Carbonic Anhydrase: Mechanism, Regulation, Links to Disease, and Industrial Applications

SUBCELLULAR BIOCHEMISTRY

SERIES EDITOR

J. ROBIN HARRIS, University of Mainz, Mainz, Germany

ASSISTANT EDITORS

B.B. BISWAS, University of Calcutta, Calcutta, India

P. QUINN, King's College London, London, UK

Recent Volumes in this Series

Volume 53	**Endotoxins: Stricture, Function and Recognition** Edited by Xiaoyuan Wang and Peter J. Quinn
Volume 54	**Conjugation and Deconjugation of Ubiquitin Family Modifiers** Edited by Marcus Groettrup
Volume 55	**Purinergic Regulation of Respiratory Diseases** Edited by Maryse Picher and Richard C. Boucher
Volume 56	**Water Soluble Vitamins** Edited by Olaf Stanger
Volume 57	**Aging Research in Yeast** Edited by Michael Breitenbach, Michal S. Jazwinski and Peter Laun
Volume 58	**Phosphoinositides I: Enzymes of Synthesis and Degradation** Edited by Tamas Balla, Matthias Wymann and John D. York
Volume 59	**Phosphoinositides II: The Diverse Biological Functions** Edited by Tamas Balla, Matthias Wymann and John D. York
Volume 60	**Adherens Junctions: From Molecular Mechanisms to Tissue Development and Disease** Edited by Tony Harris
Volume 61	**Epigenetics: Development and Disease** Edited by Kundu and Tapas Kumar
Volume 62	**The Eukaryotic Replisome: a Guide to Protein Structure and Function** Edited by MacNeill and Stuart
Volume 63	**vGPCR Signalling Complexes – Synthesis, Assembly, Trafficking and Specificity** Edited by Dupré J. Denis, Hébert E. Terence and Jockers Ralf
Volume 64	**Reprogramming Microbial Metabolic Pathways** Edited by Wang Xiaoyuan, Chen Jian and Quinn Peter
Volume 65	**Protein Aggregation and Fibrillogenesis in Cerebral and Systemic Amyloid Disease** Edited by Harris J. Robin
Volume 66	**Regulated Proteolysis in Microorganisms** Edited by David A. Dougan
Volume 68	**Structure and Physics of Viruses** Edited by Mauricio G. Mateu
Volume 69	**Peroxisomes and their Key Role in Cellular Signaling and Metabolism** Edited by Luis A. del Rio

For further volumes:
http://www.springer.com/series/6515

Chapter 1
A Concise Overview of Sources, Impacts, and Consequences of Micropollutants on the Environment

T. S. Preetham and T. V. Karthik

Introduction

Micropollutants typically consist of miniature particles, often measuring between nanometres and micrometres. Micropollutants can occur naturally in the environment even though their concentrations may be altered by human activities, these include heavy metals such as mercury, arsenic and lead that occur naturally in complex ores like Cinnabar, Tiburonia, Livingstonite, Arsenopyrite, Galena and Anglesite. Also Natural compounds such as plant toxins or biogenic chemicals can also act like micropollutants under specific conditions. Anthropogenic micropollutants exhibit significant diversity, resist natural degradation, and accumulate in ecosystems, posing a serious threat to environmental health and human well-being (Das et al., 2024a, 2024b, 2024c; Mohanty et al., 2024a; Sahu et al., 2024). Major types include pharmaceutical and personal products, insecticides, herbicides, flame retardants, micro plastics, heavy metals, industrial chemicals, and microbeads (Mishra et al., 2023a, 2023b). Micropollutants can alter the functionality of ecosystems and cause ecological imbalance. Many micropollutants particularly pesticides and industrial chemicals pollute both soil and water quality, as they accumulate, they reduce the capacity of carbon absorption of ecosystems. This in turn will accelerate changes in ecosystems and influence the production of greenhouse gases, either by enhancing or inhibiting natural processes on land and water ecosystems. Emerging Micropollutants are a serious concern to both the environment and human life. Additionally, micropollutants are very difficult to regulate due to their vast range and the complexity of detecting and quantifying them (John & Das, 2024; Mohanty et al., 2024b; Tripathy et al., 2022). Current environmental standards and governmental regulations often fail to address these problems on emerging pollutants, as a result, the persistent spread

T. S. Preetham · T. V. Karthik (✉)
Department of Chemical Engineering, IIT Guwahati, Guwahati, Assam, India
e-mail: t.karthik@iitg.ac.in

I. D. Behera et al. (eds.), *Emerging Micropollutants*, Sustainable Environmental Waste Management Strategies, https://doi.org/10.1007/978-3-032-03530-1_1

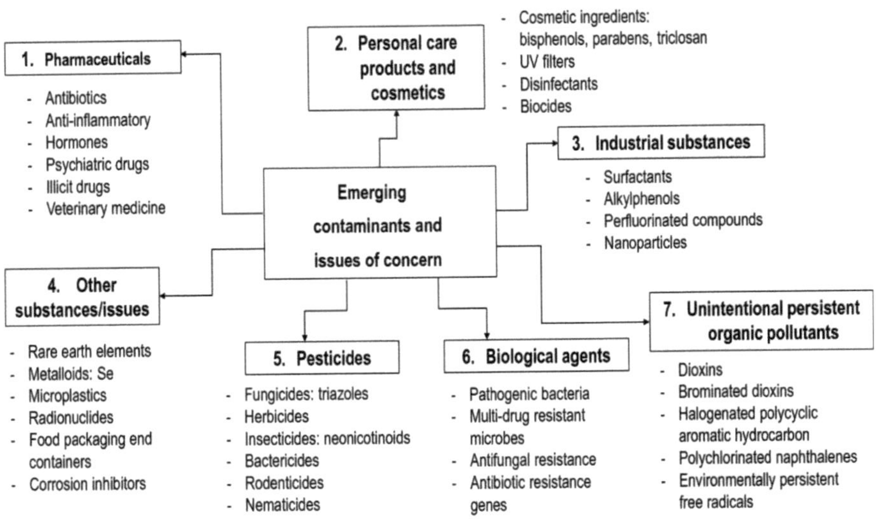

Fig. 1.1 Classification of major emerging contaminants (Morin-Crini et al., 2022)

of micropollutants continues to pose a growing threat to biodiversity, human well-being and the sustainability of environmental resources (Morin-Crini et al., 2022). Underscoring the urgent need for more effective monitoring regulation for pollution control measures. Mitigating the effects of micropollutants requires a multi-pronged approach including improved wastewater treatment, usage of greener alternative chemicals which are more environmentally friendly and biodegradable (Mishra et al., 2021; Sahu & Das, 2024; Tripathy et al., 2024a, 2024b). As the population density is increasing gradually, educating and encouraging the public about the harmful effects of chemical products that are used in daily basis, such as surfactants, exhaust emissions, medications, cosmetics, dyes, food additives, cleaning products and plastic microfibers will lead to more responsible disposal practises, which in turn reduces the overall impact of these contaminants on humans and environment (Barzman et al., 2015) (Fig. 1.1).

Sources of Micro Pollutants

Industrial Sources

Wastewater discharge, chemical production processes, and the release of medications and personal care products are the major sources of industrial micro-pollution. Hazardous materials, including heavy metals and persistent pollutants, can leak into water systems as a result of chemical industrial processes. Moreover, industrial effluents frequently contain PPCPs, including medications like hormones and antibiotics. These chemicals have an impact on human and animal health and can build up in aquatic environments, which can lead to problems like antibiotic resistance. (Barnes et al., 2004; Barzman et al., 2015; Beckers et al., 2018).

Agricultural Sources

Given that agricultural practices use fungicides, insecticides, herbicides, chemical fertilizers, and veterinary medications, they are a key source of micro-pollution in soil and water. These compounds have the potential to leak into nearby water bodies and contaminate water systems with dangerous substances, especially when used excessively. Aquatic organisms are at risk from pesticides such as glyphosate, herbicides, and veterinary antibiotics, which can disturb ecosystems and possibly lead to the emergence of bacteria resistant to antibiotics (Priyadarshini et al., 2024a, 2024b; Tripathy et al., 2024a, 2024b). Furthermore, eutrophication from nutrient-rich fertilizers can worsen the quality of the water (Barzman et al., 2015).

Urban and Domestic Sources

Water systems become contaminated when household goods that contain dangerous chemicals like phthalate and surfactant are frequently washed down the drain. Rainwater runoff from urban areas, personal care items, and household cleaning goods are examples of domestic and urban sources of micro-pollution. Preservatives and microplastics present in personal hygiene products also pollute the environment. The problem of plastic debris is further exacerbated by storm water drainage systems carrying plastics, heavy metals, and oil from urban runoff into rivers and oceans (Ahel & Jeličić, 2001; Beckers et al., 2018; Heberer & Feldmann, 2005; Rudel et al., 1998).

Environmental Sources

Through atmospheric deposition, contaminants like microplastics, heavy metals, and organic pollutants are introduced into ecosystems. Atmospheric Landfill leachates, recycling procedures, waste management systems, and air deposition are examples of environmental sources of micro-pollution. Micro pollutants leak into the environment from poorly maintained landfills and waste management systems. Toxic metals and chemicals may be released during recycling procedures, especially when recycling electronic waste (Arnold et al., 2013; Lishman et al., 2006) (Table 1.1 and Fig. 1.2)

Table 1.1 Types of micro pollutants, sources and impact on environments and human life (Gupta et al., 2024; Liu et al., 2020; Mishra et al., 2023a, 2023b; Sondhi et al., 2023; Tiwari et al., 2021)

Contaminant Category	Examples	Description	Sources	Effects on Human Health	Effects on Environment
Heavy Metals	Cadmium, mercury, lead, arsenic, copper, chromium, zinc, nickel	Toxic metals that accumulate in organisms and can disrupt cellular functions, damaging organs	Industrial discharge, mining, vehicle emissions and waste incineration	Neurological damage, developmental delays, kidney damage and cardiovascular issues	Toxic and bio-accumulate in the food chain, cause ecosystem imbalance, it also affects soil and water quality
Pharma and cosmetic Products	Antibiotics, hormones, drugs	Chemicals from medications and personal care products that persist in water sources	Wastewater effluent, improper disposal and agricultural runoff	Antibiotic resistance, endocrine disruption and liver/kidney damage	Personal care products can disrupt aquatic ecosystems, affecting growth and breeding of aquatic species
Endocrine Disruptors	Bisphenol A (BPA), phthalates, dioxins, PCBs	Chemicals that interfere with hormone systems, affecting the development of organisms	Plastic products, industrial chemicals, food packaging, pesticides	Reproductive abnormalities, developmental defects, cancer and hormone disruption	Wildlife, especially aquatic organisms, suffer from altered reproduction, hormonal changes, and population decline
Nano-materials	Metal and their oxides, nano-particles, carbon nanotubes	Engineered materials with sizes between 1–100 nm used in consumer and industrial products	Cosmetics, electronics, pharmaceuticals, food packaging, textiles	Potential lung toxicity, skin irritation, antimicrobial effects, potential for toxicity in high concentrations	Nanomaterials may cause toxicity in aquatic and terrestrial organisms, disrupting microbial communities and ecosystem function

(continued)

implying convergent evolution of a biochemical reaction essential for life processes. These ubiquitous enzymes equilibrate the reaction between three simple chemical molecules: CO_2, bicarbonate, and protons; hence, they have important roles in ion transport, acid–base regulation, gas exchange, photosynthesis, and CO_2 fixation. Furthermore, CAs are expressed variably in different species, tissues, and cellular compartments which differ in temperature, acid–base status and metabolic rate, that defined the contributions of catalyzed CO_2 reactions in many physiological processes. Drs. Robert McKenna and David Silverman, with colleagues Christopher Boone and Melissa Pinard, discuss the structure and mechanism of the α-carbonic anhydrases in Chap. 3; Dr. Roger Rowlett discusses the structure and function of the β-carbonic anhydrases in Chap. 4; Drs. Kumar and Ferry reveal the biology of the prokaryotic family of CAs in Chap. 5; and Dr. Matt Kimber describes the carboxysomal CAs in Chap. 6. Finally, Drs. Vincenzo Alterio, Simona Monti, and Giuseppina De Simone describe the thermal-stable CAs from a structural perspective in Chap. 19.

Multiple chapters are devoted to the α-class. By way of overview, Dr. Susan Frost describes the physiological significance of the catalytically active α class family in Chap. 2 while Drs. Ashok Aspatwar, Martti Tovanen, Csaba Ortutay, and Seppo Parkkila, describe the molecular biology and evolution of the "inactive" α class isoforms that are collectively called carbonic anhydrase related proteins or CARPs in Chap. 8. The importance of the α CAs in neuronal excitability is discussed by Drs. Eva Ruusuvuori and Kai Kaila in Chap. 14. Dr. Erik Swenson describes the impact of the α CAs on altitude sickness in Chap. 19. Multiple authors have contributed chapters on the membrane-associated CAs. Drs. William Sly and Abdul Waheed describe the history of the discovery of CA IV in Chap. 9. Specific emphasis is given to the tumor-associated CA IX by Dr. Egbert Oosterwijk in Chap. 10, Drs. Martin Benej, Silvia Pastorekova, and Jaromir Pastorek in Chap. 11, Drs. Narges Tafreshi Mark Lloyd, Marilyn Bui, Robert Gillies, and David Morse in Chap. 12, and Drs. Paul McDonald and Shoukat Dedhar in Chap. 13. The functional significance of these membrane associated CAs, in relation to acid–base transporters, is described in Chap. 7 by Drs. Holger Becker, Michael Klier, and Joachim Deitmer.

Inhibition of CAs has a long pharmacological history in many fields, being involved in various physiological reactions including respiration, pH regulation, Na^+ retention, calcification, tumorigenesis, electrolyte secretion, gluconeogenesis, ureagenesis, and lipogenesis. Hence CA inhibitors (CAIs) have long been used as systemic anticonvulsants, topically acting anti-glaucoma agents, and agents for treating altitude sickness, and have recently shown promising results as anti-obesity, anti-pain, and anti-tumor treatments. In addition to the treatment of human aliments, CAIs' uses are also emerging as anti-fungal and bacterial agents. The design of CA inhibitors is deliberated by Dr. Robert McKenna and Claudiu Supuran in Chap. 15 and identification of natural product inhibitors is discussed in Chap. 16 by Drs. Sally-Ann Poulsen and Rohan Davis. The application of these inhibitors are discussed by Drs. Andrea Scozzafava and Claudiu Supuran in Chap. 17 (glaucoma) and Dr. Erik Swenson in Chap. 18 (altitude sickness).

Chapter 1
Overview of the Carbonic Anhydrase Family

Robert McKenna and Susan C. Frost

Abstract The purpose of this collection of chapters is to provide a glimpse of where the carbonic anhydrase (CA) field is. This book is by no means fully inclusive, as only a few of the lead researchers around the world contributed; it serves only to show that the CA field is still pushing the boundaries of research as it has done since its discovery, and will do for a long time to come.

Keywords Carbonic anhydrase • Anhydrase family • Metalloenzymes • Structure • Function • Kinetics • Pathology • Biocatalyst • Ion transport • Acid-base balance • Regulation • Bicarbonate • CO_2 • Proton

The carbonic anhydrases (CAs; EC 4.2.1.1) are a family of metalloenzymes that catalyze the reversible hydration of carbon dioxide and bicarbonate. Since their discovery 80 years ago, in 1933, CAs have been at the forefront of scientific discovery; from the understanding of enzymatic reactions to structural biology, molecular dynamics, drug discovery and clinical medicine. The CAs are categorized into five distinct classes (α, β, γ, δ and ζ). The α-class is found primarily in vertebrates (the only class of CA in mammals). This class is also the most well characterized of the five classes. β-carbonic anhydrase is observed in higher plants and some prokaryotes, γ is present only in archaebacteria, whereas the δ and ζ classes have only been observed in diatoms. These classes of enzymes are distinct from each other in primary amino acid sequence and 3-D tertiary structure (α, β, γ),

---

Susan C. Frost and Robert McKenna (eds.). Carbonic Anhydrase: Mechanism, Regulation, Links to Disease, and Industrial Applications

R. McKenna (✉) • S.C. Frost
Department of Biochemistry and Molecular Biology, University of Florida, Gainesville, FL, USA
e-mail: rmckenna@ufl.edu; sfrost@ufl.edu

S.C. Frost and R. McKenna (eds.), *Carbonic Anhydrase: Mechanism, Regulation, Links to Disease, and Industrial Applications*, Subcellular Biochemistry 75,
DOI 10.1007/978-94-007-7359-2_1, © Springer Science+Business Media Dordrecht 2014

Part I
Introduction

Roger S. Rowlett Department of Chemistry, Colgate University, Hamilton, NY, USA

Eva Ruusuvuori Department of Biosciences, University of Helsinki, Helsinki, Finland

Andrea Scozzafava Laboratorio di Chimcia Bioinorganica, Universita degli Studi di Firenze, Florence, Italy

David Silverman Department of Pharmacology and Therapeutics, University of Florida, Gainesville, FL, USA

William S. Sly Edward A. Doisy Department of Biochemistry and Molecular Biology, Saint Louis University, School of Medicine, St. Louis, MO, USA

Claudiu T. Supuran NEUROFARBA Department, Pharmaceutical Sciences Section, Universita degli Studi di Firenze, Polo Scientifico, Florence, Italy

Eric R. Swenson VA Puget Sound Health Care System and Department of Medicine, University of Washington, Seattle, WA, USA

Narges K. Tafreshi Department of Cancer Imaging and Metabolism, H. Lee Moffitt Cancer Center and Research Institute, Tampa, FL, USA

Martti E.E. Tolvanen Institute of Biomedical Technology, University of Tampere and BioMediTech, Tampere, Finland

Abdul Waheed Edward A. Doisy Department of Biochemistry and Molecular Biology, Saint Louis University, School of Medicine, St. Louis, MO, USA

Robert J. Gillies Department of Cancer Imaging and Metabolism, H. Lee Moffitt Cancer Center and Research Institute, Tampa, FL, USA

Javier M. González Bioscience Division, Los Alamos National Laboratory, Los Alamos, NM, USA

Kai Kaila Department of Biosciences and Neuroscience Center, University of Helsinki, Helsinki, Finland

Matthew S. Kimber Department of Molecular and Cellular Biology, University of Guelph, Guelph, ON, Canada

Michael Klier Division of Biology/Membrane Transport and Division of General Zoology, Department of Biology, University of Klaiserslautern, Kaiserslautern, Germany

R. Siva Sai Kumar Department of Biochemistry and Molecular Biology, Ebery College of Science, The Pennsylvania State University, University Park, PA, USA

Mark C. Lloyd Analytic Microscopy Core Facility, H. Lee Moffitt Cancer Center and Research Institute, Tampa, FL, USA

Paul C. McDonald Department of Integrative Oncology, British Columbia Cancer Agency Centre, Vancouver, BC, Canada

Robert McKenna Department of Biochemistry and Molecular Biology, University of Florida, Gainesville, FL, USA

Simona Maria Monti Istituto di Biostrutture e Bioimmagini-CNR, Naples, Italy

David L. Morse Department of Cancer Imaging and Metabolism, H. Lee Moffitt Cancer Center and Research Institute, Tampa, FL, USA

Egbert Oosterwijk Department of Urology, University Medical Center St Radboud, Nijmegen, The Netherlands

Csaba Ortutay Institute of Biomedical Technology, University of Tampere and BioMediTech, Tampere, Finland

Seppo Parkkila Institute of Biomedical Technology and School of Medicine, University of Tampere and BioMediTech, Tampere, Finland

Jaromir Pastorek Department of Molecular Medicine, Institute of Virology, Slovak Academy of Sciences, Bratislava, Slovakia

Silvia Pastorekova Department of Molecular Medicine, Institute of Virology, Slovak Academy of Sciences, Bratislava, Slovakia

Melissa Pinard Department of Biochemistry and Molecular Biology, University of Florida, Gainesville, FL, USA

Sally-Ann Poulsen Eskitis Institute for Drug Discovery, Griffith University, Brisbane, QLD, Australia

Contributors

Vincenzo Alterio Istituto di Biostrutture e Bioimmagini-CNR, Naples, Italy

Ashok Aspatwar Institute of Biomedical Technology and School of Medicine, University of Tampere and BioMediTech, Tampere, Finland

Holger M. Becker Division of Biology/Membrane Transport, Department of Biology, University of Klaiserslautern, Kaiserslautern, Germany

Martin Benej Department of Molecular Medicine, Institute of Virology, Slovak Academy of Sciences, Bratislava, Slovakia

Christopher D. Boone Department of Biochemistry and Molecular Biology, University of Florida, Gainesville, FL, USA

Marilyn M. Bui Analytical Microscopy Core Facility and Department of Anatomic Pathology, H. Lee Moffitt Cancer Center and Research Institute, Tampa, FL, USA

Rohan A. Davis Eskitis Institute for Drug Discovery, Griffith University, Brisbane, QLD, Australia

Giuseppina De Simone Istituto di Biostrutture e Bioimmagini-CNR, Naples, Italy

Shoukat Dedhar Department of Biochemistry and Molecular Biology, University of British Columbia, Vancouver, BC, Canada

Joachim W. Deitmer Division of General Zoology, Department of Biology, University of Klaiserslautern, Kaiserslautern, Germany

James G. Ferry Department of Biochemistry and Molecular Biology, Eberly College of Science, The Pennsylvania State University, University Park, PA, USA

S. Zoë Fisher Bioscience Division, Los Alamos National Laboratory, Los Alamos, NM, USA

Susan C. Frost Department of Biochemistry and Molecular Biology, University of Florida, Gainesville, FL, USA

Contents

Editors
Susan C. Frost
Department of Biochemistry
and Molecular Biology
University of Florida
Gainesville, FL, USA

Robert McKenna
Department of Biochemistry
and Molecular Biology
University of Florida
Gainesville, FL, USA

ISSN 0306-0225
ISBN 978-94-007-7358-5 ISBN 978-94-007-7359-2 (eBook)
DOI 10.1007/978-94-007-7359-2
Springer Dordrecht Heidelberg New York London

Printed on acid-free paper

Springer is part of Springer Science+Business Media (www.springer.com)

Susan C. Frost • Robert McKenna
Editors

Carbonic Anhydrase: Mechanism, Regulation, Links to Disease, and Industrial Applications

Springer

CAs have also been gaining industrial interest as bio-catalysts for carbon sequestration of flue-gas from coal-fired power plants and in exploiting CAs in algae as a way to capture CO_2 and convert it into biofuels or other valuable products. The favorable properties of CAs such as fast kinetics, easy expression, high solubility and intermediate heat resistance have made them an attractive candidate for numerous industrial and medical applications; for example, to aid in CO_2 gas exchange in artificial lungs. Drs. Zoë Fisher and Javier González discuss these issues in detail in Chap. 20.

The editors would like to thank all the authors for their patience with our many requests, and their outstanding contributions that show the full gamut of CA research. A special word of thanks goes to Mayank Aggarwal, Christopher Boone, Melissa Pinard, and Brian Mahon for their help in the preparation of this book.

Part II
Carbonic Anhydrases: Ancient but Relevant

Chapter 2
Physiological Functions of the Alpha Class of Carbonic Anhydrases

Susan C. Frost

Abstract Carbonic anhydrases are ubiquitous enzymes that catalyze the reversible hydration of carbon dioxide. These enzymes are of ancient origin as they are found in the deepest of branches of the evolutionary tree. Of the five different classes of carbonic anhydrases, the alpha class has perhaps received the most attention because of its role in human pathology. This review focuses on the physiological function of this class of carbonic anhydrases organized by their cellular location.

Keywords Adipogenesis • Adipose • Kidney • Red blood cells • Erythrocytes • Skeletal muscle • PPARγ2 • Gluconeogenesis • Lipogenesis • Ureagenesis • Taste perception • Carbonic anhydrase I • Carbonic anhydrase II • Carbonic anhydrase III • Carbonic anhydrase IV • Carbonic anhydrase V • Carbonic anhydrase VI • Carbonic anhydrase VII • Carbonic anhydrase IX • Carbonic anhydrase XII • Carbonic anhydrase XIII • Carbonic anhydrase XIV • Mitochondria • Carbonyl phosphate synthetase • Pyruvate carboxylase • Acetyl CoA carboxylase • Sulfonamides • Saliva • Gustin • Bicarbonate transporters • Tumor-associated CAs • Retinal pigment epitheilium • Topiramate • Zonisamide

---

Susan C. Frost and Robert McKenna (eds.). Carbonic Anhydrase: Mechanism, Regulation, Links to Disease, and Industrial Applications

S.C. Frost (✉)
Department of Biochemistry and Molecular Biology, University of Florida, Gainesville, FL, USA
e-mail: sfrost@ufl.edu

S.C. Frost and R. McKenna (eds.), *Carbonic Anhydrase: Mechanism, Regulation, Links to Disease, and Industrial Applications*, Subcellular Biochemistry 75,
DOI 10.1007/978-94-007-7359-2_2,

1 Introduction

Carbonic anhydrases catalyze the reversible hydration of CO_2 ($CO_2 + H_2O \leftrightarrow HCO_3^- + H^+$), which allows this enzyme to regulate intra- and extra-cellular concentrations of CO_2, H^+, and HCO_3^-. Decades of research have implicated CA in a broad range of physiological processes including gas exchange at the air water interface, transport of CO_2 and HCO_3^- across membranes, biosynthetic reactions in metabolically active tissue, acid–base balance, secretion, calcification, signal transduction, oncogenesis, proliferation, among the many that have been reported [1–16]. These seemingly disconnected functions are mediated by specific isoforms in the α-CA family. Sixteen members of this family have been identified which have distinct tissue-specific expression, kinetic properties, and sensitivity to inhibitors [17]. It appears unlikely that this family will be expanded further as searches of genomic databases have not identified any additional CA sequences [18]. Among those identified, there are eight cytosolic proteins (CA I, CA II, CA III, CA VII, CA VIII, CA X, CA XI, CA XIII), two mitochondrial matrix proteins (CA VA, CA VB), one secreted protein (CA VI), two glycosylphosphatidylinositol (GPI)-anchored proteins (CA IV and CA XV), and three transmembrane proteins (CA IX, CA XII, CA XIV). Three of the cytosolic isoforms (VIII, X, and XI) have no activity as they lack one or more of the histidine residues that coordinate the zinc ion in the catalytic pocket. As a group, these are called CA-related proteins and appear to be expressed exclusively in the brain [19]. The other isoforms have varied activities based on the efficiency of proton transfer, differences in active site residues, quaternary structure, and potentially localization [17, 20–22]. In this chapter, the physiological role of the catalytically active forms of CA will be discussed from the perspective of location: cytosolic, mitochondrial, secretory, and membrane-associated.

2 Cytosolic CAs

The role of carbonic anhydrase in CO_2 excretion is well known. In red blood cells (RBCs), CA activity accelerates the rate of conversion between molecular CO_2, which easily diffuses across membranes, and HCO_3^-, the form in which the majority of CO_2 is transported in the circulation. CO_2 produced by tissues diffuse into RBCs where it is hydrated to form bicarbonate ions that are transported via the band 3 anion exchanger and protons that are buffered by hemoglobin. The reverse occurs at gas exchange organs where HCO_3^- is dehydrated producing CO_2 that then diffuses across the water/air interface down its partial pressure gradient. RBC CA is indirectly related to O_2 loading and unloading through the Bohr effect [23] [reviewed in [24, 25]]. Mammalian RBCs express both CA II and CA I [25, 26]. It is thought that CA II activity dominates because of its fast kinetics, although the intracellular microenvironment may influence how these enzymes operate in vivo. On the other hand, it is widely accepted that CO_2 excretion in vertebrates is not limited by RBC CA activity [1]. Further details on this topic can be found in Chap. 18 (Swenson).

Hemolytic anemia is a disease in which RBCs are destroyed prematurely which leads to anemia. Glucose-6 phosphate dehydrogenase deficiency induces hemolytic anemia [27]. These patients have significantly lower CA I expression compared to control patients [10]. It is postulated that this is related to the rate of synthesis of CA I relative to hemoglobin since data are normalized to hemoglobin content. That said, CA II expression is increased, as is total CA activity. While CA I has substantially lower activity than CA II, which is the more physiologically relevant isoform, CA I expression may serve as a marker for hemolytic anemia.

CA II is also highly expressed in kidney intercalated cells and at lower levels in the proximal tubules, loop of Henle, and collecting duct principal cells [28, 29] where CA II regulates bicarbonate flux. CA II deficiency is an autosomal recessive trait characterized by renal tubular acidosis, osteopetrosis, cerebral calcification, and growth retardation [30]. A mouse model has been developed which partially mimics the human disease [31]. Kidneys of these mice are virtually devoid of medullary collecting duct intercalated cells [32] where CA II expression is normally high. Interestingly, these cells are present at birth, but at some point during post-natal development, intercalated cells are selectively removed in the medullary collect ducts and replaced by principal cells. This suggests that CA II may play a role in regulating cell-type diversity in kidney collecting ducts. Indeed, chronic acetazolamide treatment of adult rats causes significant remodeling of the cellular profile of collecting ducts [33]. This may represent an adaptive process to correct or stabilize the metabolic acidosis that would otherwise ensue following loss of CA II function.

In addition to its ability to mediate the reversible hydration of CO_2, CA II appears to interact with a variety of membrane-bound carriers to balance cytoplasmic pH. Examples of these include the chloride/bicarbonate exchanger AEI [34, 35], the sodium bicarbonate cotransporter NBC1 [36, 37], and the sodium/hydrogen exchanger NHE1 [38]. These interactions increase the activity of the transporters and have been coined "transport metabolons" [35]. Specific amino acid motifs, along with individual residues, have been identified that are required for the binding of the metabolon partners [39, 40]. Post-translational modifications have also been implicated in these interactions. For example, phosphorylation of NHE1 in the C-terminal cytoplasmic tail significantly increases the interaction with CA II and thus its activity [39]. Metabolons may also play a role in human pathologies. For instance, the interactions between CA II and NHE1 and AE3 have been implicated in cardiomyocyte hypertrophy [7]. In addition to the above transporters, CA II also interacts with members of the monocarboxylate transporter family (MCT1 and MCT4) and increases their activity, leading to enhanced export of lactate from Xenopus oocytes [41, 42] and astrocytes [43]. Protons, provided by CA II, are cotransported by the MCTs leading to the hypothesis that CA II acts as a "proton collecting antenna" [44]. In contrast to other transport metabolons, the interaction between CAII and the MCTs does not require the catalytic activity of CA II but rather its ability to shuttle protons via the proton wire, with residue His64 playing a central role [44, 45]. These features are described in more detail in Chap. 7 (Becker et al.).

CA III has several characteristics that distinguish it from other isozymes. Expression of CA III is remarkably high in skeletal muscle [46] and adipose, both white [47] and brown [48]. However, CA III activity is low, at only 3 % of that of CA II [17]. This difference in activity has led to the idea that CA III may play a different role in cellular function beyond is catalytic activity. CA III has two reactive sulfhydryl groups that can reversibly bind to glutathione through disulfide bonds [49, 50]. This reaction would likely protect cells from irreversible protein oxidation [51]. Indeed, overexpression of CA III in cells protects them from H_2O_2-induced apoptosis [52]. Further, aged rats showed increased tissue levels of irreversibly oxidized CA III associated with decreased glutathione concentrations [53]. These data suggest that CA III might protect cells from oxidative damage [54]. However, muscle tissue in the CA III global knockout mouse responded no differently than muscle in wild type mice in response to hyperoxic challenge or muscle fatigability [55]. In fact, these authors showed that CA III expression is not required for normal growth, development, or life span of the mouse.

Adipose tissue stores fat and is central to energy homeostasis [56]. New adipocytes arise from precursors called adipose-derived stems cells or pre-adipocytes in a process called adipogenesis. Light and electron microscopy have revealed that these cells arise from perivascular sites [57–59]. Several studies have now shown that perivascular cells isolated from adipose have the ability to differentiate [60–65]. Importantly, the nuclear hormone receptor peroxisome proliferator-activated receptor $\gamma 2$ (PPAR$\gamma 2$) that acts as the master regulator of adipogenesis is found in these precursor cells [66]. CA III expression is induced during adipogenesis [67] and possibly provides HCO_3^- to acetyl CoA carboxylase [68], the rate determining step in fatty acid biosynthesis. However, CA III is down-regulated in obese states [69] in the face of enhanced fatty acid biosynthesis [70]. This questions a role in substrate metabolism. Yet, recent data reveals that CA III regulates adipogenesis at the level of PPAR$\gamma 2$ gene expression [71]. While no changes in adipose content were noted in the CA III knockout mouse [see above [55]], Mitterberger et al. have shown that adipogenesis is enhanced in mouse embryonic fibroblasts (MEFs) isolated from CA III knockout mice [71]. This was associated with a 1000-fold increase in PPAR$\gamma 2$ expression. This suggests that CA III expression exerts a negative effect on PPAR$\gamma 2$ expression. Despite the fact the CA III expression is increased during adipogenesis as mentioned [67], it apparently is not required for the normal terminal differentiation of adipose tissue. Rather, it appears that CA III controls steps early in the differentiation process. Still unknown is the mechanism by which CA III regulates PPAR$\gamma 2$ expression and whether it serves a similar role in muscle during development. As noted above, CA III may play a protective role in oxidative damage. PPAR$\gamma 2$ also plays a role in oxidative stress. It has been shown that pharmacological activation of PPAR$\gamma 2$ attenuates the production of reactive oxygen species (ROS) in 3 T3-L1 adipocytes and in the insulin-resistant leptin deficient ob/ob mouse [72]. Thus, CA III may provide long term regulation during adipogenesis and protection in response to oxidative stress.

Carbonic anhydrase VII is one of the least characterized of the CA family. The human form was identified through genomic screening [73]. While it was predicted to have CO_2-hydrase activity, this was proven later using mouse [74] and human [75] recombinant proteins. The human form has catalytic activity that is close to that of CA II [75]. It also has the highest esterase activity among the CA family members [75]. There are two forms of this protein: the long form is the predominant form and the shorter form is missing 56 residues at the N-terminus [76]. Based on western blotting, the protein is primarily expressed in colon, liver, and skeletal muscle, although it is also noted in brain [76]. Similar to CA III, it has two reactive cysteines and can be glucothionylated [75] suggesting a role as an oxygen free radical scavenger. CA VII has also been implicated in neuronal excitation by providing HCO_3^- which can mediate current through channels coupled to $GABA_A$ receptors [77]. This activity is suppressed when treated with membrane-permeant sulfonamides, supporting the hypothesis that CA VII plays a role in neuronal excitation and seizures [78]. Kaila and Ruusuvuori discuss this in further detail in Chapter 14. In addition, CA VII may play a role in neuropathic pain as acetazolamide in combination with midazolam treatment synergistically reduces neuropathic allodynia after spinal nerve damage [79]. In that regard, CA VII may represent a new drug target for managing neuropathic pain.

Human CA XIII isozyme was identified and characterized in 2004 [80]. In this study, the authors showed that CO_2 hydration activity is similar to that of CA I and CA V, each of which are characterized as having moderate catalytic activity. Inhibition profiles are similar to CA II [81]. CA XIII was localized to several tissues including the thymus, kidney, submandibular gland, small intestine, and notably in reproductive organs of both sexes [80]. Since pH and ion balance are likely to be tightly regulated in reproductive organs to ensure normal fertilization [80], it is surmised that CA XIII may contribute to reproductive processes by controlling optimal HCO_3^- concentration and pH homeostasis for the maintenance of sperm mobility. One could also postulate that CA XIII might contribute to normal fertilization process by producing the appropriate bicarbonate concentration to alkalinize the cervical and endometrial mucus [82]. CA XIII deficient animals are not yet available so testing these hypotheses must wait. However, there are data regarding a role of CA IV in bicarbonate-mediated activation of mouse and human sperm [83], an enzyme that will be discussed later in this chapter and Chap. 9 (Sly and Waheed). With renewed interest in tumor-associated CAs, Kummola et al. have demonstrated that CA XIII, along with two other cytosolic CAs (CA I and CA II), is down-regulated in colorectal cancer [82]. Because these three CAs genes are closely linked on chromosome 8, these authors suggest that down-regulation is related to reduced levels of a common transcription factor. The physiological reasons for down-regulation are left to speculation at this point.

3 Mitochondrial CAs

Chappell and Crofts demonstrated that HCO_3^- was impermeant to the inner mitochondria membrane [84]. While Elder initially proposed that HCO_3^- could provide the counter ion for energy-dependent Ca^{2+} transport [85], shortly thereafter it was shown that CO_2, not HCO_3^-, served this function [86]. With the advent of molecular technology, we now know that the bicarbonate transporter family (SLC4A) includes 11 members (see http://slc.bioparadigms.org), none of which are located in the inner mitochondrial membrane. Thus, de novo synthesis of HCO_3^- within the mitochondrial compartment is required for providing substrate for pyruvate carboxylase in the gluconeogenic pathway and carbamoyl phosphate synthetase I in ureagenesis in the liver [87, 88].

The first mitochondrial CA was isolated from guinea pig liver and called CA V [89]. It was subsequently identified in mouse, rat, and human through molecular cloning [90–92]. That the transcript for mouse CA V was only identified in liver [90] while a wider distribution was suggested by western blotting [91], led to a search of the EST database revealing that there were two mouse mitochondrial CA sequences. Ultimately, these sequences were named CA VA and CA VB, respectively, and northern and western blotting revealed a significantly different tissue-specific distribution pattern between the two [93]. Interestingly, the human ortholog for the *CA5B*, which also has broad tissue expression, has been mapped to chromosome Xp22.1 [94] while *CA5A* was mapped to chromosome 16q24 [95].

Carbamoyl phosphate synthetase I utilizes HCO_3^- rather than CO_2 for the synthesis of carbamoyl phosphate [96]. This is the committed step in ureagenesis. Ornithine transcarbamylase utilizes carbamoyl phosphate as a co-substrate in the synthesis of citrulline [97–99], which is the first intermediate of the urea cycle. Dodgson et al. demonstrated that the synthesis of citrulline could be blocked by acetazolamide in guinea pig liver mitochondria [100]. Indeed, the inhibition curve for citrulline synthesis was identical to the inhibition curve for mitochondrial CA (CA VA). This was the first physiological evidence that carbonic anhydrase enhances access of HCO_3^- to the synthetase reaction, so CA must be considered a participant in ureagenesis. These studies raised the possibility that HCO_3^- created in the CA V reaction could drive other biosynthetic reactions, particularly that of the carboxylase family of enzymes. Pyruvate carboxylase mediates the first reaction in gluconeogenesis from pyruvate. Dodgson and Forester showed that pyruvate carboxylase activity was blocked by ethoxzolamide, a membrane permeant sulfonamide, in mitochondria isolated from liver from starved guinea pigs [87]. While earlier studies had suggested that sulfonamides inhibit pyruvate carboxylase directly [101], Dodgson and Forester showed that the inhibitory effect of ethoxzolamide on pyruvate carboxylase activity was lost in experiments where guinea pig liver mitochondria were pretreated with digitonin, in the presence of high bicarbonate, to compromise membrane integrity. These and other data suggest that the effect of ethoxzolamide is on mitochondrial CA, not pyruvate carboxylase. Dodgson and Forester also demonstrated that glucose production in hepatocytes was blocked by ethoxzolamide, further implicating the dependence of the anapleurotic reaction mediated by pyruvate carboxylase on CA VA.

As mentioned, carbon fixation at pyruvate carboxylase increases the concentration of mitochondrial intermediates for other biosynthetic reactions. For gluconeogenesis, it is malate that is drawn from the cycle. For lipogenesis, it is citrate that is drawn off the cycle. Citrate is made from the condensation of acetyl

CoA and oxaloacetate, the product of the pyruvate carboxylase reaction. Citrate can be transported out of the mitochondria where it is cleaved to re-form oxaloacetate and acetyl CoA, the latter of which is the substrate for cytoplasmic acetyl CoA carboxylase, the rate-limiting step in de novo lipogenesis. Hazen et al. showed that ethoxzolamide inhibits lipogenesis from pyruvate in 3T3-L1 adipocytes, a mouse adipocyte model [102]. Acetyl CoA carboxylase, like pyruvate carboxylase, utilizes HCO_3^- as a substrate, in this case for the carboxylation of acetyl CoA. That acetyl CoA carboxylase was not the target of sulfonamide inhibition was demonstrated by lack of sulfonamide inhibition of lipogenesis from glutamate, another anapleurotic substrate that increases the concentrations of Krebs cycle intermediates but independently from pyruvate carboxylation. ^{13}C-NMR studies, reported in 2009, support these conclusions [103]. Together, these data suggest that carboxylation of pyruvate by CA VB in the mitochondria of adipocytes is required for lipogenesis and by extension CA VA in liver mitochondria [104].

While mitochondrial diseases are often associated with defects in the oxidative phosphorylation [105], the above data suggest the possibility that the mitochondrial CAs could serve as targets for modulating gluconeogenesis and lipogenesis, both of which are dysregulated in obesity and insulin resistance. Interestingly, an adverse effect of sulfonamide- and sulfamate-containing anti-epileptic drugs is weight loss in obese patients [106]. Indeed, a randomized trial in 2003 demonstrated significant weight loss in a study of 60 non-epileptic obese patients given Zonisamide, a marketed anti-epileptic aliphatic sulfonamide with known serotonergic and dopaminergic activity in addition to blocking sodium and calcium channels [107]. Furthermore, Topiramate, a sulfamate-substituted saccharide, was approved for weight loss by the FDA in 2012 to be used in conjunction with phentermine treatment (which decreases appetite). While the mechanism for this effect is currently unknown, De Simone et al. have shown that Zonisamide strongly inhibits recombinant CA VA ($K_i = 20$ nM) [108]. Like Zonisamide, Topiramate inhibits CA VA, although with somewhat less efficacy ($K_i = 63$ nM) [13]. As an aside, Topiramate is also a strong inhibitor of CA VB (Ki = 30 nM), unlike Zonisamide which is relatively poor inhibitor ($K_i = 6.0$ μM). However, both drugs block CA II in low nM range which raises questions regarding the in vivo target. That said, Topiramate has been shown to be block lipogenesis from pyruvate, not acetate, in 3T3-L1 adipocytes [103]. Presently, each of these isoforms is being pursued as novel anti-obesity targets [109–111].

4 Secreted CAs

CA VI is the only secreted isoform among the α-carbonic anhydrase family [reviewed in [112]]. The existence of CA activity in saliva has been known for decades, but it was not until 1979 when it was realized that the activity was unique from that of erythrocyte CA activity (CAII) [113]. Feldstein and Silverman provided the initial biochemical and kinetic characterization revealing that rat salivary CA

VI had a molecular weight of 42 kDa and was glycosylated which predicted a secretory protein [114]. While the kinetic parameters were similar to that of CAII, CA VI exhibited a somewhat lower affinity for sulfonamide inhibitors. Murakami and Sly reported comparable data for CA isolated from human saliva, at which point the name CA VI was adopted [115]. Interestingly Parkkila et al. have shown that salivary CA VI secretion follows circadian rhythm [116], low during sleep and rising in concentration at awakening and breakfast. Subsequently, CA VI has been found in milk [117], tears [118], respiratory airways [119], epithelial lining of the alimentary canal [120], and enamel organs [121]. It has also been found in human serum [122]. Although the physiological function of CA VI is not fully established, there are clues that it regulates against acidic environments.

Saliva plays a critical role in oral homeostasis and decreased rates of secretion increases the risk of oral infections and dental caries [123]. The buffering capacity of salivary secretions depends primarily on bicarbonate ions and provides protection against enamel erosion [124]. Several studies have shown that CA VI is responsible for acid neutralization in dental biofilm, originating from bacterial metabolism. For example, Kimoto et al. showed that patients who rinsed their mouths with sucrose in the presence of acetazolamide had significantly higher salivary pH values than patients who rinsed with only sucrose [125]. In this study, CA activity associated with plaque was specifically identified as CA VI, not CA I or CA II. In another study, Kivela et al. demonstrated that a low concentration of CAVI in saliva is associated with a higher incidence of dental caries [126]. However, a study by Frasseto et al. revealed that CA VI activity in the oral cavity of children with dental caries was higher than that found in children who were caries-free, although the statistical significance of this observation was border-line [4]. Additionally, the variation in CA VI activity in saliva, before and after a sucrose wash, was significantly greater in children with dental caries than those without. Given that there did not appear to be differences in the concentration of CA VI, the authors suggest that genetic polymorphisms may be related to the differences in CA VI activity seen across these two patient populations. Indeed, polypmorphisms have been described that are associated with higher buffering activity but, interestingly, buffering capacity is decreased in healthy children [127]. These authors have suggested that polymorphisms in the coding region may affect secondary structure to alter CA VI function. Others have shown that both pH and buffering capacity of saliva is lower in diabetics compared to normal controls [128]. While CA activity was positively correlated with frequency of polymorphisms, there was no correlation between polymorphism frequency and pH or buffering capacity. These data suggest that there is still no consensus regarding the role of CA VI and pH control in the oral cavity.

CA VI is also known as gustin [129]. It has been shown that gustin is decreased in parotid saliva of patients who experience loss of taste [130–132]. This phenomenon was associated with aberrant taste bud morphology [133], consistent with apoptosis. Return of taste function has been demonstrated by exposure to exogenous zinc [132]. Interestingly, Topiramate and other CA sulfonamide inhibitors cause taste perversion [134] perhaps targeting CA VI. Together, these data suggest a role for CA VI in taste perception. Perhaps weight loss in patients given Topiramate (see above) is in part related to the loss of food appreciation!

The crystal structure of CA VI has been solved revealing a prototypical mammalian CA fold, but with a novel dimeric arrangement as compared to previously reported CA structures [135]. The active site cavity contains a cluster of non-conserved residues that may be involved in ligand binding. This discovery may open opportunities for developing an isoform-specific inhibitor, which has been difficult because of the conservation in the catalytic site across most CA isoforms.

5 Membrane-Associated CAs

The human membrane-associated CAs include CA IV, CAIX, CA XII, and CA XIV. CA XV, like CA IV, is a GPI-anchored form of CA but is not expressed in humans or chimpanzees [136]. These enzymes are poised to reversibly hydrate CO_2 in the extracellular space. Several of these family members will be discussed in depth in later chapters in this book (Sly and Waheed, Chap. 9; Oosterwijk, Chap. 10; Benej et al., Chap. 11; Tafreshi et al., Chap. 12; and McDonald and Dedhar, Chap. 13), so please refer to those chapters as well.

A membrane-bound form of CA was initially purified from lung and tentatively called CA IV [137]. A "second" membrane-bound form was ultimately purified from human kidney [138]. Subsequently, Zhu and Sly reported a more efficient purification that allowed them to show that lung and kidney expressed the same membrane-bound form of CA [139]. These authors also showed that about 50 % of the enzyme could be released from the membrane by treatment with phosphoinositide-specific phospholipase C, suggesting that the enzyme is attached to the membrane by a GPI linkage. Human CA IV was cloned in 1992 by Okuyama et al. [140]. The deduced amino acids included an 18-amino acid signal sequence, a 260 amino acid stretch that show similarity to the catalytic regions of CA I, CA II, and CA III, with an additional 27 amino acid C-terminus containing a hydrophobic domain in the last 21 amino acids. Expression of CA IV cDNA in COS cells generated a 35 kDa membrane-bound protein. Baird et al. reported that CA IV is a high-activity isozyme showing pH independence in the hydration direction [141]. In the dehydration direction, the catalytic rate is even higher than that observed in CA II, although the esterase activity is lower. CA IV has also been localized to heart [142], brain [143], capillary bed of the eye [144], and erythrocytes [145].

Because kidney expresses both CA II and CA IV, the question arose as to whether the cytosolic CA or the membrane-bound CA was responsible for the CO_2 hydration that leads to the acidification of urine and reabsorption of filtered bicarbonate. In 1996, Conroy et al. designed a pegylated sulfonamide (F-3500) that inhibited CA activity, but was impermeant to cells [146]. This allowed the investigators to distinguish between intracellular and membrane-associated CA activity, and specifically that of CA II and CA IV in kidney [147]. Low molecular weight CA inhibitors, like acetazolamide, produce urine with a concentration of 100–200 mM HCO_3^- in all mammalian species tested [148]. Under these circumstances, both intracellular and membrane-associated CA activity will be inhibited. In contrast, rats treated with F-3500 produced urine containing only 40 mM HCO_3^- that is taken as the effect of inhibiting CA IV while retaining CA II activity. These data support the hypothesis that both CA II and CA IV are important in bicarbonate reabsorption. These results agree with studies in humans lacking CA II where bicarbonate concentration became elevated in response to acetazolamide [149]. We now know that another membrane-associated CA (CA XII) is expressed in kidney [150, 151]. At this point in time, we cannot distinguish between CA IV and CA XII function for lack of isoform-specific inhibitors so the studies above could support the involvement of both CA IV and CA XII in bicarbonate reabsorption in the kidney.

Like CA II, CA IV interacts with Cl^-/HCO_3^- transporters [152]. In this study, Sterling et al. demonstrated that CA IV interacts directly with the 4th extracellular loop of AE1. This interaction increases the activity of bicarbonate transport. CA IV also creates a functional complex with the Na^+/bicarbonate co-transporter (NBC1) [153]. This latter study showed that this interaction is required for maintaining appropriate pH balance within the environment of the retina and retinal pigment epithelium, although neither CA IV nor NBC1 are expressed in the retinal or retinal pigment epithelium. This requirement is based on the finding that mutant forms of CA IV appear to be responsible for an autosomal dominant form of retinitis pigmentosa [154] causing rod and cone photoreceptor degeneration [153]. These mutations are associated with a loss of CA activity or the inability of CA IV to interact with NCB1, in choriocapillaris leading to impaired pH homeostasis [153]. Based on the importance of CA IV in the survival of photoreceptor cells, this raises a flag for long-term use of CA inhibitors, particularly in the treatment of glaucoma, which may adversely affect vision.

CA IX and CA XII are specifically tumor-related [14, 155]. CA IX has garnered more interest because of its limited normal expression [156, 157], and its apparent role in cell proliferation and migration [158, 159], cell adhesion [160], tumorigenesis [161], and pH control [162–165]. CA IX is a transmembrane glycoprotein whose catalytic domain is oriented toward the extracellular milieu [166]. CAIX is expressed as a 49.7 kDa protein but is truncated to the mature form during processing [167]. This mature form contains an N-terminal "exofacial" proteoglycan-like domain and catalytic domain (homologous to CA II) that is attached via a transmembrane segment to a cytoplasmic tail. CA IX exists primarily as a dimer stabilized by disulfide bonds [168–170]. In rat cardiomyocytes, CA IX interacts with the NBCe1 Na^+/HCO_3^- cotransporter enhancing bicarbonate influx [171]. The cytoplasmic tail of CA IX contains the phosphorylation motif for protein kinase A (PKA) that is important for catalytic activity [15]. Recombinant CA IX, containing the catalytic domain has activity similar to that of CA II [168, 172]. CA IX is regulated by hypoxia [173], and in general predicts poor patient outcome [174–177]. The secondary structure and orientation of CA XII is similar to that of CA IX, but is a monomer, lacks the proteoglycan-like domain, and is missing the PKA motif [150]. Its catalytic activity is lower than that of CA IX [13, 178, 179] which may influence its role in pH control in tissues compared to CA IX. Northern blot analysis initially revealed CA XII expression in kidney and colon [150]. Western blotting and immunohistochemistry show a much wider tissue distribution including kidney, lung, prostate, ovaries, uterine endometrium, breast, and the basolateral membrane of epithelial cells of the gut [155, 180–182]. In contrast to CA IX, CA XII is regulated by estrogen [183, 184]. In breast cancer patients, CA XII expression correlates with positive prognosis [183–186]. In other cancers, CA XII expression can be positive, negative, or neutral as a predictor of patient outcome [177, 182, 187, 188]. Interestingly, CA XII has been associated with metabolic acidosis in patients receiving carbonic anhydrase inhibitors, specifically Topiramate or Zonisamide (see Sects. 3 and 4, this chapter) [189]. Patients sensitive to these drugs typically have serum bicarbonate concentrations of less than 20 mM. Low bicarbonate was associated with polymorphisms in CA XII (rs2306719 and rs4984241). While these data warrant further investigation, this indicates a role of CA XII in renal function. In addition, a Glu143Lys mutation in CA XII has been linked to individuals with failure to thrive, hypoatremic dehydration and hyperkalemica with isolated sweat salt wasting [190]. This autosomal recessive mutation behaves similarly to the excessive salt loss from sweat glands observed in pseudohypoaldosteronism type 1 which arises from mutations in genes encoding epithelial Na^+ channel (ENaC) subunits. These data demonstrate the importance of bicarbonate anion and proton production on salt concentration in sweat and its significance for sodium homeostasis, and implies a specific role for CA XII.

The least studied of the human CAs is CA XIV which was cloned in 1999 [191] and bears strongest sequence similarity to CA XII. CA XIV mRNA shows strong expression in most parts of the brain with weaker signals in colon, small intestine, urinary bladder and kidney. RT-PCR analysis revealed an intense signal in liver and spinal cord, but much weaker in kidney. However, by western blot and immunohistochemistry, CA XIV shows significant luminal co-localization with CA IV (but not CA XII) in regions that are involved with urinary acidification [192]. This suggests functional overlap between CA IV and CA XIV. CA XIV has also been implicated in acid–base balance in muscle and erythrocytes in an adaptive response to chronic hypoxia as observed at high elevation [193]. Like other membrane-bound CAs, CA XIV interacts with bicarbonate transporters [194]. In heart myocardium, it has been demonstrated that CA XIV interacts with AE3. In hypertrophic hearts from hypertensive rats, CA XIV expression is elevated along with AE-mediated bicarbonate transporter. This suggests a role for CA XIV in AE3 hyperactivity. Finally, CA XIV, in contrast to CA IV, has been localized to the apical and basal membranes of the retinal pigment epithelium, along with the plasma membrane of Müller cells [195]. Because CA II is also found in Müller cells [196], this implies that CA II and CA XIV have specific and unique functions in the context of acid based balance in the retina.

6 Concluding Remarks

While the physiological functions of some of the mammalian isozymes of CA are still uncertain, it is clear that the CAs are important in many physiological processes in both normal and pathological states. CA inhibitors are now widely used in the clinic to treat a number of diseases including glaucoma, epilepsy, mountain sickness, ulcers, osteoporosis, and obesity. Yet, these inhibitors collectively target enzymatic activity, which limits their targeting specificity because of the structural similarity of the CA catalytic pockets. Thus, our greatest challenge is to develop CA-specific inhibitors to further our understanding of function, develop diagnostic tools, and treat diseases in a selective fashion. This requires a better understanding of the CA structures to facilitate the design of novel drugs. In addition, it may be possible to use surface motifs for docking a catalytic inhibitor to provide specificity. Also encouraging are the membrane-impermeant compounds, which block only membrane-associated CA isoforms (discussed in Chap. 15, McKenna and Supuran). These should increase our ability to target cancer-related CAs, like CA IX and CA XII. One can also imagine nanoparticle delivery systems that use cell surface epitopes for tissue-specific drug targeting. While we have a long road ahead in the discovery process, it is clear that the stakes are high in exploiting the secrets of this ancient but critical enzyme.

References

1. Gilmour KM (2010) Perspectives on carbonic anhydrase. Comp Biochem Physiol A Mol Integr Physiol 157:193–197
2. Henry RP (1996) Multiple roles of carbonic anhydrase in cellular transport and metabolism. Annu Rev Physiol 58:523–538
3. Henry RP, Swenson ER (2000) The distribution and physiological significance of carbonic anhydrase in vertebrate gas exchange organs. Respir Physiol 121:1–12
4. Frasseto F, Parisotto TM, Peres RC, Marques MR, Line SR, Nobre Dos Santos M (2012) Relationship among salivary carbonic anhydrase VI activity and flow rate, biofilm pH and caries in primary dentition. Caries Res 46:194–200
5. Chaput CD, Dangott LJ, Rahm MD, Hitt KD, Stewart DS, Wayne Sampson H (2012) A proteomic study of protein variation between osteopenic and age-matched control bone tissue. Exp Biol Med (Maywood) 237:491–498
6. Biswas UK, Kumar A (2012) Study on the changes of carbonic anhydrase activity in insulin resistance and the effect of methylglyoxal. J Pak Med Assoc 62:417–421
7. Brown BF, Quon A, Dyck JR, Casey JR (2012) Carbonic anhydrase II promotes cardiomyocyte hypertrophy. Can J Physiol Pharmacol 90:1599–1610
8. Sterling D, Reithmeier RA, Casey JR (2001) Carbonic anhydrase: in the driver's seat for bicarbonate transport. JOP 2:165–170
9. Purkerson JM, Schwartz GJ (2007) The role of carbonic anhydrases in renal physiology. Kidney Int 71:103–115
10. Kuo WH, Yang SF, Hsieh YS, Tsai CS, Hwang WL, Chu SC (2005) Differential expression of carbonic anhydrase isoenzymes in various types of anemia. Clin Chim Acta 351:79–86

11. Thiry A, Dogne JM, Masereel B, Supuran CT (2006) Targeting tumor-associated carbonic anhydrase IX in cancer therapy. Trends Pharmacol Sci 27:566–573
12. Whittington DA, Waheed A, Ulmasov B, Shah GN, Grubb JH, Sly WS, Christianson DW (2001) Crystal structure of the dimeric extracellular domain of human carbonic anhydrase XII, a bitopic membrane protein overexpressed in certain cancer tumor cells. Proc Natl Acad Sci U S A 98:9545–9550
13. Supuran CT (2008) Carbonic anhydrase: novel therapeutic applications for inhibitors and activators. Nat Rev Drug Discov 7:168–181
14. Pastorekova S, Pastorek J (2004) Cancer-related carbonic anhydrase isozymes and their inhibition. In: Supuran CT, Scozzafava A, Conway J (eds) Carbonic anhydrase: its inhibitors and activators. CRC Press, Bocal Raton, pp 255–281
15. Ditte P, Dequiedt F, Svastova E, Hulikova A, Ohradanova-Repic A, Zatovicova M, Csaderova L, Kopacek J, Supuran CT, Pastorekova S, Pastorek J (2011) Phosphorylation of carbonic anhydrase IX controls its ability to mediate extracellualr acidification in hypoxic tumors. Cancer Res 71:7558–7567
16. Chegwidden WR, Dodgson SJ, Spencer IM (2000) The roles of carbonic anhydrase in metabolism, cell growth and cancer in animals. EXS 90:343–363
17. Supuran CT (2008) Carbonic anhydrases–an overview. Curr Pharm Des 14:603–614
18. Hilvo M, Innocenti A, Monti SM, De Simone G, Supuran CT, Parkkila S (2008) Recent advances in research on the most novel carbonic anhydrases, CA XIII and XV. Curr Pharm Des 14:672–678
19. Aspatwar A, Tolvanen ME, Parkkila S (2010) Phylogeny and expression of carbonic anhydrase-related proteins. BMC Mol Biol 11:25
20. Silverman DN, Lindskog S (1988) The catalytic mechanism of carbonic anhydrase – implication of a rate limiting protolysis of water. Acc Chem Res 21:30–36
21. Jewell DA, Tu C, Paranawithana SR, Tanhauser SM, LoGrasso PV, Laipis PJ, Silverman DN (1991) Enhancement of the catalytic properties of human carbonic anhydrase III by site-directed mutagenesis. Biochemistry 30:1484–1490
22. Sly WS (2000) The membrane carbonic anhydrases: from CO2 transport to tumor markers. EXS 90:95–104
23. Maren TH, Swenson ER (1980) A comparative study of the kinetics of the Bohr effect in vertebrates. J Physiol 303:535–547
24. Esbaugh AJ, Tufts BL (2006) The structure and function of carbonic anhydrase isozymes in the respiratory system of vertebrates. Respir Physiol Neurobiol 154:185–198
25. Swenson ER (2000) Respiratory and renal roles of carbonic anhydrase in gas exchange and acid–base regulation. EXS 90:281–341
26. Chegwidden WR, Carter ND (2000) Introduction to the carbonic anhydrases. EXS 90:14–28
27. Chiang WL, Lai JC, Yang SF, Chiou HL, Hsieh YS (2001) Alternations in quantity and activities of erythrocyte cytosolic carbonic anhydrase isoenzymes in glucose-6-phosphate dehydrogenase individuals. Clin Chim Acta 314:195–201
28. Brown D, Kumpulainen T, Roth J, Orci L (1983) Immunohistochemical localization of carbonic anhydrase in postnatal and adult rat kidney. Am J Physiol 245:F110–F118
29. Lonnerholm G, Wistrand PJ, Barany E (1986) Carbonic anhydrase isoenzymes in the rat kidney. Effects of chronic acetazolamide treatment. Acta Physiol 126:51–60
30. Sly WS, Hewett-Emmett D, Whyte MP, Yu YS, Tashian RE (1983) Carbonic anhydrase II deficiency identified as the primary defect in the autosomal recessive syndrome of osteopetrosis with renal tubular acidosis and cerebral calcification. Proc Natl Acad Sci U S A 80:2752–2756
31. Lewis SE, Erickson RP, Barnett LB, Venta PJ, Tashian RE (1988) N-ethyl-N-nitrosourea-induced null mutation at the mouse Car-2 locus: an animal model for human carbonic anhydrase II deficiency syndrome. Proc Natl Acad Sci U S A 85:1962–1966
32. Breton S, Alper SL, Gluck SL, Sly WS, Barker JE, Brown D (1995) Depletion of intercalated cells from collecting ducts of carbonic anhydrase II-deficient (CAR2 null) mice. Am J Physiol 269:F761–F774

33. Bagnis C, Marshansky V, Breton S, Brown D (2001) Remodeling the cellular profile of collecting ducts by chronic carbonic anhydrase inhibition. Am J Physiol Renal Physiol 280:F437–F448
34. Vince JW, Reithmeier RAF (1998) Carbonic anhydrase II binds to the carboxyl-terminus of human band 3, the erythrocyte Cl^{-}/HCO_3^{-} Exchanger. J Biol Chem 273:28430–28437
35. McMurtrie HL, Cleary HJ, Alvarez BV, Loiselle FB, Sterling D, Morgan PE, Johnson DE, Casey JR (2004) The bicarbonate transport metabolon. J Enzyme Inhib Med Chem 19:231–236
36. Pushkin A, Abuladze N, Gross E, Newman D, Tatishchev S, Lee I, Fedotoff O, Bondar G, Azimov R, Ngyuen M, Kurtz I (2004) Molecular mechanism of kNBC1-carbonic anhydrase II interaction in proximal tubule cells. J Physiol 559:55–65
37. Becker HM, Deitmer JW (2007) Carbonic anhydrase II increases the activity of the human electrogenic Na^{+}/HCO_3^{-} cotransporter. J Biol Chem 282:13508–13521
38. Li X, Alvarez B, Casey JR, Reithmeier RA, Fliegel L (2002) Carbonic anhydrase II binds to and enhances activity of the Na^{+}/H^{+} exchanger. J Biol Chem 277:36085–36091
39. Li X, Liu Y, Alvarez BV, Casey JR, Fliegel L (2006) A novel carbonic anhydrase II binding site regulates NHE1 activity. Biochemistry 45:2414–2424
40. Vince JW, Carlsson U, Reithmeier RAF (2002) Localization of the Cl^{-}/HCO_3^{-} anion exchanger binding site to the amino-terminal region of carbonic anhydrase II. Biochemistry 39:13344–13349
41. Becker HM, Hirnet D, Fecher-Trost C, Sultemeyer D, Deitmer JW (2005) Transport activity of MCT1 expressed in Xenopus oocytes is increased by interaction with carbonic anhydrase. J Biol Chem 280:39882–39889
42. Becker HM, Klier M, Deitmer JW (2010) Nonenzymatic augmentation of lactate transport via monocarboxylate transporter isoform 4 by carbonic anhydrase II. J Membr Biol 234:125–135
43. Stridh MH, Alt MD, Wittmann S, Heidtmann H, Aggarwal M, Riederer B, Seidler U, Wennemuth G, McKenna R, Deitmer JW, Becker HM (2012) Lactate flux in astrocytes is enhanced by a non-catalytic action of carbonic anhydrase II. J Physiol 590:2333–2351
44. Becker HM, Klier M, Schuler C, McKenna R, Deitmer JW (2011) Intramolecular proton shuttle supports not only catalytic but also noncatalytic function of carbonic anhydrase II. Proc Natl Acad Sci U S A 108:3071–3076
45. Becker HM, Deitmer JW (2008) Nonenzymatic proton handling by carbonic anhydrase II during H + —lactate cotransport via monocarboxylate transporter 1. J Biol Chem 283:21655–21667
46. Carter ND (1991) Hormonal and neuronal control of carbonic anhydrase III gene expression in skeletal muscle. In: Dodgson SJ, Tashian RE, Gross G, Carter ND (eds) The carbonic anhydrases: cellular physiology and moledular genetics. Plenum Publishing, New York, pp 247–256
47. Stanton LW, Ponte PA, Coleman RT, Snyder MA (1991) Expression of CA III in rodent models of obesity. Mol Endocrinol 5:860–866
48. Lyons GE, Buckingham ME, Tweedie S, Edwards YH (1991) Carbonic anhydrase III, an early mesodermal marker, is expressed in embryonic mouse skeletal muscle and notochord. Development 111:233–244
49. Chai YC, Jung CH, Lii CK, Ashraf SS, Hendrich S, Wolf B, Sies H, Thomas JA (1991) Identification of an abundant S-thiolated rat liver protein as carbonic anhydrase III; characterization of S-thiolation and dethiolation reactions. Arch Biochem Biophys 284:270–278
50. Lii CK, Chai YC, Zhao W, Thomas JA, Hendrich S (1994) S-thiolation and irreversible oxidation of sulfhydryls on carbonic anhydrase III during oxidative stress: a method for studying protein modification in intact cells and tissues. Arch Biochem Biophys 308:231–239
51. Thomas JA, Poland B, Honzatko R (1995) Protein sulfhydryls and their role in the antioxidant function of protein S-thiolation. Arch Biochem Biophys 319:1–9
52. Raisanen SR, Lehenkari P, Tasanen M, Rahkila P, Harkonen PL, Vaananen HK (1999) Carbonic anhydrase III protects cells from hydrogen peroxide-induced apoptosis. FASEB J 13:513–522

53. Mallis RJ, Hamann MJ, Zhao W, Zhang T, Hendrich S, Thomas JA (2002) Irreversible thiol oxidation in carbonic anhydrase III: protetion by S-glutathiolation and detection in aging rats. Biol Chem 383:649–662
54. Thomas JA, Mallis RJ (2001) Aging and oxidation of reactive protein sulfhydryls. Exp Gerontol 36:1519–1526
55. Kim G, Lee TH, Wetzel P, Geers C, Robinson MA, Myers TG, Owens JW, Wehr NB, Eckhaus MW, Gros G, Wynshaw-Boris A, Levine RL (2004) Carbonic anhydrase III is not required in the mouse for normal growth, development, and life span. Mol Cell Biol 24:9942–9947
56. Rosen ED, Spiegelman BM (2006) Adipocytes as regulators of energy balance and glucose homeostasis. Nature 444:847–853
57. Barnard T (1969) The ultrastructural differentiation of brown adipose tissue in the rat. J Ultrastruct Res 29:311–322
58. Cinti S, Cigolini M, Bosello O, Bjorntorp P (1984) A morphological study of the adipocyte precursor. J Submicrosc Cytol 16:243–251
59. Hausman GJ, Campion DR, Martin RJ (1980) Search for the adipocyte precursor cell and factors that promote its differentiation. J Lipid Res 21:657–670
60. Amos PJ, Shang H, Bailey AM, Taylor A, Katz AJ, Peirce SM (2008) IFATS collection: the role of human adipose-derived stromal cells in inflammatory microvascular remodeling and evidence of a perivascular phenotype. Stem Cells 26:2682–2690
61. Crisan M, Yap S, Casteilla L, Chen CW, Corselli M, Park TS, Andriolo G, Sun B, Zheng B, Zhang L, Norotte C, Teng PN, Traas J, Schugar R, Deasy BM, Badylak S, Buhring HJ, Giacobino JP, Lazzari L, Huard J, Peault B (2008) A perivascular origin for mesenchymal stem cells in multiple human organs. Cell Stem Cell 3:301–313
62. Traktuev DO, Merfeld-Clauss S, Li J, Kolonin M, Arap W, Pasqualini R, Johnstone BH, March KL (2008) A population of multipotent CD34-positive adipose stromal cells share pericyte and mesenchymal surface markers, reside in a periendothelial location, and stabilize endothelial networks. Circ Res 102:77–85
63. Zannettino AC, Paton S, Arthur A, Khor F, Itescu S, Gimble JM, Gronthos S (2008) Multipotential human adipose-derived stromal stem cells exhibit a perivascular phenotype in vitro and in vivo. J Cell Physiol 214:413–421
64. Zimmerlin L, Donnenberg VS, Pfeifer ME, Meyer EM, Peault B, Rubin JP, Donnenberg AD (2010) Stromal vascular progenitors in adult human adipose tissue. Cytometry A 77:22–30
65. Gupta RK, Mepani RJ, Kleiner S, Lo JC, Khandekar MJ, Cohen P, Frontini A, Bhowmick DC, Ye L, Cinti S, Spiegelman BM (2012) Zfp423 expression identifies committed preadipocytes and localizes to adipose endothelial and perivascular cells. Cell Metab 15:230–239
66. Tang W, Zeve D, Suh JM, Bosnakovski D, Kyba M, Hammer RE, Tallquist MD, Graff JM (2008) White fat progenitor cells reside in the adipose vasculature. Science 322:583–586
67. Lynch CJ, Hazen SA, Horetsky RL, Carter ND, Dodgson SJ (1993) Differentiation-dependent expression of carbonic anhydrase II and III in 3T3 adipocytes. Am J Physiol 265:C234–C243
68. Cao TP, Rous S (1978) Inhibitory effect of acetazolamide on the activity of acetyl CoA carboxylase of mouse liver. Life Sci 22:2067–2072
69. Lynch CJ, Brennan WA Jr, Vary TC, Carter N, Dodgson SJ (1993) Carbonic anhydrase III in obese Zucker rats. Am J Physiol 264:E621–E630
70. Bray GA, York DA (1979) Hypothalamic and genetic obesity in experimental animals: an autonomic and endocrine hypothesis. Physiol Rev 59:719–809
71. Mitterberger MC, Kim G, Rostek U, Levine RL, Zwerschke W (2012) Carbonic anhydrase III regulates peroxisome proliferator-activated receptor-gamma2. Exp Cell Res 318:877–886
72. Houstis N, Rosen ED, Lander ES (2006) Reactive oxygen species have a causal role in multiple forms of insulin resistance. Nature 440:944–948
73. Montgomery JC, Venta PJ, Eddy RL, Fukushima YS, Shows TB, Tashian RE (1991) Characterization of the human gene for a newly discovered carbonic anhydrase, CA VII, and its localization to chromosome 16. Genomics 11:835–848
74. Lakkis MM, Bergenhem NC, Tashian RE (1996) Expression of mouse carbonic anhydrase VII in E. coli and demonstration of its CO_2 hydrase activity. Biochem Biophys Res Commun 226:268–272

75. Truppo E, Supuran CT, Sandomenico A, Vullo D, Innocenti A, Di Fiore A, Alterio V, De Simone G, Monti SM (2012) Carbonic anhydrase VII is S-glutathionylated without loss of catalytic activity and affinity for sulfonamide inhibitors. Bioorg Med Chem Lett 22:1560–1564
76. Bootorabi F, Janis J, Smith E, Waheed A, Kukkurainen S, Hytonen V, Valjakka J, Supuran CT, Vullo D, Sly WS, Parkkila S (2010) Analysis of a shortened form of human carbonic anhydrase VII expressed in vitro compared to the full-length enzyme. Biochimie 92:1072–1080
77. Thiry A, Dogne JM, Supuran CT, Masereel B (2007) Carbonic anhydrase inhibitors as anticonvulsant agents. Curr Top Med Chem 7:855–864
78. Ruusuvuori E, Li H, Huttu K, Palva JM, Smirnov S, Rivera C, Kaila K, Voipio J (2004) Carbonic anhydrase isoform VII acts as a molecular switch in the development of synchronous gamma-frequency firing of hippocampal CA1 pyramidal cells. J Neurosci 24:2699–2707
79. Asiedu M, Ossipov MH, Kaila K, Price TJ (2010) Acetazolamide and midazolam act synergistically to inhibit neuropathic pain. Pain 148:302–308
80. Lehtonen J, Shen B, Vihinen M, Casini A, Scozzafava A, Supuran CT, Parkkila A, Saarnio J, Kivela AJ, Waheed A, Sly WS, Parkkila S (2004) Characterization of CA XIII, a novel member of the carbonic anhydrase isozyme family. J Biol Chem 279:2719–2727
81. Lehtonen JM, Parkkila S, Vullo D, Casini A, Scozzafava A, Supuran CT (2004) Carbonic anhydrase inhibitors. Inhibition of cytosolic isozyme XIII with aromatic and heterocyclic sulfonamides: a novel target for the drug design. Bioorg Med Chem Lett 14:3757–3762
82. Kummola L, Hamalainen JM, Kivela J, Kivela AJ, Saarnio J, Karttunen T, Parkkila S (2005) Expression of a novel carbonic anhydrase, CA XIII, in normal and neoplastic colrectal mucosa. BMC Cancer 4:1–7
83. Wandernoth PM, Raubuch M, Mannowetz N, Becker HM, Deitmer JW, Sly WS, Wennemuth G (2010) Role of carbonic anhydrase IV in the bicarbonate-mediated activation of murine and human sperm. PLoS One 5:e15061
84. Chappell JB, Crofts AR (1966) Ion transport and reversible volume changes of isolated mitochondria. In: Tager JM, Papa S, Qualiariello E, Slater EC (eds) Regulation of metabolic processes in mitochondria. Elsevier, Amsterdam, pp 293–316
85. Elder JA (1972) Energy-linked accumulation of bicarbonate by rat liver mitochondria. FASEB J 31:856
86. Elder JA, Lehninger AL (1973) Respiration-dependent transport of carbon dioxide into rat liver mitochondria. Biochemistry 12:976–982
87. Dodgson SJ, Forster RE 2nd (1986) Inhibition of CA V decreases glucose synthesis from pyruvate. Arch Biochem Biophys 251:198–204
88. Dodgson SJ, Forster RE 2nd (1986) Carbonic anhydrase: inhibition results in decreased urea production by hepatocytes. J Appl Physiol 60:646–652
89. Dodgson SJ (1987) Inhibition of mitochondrial carbonic anhydrase and ureagenesis: a discrepancy examined. J Appl Physiol 63:2134–2141
90. Amor-Gueret M, Levi-Strauss M (1990) Nucleotide and derived amino-acid sequence of a cDNA encoding a new mouse carbonic anhydrase. Nucleic Acids Res 18:1646
91. Nagao Y, Srinivasan M, Platero JS, Svendrowski M, Waheed A, Sly WS (1994) Mitochondrial carbonic anhydrase (isozyme V) in mouse and rat: cDNA cloning, expression, subcellular localization, processing, and tissue distribution. Proc Natl Acad Sci U S A 91:10330–10334
92. Nagao Y, Platero JS, Waheed A, Sly WS (1993) Human mitochondrial carbonic anhydrase: cDNA cloning, expression, subcellular localization, and mapping to chromosome 16. Proc Natl Acad Sci U S A 90:7623–7627
93. Shah GN, Hewett-Emmett D, Grubb JH, Migas MC, Fleming RE, Waheed A, Sly WS (2000) Mitochondrial carbonic anhydrase CA VB: differences in tissue distribution and pattern of evolution from those of CA VA suggest distinct physiological roles. Proc Natl Acad Sci U S A 97:1677–1682
94. Fujikawa-Adachi K, Nishimori I, Taguchi T, Onishi S (1999) Human mitochondrial carbonic anhydrase VB. cDNA cloning, mRNA expression, subcellular localization, and mapping to chromosome x. J Biol Chem 274:21228–21233

95. Nagao Y, Batanian JR, Clemente MF, Sly WS (1995) Genomic organization of the human gene (CA5) and pseudogene for mitochondrial carbonic anhydrase V and their localization to chromosomes 16q and 16p. Genomics 28:477–484
96. Lusty CJ (1978) Carbamyl phosphate synthetase. Bicarbonate-dependent hydrolysis of ATP and potassium activation. J Biol Chem 253:4270–4278
97. Cohen PP (1981) The ornithine-urea cycle: biosynthesis and regulation of carbamyl phosphate synthetase I and ornithine transcarbamylase. Curr Top Cell Regul 18:1–19
98. McGivan JD, Bradford NM, Mendes-Mourao J (1976) The regulation of carbamoyl phosphate synthase activity in rat liver mitochondria. Biochem J 154:415–421
99. Lusty CJ (1978) Carbamoylphosphate synthetase I of rat-liver mitochondria. Purification, properties, and polypeptide molecular weight. Eur J Biochem/FEBS 85:373–383
100. Dodgson SJ, Forster RE 2nd, Schwed DA, Storey BT (1983) Contribution of matrix carbonic anhydrase to citrulline synthesis in isolated guinea pig liver mitochondria. J Biol Chem 258:7696–7701
101. Cao TP, Rous S (1978) Action of acetazolamide on liver pyruvate carboxylase activity, glycogenolysis and gluconeogenesis of mice. Int J Biochem 9:603–605
102. Hazen SA, Waheed A, Sly WS, LaNoue KF, Lynch CJ (1996) Differentiation-dependent expression of CA V and the role of carbonic anhydrase isozymes in pyruvate carboxylation in adipocytes. FASEB J 10:481–490
103. Mohammadi A, Leibfritz D (2009) Inhibitory effect of carbonic anhydrase inhibitors on the de novo lipogenesis. A study with 13C-NMR spectroscopy. Proc Int Soc Magn Reson Med 17:2374
104. Lynch CJ, Fox H, Hazen SA, Stanley BA, Dodgson S, Lanoue KF (1995) Role of hepatic carbonic anhydrase in de novo lipogenesis. Biochem J 310(Pt 1):197–202
105. Wallace DC (1999) Mitochondrial diseases in man and mouse. Science 283:1482–1488
106. Oommen KJ, Mathews S (1999) Zonisamide: a new antiepileptic drug. Clin Neuropharmacol 22:192–200
107. Gadde KM, Franciscy DM, Wagner HR II, Krishnan KR (2003) Zonisamide for weight loss in obese adults: a randomized controlled trial. JAMA 289:1820–1825
108. De Simone G, Di Fiore A, Menchise V, Pedone C, Antel J, Casini A, Scozzafava A, Wurl M, Supuran CT (2005) Carbonic anhydrase inhibitors. Zonisamide is an effective inhibitor of the cytosolic isozyme II and mitochondrial isozyme V: solution and X-ray crystallographic studies. Bioorg Med Chem Lett 15:2315–2320
109. Poulsen SA, Wilkinson BL, Innocenti A, Vullo D, Supuran CT (2008) Inhibition of human mitochondrial carbonic anhydrases VA and VB with para-(4-phenyltriazole-1-yl)-benzenesulfonamide derivatives. Bioorg Med Chem Lett 18:4624–4627
110. Arechederra RL, Waheed A, Sly WS, Supuran CT, Minteer SD (2013) Effect of sulfonamides as carbonic anhydrase VA and VB inhibitors on mitochondrial metabolic energy conversion. Bioorg Med Chem 21:1544–1548
111. Nishimori I, Vullo D, Innocenti A, Scozzafava A, Mastrolorenzo A, Supuran CT (2005) Carbonic anhydrase inhibitors. The mitochondrial isozyme VB as a new target for sulfonamide and sulfamate inhibitors. J Med Chem 48:7860–7866
112. Kivela J, Parkkila S, Parkkila AK, Leinonen J, Rajaniemi H (1999) Salivary carbonic anhydrase isoenzyme VI. J Physiol 520(Pt 2):315–320
113. Fernley RT, Wright RD, Coghlan JP (1979) A novel carbonic anhydrase from ovine parotid glands. FEBS Lett 105:299–302
114. Feldstein JB, Silverman DN (1984) Purification and characterization of carbonic anhydrase from the saliva of the rat. J Biol Chem 259:5447–5453
115. Murakami H, Sly WS (1987) Purification and characterization of human salivary carbonic anhydrase. J Biol Chem 262:1382–1388
116. Parkkila S, Parkkila AK, Rajaniemi H (1995) Circadian periodicity in salivary carbonic anhydrase VI concentration. Acta Physiol Scand 154:205–211
117. Karhumaa P, Lienonen J, Parkkila S, Kaunisto K, Tapanainen J, Rajanemi H (2001) The identification of secreted carbonic anydrase VI as a constitutive glycoprotein of human and rat milk. Proc Natl Acad Sci U S A 98:11604–11608

118. Ogawa Y, Matsumoto K, Maeda T, Tamai R, Suzuki T, Sasano H, Fernley RT (2002) Characterization of lacrimal gland carbonic anhydrase VI. J Histochem Cytochem 50:821–827
119. Leinonen JS, Saari KA, Seppanen JM, Myllyla HM, Rajaniemi HJ (2004) Immunohistochemical demonstration of carbonic anhydrase isoenzyme VI (CA VI) expression in rat lower airways and lung. J Histochem Cytochem 52:1107–1112
120. Kaseda M, Ichihara N, Nishita T, Amasaki H, Asari M (2006) Immunohistochemistry of the bovine secretory carbonic anhydrase isozyme (CA-VI) in bovine alimentary canal and major salivary glands. J Vet Med Sci 68:131–135
121. Smith CE, Nanci A, Moffatt P (2006) Evidence by signal peptide trap technology for the expression of carbonic anhydrase 6 in rat incisor enamel organs. Eur J Oral Sci 114(Suppl 1):147–153
122. Kivela J, Parkkila S, Waheed A, Parkkila AK, Sly WS, Rajaniemi H (1997) Secretory carbonic anhydrase isoenzyme (CA VI) in human serum. Clin Chem 43:2318–2322
123. Ship JA (2003) Diabetes and oral health: an overview. J Am Dent Assoc 134 Spec No:4S–10S
124. Dowd FJ (1999) Saliva and dental caries. Dent Clin North Am 43:579–597
125. Kimoto M, Kishino M, Yura Y, Ogawa Y (2006) A role of salivary carbonic anhydrase VI in dental plaque. Arch Oral Biol 51:117–122
126. Kivela J, Parkkila S, Parkkila AK, Rajaniemi H (1999) A low concentration of carbonic anhydrase isoenzyme VI in whole saliva is associated with caries prevalence. Caries Res 33:178–184
127. Peres RC, Camargo G, Mofatto LS, Cortellazzi KL, Santos MC, Nobre-dos-Santos M, Bergamaschi CC, Line SR (2010) Association of polymorphisms in the carbonic anhydrase 6 gene with salivary buffer capacitiy, dental plaque pH, and caries index in children aged 7–9 years. Pharmacogenomics J 10:114–119
128. Ozturk K, Ulucan K, Akyuz S, Furuncuoglu H, Bayer H, Yarat A (2012) The investigation of genetic polymorphisms in the carbonic anhydrase VI gene exon 2 and salivary parameters in type 2 diabetic patients and healthy adults. Mol Biol Rep 39:5677–5682
129. Henkin RI, Martin BM, Agarwal RP (1999) Decreased parotid saliva gustin/carbonic anhydrase VI secretion: an enzyme disorder manifested by gustatory and olfactory dysfunction. Am J Med Sci 318:380–391
130. Henkin RI, Lippoldt RE, Bilstad J, Edelhoch H (1975) A zinc protein isolated from human parotid saliva. Proc Natl Acad Sci U S A 72:488–492
131. Shatzman AR, Henkin RI (1980) Metal-binding characteristics of the parotid salivary protein gustin. Biochim Biophys Acta 623:107–118
132. Shatzman AR, Henkin RI (1981) Gustin concentration changes relative to salivary zinc and taste in humans. Proc Natl Acad Sci U S A 78:3867–3871
133. Henkin RI, Schechter PJ, Hoye R, Mattern CF (1971) Idiopathic hypogeusia with dysgeusia, hyposmia, and dysosmia. A new syndrome. JAMA 217:434–440
134. Ortho-McNeil-Janssen Pharmaceuticals I (2013) Topomax: drug summary. Physicians' desk reference http://www.pdr.net
135. Pilka ES, Kochan G, Oppermann U, Yue WW (2012) Crystal structure of the secretory isozyme of mammalian carbonic anhydrases CA VI: implications for biological assembly and inhibitor development. Biochem Biophys Res Commun 419:485–489
136. Hilvo M, Tolvanen M, Clark A, Shen B, Shah GN, Waheed A, Halmi P, Hanninen M, Hamalainen JM, Vihinen M, Sly WS, Parkkila S (2005) Characterization of CA XV, a new GPI-anchored from of carbonic anhydrase. Biochem J 392:83–92
137. Whitney PL, Briggle TV (1982) Membrane-associated carbonic anhydrase purified from bovine lung. J Biol Chem 257:12056–12059
138. Wistrand PJ (1984) Properties of membrane-bound carbonic anhydrase. Ann N Y Acad Sci 429:195–206
139. Zhu XL, Sly WS (1990) Carbonic anhydrase IV from human lung. Purification, characterization, and comparison with membrane carbonic anhydrase from human kidney. J Biol Chem 265:8795–8801

140. Okuyama T, Sato S, Zhu XL, Waheed A, Sly WS (1992) Human carbonic anhydrase IV: cDNA cloning, sequence comparison, and expression in COS cell membranes. Proc Natl Acad Sci U S A 89:1315–1319
141. Baird TT Jr, Waheed A, Okuyama T, Sly WS, Fierke CA (1997) Catalysis and inhibition of human carbonic anhydrase IV. Biochemistry 36:2669–2678
142. Sender S, Decker B, Fenske CD, Sly WS, Carter ND, Gros G (1998) Localization of carbonic anhydrase IV in rat and human heart muscle. J Histochem Cytochem 46:855–861
143. Brion LP, Suarez C, Zhang H, Cammer W (1994) Up-regulation of carbonic anhydrase isozyme IV in CNS myelin of mice genetically deficient in carbonic anhydrase II. J Neurochem 63:360–366
144. Hageman GS, Zhu XL, Waheed A, Sly WS (1991) Localization of carbonic anhydrase IV in a specific capillary bed of the human eye. Proc Natl Acad Sci U S A 88:2716–2720
145. Wistrand PJ, Carter ND, Conroy CW, Mahieu I (1999) Carbonic anhydrase IV activity is localized on the exterior surface of human erythrocytes. Acta Physiol Scand 165:211–218
146. Conroy CW, Wynns GC, Maren TH (1996) Synthesis and properties of two new membrane-impermeant high-molecular-weight carbonic anhydrase inhibitors. Bioorg Chem 24:262–272
147. Maren TH, Conroy CW, Wynns GC, Godman DR (1997) Renal and cerbrospinal fluid formation pharmacology of a high molecular weight carbonic anhydrase inhibitor. J Pharmacol Exp Ther 280:98–104
148. Maren TH (1969) Renal carbonic anhydrase and the pharmacology of sulfonamide inhibitors. Springer-Verlag, Berlin
149. Sly WS, Whyte MP, Krupin T, Sundaram V (1985) Positive renal response to intravenous acetazolamide in patients with carbonic anhydrase II deficiency. Pediatr Res 19:1033–1036
150. Tureci O, Sahin U, Vollmar E, Siemer S, Gottert E, Seitz G, Parkkila A, Shah GN, Grubb JH, Pfreundschuh M, Sly WS (1998) Human carbonic anhydrase XII: cDNA cloning, expression, and chromosomal location of a carbonic anhydrase gene that is overexpressed in some renal cancers. Proc Natl Acad Sci U S A 93:7608–7613
151. Schwartz GL, Kittelberger AM, Watkins RH, O'Reilly MA (2003) Carbonic anhydrase XII mRNA encodes a hydratase that is differentially expressed along the rabbit nephron. Am J Physiol 284:F399–F410
152. Sterling D, Alvarez BV, Casey JR (2002) The extracellular component of a transport metabolon: extracellular loop 4 of the human AE1 Cl^-/HCO_3^- exchanger binds carbonic anhydrase IV. J Biol Chem 277:25239–25246
153. Yang Z, Alvarez B, Chakarova C, Jiang L, Karan G, Frederick JM, Zhao Y, Sauve Y, Zrenner E, Wissinger B, Den Hollander AI, Katz B, Baehr W, Cremers FP, Casey JR, Bhattacharya SS, Zhang K (2005) Mutant carbonic anhydrase 4 impairs pH regulation and causes retinal photoreceptor degeneration. Hum Mol Genet 14:255–265
154. Rebello G, Ramesar R, Vorster A, Roberts L, Ehrenreich L, Oppon E, Gama D, Bardien S, Greenberg J, Bonapace G, Waheed A, Shah GN, Sly WS (2004) Apoptosis-inducing signal sequence mutation in carbonic anhydrase IV identified in patients with the RP17 form of retinitis pigmentosa. Proc Natl Acad Sci U S A 101:6617–6622
155. Ivanov S, Liao SY, Ivanova A, Danilkovitch-Miagkova A, Tarasova N, Weirich G, Merrill MJ, Proescholdt MA, Oldfield EH, Lee J, Zavada J, Waheed A, Sly W, Lerman MI, Stanbridge EJ (2001) Expression of hypoxia-inducible cell-surface transmembrane carbonic anhydrases in human cancer. Am J Pathol 158:905–919
156. Pastorekova S, Parkkila S, Parkkila A, Opavsky R, Zelnik V, Saarnio J, Pastorek J (1997) Carbonic anhydrase IX, MN/CAIX: analysis of stomach complementary DNA sequence and expression in human and rat alimentary tracts. Gastroenterology 112:398–408
157. Saarnio J, Parkkila S, Parkkila AK, Waheed A, Casey MC, Zhou XY, Pastorekova S, Pastorek J, Karttunen T, Haukipuro K, Kairaluoma MI, Sly WS (1998) Immunohistochemistry of carbonic anhydrase isozyme IX (MN/CA IX) in human gut reveals polarized expression in epithelial cells with the highest proliferative capacity. J Histochem Cytochem 46:497–504

158. Parkkila S, Rajaniemi H, Parkkila A, Kivela J, Waheed A, Pastorekova S, Pastorek J, Sly WS (2000) Carbonic anhydrase inhibitor suppresses invasion of renal cancer cells in vitro. Proc Natl Acad Sci U S A 97:2220–2224
159. Robertson N, Potter C, Harris AL (2004) Role of carbonic anhydrase IX in human tumor cell growth, survival, and invasion. Cancer Res 64:6160–6165
160. Svastova E, Zilka N, Zatovicova M, Gibadulinova A, Ciampor F, Pastorek J, Pastorekova S (2003) Carbonic anhydrase IX reduces E-cadherin-mediated adhesion of MDCK cells via interaction with á-catenin. Exp Cell Res 290:332–345
161. Lou Y, McDonald PC, Oloumi A, Chia S, Ostlund C, Ahmadi A, Kyle A, auf dem Keller U, Leung S, Huntsman D, Clarke B, Sutherland BW, Waterhouse D, Bally M, Roskelley C, Overall CM, Minchinton A, Pacchiano F, Carta F, Scozzafava A, Touisni N, Winum J, Supuran CT, Dedhar S (2011) Targeting tumor hypoxia: suppression of breast tumor growth and metastasis by novel carbonic anhydrase IX inhibitors. Cancer Res 71:3364–3376
162. Swietach P, Hulikova A, Vaughan-Jones RD, Harris AL (2010) New insights into the physiological role of carbonic anhydrase IX in tumour pH regulation. Oncogene 29:6509–6521
163. Chiche J, Ilc K, Brahimi-Horn MC, Pouyssegur J (2010) Membrane-bound carbonic anhydrases are key pH regulators controlling tumor growth and cell migration. Adv Enzyme Regul 50:20–33
164. Svastova E, Hulikova A, Rafajova M, Zatovicova M, Gibadulinova A, Casini A, Cecchi A, Scozzafava A, Supuran CT, Pastorek J, Pastorekova S (2004) Hypoxia activates the capacity of tumor-associated carbonic anhydrase IX to acidify extracellular pH. FEBS Lett 577: 439–445
165. Li Y, Tu C, Wang H, Silverman DN, Frost SC (2011) Catalysis and pH control by membrane-associated carbonic anhydrase IX in MDA-MB-231 breast cancer cells. J Biol Chem 286:15789–15796
166. Pastorek J, Pastorekova S, Callebaut I, Mornon JP, Zeln¡k V, Opavsky R, Zat'ovicov M, Liao S, Portetelle D, Stanbridge EJ, Zá-vada J, Burny A, Kettmann R (1994) Cloning and characterization of MN, a human tumor-associated protein with a domain homologous to carbonic anhydrase and a putative helix-loop-helix DNA binding segment. Oncogene 9:2877–2888
167. Opavsky R, Pastorekova S, Zeln¡k V, Gibadulinov A, Stanbridge EJ, Zá-vada J, Kettmann R, Pastorek J (1996) Human MN/CA9 Gene, a novel member of the carbonic anhydrase family: structure and exon to protein domain relationships. Genomics 33:480–487
168. Hilvo M, Baranauskiene L, Salzano AM, Scaloni A, Matulis D, Innocenti A, Scozzafava A, Monti SM, Di Fiore A, De Simone G, Lindfors M, Janis J, Valjakka J, Pastorekova S, Pastorek J, Kulomaa MS, Mordlund HR, Supuran CT, Parkkila S (2008) Biochemical characterization of CA IX, one of the most active carbonic anhydrase isozymes. J Biol Chem 283:27799–27809
169. Li Y, Wang H, Tu C, Shiverick KT, Silverman DN, Frost SC (2011) Role of hypoxia and EGF on expression, activity, localization, and phosphorylation of carbonic anhydrase IX in MDA-MB-231 breast cancer cells. Biochim Biophys Acta 1813:159–167
170. Alterio V, Hilvo M, Di Fiore A, Supuran CT, Pan P, Parkkila S, Scaloni A, Pastorek J, Pastorekova S, Pedone C, Scozzafava A, Monti SM, De Simone G (2009) Crystal structure of the catalytic domain of the tumor-associated human carbonic anhydrase IX. Proc Natl Acad Sci U S A 106:16233–16238
171. Orlowski A, De Giusti VC, Morgan PE, Aiello EA, Alvarez BV (2012) Binding of carbonic anhydrase IX to extracellular loop 4 of the NBCe1 Na^+/HCO_3^- cotransporter enhances NBCe1-medicated HCO_3^- influx in the heart. Am J Physiol Cell Physiol 303:C69–C80
172. Wingo T, Tu C, Laipis PJ, Silverman DN (2001) The catalytic properties of human carbonic anhydrase IX. Biochem Biophys Res Commun 288:666–669
173. Wykoff CC, Beasley NJP, Watson PH, Turner KJ, Pastorek J, Sibtain A, Wilson GD, Turley H, Talks KL, Maxwell PH, Pugh CW, Ratcliffe PJ, Harris AL (2000) Hypoxia-inducible expression of tumor-associated carbonic anhydrase. Cancer Res 60:7075–7083

174. Chia SK, Wykoff CC, Watson PH, Han C, Leek RD, Pastorek J, Gatter KC, Ratcliffe P, Harris AL (2001) Prognostic significance of a novel hypoxia-regulated marker, carbonic anhydrase IX, in invasive breast cancer. J Clin Oncol 19:3660–3668
175. Generali D, Fox SB, Berruti A, Brizzi MP, Campo L, Bonardi S, Wigfield SM, Bruzzi P, Bersiga A, Allevi G, Milani M, Aguggini S, Dogliotti L, Bottini A, Harris AL (2006) Role of carbonic anhydrase IX expression in prediction of the efficacy and outcome of primary epirubicin/tamoxifen therapy for breast cancer. Endocr Relat Cancer 13:921–930
176. Span PM, Bussink J, Manders P, Beex LVAM, Sweep CGJ (2003) Carbonic anhydrase-9 expression levels and prognosis in human breast cancer: association with treatment outcome. Br J Cancer 89:271–276
177. Nordfors K, Haapasalo J, Korja M, Niemela A, Laine J, Parkkila A, Pastorekova S, Pastorek J, Waheed A, Sly WS, Parkkila S, Haapasalo H (2010) The tumour-associated carbonic anhydrases CA II, CA IX and CA XII in a group of medulloblastomas and supratentorial primitive neuroectodermal tumours: an association of CA IX with poor prognosis. BMC Cancer 10:148
178. Ulmasov B, Waheed A, Shah GN, Grubb JH, Sly WS, Tu C, Silverman DN (2000) Purification and kinetic analysis of recombinant CAXII, a membrane carbonic anhydrase overexpressed in certain cancers. Proc Natl Acad Sci U S A 97:14212–14217
179. Vullo D, Innocenti A, Nishimori I, Pastorek J, Scozzafava A, Pastorekova S, Supuran CT (2005) Carbonic anhydrase inhibitors: inhibition of the transmembrane isozyme XII with sulfonamides – a new target for the design of antitumor and antiglaucoma drugs. Bioorg Med Chem Lett 15:963–969
180. Parkkila S, Parkkila AK, Saarnio J, Kivela J, Karttunen TJ, Kaunisto K, Waheed A, Sly WS, Tureci O, Virtanen I, Rajaniemi H (2000) Expression of the membrane-associated carbonic anhydrase isozyme XII in the human kidney and renal tumors. J Histochem Cytochem 48:1601–1608
181. Hynninen P, Parkkila S, Huhtala H, Pastorekova S, Pastorek J, Wahl RL, Sly WS, Tomas E (2011) Carbonic anhydrase isozymes II, IX and XII in uterine tumors. Acta Physiol Microbiol Immunol Scand 120:117–129
182. Kivela A, Parkkila S, Saarnio J, Karttunen TJ, Kivela J, Parkkila A, Waheed A, Sly WS, Grubb JH, Shah G, Tureci O, Rajaniemi H (2000) Expression of a novel transmembrane carbonic anhydrase XII in normal human gut and colorectal tumors. Am J Pathol 156:577–584
183. Creighton CJ, Cordero KE, Larios JM, Miller RS, Johnson MD, Chinnaiyan AR, Lippman ME, Rae JM (2006) Genes regulated by estrogen in breast tumor cells in vitro are similarly regulated in vivo in tumor xernografts and human breast tumors. Genome Biol 7(R28):1–13
184. Barnett DH, Sheng S, Charn TH, Waheed A, Sly WS, Lin CY, Liu ET, Katzenellenbogen BS (2008) Estrogen receptor regulation of carbonic anhydrase XII through a distal enhancer in breast cancer. Cancer Res 68:3505–3515
185. Wykoff CC, Beasley N, Watson PH, Campo L, Chia SK, English R, Pastorek J, Sly WS, Ratcliffe P, Harris AL (2001) Expression of hypoxia-inducible and tumor-associated carbonic anhydrases in ductal carcinoma in situ of the breast. Am J Pathol 158:1011–1019
186. Watson PH, Chia SK, Wykoff CC, Han C, Leek RD, Sly WS, Gatter KC, Ratcliffe P, Harris AL (2003) Carbonic anhydrase XII is a marker of good prognosis in invasive breast carcinoma. Br J Cancer 88:1065–1070
187. Ilie MI, Hofman V, Ortholan C, El Ammadi R, Bonnetaud C, Havet K, Venissac N, Mouroux J, Mazure NM, Pouyssegur J, Hofman P (2011) Overexpression of carbonic anhydrase XII in tissues from resectable non-small cell lung cancers is a biomarker of good prognosis. Int J Cancer 128:1614–1623
188. Chien MH, Ying TH, Hsieh YH, Lin CH, Shih CH, Wei LH, Yang SF (2012) Tumor-associated carbonic anhydrase XII is linked to the growth of primary oral squamous cell carcinoma and its poor prognosis. Oral Oncol 48:417–423
189. Mirza NS, Alfirevic A, Jorgensen A, Marson AG, Pirmohamed M (2011) Metabolic acidosis with topiramate and zonisamide: an assessment of its severity and predictors. Pharmacogenet Genomics 21:297–302

190. Muhammad E, Leventhal N, Parvari G, Hanukoglu A, Hanukoglu I, Chalifa-Caspi V, Feinstein Y, Weinbrand J, Jacoby H, Manor E, Nagar T, Beck JC, Sheffield VC, Hershkovitz E, Parvari R (2011) Autosomal recessive hyponatremia due to isolated salt wasting in sweat associated with a mutation in the active site of carbonic anhydrase 12. Hum Genet 129:397–405
191. Fujikawa-Adachi K, Nishimori I, Taguchi T, Onishi S (1999) Human carbonic anhydrase XIV (CA14): cDNA cloning, mRNA expression, and mapping to chromosome 1. Genomics 61:81
192. Kaunisto K, Parkkila S, Rajaniemi H, Waheed A, Grubb J, Sly WS (2002) Carbonic anhydrase XIV: luminal expression suggests key role in renal acidification. Kidney Int 61:2111–2118
193. Juel C, Lundby C, Sander M, Calbet JA, Hall G (2003) Human skeletal muscle and erythrocyte proteins involved in acid–base homeostasis: adaptations to chronic hypoxia. J Physiol 548:639–648
194. Vargas LA, Alvarez BV (2012) Carbonic anhydrase XIV in the normal and hypertrophic myocardium. J Mol Cell Cardiol 52:741–752
195. Ochrietor JD, Clamp MF, Moroz TP, Grubb JH, Shah GN, Waheed A, Sly WS, Linser PJ (2005) Carbonic anhydrase XIV identified as the membrane CA in mouse retina: strong expression in Muller cells and the RPE. Exp Eye Res 81:492–500
196. Linser PJ, Mosconna AA (1984) Variable CA II compartmentalization in the vertebrate retina. Ann N Y Acad Sci 429:430–446

Chapter 3
Catalytic Mechanism of α-Class Carbonic Anhydrases: CO_2 Hydration and Proton Transfer

Christopher D. Boone, Melissa Pinard, Rob McKenna, and David Silverman

Abstract The carbonic anhydrases (CAs; EC 4.2.1.1) are a family of metalloenzymes that catalyze the reversible hydration of carbon dioxide (CO_2) and dehydration of bicarbonate (HCO_3^-) in a two-step ping-pong mechanism:

$$CO_2 + H_2O \leftrightarrow HCO_3^- + H^+$$

CAs are ubiquitous enzymes and are categorized into five distinct classes (α, β, γ, δ and ζ). The α-class is found primarily in vertebrates (and the only class of CA in mammals), β is observed in higher plants and some prokaryotes, γ is present only in archaebacteria whereas the δ and ζ classes have only been observed in diatoms.

The focus of this chapter is on α-CAs as the structure-function relationship is best understood for this class, in particular for humans. The reader is referred to other reviews for an overview of the structure and catalytic mechanism of the other CA classes. The overall catalytic site structure and geometry of α-CAs are described in the first section of this chapter followed by the kinetic studies, binding of CO_2, and the proton shuttle network.

Keywords Carbonic anhydrase • CO_2 hydration • Proton • Bicarbonate • Proton transfer • Biocatalyst

Susan C. Frost and Robert McKenna (eds.). Carbonic Anhydrase: Mechanism, Regulation, Links to Disease, and Industrial Applications

C.D. Boone • M. Pinard • R. McKenna
Department of Biochemistry and Molecular Biology, University of Florida, Gainesville, FL, USA
e-mail: cdboone@ufl.edu; mpinard@ufl.edu; rmckenna@ufl.edu

D. Silverman (✉)
Department of Pharmacology and Therapeutics, University of Florida, Gainesville, FL, USA
e-mail: silvrmn@ufl.edu

S.C. Frost and R. McKenna (eds.), *Carbonic Anhydrase: Mechanism, Regulation, Links to Disease, and Industrial Applications*, Subcellular Biochemistry 75,
DOI 10.1007/978-94-007-7359-2_3, © Springer Science+Business Media Dordrecht 2014

1 Structure of the Carbonic Anhydrase Active Site

1.1 Human Carbonic Anhydrases

The structures available for α-CAs, including most isoforms of human CA (HCA), reveal that the catalytic site is located deep within the protein structure and accessible to solvent by a large open conical cleft that is approximately 15 Å deep (volume of ~900 $Å^3$) [1–3]. As HCA II is the best-studied of all CAs, the amino acid numbering and side chains for HCA II will be referred throughout this text with any differences among other α-CAs noted.

The Zn^{2+} metal at the base of the active site is in a distorted tetrahedral coordination with three His residues [4–6] and a solvent molecule (Fig. 3.1a, b). Proper orientation of the imidazole rings to allow for correct coordination of the metal ion occurs via several hydrogen bond networks with other residues in the active site: [7] Nδ1 of His94 to Oε1 of Gln92; [8] Nδ1 of His96 to O Asn244; and [1] Nδ2 of His119 to Oε2 of Glu117 which itself is orientated via acceptance of a hydrogen from Nδ1 of His107 [9].

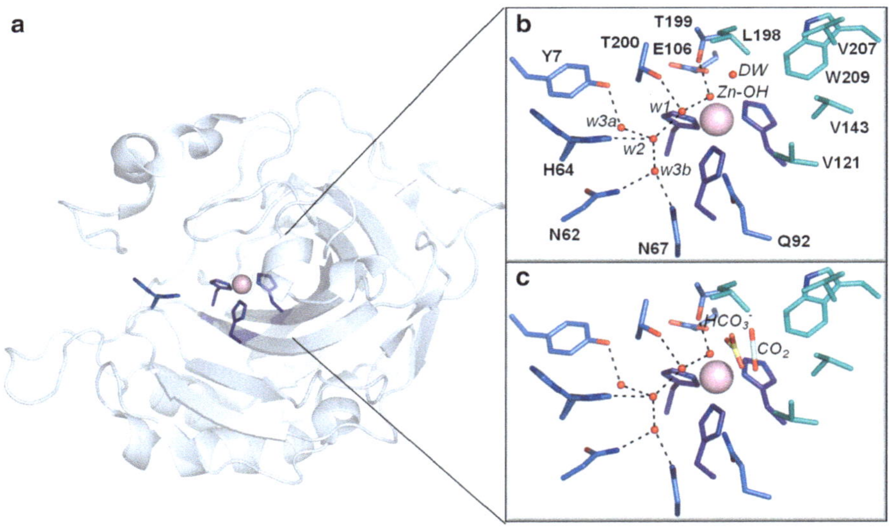

Fig. 3.1 (**a**) The structure of HCA II (PDB: 3KS3; [7]) shown in cartoon representation with the C_α backbone colored in silver, the Zn^{2+} metal ion as a pink sphere, the coordinating histidine residues as *purple sticks* and the proton shuttle residue, His64, in *blue*. (**b**) Structure of the active site showing the ordered water network (*red spheres*) hydrogen bond (*dashed line*) to their respective hydrophilic contacts (*blue sticks*). The hydrophobic residues important in coordinating CO_2 in the active site are shown as teal sticks. The C_α backbone has been omitted for clarity. (**c**) Models of the bicarbonate (*yellow*) and CO_2 (*light blue*) binding sites. Active site residues are colored and labeled as in (**b**)

The catalytic site is separated into hydrophobic and hydrophilic faces with the hydrophobic portion being further narrowed into two spatial categories: the residues which comprise the CO_2 binding site and those residues which line the entrance of the active site (Fig. 3.1b). These proximal (residues 121, 143, 198, 207 and 209) and distal (residues 131, 135, 201, 202 and 204) hydrophobic residues relative to the Zn^{2+} metal are often referred to as the primary and secondary hydrophobic binding sites, respectively [3, 10, 11]. The secondary hydrophobic binding site is important in facilitation of CO_2 into the active site and defining the binding affinities of several CA inhibitors (CAIs) [12–15].

The hydrophilic residues of the binding cavity (residues 62, 64, 67, 92, 106, 199 and 200) are highly conserved among the α-CAs and can be subdivided according to their role in catalysis [16]. The Oγ1 atom of Thr199 donates a proton to the Oε1 of Glu106 which properly aligns Thr199 to accept a hydrogen bond from the Zn-OH and orients the lone electron pair of the Zn-OH for nucleophilic attack on CO_2 [1, 3, 9, 17, 18]. The Oγ1 atom of Thr199 often acts as a hydrogen bond acceptor with several CAIs that displace the Zn-OH [13]. Tyr7, Asn62, Asn67 and Thr200 stabilize the ordered water network that transfers the proton produced during catalysis (see Eq. 3.1) out of the active site via the proton shuttle residue His64 (Fig. 3.1b) [16, 19, 20].

X-ray and neutron crystallographic models of CAs over the past 20 years in various pH and buffer conditions reveal that two rotamers of His64 exist, commonly referred to as the 'in' and 'out' conformations relative to the Zn^{2+} metal [3, 16, 21, 22]. There are theories that proton transfer from His64 into the bulk solvent arise from either a flipping mechanism [16] or tautomerization of the histidine ring [23]. It has been seen that His64 assumes an outward configuration at pH 5.7, presumably in a protonated state, whereas in basic conditions it is seen in the inward conformation and deprotonated [1, 8, 9, 21, 22, 24, 25]. The pK_a of His64 has been shown to be sensitive to the surrounding residues and can have an effect on catalytic rates [26–31].

1.2 Other α-CA Active Sites

The presence of His64 in most of the catalytically active isoforms of CAs in humans suggests the importance of this residue in proton shuttling [16]. However, HCA III and murine CA V (MCA V) contain a lysine and tyrosine residue at position 64 (HCA II numbering), respectively. These substitutions could provide an explanation to their observed lowered catalytic efficiency but there are other factors that influence kinetic parameters including the pK_a of the zinc-bound water, the number of water molecules involved in the proton transfer, and the distance between the proton donor and acceptor [22]. Additionally, reduction of the active site volume of HCA III is due to bulky side chains at residue position 67 and 198 (N67R and L198F; HCA II to HCA III active site variant), which has been proposed to also have an effect on the lowered catalytic rate [32, 33]. This bulky residue substitution is also seen in MCA V at positions 64 and 65 with the mutations H64Y and A65F. The hydrophobic nature of these residues make them unlikely to be involved in proton transfer which may actually occur via either Tyr131 or Lys132 [34]. X-ray crystallographic, catalytic and inhibition studies of several prokaryotic α-CAs show similar active site structures and kinetics compared to that seen in mammals and among other vertebrates [35–49], sponges [50], coral [51, 52] and nematodes [53, 54].

Interestingly, His64 is also conserved in the acatalytic isoform of CA termed CA-Related Protein VIII (CA-RP VIII) that is inactive due to replacement of one of the zinc-coordinating residues His94 with an arginine [55, 56]. HCA-RP VIII is highly expressed in the cerebellum [57] where it has been identified as a binding partner for inositol 1,4,5 triphosphate (IP3) receptor type 1 [58–60] and could play a key role in facilitating the transport activity of monocarboxylate transporters 1 and 4 (MCT1/4) [61].

2 Catalytic Mechanism

2.1 *Overview of CA Catalysis*

The kinetics and catalytic mechanism of CA has been extensively studied using HCA II as a model; however, it is thought that all α-CAs exhibit the same general scheme; that is, the nucleophilic attack of the Zn-OH^- on CO_2. The catalysis occurs via a two-step ping-pong mechanism [11, 20, 62, 63]:

$$\mathrm{EZnOH^- + CO_2 \leftrightarrow EZn(OH^-)CO_2 \leftrightarrow EZnHCO_3{}^- \leftrightarrow EZnH_2O + HCO_3{}^-} \tag{3.1}$$

$$\mathrm{EZnH_2O \leftrightarrow H^+EZnOH^- + B \leftrightarrow EZnOH^- + BH^+} \tag{3.2}$$

The binding of CO_2 in the hydrophobic region adjacent to the zinc metal promotes the nucleophilic attack by the Zn-OH^- leading to the formation of $HCO_3{}^-$ which is later displaced by the random diffusion of water in the active site. The transfer of a proton in the second step from the bound water molecule at the zinc to an acceptor in the bulk solvent (B) is needed to regenerate the hydroxide for a subsequent round of catalysis via the proton shuttle residue His64 [19, 20, 63, 64]. This transfer occurs on the order of $10^6\ s^{-1}$ for HCA II and is the rate-limiting step of the overall maximum velocity of catalysis [20, 65]. The intramolecular proton transport between the Zn-bound solvent and His64 occurs via intervening water molecules in the active site. From His64, an intermolecular transfer event delivers the proton to the bulk solvent.

Table 3.1 Steady-state constants for catalysis of the hydration of CO_2 by carbonic anhydrases in the α, β, and γ classes

Class	Organism	Isoform	k_{cat} (ms^{-1})	k_{cat}/K_m ($M^{-1}\ \mu s^{-1}$)	Reference
α-CAs	***Human***	**I**	200	50	[66, 67]
		II	1100	120	[20]
		III	10	0.3	[66, 67]
		IV	1100	51	[67]
		VA	290	29	[68]
		VB	950	98	[68]
		VI	340	49	[69]
		VII	950	83	[70]
		IX	380	55	[71]
		XII	420	35	[72]
		XIII	150	11	[73]
		XIV	310	39	[74]
	Bovine	***III***	6.4	0.4	[75]
	Murine	***IV***	1100	32	[76]
	Rat	***VI***	70	16	[77]
	Neisseria gonorrheae		1700	25	[78]
	Vibrio cholerae		823	70	[48, 79]
α-CAs	***Sulfurihydrogenibium azorense***		4400	350	[80]
	Sulfurihydrogenibium yellowstonense YO3AOP1		935	110	[38, 81]
	Drosophila melanogaster		950	110	[40, 82]
	Helicobacter pylori		250	15	[83, 84]
	Astrosclera willeyana		900	110	[50]
β-CAs	***Arabidopsis thaliana***		181	68	[85]
	Helycobacter pylori		710	48	[83]
γ-CAs	***Methanosarcina thermophila*****(zinc)**		61	3.3	[86]

The k_{cat} and k_{cat}/K_M of CO_2 hydration in HCA II have pH profiles that appear as titration curves with a $pK_a \sim 7$ and maximal activity at high pH [10, 63]. Of the active HCAs, HCA II is one of the most efficient ($k_{cat}/K_M \approx 10^8 M^{-1}s^{-1}$) and fastest ($k_{cat} \approx 10^6 s^{-1}$) isoforms and HCA III is the slowest ($k_{cat}/K_M \approx 10^5 M^{-1}s^{-1}$; $k_{cat} \approx 10^4 s^{-1}$; Table 3.1) [16, 64].

2.2 CO_2 Binding

Early studies on the binding of CO_2 relied on the use of substrate analogs and inhibitors due to high catalytic activity of HCA II [87]. Crystallographic studies of HCA II in the presence of these same substrate analogs and inhibitors reveal that sulfonamide inhibitors such as acetazolamide displace the Zn-OH molecule, keeping the same tetrahedral coordination with the Zn^{2+} metal. Structures of HCA II with anionic inhibitors, such as cyanate, revealed the displacement of the deep-water (DW) molecule, which is coordinated by the amide nitrogen of Thr199, to create a new trigonal-bipyramidal coordination sphere of the Zn^{2+} metal [88]. Furthermore, spectroscopic studies indicated that the binding of CO_2 does not require any inner-sphere coordination by the Zn^{2+} metal [89–91]. These results ultimately led to the conclusion that CO_2 must interact with residues in the primary hydrophobic binding site and that Thr199 plays a specific role in CO_2 orientation [66, 67, 92–96]. Molecular dynamic simulations revealed that the primary binding site was not the only binding site of CO_2 but that there are secondary and tertiary sites in the active site that are suggested to play in a role in rapid diffusion of CO_2 into the catalytic site [18, 97]. The binding of CO_2 in the active site of HCA II is weak (K_d is near 100 mM) and also weak ($K_d \approx 80$ mM) for bicarbonate [10, 91].

The X-ray crystallographic structure of HCA II in complex with CO_2 (Fig. 3.1c) was obtained via high-pressure pumping of room temperature CO_2 inside of a cryo-cooling apparatus above a liquid nitrogen bath [98–100]. The crystals were dipped slowly into the liquid nitrogen bath over a two minute span to trap CO_2 in the active site [96]. The high-resolution (1.1 Å) CO_2-bound HCA II structure revealed that the backbone NH of Thr199 could make a potential hydrogen bond with CO_2 at a distance of 3.5 Å. The orientation of CO_2 via Thr199 is such that the oxygens of the substrate are equidistant from the zinc-bound solvent molecule (3.0 and 3.1 Å) while the carbon atom is 2.8 Å away. However, the Zn^{2+} metal only seems to play a minor role in interacting with and orientating the CO_2. The structure of the substrate-bound, apo-HCA II revealed that CO_2 undergoes a minor rotation about its long axis towards the uncoordinated histidine residues that bind metal in the holo-enzyme. As a result CO_2 moves an oxygen closer towards the amide nitrogen of Thr199 (3.1 versus 3.5 Å for apo- and holo-HCA II, respectively).

2.3 *Bicarbonate Coordination*

The first structure of bicarbonate bound in the active site of HCA II (Fig. 3.1c) was achieved via mutation of T200H, which has a higher binding affinity for bicarbonate than the native enzyme [101]. The structure revealed that one of the oxygens of the bicarbonate has the same spatial positioning as the zinc-bound solvent molecule seen in other structures and the other two oxygens of the bicarbonate are shifted about 1.2 Å away from the coordinates seen for the CO_2-bound structure. Later studies with the mutant T199P/C206S of HCA II revealed that the T200H mutation had minimal effect on the positioning of the bicarbonate [102].

Bicarbonate was first seen in crystallographic models of native HCA II as the result of enzyme activation via X-ray damage from either water photoradiolysis and/or electron radiolysis [99]. The model showed that O3 of bicarbonate is bound 2.0 Å away from the Zn^{2+} metal and is within hydrogen bond distance of the hydroxyl group of Thr199 (2.6 Å) whereas the O2 atom of bicarbonate makes a hydrogen bond with the backbone amide of Thr199 (2.9 Å). The O1 atom of bicarbonate is 2.9 Å away from the Zn^{2+} ion. Like CO_2, bicarbonate was observed to make van der Waals contacts with the residues lining the hydrophobic pocket (Val121, Val143, Leu198 and Trp209).

A crystallographic structure of both CO_2 and bicarbonate bound in the active site of HCA II with the mutation V143I confirmed that the oxygens of the bicarbonate mimic the coordinates and geometry seen with the oxygens of the CO_2, the zinc-bound solvent molecule and previous bicarbonate-bound structures (Fig. 3.1c) [103]. The authors note that the V143I mutation has moved the CO_2 closer to the zinc metal by 0.3 Å and lowered the k_{cat}/K_M by ~15 fold as compared to wild-type HCA II. The differences seen in the position of the bicarbonate molecule in the V143I variant to that of other bicarbonate-bound HCA II structures could be due to the mutation itself but also could be an artifact of diffusing bicarbonate into crystals instead of capturing bicarbonate formation immediately after catalysis.

The structures of all substrate- and product-bound CAs correlate well with the proposed catalytic mechanism of CAs which is that of direct nucleophilic attack of the Zn-OH on CO_2. Formation of bicarbonate promotes a bidentate interaction with the zinc metal, resulting in a penta-coordinated metal ion with trigonal bipyramidal geometry. A water molecule displaces the zinc-bound bicarbonate which is then deprotonated via the proton transfer water network for the subsequent round of catalysis. A previously unidentified water molecule was observed in both the apo- and holo-CO_2 bound HCA II structures which may play the role of displacing bicarbonate [96, 98]. This water molecule is located between the side chain of Thr200 and the CO_2 molecule and could allow for easier access to the bicarbonate molecule and increased catalytic activity (as compared to random diffusion of water from the bulk solvent into the active site).

2.4 *Rate of Proton Transfer: Turnover Number*

The relevant topic here is the hydration of CO_2 and the dehydration of bicarbonate for which proton transfers can be rate-limiting steps. There is a broad documentation of other reactions catalyzed by the CAs in the α class, including the hydrolysis of aryl carboxylate esters and a variety of carbonate esters as well as the hydration of aliphatic aldehydes [104]. However, the rates of these catalyses are slower than catalysis of CO_2 hydration by orders of magnitude, and there has not been convincing evidence, for example H/D kinetic isotope effects, to conclude that proton transfer (PT) is limiting for these other catalyses.

There are two rate-contributing proton transfers that have been extensively studied in the catalysis of the hydration of CO_2, one intra- and one intermolecular. The intermolecular PT is reviewed in a subsequent section. The first evidence for an intramolecular PT was presented from the lab of Sven Lindskog [20] in the discovery that k_{cat} for catalysis of CO_2 hydration by HCA II had a H/D kinetic isotope effect (KIE) near 3.8 and k_{cat}/K_M had an isotope effect near unity, with similar isotope effects observed for the dehydration direction. Since these observations were made using solutions containing ample buffer (50 mM) precluding a rate-contributing intermolecular proton transfer, the conclusion from the KIE on k_{cat} was that the maximal velocity was limited in rate by an intramolecular proton transfer. Another conclusion drawn from the KIE on k_{cat}/K_m was that the interconversion of CO_2 and bicarbonate had no rate-contributing proton transfer. This work was definitive and the basis for many years of studies of PT in catalysis. Similar H/D KIEs were later observed for HCA II activity measured using ^{18}O exchange [105] and of bovine CA II [106]. Subsequent measurements of the H/D KIE on k_{cat} with values 2.0–2.6 were found for the β-class CA from spinach [107] and for the γ-class CA from *Methanosarcina thermophila* [86], suggesting catalytic mechanisms similar to that of the CAs in the α class. Values of the steady-state constants for hydration of CO_2 catalyzed by examples of CAs in the α, β, and γ classes are given in Table 3.1.

These considerations were the initial basis for the proposed ping-pong mechanism of carbonic anhydrase, evidence for which extends to the α, β, and γ classes. In this mechanism, rate-contributing PT occurs only in the second stage, as in Eq. 3.2. Here H^+ before $EZnOH^-$ indicates the protonated residue that acts as a proton shuttle between the zinc-bound solvent molecule and buffer in solution, B. As shown in Table 3.1, the rate of intramolecular PT in k_{cat} occurs as fast as 1,000 ms^{-1} for HCA II. This is very rapid for enzymic catalysis; for example, the value of k_{cat} for the very efficient manganese superoxide dismutase, which has a H/D KIE near 2, is 40 ms^{-1} [108, 109].

2.5 *Proton Acceptors and Donors*

There is strong evidence for a catalytic role of a zinc-bound solvent molecule [110] in catalysis by HCA II, and the pH profile of k_{cat} suggests a PT group with pK_a near 7. Lindskog and colleagues concluded that a group in the active site cavity with pK_a near 7 was involved in the PT and suggested that this was His64, the only group in the active site cavity with the appropriate pK_a [20]. This was an excellent suggestion that was supported in a number of studies such as examination of CA III which has Lys64 at this position and showed a value of k_{cat} near 10 ms^{-1} and is independent of pH in the region of pH 6–9 [4]. NMR studies confirmed that the imidazole ring of His64 in HCA II titrated with a pK_a near 7 [111]. However, confirmation of the proton shuttle role of His64 came with the observation that catalysis by the site-directed variant H64A HCA II had values of k_{cat} as low as 10 ms^{-1} and were rescued to nearly full activity in a manner that followed Michaelis kinetics by small-molecule proton donors such as imidazolium ion [5, 19]. This proton migration between His64 and the zinc-bound solvent molecule in CA is of great current interest by theorists and is under study using computational methods of proton motion in the labs of Greg Voth (University of Chicago [25]), Arieh Warshel (University of Southern California [112]), and Qiang Cui (University of Wisconsin [113]) among others.

The imidazole side chain of His64 in HCA II extends into the active-site cavity in an inward and outward conformation, observed as superimposed structures in crystallography (Fig. 3.1b) [21, 22]. In the inward conformation, the imidazole side chain is about 7 Å from the zinc, and an array of ordered water molecules extends between them that is likely the pathway for intramolecular proton transfer. The initial evidence for the role of such a water chain was obtained by the dependence of k_{cat} for hydration catalyzed by CA II on the atom fraction of deuterium in solvent [114]. The results were consistent with the motion of two or three protons in the transition state, as would be observed with PT through water bridges. The main feature in explaining the rate constants for PT appears to be the ordered water structure (see next section). The orientation of His64 in crystal structures has not been shown to affect the rate constant for PT according to much data [see for example [23, 115, 116]]. Moreover, reports show that the pH dependence of the rate constant for intramolecular PT is rather flat in the vicinity of $\Delta pK_a(pK_{aZnH2O} - pK_{a\ His64})$ near zero [5, 117].

These considerations extend to the β CA from *Arabidopsis thaliana* for which there is evidence that His216 acts as a proton shuttle and in the γ CA from *M. thermophila* for which Glu84 is a proton shuttle. In each case the replacement of these residues reduced rates of PT in the catalysis, and the activity of the variant enzymes (H216N for the β CA and E84A for the γ CA) was partially recovered by addition of proton donors in solution, substituted imidazolium and pyridinium ions [118]. Moreover, the activation follows Michaelis kinetics in which the external proton donor acts as a second substrate in the catalysis. The proposed proton shuttle Glu84 in the β CA is also observed to have two distinct side-chain conformations in the crystal structure, with one pointing toward the metal of the active site [119]. There is a proposed advantage to have conformational mobility in the side chain of the proton shuttle; that is, the shuttle in accepting a proton from the zinc-bound water can flip conformation and deliver it to solution in an outward orientation [65, 120]; however, such a mechanism has not been definitively shown.

There is yet another aspect of the stereochemistry of the active site that may promote proton transfer. Crystal structures of HCA II at 0.9 Å resolution have revealed a short, strong hydrogen bond between a deeply buried water molecule in the active site and the zinc-bound hydroxide [7]. This hydrogen bond may contribute to the protolysis of the zinc-bound water and contribute to its low pK_a near 7 by providing the energy of the short, strong hydrogen bond. In this aspect, the role of the strong hydrogen bond is similar to that proposed for liver alcohol dehydrogenase for which a similar bond promotes the removal of a proton from zinc-bound alcohol [121].

2.6 *Proton Transfer Through Intervening Water Molecules*

Dr. Zoë Fisher at Los Alamos National Lab and colleagues have reported three groundbreaking neutron diffraction structures of HCA II at pH 7.8 [9] and pH 10.0 [8] and in complex with the clinically used inhibitor acetazolamide [122]. These data

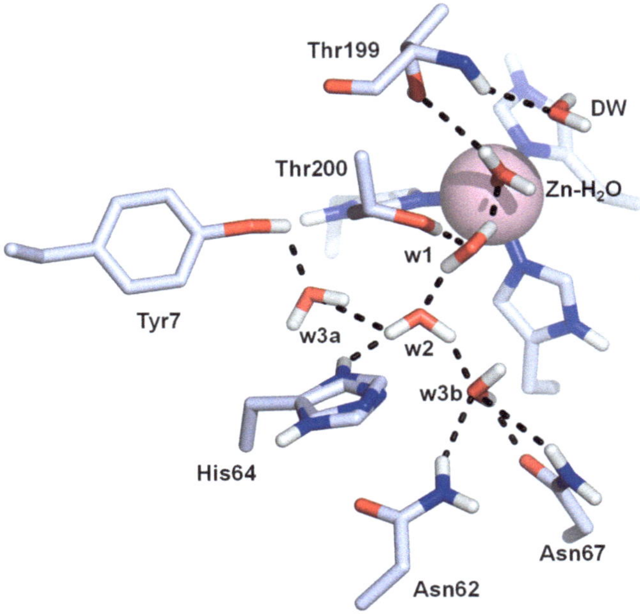

Fig. 3.2 The orientation of hydrogen bonds in the active site of HCA II determined by neutron diffraction at pH 7.8 and room temperature [8]. The sample had been vapor exchanged with D_2O to place deuterium atoms at exchangeable sites and enhance diffraction of neutrons

allowed observation for the first time of the zinc-bound water molecule with explicit information on hydrogen atom positions. Moreover, observation of the hydrogen positions showed the ionization state of the imidazole ring of the proton shuttle His64 (unionized) and other residues near the active site, such as Tyr7 which was unionized at pH 7.8 (Fig. 3.2). A major advantage was to be able to observe the orientation of the hydrogen bonds in the water structure in the active site and how they change conformations as a function of pH [21]. At high pH the orientation of these hydrogen bonds was not appropriate for PT; that is, breaking of hydrogen bonds and reorientation of water molecules would be necessary for PT. Neutron crystal structures subsequently determined at pH 7.8 showed a reorganization of the hydrogen-bonded water network more consistent with PT. These fine detail structural changes cannot be observed by X-ray crystallography alone, and only by a joint X-ray and neutron study. Such reorganization steps probably contribute significantly to the observed free energy barrier to PT. This is the first examination by neutron diffraction of a PT pathway at different conditions of pH. Structures at lower pH are being pursued, some are currently under analysis, and will provide new information on the biophysics of PT.

Computational studies have elucidated a number of possible proton pathways in HCA II between the zinc-bound solvent, His64 and external solution [123]. Examination of the mechanism of the steps of PT by Voth and colleagues using molecular simulations was consistent with proton motion from an Eigen cation ($H_9O_4^+$) on w1 through a Zundel cation ($H_5O_2^+$) containing w2, to an Eigen cation on w3a, and finally to His64 [124, 125] (see Fig. 3.1b for labeling of water molecules). The barrier making the largest contribution to the rate is a high free energy Eigen cation formed by the excess proton residing on w3a with a small amplitude contribution from w2 and His64. This establishes a barrier with a free energy of 10.0 kcal/mol and a rate constant for PT near 1.0 μs^{-1}. Water structures with more than three water molecules represent clusters that may be branched, and therefore are more likely to form an Eigen cation which has been shown to elevate the PT barrier when compared to smaller unbranched water wires [126]. The results of Voth and colleagues indicate that the Y7F variant of HCA II possesses a significantly elevated probability of forming smaller water clusters of size 3; this favors an unbranched water structure and a higher probability of PT that is likely to account for the elevated turnover number of 4–7 μs^{-1} reported by kinetic studies [127]. Wild-type HCA II with a turnover number near 1 μs^{-1}, while possessing a moderate number of size 3 water clusters, favors water clusters of size 4 which is more likely to involve an Eigen cation and hence a higher overall barrier to proton transfer. Another factor is that the Y7F mutant exhibits lifetimes for water structure greater than twice that of the WT water wires with lifetimes of 5.0 ps for the mutant. Part of the explanation for longer lifetimes of specific water structures in the Y7F mutant may be that the enzyme can expand and contract without placing strain on the water wire responsible for proton transport [127, 128]. These computational studies have offered a strong complementary approach to models based on observations in solution and in crystal structures.

One hypothesis suggests a picture in which separate components of the water structure must come together to form a continuous hydrogen bonded network of water molecules through which PT proceeds. Based on the computational results and correlations with kinetics and structural data, we constructed mutants with enhanced PT rates. Specifically, HCA II Y7F-N67Q has a rate constant for PT from His64 to the zinc-bound hydroxide in catalysis of 9 μs^{-1}, compared with a value near 1 μs^{-1} for wild-type HCA II [28]. These higher values observed for Y7F-N67Q HCA II could not be explained by differences in the values of the pK_a of the proton donor (His64) and acceptor (zinc-bound hydroxide) or by orientation of the side chain of the proton shuttle residue His64. They appeared to be associated with reduced branching in the active site water networks as observed in crystal structures. Moreover, Y7F-N67Q HCA II is unique among the variants studied in having a direct, hydrogen-bonded chain of water molecules between the zinc-bound solvent and N^{ε} of His64 [28]. This study provides the clearest example to date of the relevance of ordered water structure to rate constants for PT in catalysis by carbonic anhydrase.

2.7 Intermolecular Proton Transfer

There arose in the 1970s a dilemma concerning catalysis by HCA II. The issue was whether carbonic acid (H_2CO_3) or bicarbonate and a proton were substrates for the catalysis [129]. The problem with carbonic acid as substrate is that in the range of physiological pH carbonic acid is a very small fraction [2] of all species of CO_2 [130]. At such a low concentration, rates exceeding the diffusion-controlled limit would be required to explain the measured value of k_{cat}/K_M of 10^8 M^{-1} s^{-1} [64, 129]. On the other hand, requiring bicarbonate and a proton be substrates faces the problem that at physiological pH the concentration of protons near 10^{-7} M would also require rates exceeding diffusion control to explain catalysis by HCA II. Water itself is too poor a proton donor or acceptor to participate significantly.

An answer to this dilemma was proposed much earlier by Alberty who stated that although the concentration of H^+ in solution was low the concentration of protonated buffers was considerably larger and should be considered [131]. This idea was suggested as pertinent to catalysis by carbonic anhydrase in 1973 [110, 129, 132] since measurements of the rapid turnover of CA II near 1 μs^{-1} were measured in solutions containing at least 10 mM buffer. However, lowering the buffer concentration to a level in which intermolecular PT becomes rate limiting was a problem since the assays available at the time relied on buffer to control pH; without buffers the pH would vary widely in this catalysis which generates or consumes a proton. Khalifah [129] suggested measuring the catalysis by isotope exchange at chemical equilibrium for which pH control would not be so large a problem. This experiment was achieved by Silverman and Tu [6, 133] who utilized carbonic anhydrase catalyzed exchange of ^{18}O between CO_2 and water [134]. The experiment showed that this stable isotope exchange measured by mass spectrometry decreased as the buffer concentration was reduced below about 10 mM.

Buffers in the imidazole and pyridine class activated catalysis in these studies, whereas compounds similar in structure but not PT agents, such as 1,3-dimethylimidazole were not activators [133]. The rate constant for PT between such buffers and the enzyme followed a Brønsted plot with a maximal value at 10^9 M^{-1} s^{-1} that is near diffusion control [135]. The shape of the Brønsted plot indicated a proton acceptor group on the enzyme with a pK_a near 7, consistent with PT either to His64 or to the zinc-bound solvent molecule.

2.8 Chemical Rescue and Catalysis

The activation of catalysis by derivatives of imidazole and pyridine has been helpful in understanding the mechanism of proton transfer. This was achieved in a series of studies using variants of carbonic anhydrase in which the proton shuttle residue was replaced with a residue having no PT capabilities. Such studies were reported for the variant H64A HCA II in the α class [5, 19], for H216N CA from *Arabidopsis thaliana* in the β class, and E84A CA from *Methanosarcina thermophila* from the γ class [118], as mentioned earlier. The activation usually by a derivative of imidazole or pyridine followed Michaelis kinetics with the exogenous proton donor acting as second substrate in the ping-pong mechanism. Generally, the observed value of K_M for the activation by buffer molecules was large, in the range of 10–50 mM. In some cases, the maximal rescued catalytic activity at large concentrations of proton donor were near in magnitude to the catalytic activity of the wild-type enzyme [20]. Herschlag and colleagues have presented valuable comments on the inherent limitations of such chemical rescue studies; although replacement of a side chain is often viewed as a conservative change, the energetic effects of the replacement depend on the molecular properties of the cavity created and the surrounding environment of the protein [136].

Attempts to locate a binding site for the exogenous proton donor within the active site cavity have not been successful. Thus, although crystallography showed binding of 4-methylimidazole (4-MI) in a π-stacking interaction with the indole ring of Trp5, the activation of catalysis by 4-MI was identical for H64A HCA and W5A-H64A HCA II [5]. From this observation it was concluded that the binding site for 4-MI on Trp5 was not productive for PT in catalysis. Using ^{1}H NMR relaxivities, Elder et al. [137] found 4-MI bound as a second-shell ligand of the tetrahedrally coordinated cobalt in Co(II)-substituted H64A HCA II, with 4-MI estimated to be about 4.5 Å from the metal. However, this site was determined to be inhibitory. Examination of chemical rescue of H64A HCA II by empirical valence-bond simulations showed that 4-MI could act as an activator by direct PT with the active site, or by acting as an intermediary transferring protons between the active site and ionizable residues near the rim of the active-site cavity [138]. This study confirmed the weak binding of 4-MI in the active site and suggested that a firm binding site was not needed to explain the activation, rather 4-MI could diffuse around the active site with no preferred binding site and when 4-MI is located < 8.5 Å from the zinc a stable hydrogen-bonded chain has a probability of forming that would transfer the proton [138].

The dependence of catalytic rate on the pK_a of the proton donor provides information on the properties of the PT by construction of a Brønsted plot and comparison with bimolecular PT between small molecules. Such a Brønsted plot was originally reported for intramolecular PT in HCA III in which the pK_a of the zinc-bound water was varied by replacement of the nearby residue Phe198 and other residues [117]. The data plotted onto a highly curved free-energy plot that was interpreted in terms of a classic, two-state Marcus theory [117]. Such experiments were subsequently reported using variants in the α, β, and γ classes and activating catalysis with derivatives of imidazole and pyridine having a range of pK_a values from 5 to 9 [5, 118]. The free-energy curve derived from activation of H216N CA from *A. thaliana* (β class) is shown in Fig. 3.3.

Application of Marcus theory to catalysis by the different classes of carbonic anhydrase yield similar results: the intrinsic Marcus barrier $\Delta G^{\ddagger}_o$ is low, near or

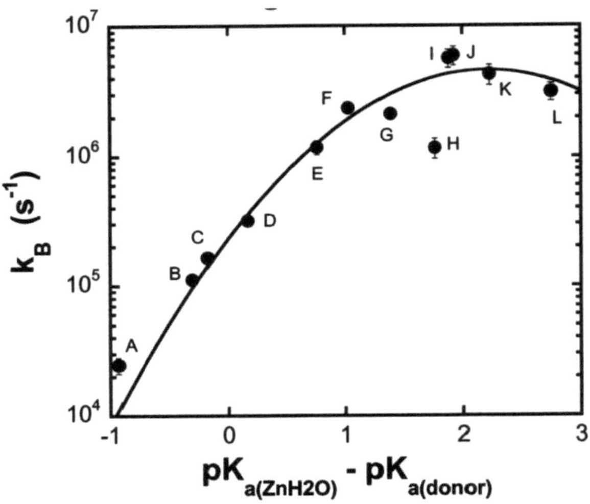

Fig. 3.3 Dependence of k_B (s^{-1}), a rate constant for intermolecular proton transfer, on (pKa(ZnH2O) – pKa (donor)) for catalysis of ^{18}O exchange by H216N CA from *A. thaliana*. The value of pKa(ZnH2O) is taken as 8.0. The proton donors are listed with solution values of pK_a in parentheses: *A*, purine (8.9); *B*, 1,2-dimethylimidazole (8.3); *C*, 2-methylimidazole (8.2); *D*, 4-methylimidazole (7.9); *E*, 1-methylimidazole (7.3); *F*, imidazole (7.0); *G*, 2,3-dimethylpyridine (6.6); *H*, 3,5-dimethylpyridine (6.3); *I*, 4-methylpyridine (6.1); *J*, 2-methylpyridine (6.1); *K*, 3-methylpyridine (5.8); *L*, pyridine (5.2). In each case the pH of the experiment was adjusted approximately to the pK_a of the proton donor. The *solid line* is a least-squares fit to Marcus theory as described in Ref. [118]

less than about 2 kcal/mol, and the work functions w^r and w^p are large, generally 8–10 kcal/mol [5, 139]. These results reflect a lowering of the barrier to PT, reflected in $\Delta G^{\ddagger}{}_o$, just as the enzyme has lowered the barrier for the CO_2 hydration step. The large value of the work functions is suggested to contain solvent reorganizations (as well as possibly conformational changes in side chains). The breaking of hydrogen bonds and rotation of water molecules is necessary to achieve PT from zinc-bound water to His64 in the neutron diffraction structure of Fig. 3.2. This could be a significant part of w^r and w^p. These data give us some confidence that these parameters based on application of Marcus theory are measuring a real phenomenon associated with the catalysis of PT and not an artifact of the experiment. It should be mentioned that computational groups have varying opinions about this application of classical Marcus theory. Warshel and colleagues have described an alternative approach for these PTs applying empirical valence bond methods to a three-state description of PT in CA III [112, 140]. In this analysis, the intrinsic barrier to PT is large, near 10 kcal/mol, and the associated work functions describing the solvation pre-equilibrium are very small. This is the opposite conclusion from the application of the classical Marcus formalism. Further experiments may elucidate this topic.

The quadratic form of the Marcus equation makes the well-known prediction that as the PT becomes more and more favorable thermodynamically the observed activation barrier $\Delta G^{\ddagger}$ decreases until a point is reached at $\Delta G^o = -4\Delta G_o^{\ddagger}$ beyond which further deceases in ΔG^o give increasing activation barriers. That is, the parabolic form of Marcus plots predicts that in the inverted region as the transfer becomes more and more favorable thermodynamically, the PT rate constant is decreased. This prediction of reaction rates forming an inverted region as exoergicity is increased has been observed for electron transfers [141]. Such an inverted region has been observed for PT within benzophenone/N,N-dimethylaniline contact radical ion pairs in a nonopolar environment [142], but there is no evidence that this would extend to PT in an aqueous environment.

A problem in the application of the classic Marcus treatment of reactivity to proton transfers is the assumption that all the distortions that convert the reactant to the transition state, and the transition state to product, can be represented by changes in the value of a single collective coordinate ΔG^o. Real potential surfaces are hyperdimensional, and the distortions leading to the transition state are not all coordinated. The inverted region might be avoided for PT because the transition state predicted for the inverted region will shift to another place on the surface where the energy barrier is more favorable. Because of the richness of the vibrational spectrum there are many other modes that can be excited, and may influence the energy of proton transfer, so that in a protein with many breathing modes the problem could be amplified. This has been discussed by Hammes-Schiffer and colleagues [143]. Nevertheless, the basic hypotheses upon which these statements are based need to be tested further.

References

1. Eriksson AE, Jones TA, Liljas A (1988) Refined structure of human carbonic anhydrase II at 2.0 A resolution. Proteins 4:274–282
2. Pocker Y, Sarkanen S (1978) Carbonic anhydrase: structure, catalytic versatility, and inhibition. Adv Enzymol Relat Areas Mol Biol 47:149–274
3. Krishnamurthy VM, Kaufman GK, Urbach AR, Gitlin I, Gudiksen KL, Weibel DB, Whitesides GM (2008) Carbonic anhydrase as a model for biophysical and physical-organic studies of proteins and protein-ligand binding. Chem Rev 108:946–1051
4. Tu C, Sanyal G, Wynns GC, Silverman DN (1983) The pH dependence of the hydration of CO2 catalyzed by carbonic anhydrase III from skeletal muscle of the cat. Steady state and equilibrium studies. J Biol Chem 258:8867–8871
5. An H, Tu C, Duda D, Montanez-Clemente I, Math K, Laipis PJ, McKenna R, Silverman DN (2002) Chemical rescue in catalysis by human carbonic anhydrases II and III. Biochemistry 41:3235–3242
6. Silverman DN, Tu CK (1975) Buffer dependence of carbonic-anhydrase catalyzed oxygen-18 exchange at equilibrium. J Am Chem Soc 97:2263–2269
7. Avvaru BS, Kim CU, Sippel KH, Gruner SM, Agbandje-McKenna M, Silverman DN, McKenna R (2010) A short, strong hydrogen bond in the active site of human carbonic anhydrase II. Biochemistry 49:249–251
8. Fisher Z, Kovalevsky AY, Mustyakimov M, Silverman DN, McKenna R, Langan P (2011) Neutron structure of human carbonic anhydrase II: a hydrogen-bonded water network "switch" is observed between pH 7.8 and 10.0. Biochemistry 50:9421–9423

9. Fisher SZ, Kovalevsky AY, Mustyakimov M, McKenna R, Silverman DN, Langan P (2010) The neutron structure of human carbonic anhydrase II: Implications for proton transfer. Biochemistry 49:415–421
10. Lindskog S, Silverman DN (2000) The catalytic mechanism of mammalian carbonic anhydrases. In: Chegwidden WR, Carter ND, Edwards YH (eds) The carbonic anhdyrases: new horizons. Birkhäuser Verlag, Boston, pp 175–195
11. Christianson DW, Fierke CA (1996) Carbonic anhydrase: evolution of the zinc binding site by nature and by design. Acc Chem Res 29:331–339
12. Aggarwal M, Kondeti B, McKenna R (2013) Insights towards sulfonamide drug specificity in alpha-carbonic anhydrases. Bioorg Med Chem 21:1526–1533
13. Aggarwal M, McKenna R (2012) Update on carbonic anhydrase inhibitors: a patent review (2008–2011). Expert Opin Ther Pat 22:903–915
14. Alterio V, Di Fiore A, D'Ambrosio K, Supuran CT, De Simone G (2012) Multiple binding modes of inhibitors to carbonic anhydrases: how to design specific drugs targeting 15 different isoforms? Chem Rev 112:4421–4468
15. Boriack PA, Christianson DW, Kingery-Wood J, Whitesides GM (1995) Secondary interactions significantly removed from the sulfonamide binding pocket of carbonic anhydrase II influence inhibitor binding constants. J Med Chem 38:2286–2291
16. Aggarwal M, Boone CD, Kondeti B, McKenna R (2013) Structural annotation of human carbonic anhydrases. J Enzyme Inhib Med Chem 28:267–277
17. Håkansson K, Carlsson M, Svensson LA, Liljas A (1992) Structure of native and apo carbonic anhydrase II and structure of some of its anion-ligand complexes. J Mol Biol 227:1192–1204
18. Merz KM (1991) Carbon dioxide binding to human carbonic anhydrase II. J Am Chem Soc 113:406–411
19. Tu CK, Silverman DN, Forsman C, Jonsson BH, Lindskog S (1989) Role of histidine 64 in the catalytic mechanism of human carbonic anhydrase II studied with a site-specific mutant. Biochemistry 28:7913–7918
20. Steiner H, Jonsson BH, Lindskog S (1975) Catalytic mechanism of carbonic-anhydrase – hydrogen-isotope effects on kinetic-parameters of human C isoenzyme. Eur J Biochem 59:253–259
21. Nair SK, Christianson DW (1991) Unexpected pH-dependent conformation of His-64, the proton shuttle of carbonic anhydrase-II. J Am Chem Soc 113:9455–9458
22. Fisher Z, Hernandez Prada JA, Tu C, Duda D, Yoshioka C, An H, Govindasamy L, Silverman DN, McKenna R (2005) Structural and kinetic characterization of active-site histidine as a proton shuttle in catalysis by human carbonic anhydrase II. Biochemistry 44:1097–1105
23. Shimahara H, Yoshida T, Shibata Y, Shimizu M, Kyogoku Y, Sakiyama F, Nakazawa T, Tate S, Ohki SY, Kato T, Moriyama H, Kishida K, Tano Y, Ohkubo T, Kobayashi Y (2007) Tautomerism of histidine 64 associated with proton transfer in catalysis of carbonic anhydrase. J Biol Chem 282:9646–9656
24. Nair SK, Ludwig PA, Christianson DW (1994) Two-site binding of phenol in the active site of human carbonic anhydrase II: structural implications for substrate association. J Am Chem Soc 116:3659–3660
25. Fisher SZ, Maupin CM, Budayova-Spano M, Govindasamy L, Tu C, Agbandje-McKenna M, Silverman DN, Voth GA, McKenna R (2007) Atomic crystal and molecular dynamics simulation structures of human carbonic anhydrase II: insights into the proton transfer mechanism. Biochemistry 46:2930–2937
26. Fisher SZ, Tu C, Bhatt D, Govindasamy L, Agbandje-McKenna M, McKenna R, Silverman DN (2007) Speeding up proton transfer in a fast enzyme: kinetic and crystallographic studies on the effect of hydrophobic amino acid substitutions in the active site of human carbonic anhydrase II. Biochemistry 46:3803–3813
27. Mikulski R, Avvaru BS, Tu C, Case N, McKenna R, Silverman DN (2011) Kinetic and crystallographic studies of the role of tyrosine 7 in the active site of human carbonic anhydrase II. Arch Biochem Biophys 506:181–187

28. Mikulski R, West D, Sippel KH, Avvaru BS, Aggarwal M, Tu C, McKenna R, Silverman DN (2013) Water networks in fast proton transfer during catalysis by human carbonic anhydrase II. Biochemistry 52:125–131
29. Mikulski R, Domsic JF, Ling G, Tu C, Robbins AH, Silverman DN, McKenna R (2011) Structure and catalysis by carbonic anhydrase II: role of active-site tryptophan 5. Arch Biochem Biophys 516:97–102
30. Mikulski RL, Silverman DN (2010) Proton transfer in catalysis and the role of proton shuttles in carbonic anhydrase. Biochim Biophys Acta 1804:422–426
31. Domsic JF, Williams W, Fisher SZ, Tu C, Agbandje-McKenna M, Silverman DN, McKenna R (2010) Structural and kinetic study of the extended active site for proton transfer in human carbonic anhydrase II. Biochemistry 49:6394–6399
32. Duda DM, Tu C, Fisher SZ, An H, Yoshioka C, Govindasamy L, Laipis PJ, Agbandje-McKenna M, Silverman DN, McKenna R (2005) Human carbonic anhydrase III: structural and kinetic study of catalysis and proton transfer. Biochemistry 44:10046–10053
33. Harju AK, Bootorabi F, Kuuslahti M, Supuran CT, Parkkila S (2012) Carbonic anhydrase III: a neglected isozyme is stepping into the limelight. J Enzyme Inhib Med Chem 28:231–239
34. Boriack PA, Heck RW, Laipis PJ, Silverman DN, Christianson DW (1995) Structure determination of murine mitochondrial carbonic anhydrase V at 2.45-A resolution: implications for catalytic proton transfer and inhibitor design. Proc Natl Acad Sci U S A 92:10949–10953
35. Liu Z, Bartlow P, Dilmore RM, Soong Y, Pan Z, Koepsel R, Ataai M (2009) Production, purification, and characterization of a fusion protein of carbonic anhydrase from *Neisseria gonorrhoeae* and cellulose binding domain from *Clostridium thermocellum*. Biotechnol Prog 25:68–74
36. Mårtensson L-G, Karlsson M, Carlsson U (2002) Dramatic stabilization of the native state of human carbonic anhydrase II by an engineered disulfide bond. Biochemistry 41:15867–15875
37. De Luca V, Vullo D, Scozzafava A, Carginale V, Rossi M, Supuran CT, Capasso C (2012) Anion inhibition studies of an alpha-carbonic anhydrase from the thermophilic bacterium Sulfurihydrogenibium yellowstonense YO3AOP1. Bioorg Med Chem Lett 22:5630–5634
38. Vullo D, De Luca V, Scozzafava A, Carginale V, Rossi M, Supuran CT, Capasso C (2012) The first activation study of a bacterial carbonic anhydrase (CA). The thermostable alpha-CA from Sulfurihydrogenibium yellowstonense YO3AOP1 is highly activated by amino acids and amines. Bioorg Med Chem Lett 22:6324–6327
39. Luca VD, Vullo D, Scozzafava A, Carginale V, Rossi M, Supuran CT, Capasso C (2013) An alpha-carbonic anhydrase from the thermophilic bacterium Sulphurihydrogenibium azorense is the fastest enzyme known for the CO(2) hydration reaction. Bioorg Med Chem 21:1465–1469
40. Syrjanen L, Tolvanen ME, Hilvo M, Vullo D, Carta F, Supuran CT, Parkkila S (2013) Characterization, bioinformatic analysis and dithiocarbamate inhibition studies of two new alpha-carbonic anhydrases, CAH1 and CAH2, from the fruit fly *Drosophila melanogaster*. Bioorg Med Chem 21:1516–1521
41. Demirdag R, Comakli V, Senturk M, Ekinci D, Irfan Kufrevioglu O, Supuran CT (2013) Purification and characterization of carbonic anhydrase from sheep kidney and effects of sulfonamides on enzyme activity. Bioorg Med Chem 21:1522–1525
42. Maresca A, Vullo D, Scozzafava A, Supuran CT (2013) Inhibition of the alpha- and beta-carbonic anhydrases from the gastric pathogen *Helycobacter pylori* with anions. J Enzyme Inhib Med Chem 28:388–391
43. Demirdag R, Yerlikaya E, Senturk M, Kufrevioglu OI, Supuran CT (2013) Heavy metal ion inhibition studies of human, sheep and fish alpha-carbonic anhydrases. J Enzyme Inhib Med Chem 28:278–282
44. Kolayli S, Karahalil F, Sahin H, Dincer B, Supuran CT (2011) Characterization and inhibition studies of an alpha-carbonic anhydrase from the endangered sturgeon species *Acipenser gueldenstaedti*. J Enzyme Inhib Med Chem 26:895–900

45. Cincinelli A, Martellini T, Innocenti A, Scozzafava A, Supuran CT (2011) Purification and inhibition studies with anions and sulfonamides of an alpha-carbonic anhydrase from the Antarctic seal *Leptonychotes weddellii*. Bioorg Med Chem 19:1847–1851
46. Ekinci D, Ceyhun SB, Senturk M, Erdem D, Kufrevioglu OI, Supuran CT (2011) Characterization and anions inhibition studies of an alpha-carbonic anhydrase from the teleost fish *Dicentrarchus labrax*. Bioorg Med Chem 19:744–748
47. Innocenti A, Supuran CT (2010) Paraoxon, 4-nitrophenyl phosphate and acetate are substrates of alpha- but not of beta-, gamma- and zeta-carbonic anhydrases. Bioorg Med Chem Lett 20:6208–6212
48. Del Prete S, Isik S, Vullo D, De Luca V, Carginale V, Scozzafava A, Supuran CT, Capasso C (2012) DNA cloning, characterization, and inhibition studies of an alpha-carbonic anhydrase from the pathogenic bacterium *Vibrio cholerae*. J Med Chem 55:10742–10748
49. Krungkrai J, Supuran CT (2008) The alpha-carbonic anhdyrase from the malaria parasite and its inhibition. Curr Pharm Des 14:631–640
50. Ohradanova A, Vullo D, Pastorekova S, Pastorek J, Jackson DJ, Worheide G, Supuran CT (2012) Cloning, characterization and sulfonamide inhibition studies of an alpha-carbonic anhydrase from the living fossil sponge *Astrosclera willeyana*. Bioorg Med Chem 20:1403–1410
51. Bertucci A, Innocenti A, Scozzafava A, Tambutte S, Zoccola D, Supuran CT (2011) Carbonic anhydrase inhibitors. Inhibition studies with anions and sulfonamides of a new cytosolic enzyme from the scleractinian coral *Stylophora pistillata*. Bioorg Med Chem Lett 21:710–714
52. Bertucci A, Tambutte S, Supuran CT, Allemand D, Zoccola D (2011) A new coral carbonic anhydrase in *Stylophora pistillata*. Mar Biotechnol (NY) 13:992–1002
53. Guzel O, Innocenti A, Hall RA, Scozzafava A, Muhlschlegel FA, Supuran CT (2009) Carbonic anhydrase inhibitors. The nematode alpha-carbonic anhydrase of *Caenorhabditis elegans* CAH-4b is highly inhibited by 2-(hydrazinocarbonyl)-3-substituted-phenyl-1H-indole-5-sulfonamides. Bioorg Med Chem 17:3212–3215
54. Crocetti L, Maresca A, Temperini C, Hall RA, Scozzafava A, Muhlschlegel FA, Supuran CT (2009) A thiabendazole sulfonamide shows potent inhibitory activity against mammalian and nematode alpha-carbonic anhydrases. Bioorg Med Chem Lett 19:1371–1375
55. Bergenhem NC, Hallberg M, Wisén S (1998) Molecular characterization of the human carbonic anhydrase-related protein (HCA-RP VIII). Biochim Biophys Acta 1384:294–298
56. Nishimori I, Vullo D, Minakuchi T, Scozzafava A, Capasso C, Supuran CT (2013) Restoring catalytic activity to the human carbonic anhydrase (CA) related proteins VIII, X and XI affords isoforms with high catalytic efficiency and susceptibility to anion inhibition. Bioorg Med Chem Lett 23:256–260
57. Lakkis MM, Bergenhem NC, O'Shea S, Tashian RE (1997) Expression of the acatalytic carbonic anhdyrase VIII gene, *Car8*, during mouse embryonic development. Histochem J 29:135–141
58. Hirota J, Ando H, Hamada K, Mikoshiba K (2003) Carbonic anhydrase-related protein is a novel binding protein for inositol 1,4,5-trisphosphate receptor type 1. Biochem J 372:435–441
59. Bosanac I, Alattia JR, Mal TK, Chan J, Talarico S, Tong FK, Tong KI, Yoshikawa F, Furuichi T, Iwai M, Michikawa T, Mikoshiba K, Ikura M (2002) Structure of the inositol 1,4,5-trisphosphate receptor binding core in complex with its ligand. Nature 420:696–700
60. Jiao Y, Yan J, Zhao Y, Donahue LR, Beamer WG, Li X, Roe BA, Ledoux MS, Gu W (2005) Carbonic anhydrase-related protein VIII deficiency is associated with a distinctive lifelong gait disorder in waddles mice. Genetics 171:1239–1246
61. Becker HM, Klier M, Schuler C, McKenna R, Deitmer JW (2011) Intramolecular proton shuttle supports not only catalytic but also noncatalytic function of carbonic anhydrase II. Proc Natl Acad Sci U S A 108:3071–3076
62. Lindskog S (1997) Structure and mechanism of carbonic anhydrase. Pharmacol Ther 74:1–20
63. Silverman DN, Lindskog S (1988) The catalytic mechanism of carbonic anhydrase: implications of a rate-limiting protolysis of water. Acc Chem Res 21:30–36
64. Khalifah RG (1971) The carbon dioxide hydration activity of carbonic anhydrase I. Stop-flow kinetic studies on the native human isoenzyme B and C. J Biol Chem 246:2561–2573

65. Silverman DN, McKenna R (2007) Solvent-mediated proton transfer in catalysis by carbonic anhydrase. Acc Chem Res 40:669–675
66. Supuran CT (2008) Carbonic anhydrases: novel therapeutic applications for inhibitors and activators. Nat Rev Drug Discov 7:168–181
67. Supuran CT, Scozzafava A, Casini A (2003) Carbonic anhydrase inhibitors. Med Res Rev 23:146–189
68. Nishimori I, Vullo D, Innocenti A, Scozzafava A, Mastrolorenzo A, Supuran CT (2005) Carbonic anhydrase inhibitors. The mitochondrial isozyme VB as a new target for sulfonamide and sulfamate inhibitors. J Med Chem 48:7860–7866
69. Nishimori I, Minakuchi T, Onishi S, Vullo D, Scozzafava A, Supuran CT (2007) Carbonic anhydrase inhibitors. DNA cloning, characterization, and inhibition studies of the human secretory isoform VI, a new target for sulfonamide and sulfamate inhibitors. J Med Chem 50:381–388
70. Vullo D, Voipio J, Innocenti A, Rivera C, Ranki H, Scozzafava A, Kaila K, Supuran CT (2005) Carbonic anhydrase inhibitors. Inhibition of the human cytosolic isozyme VII with aromatic and heterocyclic sulfonamides. Bioorg Med Chem Lett 15:971–976
71. Vullo D, Franchi M, Gallori E, Pastorek J, Scozzafava A, Pastorekova S, Supuran CT (2003) Carbonic anhydrase inhibitors: inhibition of the tumor-associated isozyme IX with aromatic and heterocyclic sulfonamides. Bioorg Med Chem Lett 13:1005–1009
72. Whittington DA, Waheed A, Ulmasov B, Shah GN, Grubb JH, Sly WS, Christianson DW (2001) Crystal structure of the dimeric extracellular domain of human carbonic anhydrase XII, a bitopic membrane protein overexpressed in certain cancer tumor cells. Proc Natl Acad Sci U S A 98:9545–9550
73. Lehtonen J, Shen B, Vihinen M, Casini A, Scozzafava A, Supuran CT, Parkkila AK, Saarnio J, Kivela AJ, Waheed A, Sly WS, Parkkila S (2004) Characterization of CA XIII, a novel member of the carbonic anhydrase isozyme family. J Biol Chem 279:2719–2727
74. Whittington DA, Grubb JH, Waheed A, Shah GN, Sly WS, Christianson DW (2004) Expression, assay, and structure of the extracellular domain of murine carbonic anhydrase XIV: implications for selective inhibition of membrane-associated isozymes. J Biol Chem 279:7223–7228
75. Ren X, Jonsson B-H, Millqvist E, Lindskog S (1988) A comparison of the kinetic properties of native bovine muscle carbonic anhydrase and an activated derivative with modified thiol groups. Biochim Biophys Acta (BBA) Protein Struct Mol Enzymol 953:79–85
76. Hurt JD, Tu C, Laipis PJ, Silverman DN (1997) Catalytic properties of murine carbonic anhydrase IV. J Biol Chem 272:13512–13518
77. Feldstein JB, Silverman DN (1984) Purification and characterization of carbonic anhdyrase from the saliva of the rat. J Biol Chem 259:5447–5453
78. Elleby B, Chirica LC, Tu C, Zeppezauer M, Lindskog S (2001) Characterization of carbonic anhydrase from *Neisseria gonorrhoeae*. Eur J Biochem 268:1613–1619
79. Del Prete S, De Luca V, Scozzafava A, Carginale V, Supuran CT, Capasso C (2013) Biochemical properties of a new alpha-carbonic anhydrase from the human pathogenic bacterium, *Vibrio cholerae*. J Enzyme Inhib Med Chem. doi:10.3109/14756366.2012.747197
80. Akdemir A, Vullo D, De Luca V, Scozzafava A, Carginale V, Rossi M, Supuran CT, Capasso C (2013) The extremo-alpha-carbonic anhydrase (CA) from Sulfurihydrogenibium azorense, the fastest CA known, is highly activated by amino acids and amines. Bioorg Med Chem Lett 23:1087–1090
81. Capasso C, De Luca V, Carginale V, Cannio R, Rossi M (2012) Biochemical properties of a novel and highly thermostable bacterial alpha-carbonic anhydrase from Sulfurihydrogenibium yellowstonense YO3AOP1. J Enzyme Inhib Med Chem 27:892–897
82. Syrjanen L, Tolvanen M, Hilvo M, Olatubosun A, Innocenti A, Scozzafava A, Leppiniemi J, Niederhauser B, Hytonen VP, Gorr TA, Parkkila S, Supuran CT (2010) Characterization of the first beta-class carbonic anhydrase from an arthropod (*Drosophila melanogaster*) and phylogenetic analysis of beta-class carbonic anhydrases in invertebrates. BMC Biochem 11:28

83. Nishimori I, Minakuchi T, Kohsaki T, Onishi S, Takeuchi H, Vullo D, Scozzafava A, Supuran CT (2007) Carbonic anhydrase inhibitors: the beta-carbonic anhydrase from *Helicobacter pylori* is a new target for sulfonamide and sulfamate inhibitors. Bioorg Med Chem Lett 17:3585–3594
84. Nishimori I, Minakuchi T, Morimoto K, Sano S, Onishi S, Takeuchi H, Vullo D, Scozzafava A, Supuran CT (2006) Carbonic anhydrase inhibitors: DNA cloning and inhibition studies of the alpha-carbonic anhydrase from *Helicobacter pylori*, a new target for developing sulfonamide and sulfamate gastric drugs. J Med Chem 49:2117–2126
85. Rowlett RS, Tu C, McKay MM, Preiss JR, Loomis RJ, Hicks KA, Marchione RJ, Strong JA, Donovan GS Jr, Chamberlin JE (2002) Kinetic characterization of wild-type and proton transfer-impaired variants of beta-carbonic anhydrase from *Arabidopsis thaliana*. Arch Biochem Biophys 404:197–209
86. Alber BE, Colangelo CM, Dong J, Stalhandske CM, Baird TT, Tu C, Fierke CA, Silverman DN, Scott RA, Ferry JG (1999) Kinetic and spectroscopic characterization of the gamma-carbonic anhydrase from the methanoarchaeon *Methanosarcina thermophila*. Biochemistry 38:13119–13128
87. Eriksson AE, Kylsten PM, Jones TA, Liljas A (1968) Crystallographic studies of inhibitor binding sites in human carbonic anhydrase II: a pentacoordinated binding of the SCN-ion to the zinc at high pH. Proteins 4:283–293
88. Lindahl M, Svensson LA, Liljas A (1993) Metal poison inhibition of carbonic anhydrase. Proteins 15:177–182
89. Williams TJ, Henkens RW (1985) Dynamic carbon-13 NMR investigations of substrate interaction and catalysis by cobalt(II) human carbonic anhydrase I. Biochemistry 24:2459–2462
90. Bertini I, Luchinat C, Monnanni R, Roelens S, Moratal JM (1987) Interaction of carbon dioxide and copper(II) carbonic anhydrase. J Am Chem Soc 109:7855–7856
91. Krebs JF, Rana F, Dluhy RA, Fierke CA (1993) Kinetic and spectroscopic studies of hydrophilic amino acid substitutions in the hydrophobic pocket of human carbonic anhydrase II. Biochemistry 32:4496–4505
92. Supuran CT (2008) Carbonic anhydrases–an overview. Curr Pharm Des 14:603–614
93. Supuran CT (2010) Carbonic anhydrase inhibitors. Bioorg Med Chem Lett 20:3467–3474
94. Supuran CT (2012) Inhibition of carbonic anhydrase IX as a novel anticancer mechanism. World J Clin Oncol 3:98–103
95. Supuran CT, Scozzafava A (2007) Carbonic anhydrases as targets for medicinal chemistry. Bioorg Med Chem 15:4336–4350
96. Domsic JF, McKenna R (2010) Sequestration of carbon dioxide by the hydrophobic pocket of the carbonic anhydrases. Biochim Biophys Acta 1804:326–331
97. Liang JY, Lipscomb WN (1990) Binding of substrate CO2 to the active site of human carbonic anhydrase II: a molecular dynamics study. Proc Natl Acad Sci U S A 87:3675–3679
98. Domsic JF, Avvaru BS, Kim CU, Gruner SM, Agbandje-McKenna M, Silverman DN, McKenna R (2008) Entrapment of carbon dioxide in the active site of carbonic anhydrase II. J Biol Chem 283:30766–30771
99. Sjoblom B, Polentarutti M, Djinovic-Carugo K (2009) Structural study of X-ray induced activation of carbonic anhydrase. Proc Natl Acad Sci U S A 106:10609–10613
100. Kim CU, Kapfer R, Gruner SM (2005) High-pressure cooling of protein crystals without cryoprotectants. Acta Crystallogr D Biol Crystallogr 61:881–890
101. Xue Y, Vidgren J, Svensson LA, Liljas A, Jonsson BH, Lindskog S (1993) Crystallographic analysis of Thr-200 → His human carbonic anhydrase II and its complex with the substrate, HCO3. Proteins 15:80–87
102. Huang S, Sjöblom B, Sauer-Eriksson AE, Jonsson B-H (2002) Organization of an efficient carbonic anhydrase: implications for the mechanism based on structure–function studies of a T199P/C206S mutant. Biochemistry 41:7628–7635
103. West D, Kim CU, Tu C, Robbins AH, Gruner SM, Silverman DN, McKenna R (2012) Structural and kinetic effects on changes in the CO(2) binding pocket of human carbonic anhydrase II. Biochemistry 51:9156–9163

104. Pocker Y, Sarkanen S (1978) Carbonic anhydrase: structure, catalytic versatility, and inhibition. Adv Enzymol Relat Areas Mol Biol 47:149–274
105. Tu CK, Silverman DN (1982) Solvent deuterium-isotope effects in the catalysis of O-18 exchange by human carbonic anhydrase-II. Biochemistry 21:6353–6360
106. Pocker Y, Bjorkquist DW (1977) Comparative studies of bovine carbonic anhydrase in H2O and D2O. Stopped-flow studies of the kinetics of interconversion of CO2 and HCO3. Biochemistry 16:5698–5707
107. Rowlett RS, Chance MR, Wirt MD, Sidelinger DE, Royal JR, Woodroffe M, Wang YFA, Saha RP, Lam MG (1994) Kinetic and structural characterization of spinach carbonic-anhydrase. Biochemistry 33:13967–13976
108. Hsu JL, Hsieh Y, Tu C, O'Connor D, Nick HS, Silverman DN (1996) Catalytic properties of human manganese superoxide dismutase. J Biol Chem 271:17687–17691
109. Bull C, Niederhoffer EC, Yoshida T, Fee JA (1991) Kinetic-studies of superoxide dismutases – properties of the manganese-containing protein from *Thermus thermophilus*. J Am Chem Soc 113:4069–4076
110. Lindskog S, Coleman JE (1973) Catalytic mechanism of carbonic-anhydrase. Proc Natl Acad Sci U S A 70:2505–2508
111. Campbell ID, Lindskog S, White AI (1975) A study of the histidine residues of human carbonic anhydrase C using 270 MHz proton magnetic resonance. J Mol Biol 98:597–614
112. Braun-Sand S, Strajbl M, Warshel A (2004) Studies of proton translocations in biological systems: simulating proton transport in carbonic anhydrase by EVB-based models. Biophys J 87:2221–2239
113. Riccardi D, Konig P, Prat-Resina X, Yu H, Elstner M, Frauenheim T, Cui Q (2006) "Proton holes" in long-range proton transfer reactions in solution and enzymes: a theoretical analysis. J Am Chem Soc 128:16302–16311
114. Venkatasubban KS, Silverman DN (1980) Carbon-dioxide hydration activity of carbonic-anhydrase in mixtures of water and deuterium-oxide. Biochemistry 19:4984–4989
115. Krebs JF, Fierke CA, Alexander RS, Christianson DW (1991) Conformational mobility of His-64 in the Thr200Ser mutant of human carbonic anhydrase-II. Biochemistry 30:9153–9160
116. Riccardi D, Konig P, Guo H, Cui Q (2008) Proton transfer in carbonic anhydrase is controlled by electrostatics rather than the orientation of the acceptor. Biochemistry 47:2369–2378
117. Silverman DN, Tu C, Chen X, Tanhauser SM, Kresge AJ, Laipis PJ (1993) Rate-equilibria relationships in intramolecular proton transfer in human carbonic anhydrase III. Biochemistry 32:10757–10762
118. Tu C, Rowlett RS, Tripp BC, Ferry JG, Silverman DN (2002) Chemical rescue of proton transfer in catalysis by carbonic anhydrases in the beta- and gamma-class. Biochemistry 41:15429–15435
119. Iverson TM, Alber BE, Kisker C, Ferry JG, Rees DC (2000) A closer look at the active site of gamma-class carbonic anhydrases: high-resolution crystallographic studies of the carbonic anhydrase from *Methanosarcina thermophila*. Biochemistry 39:9222–9231
120. Jude KM, Wright SK, Tu C, Silverman DN, Viola RE, Christianson DW (2002) Crystal structure of F65A/Y131C-methylimidazole carbonic anhydrase V reveals architectural features of an engineered proton shuttle. Biochemistry 41:2485–2491
121. Ramaswamy S, Park DH, Plapp BV (1999) Substitutions in a flexible loop of horse liver alcohol dehydrogenase hinder the conformational change and unmask hydrogen transfer. Biochemistry 38:13951–13959
122. Fisher SZ, Aggarwal M, Kovalesky A, Silverman DN, McKenna R (2012) Neutron-diffraction of acetazolamide-bound human carbonic anhydrase II reveals atomic details of drug binding. J Am Chem Soc 134:14726–14729
123. Roy A, Taraphder S (2007) Identification of proton-transfer pathways in human carbonic anhydrase II. J Phys Chem B 111:10563–10576
124. Maupin CM, Voth GA (2010) Proton transport in carbonic anhydrase: Insights from molecular simulation. Biochim Biophys Acta-Proteins Proteomics 1804:332–341

125. Maupin CM, McKenna R, Silverman DN, Voth GA (2009) Elucidation of the proton transport mechanism in human carbonic anhydrase II. J Am Chem Soc 131:7598–7608
126. Cui Q, Karplus M (2003) Is a "proton wire" concerted or stepwise? A model study of proton transfer in carbonic anhydrase. J Phys Chem B 107:1071–1078
127. Fisher SZ, Tu CK, Bhatt D, Govindasamy L, Agbandje-McKenna M, McKenna R, Silverman DN (2007) Speeding up proton transfer in a fast enzyme: kinetic and crystallographic studies on the effect of hydrophobic amino acid substitution in the active site of human carbonic anhydrase II. Biochemistry 42:3803–3813
128. Maupin CM, Saunders MG, Thorpe IF, McKenna R, Silverman DN, Voth GA (2008) Origins of enhanced proton transport in the Y7F mutant of human carbonic anhydrase II. J Am Chem Soc 130:11399–11408
129. Khalifah RG (1973) Carbon-dioxide hydration activity of carbonic-anhydrase – paradoxical consequences of unusually rapid catalysis. Proc Natl Acad Sci U S A 70:1986–1989
130. Butler JN (1982) Carbon dioxide equilibria and their applications. Addison-Wesley, Reading
131. Alberty RA (1962) The interpretation of steady state kinetic data on enzymatic reactions. Brookhaven Symp Biol 15:18
132. Prince RH, Woolley PR (1973) Mechanism of action of carbonic-anhydrase. Bioorg Chem 2:337–344
133. Tu CK, Silverman DN (1975) Mechanism of carbonic-anhydrase studied by C-13 and O-18 labeling of carbon-dioxide. J Am Chem Soc 97:5935–5936
134. Silverman DN (1982) Carbonic anhydrase: oxygen-18 exchange catalyzed by an enzyme with rate-contributing proton-transfer steps. Methods Enzymol 87:732–752
135. Rowlett RS, Silverman DN (1982) Kinetics of the protonation of buffer and hydration of Co-2 catalyzed by human carbonic anhydrase-II. J Am Chem Soc 104:6737–6741
136. Admiraal SJ, Meyer P, Schneider B, Deville-Bonne D, Janin J, Herschlag D (2001) Chemical rescue of phosphoryl transfer in a cavity mutant: a cautionary tale for site-directed mutagenesis. Biochemistry 40:403–413
137. Elder I, Tu CK, Ming LJ, McKenna R, Silverman DN (2005) Proton transfer from exogenous donors in catalysis by human carbonic anhydrase II. Arch Biochem Biophys 437:106–114
138. Maupin CM, Castillo N, Taraphder S, Tu CK, McKenna R, Silverman DN, Voth GA (2011) Chemical rescue of enzymes: proton transfer in mutants of human carbonic anhydrase II. J Am Chem Soc 133:6223–6234
139. Kresge AJ, Silverman DN (1999) Application of Marcus rate theory to proton transfer in enzyme-catalyzed reactions. Methods Enzymol 308:276–297
140. Schutz CN, Warshel A (2004) Analyzing free energy relationships for proton translocations in enzymes: carbonic anhydrase revisited. J Phys Chem B 108:2066–2075
141. Closs GL, Calcaterra LT, Green NJ, Penfield KW, Miller JR (1986) Distance, stereoelectronic effects, and the Marcus inverted region in intramolecular electron-transfer in organic radical-anions. J Phys Chem 90:3673–3683
142. Peters KS, Cashin A, Timbers P (2000) Picosecond dynamics of nonadiabatic proton transfer: a kinetic study of proton transfer within the contact radical ion pair of substituted benzophenones/N, N-dimethylaniline. J Am Chem Soc 122:107–113
143. Edwards SJ, Soudackov AV, Hammes-Schiffer S (2009) Driving force dependence of rates for nonadiabatic proton and proton-coupled electron transfer: conditions for inverted region behavior. J Phys Chem B 113:14545–14548

Chapter 4
Structure and Catalytic Mechanism of β-Carbonic Anhydrases

Roger S. Rowlett

Abstract The β-carbonic anhydrases (β-CAs) are a structurally distinct family of carbonic anhydrase that is widely distributed in microorganisms, algae, plants, and invertebrates. Like all carbonic anhydrases, β-CAs catalyze the reaction $CO_2 + H_2O \leftrightarrows HCO_3^- + H^+$, and is typically associated with other enzymes that produce or utilize CO_2 or HCO_3^-. β-CA is required for normal growth for many organisms. Unique among the five different families of carbonic anhydrases, β-CA is the only family of carbonic anhydrase to exhibit allostery. This chapter summarizes the structure, catalytic mechanism, and allosteric regulation of β-CA.

Keywords Allostery • β-carbonic anhydrases • Ribulose-1, 5-phosphate carboxylase • Crystal structure • Bicarbonate • Catalytic mechanism • Proton transfer • Spinach • Garden pea • Arabidopsis • *E.coli*

1 Introduction

The β-carbonic anhydrases (CAs) were first recognized as an evolutionarily distinct class of carbonic anhydrase in 1990, when the DNA sequence of the gene coding for the enzyme from *Spinacea oleraceae* was obtained [1]. This class of CAs received the β designation because it was the first convergently evolved CA (after α-CA) to have been discovered. Since then, several additional convergently evolved CAs (γ, δ, and ζ) have been discovered. Since the recognition of *S. oleraceae* CA as

Susan C. Frost and Robert McKenna (eds.). Carbonic Anhydrase: Mechanism, Regulation, Links to Disease, and Industrial Applications

R.S. Rowlett (✉)
Department of Chemistry, Colgate University, Hamilton, NY, USA
e-mail: rrowlett@colgate.edu

S.C. Frost and R. McKenna (eds.), *Carbonic Anhydrase: Mechanism, Regulation, Links to Disease, and Industrial Applications*, Subcellular Biochemistry 75,
DOI 10.1007/978-94-007-7359-2_4,

a β-CA, many additional β-CAs—putative and verified—have been discovered, and found to occur in a wide range of organisms, including algae, bacteria, yeast, archaeal species, worms, and insects. Perhaps unlike any other class of CA, β-CAs exhibit a broad structural and functional diversity, including allostery, and the ability to hydrolyze related, alternative substrates, like carbon disulfide. This chapter will focus on the structure and catalytic mechanism of the β-CAs.

2 Distribution and Significance

2.1 Distribution

CA activity was first recognized in plant chloroplasts from *Trifolium pratense* and *Onoclea sensibilis* [2]. Remarkably, plant chloroplast CA was not recognized as a β-CA until more than 50 years later, when the cDNA of the homologous enzymes from spinach (*S. oleraceae*), garden pea (*Pisium sativum*) and *Arabidopsis thaliana* were sequenced [1, 3, 4]. The bulk of what is known about the structure and mechanism of the plant β-CAs has been derived from studies of these three enzymes, which inhabit the chloroplastic stroma. However, additional β-CAs are thought to be localized in the thylakoid space, cytoplasm, and mitochondria of higher plants and algae [5].

The *CynT* gene of *E. coli* was the first recognized β-CA in bacteria. Since that time, many additional bacterial β-CAs have been identified [6–8], including enzymes from well-known human pathogens such as *Mycobacterium tuberculosis*, *Helicobacter pylori* [9], *Salmonella enterica* [10], *Brucella suis* [11], and *Streptococcus pneumoniae* [12]. β-CAs are also found in other microorganisms including the archaea (*Methanobacterium thermoautotrophicum*) [13], yeast (*Saccharomyces cereviseae*) [14], cyanobacteria (*Sychechocystis PCC6803*) [15], chemoautotrophic bacteria (*Halothiobacillus neapolitanus*) [16], and in green (*Chlamydomonas reinhardtii*) [17] and red (*Porphyridium purpureum*) [18] algae.

More recently, β-CAs have been discovered in animal species, including *Drosophila melanogaster* [19] and *Caenorhabditis elegans* [20]. Genomic sequence analysis suggests that β-CAs are widely distributed among non-chordates and invertebrates, but have not yet been demonstrated to exist in any vertebrate species [19].

2.2 Physiological Role and Significance

The physiological role(s) of β-CA are not well-known. However, for those organisms for which there is useful data, β-CA often serves in a support role for enzymes that utilize or dispose of CO_2 or HCO_3^-, such as ribulose-1,5-bisphosphate carboxylase (Rubisco), urease, cyanase, and carboxylases associated with fatty acid biosynthesis. In *P. sativum*, Rubisco and β-CA are transcriptionally linked and experience an increase in expression levels when plants are transferred from a high CO_2 environment to one with lower, atmospheric pressure of CO_2 [21]. However, it should be noted that tobacco (*Nicotiana tabacum*) plants with antisense suppression of β-CA expression show no clear decrease in carbon assimilation despite having as small as 1 % of normal chloroplastic β-CA levels [22]. β-CA is an essential component of the carboxysome, a CO_2-concentrating structure in cyanobacteria. Mutagenesis of this β-CA results in a phenotype that requires high CO_2 concentrations for normal growth in *Synechococcus* [23]. β-CA has also been determined to be essential for normal growth of *Escherichia coli* [24], *Corynebacterium glutamicum* [25], and *Saccharomyces cerevisiae* [26, 27] under aerobic conditions and atmospheric pressures of CO_2. In β-CA-deficient strains of *S. cerevisiae*, complementation with catalytically competent α- or β-CA restores the ability to grow normally under atmospheric CO_2 concentrations [27]. In β-CA deficient *Streptococcus pneumoniae*, complementation with unsaturated fatty acids restores normal growth under low CO_2 concentration conditions [28]. This suggests that β-CA in this organism, and perhaps in other bacteria and yeast, supports the activity of bicarbonate-dependent carboxylases required for fatty acid synthesis. *Helicobacter pylori* has both α- and β-CA gene, which are localized to the periplasm and cytoplasm, respectively. Deletion of the α-CA affects CO_2-HCO_3^- exchange rates and delays the urease-dependent increase in extracellular pH in *H. pylori*, whereas deletion of β-CA does not. However, β-CA deficient *H. pylori* produced strongly reduced gastric mucosal inflammation in mouse models of *H. pylori* infection [29].

In animal species, the physiological roles of β-CA are not established. The *Drosophila melanogaster* β-CA is localized to the mitochondrion, and is possibly involved in gluconeogenesis, ureagenesis, and/or lipogenesis, but this has not been directly demonstrated [19]. In *C. elegans,* no observable phenotype was observed for RNAi suppression of β-CA expression [20].

Finally, some β-CAs have adapted evolved to adopt novel metabolic functions. The *Acidianus* carbon disulfide hydrolase is clearly structurally and genetically closely related to β-CA. This enzyme cleaves CS_2 to COS and H_2S, but has no detectable CO_2 hydration activity [30].

3 Structure

The first X-ray crystallographic structures of β-CA were not determined until 2000, nearly a decade after the discovery of β-CA as an evolutionarily distinct class of CA. The first β-CA structure to be determined was for the red alga *P. purpureum* [18]. Structures for the *P. sativum* [31] and *E. coli* [32] enzymes followed quickly. As of this writing, there are a total of 12 distinct X-ray crystallographic structures of β-CAs deposited in the Protein Data Bank (Table 4.1).

Table 4.1 Structurally and kinetically characterized β-CAs

Source	PDB entry	Abbreviation	Structural subtype[a]
Pisium sativum	1EKJ	PSCA	Type I
Spinacea oleraceae	—	SOCA	Type I
Arabidopsis thaliana	—	ATCA	Type I
Methanobacterium thermoautotrophicum	1G5C	MTCA	Type I
Mycobacterium tuberculosis Rv1284	1YLK	Rv1284	Type I
Halothiobacillus neopolitanus	2FGY	HNCA	Type I
Coccomyxa	3UCO	CoCA	Type I
Saccharomyces cerevisiae	3EYX	SCCA	Type I
Streptococcus mutans	3LAS	SMCA	Type I
Porphyridium purpureum	1DDZ	PPCA	Type II
Escherichia coli	1I6P	ECCA	Type II
Haemophilus influenzae	2A8D	HICA	Type II
Mycobacterium tuberculosis Rv3588	1YM3	Rv3588	Type II
Salmonella enterica	3QY1	SECA	Type II

[a]Structural subtypes are described in Sect. 3.5

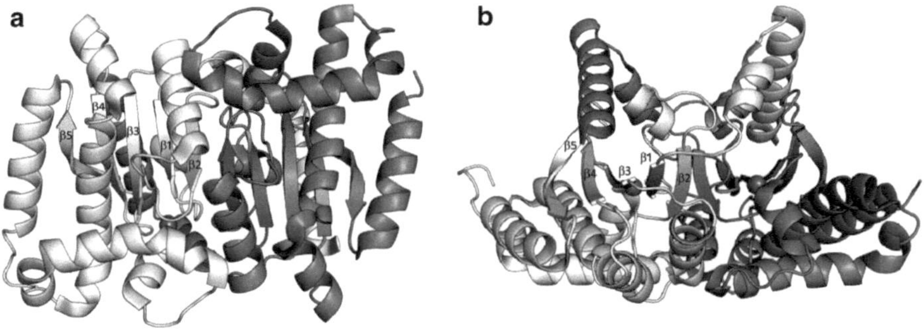

Fig. 4.1 Fundamental dimer of HICA. *Panel B* is rotated 90° about its horizontal axis relative to *panel A*. Monomers are colored *white* and *gray*. The dimerization interface lies along the *vertical axis* of panels A and B. The tetramerization interface lies along the face of *panel A* and the bottom of *panel B* (see Sect. 3.3)

3.1 *β-CAs Are Composed of Dimers*

The β-CAs are characterized by a unique α/β fold that associates to form dimers. In some β-CAs, this basic structural dimer is a single polypeptide chain composed of a tandem repeat of the a monomer sequence connected by a short polypeptide linker—a "pseudo-dimer." The β-CA monomer (or pseudo-monomer) is composed of a central core formed from a four-strand parallel β-sheet in a 2-1-3-4 arrangement (Fig. 4.1). Some β-CAs incorporate a short fifth β-strand that associates with β4 in an antiparallel arrangement. In most β-CAs the β-sheet cores of the fundamental structural dimer or pseudo-dimer are associated in an approximately antiparallel arrangement to form an 8–10 strand β-sheet that extends across the entire assembly.

More than a half-dozen α-helices pack against the β-sheet core, and with β5, when present, form the solvent-accessible surface of the dimer or pseudo-dimer unit. The fundamental dimer of β-CA is a tightly integrated structure. There are extensive interactions between monomer units in the region where the two β-sheet cores contact each other. In addition, two N-terminal helices wrap around the neighboring monomer like a clasp, where α1 of one monomer interacts with β4 and β5 from the neighboring monomer. For most β-CAs the total buried surface area [33] of the dimerization interface is approximately 7,000 $Å^2$, or 3,500 $Å^2$ per monomer unit.

HNCA departs significantly from the canonical dimer structure of β-CA. This β-CA is composed of three distinct domains: an N-terminal four-helix bundle; a catalytic domain that is similar to that of other β-CA monomers in structure; and a C-terminal domain that is similar in structure to the catalytic domain, but has lost its catalytic site, including metal ion binding residues [16]. This enzyme is so structurally distinct from other β-CAs that is was originally thought to be a member of a distinct (ε) family of CA, but is now recognized as a variant of β-CA [34].

The *Acidanius* carbon disulfide hydrolase (PDB 3TEN) is also very likely a divergently evolved β-CA [30]. It shares the same 2-1-3-4 central β-sheet core structure and dimer organization as β-CA, with 1.55 Å rmsd 3D structural homology over 70 % of the aligned sequence to *M. thermotautotrophicum* β-CA.

3.2 *β-CAs Are Zinc Metalloenzymes*

With one exception—HNCA, which is missing one active site—all β-CAs that have been structurally characterized contain one zinc ion per monomer or pseudo-monomer. The metal ion is in a pseudo-tetrahedral $Cys_2His(X)$ coordination environment where X is either an Asp residue or an exchangeable ligand such as water or an inorganic or organic anion. The zinc ion is the location of the catalytic (active) site. The pseudo-dimer of HNCA is asymmetric. In this enzyme, one of the pseudo-monomers has lost its metal-binding site during the course of evolution [16].

3.3 *Quaternary Structure of β-CAs*

The biological units of β-CAs are multiples of the fundamental dimer structure. Dimers, tetramers, and octamers are known or suggested from X-ray crystallographic data (Fig. 4.2).

3.3.1 Tetramers

The most frequently observed quaternary structure for β-CA is a tetramer. HICA, ECCA, COCA, SCCA, Rv1284 and SECA adopt a homotetrameric structure based on interface analysis of the X-ray crystal structures [33], and this has been verified

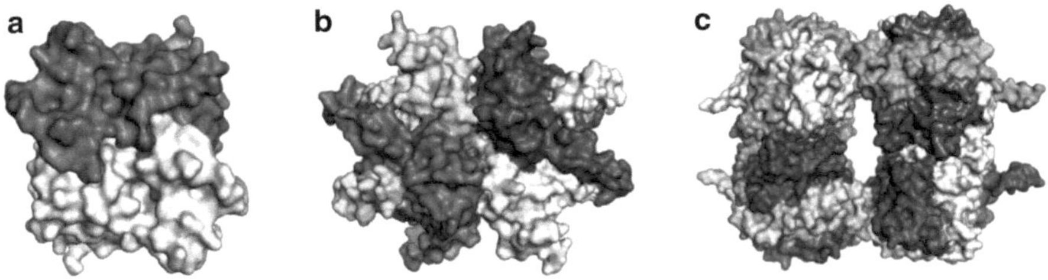

Fig. 4.2 Quaternary structures of β-CAs. (**a**) Rv3588 dimer. Dimerization interface is horizontal. (**b**) HICA tetramer. Dimerization interface is *horizontal*; tetramerization interface is *vertical*. (**c**) PSCA octamer. The tetramerization interfaces are along the *horizontal axis*; the octamerization interfaces are along the *vertical axis*

by gel exclusion chromatography for HICA [35]. PPCA [18] and HNCA [16] adopt analogous pseudo-tetrameric structures by the association of their pseudo-dimers. β-CA tetramers are best described as a dimer of dimers with 222 symmetry and two distinct oligomerization interfaces, one of which—the "dimerization" interface—has been described previously. The "tetramerization" interface is primarily composed of α-helices that form a nearly flat face on the fundamental dimer (Fig. 4.1). The total buried surface area [33] at the tetramerization interface is about 1,600–1,700 $Å^2$ for HICA, ECCA, STCA, and Rv1284, and similar values are obtained for the association of pseudo-dimers for HNCA. Thus it appears that the tetramerization interaction is perhaps only a quarter or a third as strong as the dimerization interface. The tetramerization interface of PPCA and COCA is significantly larger than the other tetrameric or pseudo-tetrameric β-CAs with a total buried surface area of ≈4,700 $Å^2$ and 3,500 $Å^2$, respectively. MTCA forms crystals in which there is no reasonable tetramerization interface in the asymmetric unit [36]. MTCA is missing most of the N- and C-terminal helices that characterize most of the other β-CAs. In particular, MTCA is missing both α1 and α2, structural features that are typically integral to the dimerization interface; instead, residues 90–125 form a pair of helices (α4 and α5) that fulfill this role. The location of helices α4 and α5 on the tetramerization interface appears to preclude tetramerization of this enzyme in the same manner as HICA or its homologs. However, gel exclusion chromatography [13] and analytical ultracentrifugation [37] suggest that the MTCA exists in solution as a tetramer. The crystal structure of MTCA has an unusual, but probably artifactual feature where the N-terminal helix of chain A appears to be domain swapped with its nearest crystal-packing neighbor, where it packs against β4 and β5, similar to that of tetrameric β-CAs. It seems likely that in vitro the N-terminal helix of chain A of MTCA is arranged similarly to that of chain B.

3.3.2 Dimers

Native Rv3588 forms crystals that contain a single monomer per asymmetric unit in a manner that can allow for the formation of a dimer, but not a tetramer with symmetry partners in the unit cell [38]. The thiocyanate complex of the same enzyme crystallizes with a dimer-of-dimers homotetramer in the asymmetric unit [39]. Dynamic light scattering studies show that the enzyme exists primarily as a tetramer at high pH, and dissociates into dimers when the pH is lowered from 8.4 to 7.5. Unlike other tetrameric β-CAs, Rv3588 has a high concentration of basic residues in the tetramerization interface that participate in salt bridges that stabilize the dimer-dimer interaction. Basic residues outnumber acidic residues in the tetramerization interface by almost two to one. It seems likely that high pH is required for tetramerization in order to neutralize excess positive charge at the tetramerization interface [39].

The *Drosophila melanogaster* β-CA has been characterized by gel exclusion chromatography and by dynamic light scattering, and suggest that it exists as a dimer in vitro [19].

3.3.3 Octamers

The biological unit of PSCA is an octamer with 222 symmetry: a dimer of a dimer of dimers [31]. The structural feature that is likely responsible for this organization is an extended C-terminus that forms a long β5 strand which engages in an antiparallel β5-β5′ interaction that allows dimers to associate into higher-order quaternary structures. The assembly of the PCSA octamer involves a "tetramerization" interface of $\approx$1,700 Å^2 total interaction area, and an "octamerization" interface with a total contacts area of $\approx$1,500 Å^2 [31]. The resulting structure looks something like a square ring. Gel exclusion chromatography studies of SOCA [40] and non-denaturing gel electrophoresis of SOCA and ATCA [41] are consistent with an octameric quaternary structure for these homologous enzymes.

The *Acidianus* carbon disulfide hydrolase adopts a quaternary structure remarkably like PSCA. However, in the case of the *Acidianus* enzyme, two octameric rings appear to interlock to form an unusual hexadodecameric catenane. Analytical ultracentrifugation and small angle X-ray scattering confirm the expected molecular weight of the catenane structure [30].

3.4 Post-translational Modification of Plant β-CAs

The genes that code for the precursor of chloroplast β-CAs contain a Ser/Thr rich chloroplastic transit peptide [1, 42–45]. Native SOCA and PSCA isolated from plants suggests that processing of the transit peptide is heterogeneous. In particular PSCA appears to exist as two discrete molecular weight forms [45] resulting from the proteolysis of the transit peptide in two sequential steps [46]. These different forms of PSCA do not appear to differ significantly in catalytic activity [45].

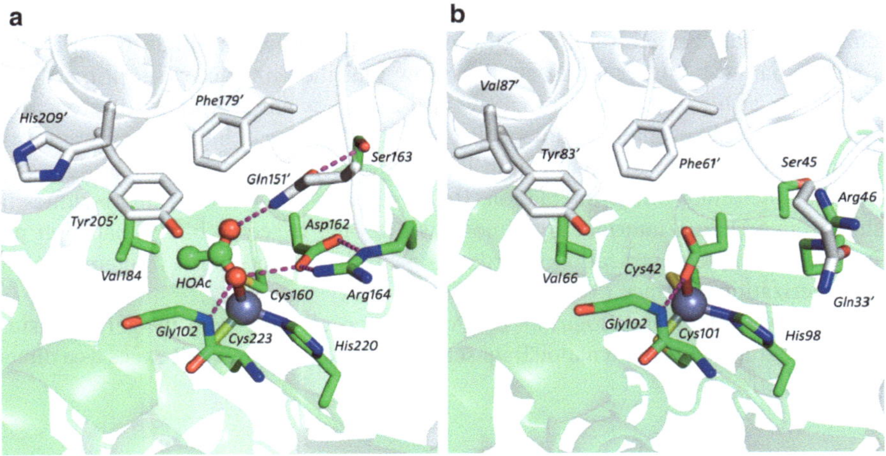

Fig. 4.3 Active site regions of archetypical Type I (PSCA, PDB 1EKJ chains G & H, *panel A*) and Type II (HICA, PDB 2A8D chains D & F, *panel B*) β-CAs. The active site monomer is *green*; the neighboring protein chain in the fundamental dimer is *gray*. Zinc is depicted as a *gray sphere*. Key residues, including acetic acid (HOAc) are depicted as *sticks* and/or *spheres*. *Dashed magenta lines* indicate hydrogen bonding interactions

3.5 Two Classes of β-CA Are Defined by Active Site Structure

X-ray crystallography of β-CAs reveal two distinct structural subclasses differentiated by the nature of the fourth metal ion ligand and a characteristic triad of residues (Trp, Tyr, and Arg) in a putative allosteric site. These two classes of β-CA—Type I and Type II—are not defined by genetic origins, but rather by active site organization and functional behavior (*vide infra*). The type classification of structurally and functionally well-characterized β-CAs is listed in Table 4.1.

3.5.1 Type I (Open) β-CAs

Type I β-CAs are characterized by three principal structural features. The first is a $Zn(Cys)_2(His)(X)$ coordination sphere in which X is an exchangeable ligand, such as water or an inorganic or organic ion. In active enzyme this fourth ligand is assumed to be a water or hydroxide ion. The second structural feature is an Asp-Arg dyad in which the Asp residue acts as a hydrogen bond acceptor from water, hydroxide, or an OH or NH group that is ligated to the metal ion. The third is a hydrogen bond donor (typically a Gln or His) that can interact with a distal oxygen of metal ion bound bicarbonate or bicarbonate analog. The defining example of a Type I β-CA is PSCA (Fig. 4.3a). The active site metal ion, as in all β-CAs, is situated on the dimerization interface. In PSCA the exchangeable ligand is acetic acid, a bicarbonate analog. In this structure, Arg164 stabilizes and orients Asp162 so that it can accept a hydrogen bond from the OH group of the bound acetic acid. Asp162 is hypothesized to act as a "gate-keeper" residue, excluding anions or other ligands from the metal ion that cannot donate a hydrogen bond from the metal-bonded atom. Gln151′ from the neighboring monomer donates a hydrogen bond to the distal oxygen atom of acetic acid. Finally, the amide NH of Gly224 donates a hydrogen bond to the proximal, metal-bound oxygen atom of acetic acid. The binding of acetic acid to PSCA is probably representative of the enzyme-bicarbonate intermediate in catalysis.

The active site tunnel leading to the metal ions lies along the dimerization interface and is highly constricted. It is lined with hydrophobic residues: Phe 179′, Val184, and Tyr205′ in PSCA. The only polar residue in the active site cleft is His209′, which lies at the exterior surface of the active site opening. For most β-CAs that have been structurally characterized by X-ray crystallography, the active site opening is too small to accommodate CO_2 or HCO_3^-. Therefore, there must be considerable rearrangement of the bulky side chains to allow for substrate/product ingress/egress.

The active site of PSCA is a remarkable example of convergent evolution: it maps nearly perfectly on a mirror image of the active site of human α-CA II (HCAII) [31], a CA whose mechanism of action is well understood. The mapping of analogous residues/atoms in PSCA onto HCAII allow the designation of catalytic roles of functional groups in β-CA. The most significant mappings include Asp162-Oδ1 in PSCA and Thr199-Oγ1 in HCAII. In HCAII, Thr199-Oγ1 donates a hydrogen bond to the side chain of Glu106, so that Thr199 is an obligate hydrogen bond acceptor from the atom bound to the metal ion. That is, Thr199 acts as the "gatekeeper" residue in HCAII, and by analogy, so does Asp162 in PSCA. The main chain amide of Thr199 in HCAII is hypothesized to donate a hydrogen bond to the distal oxygen atom of bicarbonate curing catalysis; in PSCA, the amide group of Gln151′ plays an analogous role. Notably, both Asp162 and Gln 151 (PSCA numbering) are highly conserved in β-CAs.

The carboxysomal enzyme HNCA possesses an active site similar to Type I β-CAs, but this enzyme is somewhat of an anomaly. It has only 15 % sequence identity with other β-CAs and has numerous insertions and deletions compared to the typical β-CA. Most significantly, the active site has been entirely lost from one of the pseudo-dimers. Both HNCA and Rv1284 utilize a His residue in place of the analogous Gln151′ in PSCA.

3.5.2 Type II (Closed) β-CAs

HICA is the archetype for Type II β-CAs (Fig. 4.3b). Type II β-CAs have a closed Cys_2HisAsp coordination sphere around the active site metal ion, and significant changes in other structural features that characterize Type I β-CAs (Fig. 4.3b). Specifically, the Asp-Arg dyad of Type I β-CAs is broken, and the hydrogen bond donor residue (Gln151′ in PSCA, Gln33′ in HICA) is rotated away from the active site. In PSCA, Ser163 donates a hydrogen bond to the side-chain carbonyl of

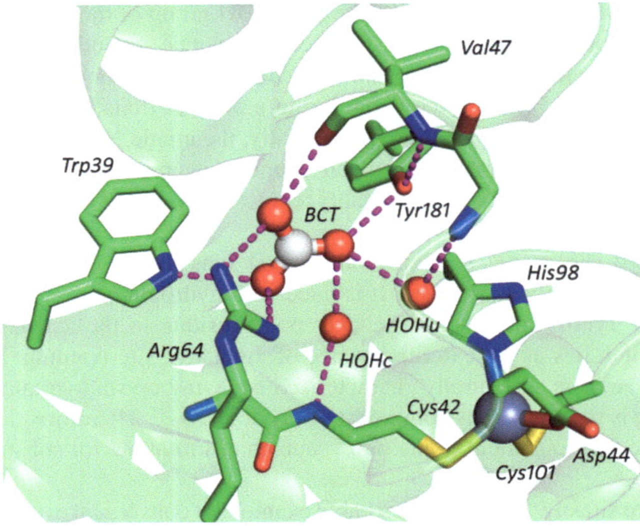

Fig. 4.4 Non-catalytic (allosteric) bicarbonate binding site in HICA (PDB 2A8D, chain D). Key residues, including bicarbonate (BCT) and the common and unique water molecules (HOHc and HOHu, respectively) are depicted as *sticks* and/or *spheres*

Gln151′, orienting it properly to donate a hydrogen bond to the distal oxygen of metal-bound acetic acid or bicarbonate. In HICA, the analogous Ser45-Gln33′ interaction is broken because the 44–48 loop has been rotated about 90° relative to the comparable region in PSCA.

Some type II β-CAs also have a non-catalytic bicarbonate binding site (Fig. 4.4) characterized by a Trp-Arg-Tyr triad (Trp39, Arg64, and Tyr181 in HICA). The non-catalytic binding site has been characterized by X-ray crystallography for ECCA and HICA [35]. It is located near the dimerization interface, opposite the tetramerization interface. A network of 7 highly specific hydrogen bonds serves to bind the bicarbonate in this site. The indole nitrogen of Trp39, two of the guanidinium NH groups of Arg64, the hydroxyl group of Tyr181, and two water molecules donate hydrogen bonds to the oxygen atoms of bicarbonate. One of these water molecules is found in many Type I and Type II β-CAs. This water accepts a hydrogen bond from the main-chain amide of Cys42. The other water molecule is unique to HICA and ECCA. It is positioned ≈3.4 Å directly above the imidazole ring of His98, where it can engage in a π-hydrogen bond, and accept a hydrogen bond from the main chain amide of Arg46. A final hydrogen bond exists between the main-chain carbonyl of Val47 and the hydroxyl group of bicarbonate. This is the only hydrogen bonding interaction in the binding site in which bicarbonate ion acts as a donor. This arrangement of hydrogen bonds is quite specific for bicarbonate: isolectronic anions such as nitrate or acetate are not able to act as a donor to the Val47 main-chain carbonyl. While HICA crystals soaked in bicarbonate ion readily bind it into the non-catalytic site, identical crystals soaked in up to 1 M acetate or nitrate are not observed to bind these ions in the bicarbonate binding site (R. Rowlett, unpublished results).

Binding of bicarbonate to the non-catalytic site stabilizes the conformation of the 44–48 loop that characterizes the structure of Type II β-CAs. The binding of the bicarbonate ion to the non-catalytic site expels the side chain of Val47, which causes a reorganization of the Asp44-Pro48 loop. This reorganization apparently breaks the Asp44-Arg46 dyad, which in turn allows Asp44 to bind directly to the metal ion, displacing the catalytically essential water molecule. Additionally, the rotation of the Asp44-Pro48 loop repositions Ser45 so that it can no longer stabilize Gln33′ in an orientation favorable for stabilizing the binding of an anionic ligand to the metal ion. The noncatalytic bicarbonate binding site is thought to mediate allosteric behavior in Type II β-CA (*vide infra*).

While the existence of the non-catalytic binding site has been shown only for HICA and ECCA, it is likely that many other type II β-CAs can bind bicarbonate in this fashion as well. SECA and PPCA have an identical triad, whereas Rv3588 substitutes an Ile residue for the Trp. SMCA, although it is classified as a Type II β-CA here, is missing most of the non-catalytic bicarbonate binding groups: it has a Val residue instead of Trp, and does not have the equivalent of Tyr181 in HICA because of a truncated C-terminus. All structurally characterized Type I β-CAs are missing one or more elements of the Trp-Arg-Tyr triad that characterizes the non-catalytic bicarbonate binding site.

4 Catalytic Mechanism

SOCA, PSCA, ATCA, MTCA, and HICA have been thoroughly kinetically characterized. Measurements of the rate of CO_2 hydration at steady state by stopped-flow spectrophotometry shows that β-CAs, like other CAs, are very fast enzymes with maximal k_{cat} values as high as 950 ms^{-1} and k_{cat}/K_m values as high as 180 $\mu M^{-1}\ s^{-1}$ (Table 4.2). MTCA and HICA are significantly slower enzymes than the eukaryotic enzymes.

The catalytic mechanism of β-CA follows a metal-hydroxide mechanism similar to that of other CAs (Fig. 4.5). This assessment is based on the close structural homology of the β-CA and α-CA active sites [31], as well as a wealth of kinetic evidence. Among the most convincing data is that β-CAs catalyze the exchange of O-18 in isotopically labeled bicarbonate in a way that can only be rationalized by deposition of the O-18 label at the active site metal ion during catalysis [35, 41, 47].

In the first step of the mechanism for CO_2 hydration, CO_2 is concentrated in the active site by association with a number of hydrophobic residues that line the active site cleft. CO_2 is known to bind to a similar hydrophobic pocket in the active site of HCAII [52]. This trapped CO_2 is then attacked by the zinc-bound hydroxide ion (step A → B in Fig. 4.5), the active species of the enzyme for CO_2 hydration. The hydroxide ion is properly oriented by the Asp-Arg dyad and the main chain

Table 4.2 Steady state kinetic parameters for CO_2 hydration for selected β-CAs

Enzyme	k_{cat} (ms^{-1})	k_{cat}/K_m ($\mu M^{-1}\ s^{-1}$)
SOCA[a]	163 ± 1	—
PSCA[b]	400	180
ATCA[c]	320 ± 20	68 ± 11
MTCA[d]	24 ± 2	8.3 ± 0.7
HICA[e]	69 ± 29	4.3 ± 0.8
SCCA[f]	940	98
DMCA[g]	950	110

[a]Maximal values at high pH [47]
[b]Values at pH 9.0 [48, 49]
[c]Maximal values at high pH [41]
[d]Values at pH 8.5 [37, 50]
[e]Maximal values at high pH [35]
[f]Values at pH 8.3 [51]
[g]*Drosophila melanogaster* enzyme; values at pH 8.3 [19]

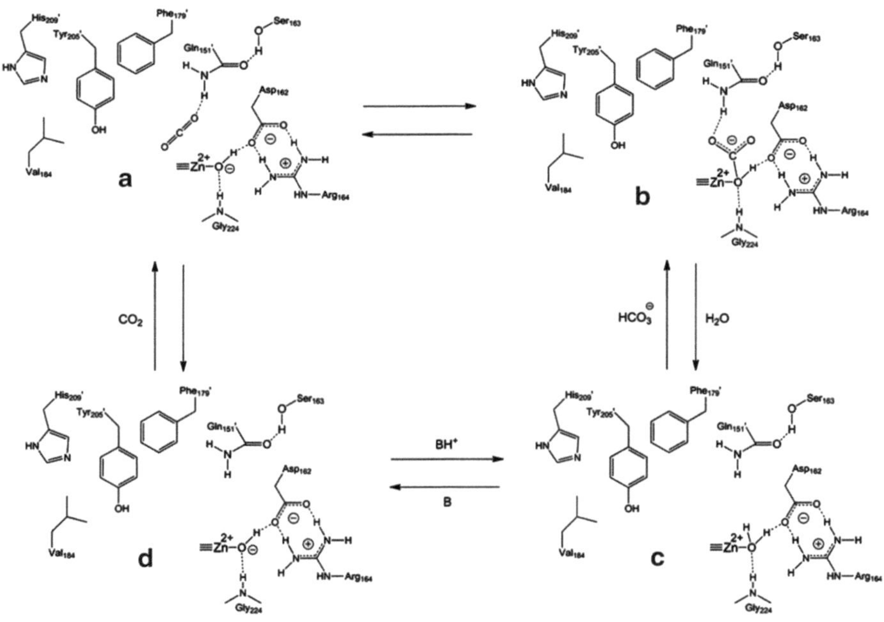

Fig. 4.5 Catalytic Mechanism of β-CA. Residue numbering system is for PSCA. "B" represents an external hydrogen ion acceptor; "BH$^+$" represents an external hydrogen ion donor. CO_2 hydration is represented by D-A-B-C-D; HCO_3^- dehydration proceeds as C-B-A-D-C. Figure adapted from [8]

amide of the active site Gly for nucleophilic attack of CO_2. The developing negative charge on the metal-bound bicarbonate ion is partly neutralized by hydrogen bond donation from the side chain amide of a Gln (His in MTCA and HTCA) from the neighboring subunit. In the third step of the mechanism (B → C in Fig. 4.5) water exchanges for bicarbonate at the metal ion, completing the conversion of CO_2 to HCO_3^-. To regenerate the original, active form of the enzyme, a hydrogen ion must be lost (C → D in Fig. 4.5). The final acceptor of this hydrogen ion are exogenous buffer molecules. However, it seems likely that one or more proton shuttle residues (PSRs) are involved, considering the distance (≈8–10 Å) between the metal-bound water molecule and the surface of the protein. Site-directed mutagenesis suggests that the active site His and Tyr residues (209 and 212 in PSCA) are important for proton transfer in PSCA [53] and ATCA [41]. In MTCA, the active site Asp34 residue has a role in proton transport as well [50]. Although it has not been experimentally tested, in the non-plant β-CAs the weakly conserved active site Tyr residue (Tyr83 in HICA) is the only polar group in the active site cleft that could act as a PSR. The equivalent of Tyr83 is missing in HTCA, MTCA, and Rv1284, and these β-CAs do not have any other likely candidates for a PSR. This is a likely reason why MTCA is among the slowest of β-CAs.

4.1 *Proton Transfer Is the Rate-Determining Step in Catalysis*

A significant body of evidence suggests that the rate-determining step for β-CA is the proton transfer step. NMR measurements of the rate of CO_2-HCO_3^- exchange (all steps except C ⇆ D in Fig. 4.5) catalyzed by SOCA show that it is ≈10× faster than the overall rate of catalysis [47]. O-18 isotope exchange experiments conducted with ATCA show that R_1, the rate of CO_2-HCO_3^- exchange at chemical equilibrium, is faster than R_{H2O}, the rate of release of O-18 labeled water from the enzyme [41]. The chemical step measured by the rate R_{H2O} requires the transfer of a proton from an exogenous or endogenous donor to the metal-bound hydroxide ion during the conversion of HCO_3^- to CO_2. For SOCA, PSCA, and MTCA, there is a significant solvent deuterium isotope effect (SDIE) of approximately 2 on k_{cat} for the CO_2 hydration reaction, which strongly suggests that there is a rate-determining hydrogen ion transfer in the mechanism. Additionally, the SDIE for k_{cat}/K_m, which includes all steps in Fig. 4.5 except C ⇆ D, is near unity: this means that the origin of the SDIE must reside in step C ⇆ D in Fig. 4.5, i.e., within an intermolecular or intramolecular proton transfer step. The observation that k_{cat} for CO_2 hydration in SOCA [47], PSCA [49], ATCA [41], and MTCA [37], depends upon exogenous buffer concentration in a saturable way suggests that intermolecular proton transfer is rate-limiting at low buffer concentrations, and that an intramolecular proton transfer step may be rate-limiting at high buffer concentrations. This is directly analogous to other classes of CA.

4.2 *pH-rate Profiles*

Both k_{cat} and k_{cat}/K_m increase with increasing pH for β-CA catalyzed CO_2 hydration [35, 37, 41, 47, 48]. This observation is entirely consistent with the mechanism of

Fig. 4.5. If intramolecular proton transfer is rate-limiting, as it is expected to be at high exogenous buffer concentrations, the pH profile of k_{cat} should be proportional to the fraction of the PSR that is in its conjugate base form. Thus the pH-rate profile of k_{cat} should provide an estimate of the macroscopic pK_a of the PSR. The pH-rate profile of k_{cat}/K_m, on the other hand, should reflect the ionization state of the metal-bound water: specifically k_{cat}/K_m should track with the relative fraction of the metal-hydroxide species present.

The pH-rate profiles of β-CA catalyzed CO_2 hydration are complex, and are likely involve the ionization of more than one species [37, 41, 48]. In addition, non-Michaelis-Menten kinetics are often observed [41, 48]. For ATCA, the pH-rate profile for k_{cat}/K_m can be satisfactorily modeled by a single ionization event with a pK_a of 7.2 ± 0.1 [41]. The H209A variant of PSCA also has a tractable pH-rate profile for k_{cat}/K_m that can be fit to a titration curve with a pK_a of 7.1 ± 0.1 [53]. These studies suggest that the pK_a for the zinc-bound water molecule is near neutrality for plant β-CAs, and is similar to the pK_a of the metal-bound water in α-CAs. This result is somewhat surprising, given the much more electron-rich ligation sphere for the metal ion in β-CA compared to α-CA. The pH-rate profile of k_{cat}/K_m for HICA is highly cooperative, but is estimated to be 7.74 ± 0.04 [35]. This is significantly higher than for plant β-CAs. The pH-rate profiles of k_{cat} are complex for all β-CAs studied so far. In ATCA, it is possible to identify two separate apparent pK_a values of 6.0 ± 0.8 and 8.8 ± 0.3 in the k_{cat} pH profiles [41]. This suggests that there may be two PSRs in plant CAs. Likely candidates are His209 and Tyr 205 (PSCA numbering).

4.3 *Site-Directed Mutagenesis Studies*

A significant number of site-directed mutagenesis studies have critically examined the mechanism of Fig. 4.5. To date, these studies have been supportive of this mechanism.

4.3.1 Active Site Variants

The role of the active site Gln residue was studied in the Q158A variant of ATCA [54]. Computational studies predict it is important in stabilizing developing negative charge on the metal-bound bicarbonate ion during catalysis [55]. This variant has a k_{cat}/K_m for CO_2 hydration that 33 % of wild type, and a maximal value of R_1/[E] that is an order of magnitude lower than the wild-type enzyme. The k_{cat} for CO_2 hydration and R_{H2O}/[E] values are not significantly altered by this variant. This result strongly suggests that the active site Gln residue is important for catalytic steps involved in the interconversion of CO_2 and HCO_3^-, but not for proton transfer, a result that is consistent with the proposed role for this residue.

The role of the Asp-Arg dyad in β-CA has also been studied. Computational studies suggest that the Asp-Arg dyad is essential for efficient catalysis by orienting the metal bound-hydroxide for nucleophilic attack [55]. The D34A variant of MTCA [50] reduces k_{cat} for CO_2 hydration by an order of magnitude in MOPS buffer, but not at all in the presence of imidazole buffer, suggesting that imidazole can chemically rescue this variant. The rate constant k_{cat}/K_m is relatively unaffected by the mutation of Asp34 to Ala. This observation suggests that the active site Asp residue may play an essential role in proton transfer. However, the replacement of the analogous Asp residue in SOCA [56] or in HICA [57] results in the nearly total loss of activity. Imidazole rescue of SOCA restores only 9 % of its activity at pH 8.0 [56]. All of these results taken together suggest that the dyad Asp residue may play important roles in both CO_2-HCO_3^- conversion and proton transfer. The replacement of Arg36 in MTCA results in the reduction of both k_{cat} and k_{cat}/K_m for CO_2 hydration to 0.1–1.0 % of wild-type [50]. This variant can be partially chemically rescued by exogenous guanidine. This result is consistent with the proposed role of the dyad Arg residue in orienting the dyad Asp residue to orient the metal-bound hydroxide for nucleophilic attack of CO_2.

4.3.2 Proton Transfer Variants

PSRs appear to be a general feature of all CAs, providing an efficient route for proton transfer from the metal-bound water molecule to exogenous proton acceptors. His64 fulfills this role in α-CA [58]; in γ-CA Glu84 performs this role [59]. Likely PSR candidates for β-CA are a highly conserved active site tyrosine residue (Tyr205 in PSCA; Tyr83 in HICA), and a histidine residue that is present in some plant β-CAs (His209 in PSCA), including PSCA, ATCA, and SOCA. The pH-rate profiles of k_{cat} for CO_2 hydration for ATCA suggests that there are two residues that participate in intramolecular proton transfer with apparent pK_a values of approximately 6.0 and 8.8 [41]. The ATCA variant H216N has a pH-rate profile of k_{cat} for CO_2 hydration, and a pH-rate profile of R_{H2O}/[E] that is significantly decreased at low pH, suggesting a role for His216 in proton transfer in this enzyme [41]. The H209A variant of PSCA behaves similarly: the pH-rate profile of k_{cat} for CO_2 hydration is significantly decreased at low, but not high pH [53]. The role of His216 as a PSR in ATCA is strongly supported by the chemical rescue of k_{cat} and R_{H2O}/[E] by exogenous imidazole [41, 60]. The role of the active site tyrosine residue as a PSR is less clear, but the Y212A variant of ATCA has a significantly diminished k_{cat} value at high pH (but not a low pH) compared to the wild-type enzyme [41].

4.4 Metallosubstitution of β-CA

Co(II)-substituted HICA (Co-HICA) can be made by overexpression of this enzyme in zinc-depleted, cobalt(II)-enriched defined growth medium [61]. Co-HICA is a bright blue color. Like most zinc-metalloenzymes, Co-HICA has catalytic activity very similar to the wild type enzyme, with a maximal k_{cat} of 120 % of the zinc enzyme. The apparent pK_a of the metal-bound water, derived from measurements of the pH-rate profile of k_{cat}/K_m for CO_2 hydration, is 7.48 ± 0.03, similar to or slightly lower than the pK_a of the metal-bound water in the zinc enzyme (7.7 ± 0.4). The visible absorption spectrum of Co-HICA at pH 8.0 is similar to that of other Co(II)-substituted CAs with *d-d* absorption bands at 560 nm ($\varepsilon = 340\ M^{-1}\ cm^{-1}$), 620 nm ($\varepsilon = 720\ M^{-1}\ cm^{-1}$), and 703 nm ($\varepsilon = 290\ M^{-1}\ cm^{-1}$), the last of which may be two overlapping features. The molar absorptivity of these bands increase markedly in a highly cooperative fashion as the pH is lowered, and there is no isosbestic point, suggesting that there than two metal coordination species present. The most likely coordination environments involved are $Co(Cys)_2(His)(Asp44)$, $Co(Cys)_2(His)OH_2$, and $Co(Cys)_2OH^-$ species, with the first predominant at low pH, the last predominant at high pH, and the middle species mixing in at intermediate pH values.

4.5 Inhibitors of β-CA

Like other CAs, β-CAs are inhibited by monovalent anions and sulfonamides [47, 48, 51, 62–68]. The mode of inhibition is probably similar to that of other CAs: that is, anions and sulfonamides (in the deprotonated anionic form) bind to the active site zinc ion, displacing the catalytically essential water molecule. Direct evidence for anion binding to β-CA comes from the X-ray crystal structure of the thiocyanate complex of Rv3588, in which the nitrogen atom of thiocyanate is observed to be directly bound to the zinc ion in place of the catalytic water [39]. The affinity of inhibitors varies greatly between various CAs, but in general, anions and sulfonamides inhibit plant β-CAs more strongly than for microbial β-CAs [8]. Plant β-CAs have anion K_i values that are comparable to that of α-CAs, but for MTCA similar anion K_i values are 1–2 orders of magnitude higher. In α-CA, acetazolamide and ethoxzolamide are potent inhibitors with K_i values in the nanomolar to picomolar range [69]. For PSCA and MTCA, these inhibitors have K_i values in the micromolar range. Phenylboronic and phenylarsonic acids are more potent inhibitors of MTCA β-CA than α-CA, with K_i values in the micromolar range [70]; these inhibitors are moderately specific for β-CA relative to α-CA.

5 Allostery

Another defining characteristic of type II β-CAs are dramatically steep pH-rate profiles. ECCA and HICA [35], SECA [10], and Rv3588 [38] all show a sudden loss of enzymatic activity below pH 8.0. pH-rate profiles of k_{cat} and R_{H2O}/[E] for HICA-catalyzed CO_2 hydration show highly cooperative behavior that correlates to the

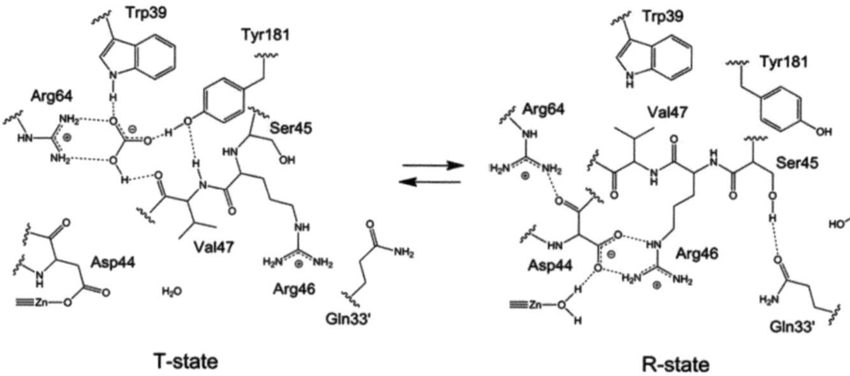

Fig. 4.6 Key residues involved in the allosteric switching of HICA between the inactive (T) and active (R) states

loss of 2 or 4 protons to generate the catalytically competent species [35]. Finally, the variation of both R_1 and R_{H2O} with $[CO_2 + HCO_3^-]$ concentration is biphasic: at low substrate concentration, these rates increase with increasing $[CO_2 + HCO_3^-]$ concentration; at higher $[CO_2 + HCO_3^-]$ concentrations R_1 and R_{H2O} decrease as the square of the $[CO_2 + HCO_3^-]$ concentration [35]. This data, when combined with the observation that bicarbonate ion binds to a non-catalytic site in HICA, suggests that HICA is an allosteric enzyme that is regulated by HCO_3^-, and that the fundamental allosteric unit is a dimer, presumably the fundamental structural dimer of β-CA. It seems likely that all type II β-CAs that have an intact Arg-Trp-Tyr triad in the non-catalytic pocket are likewise allosteric enzymes. The inactive T-state of type II β-CA is characterized by the vast majority of the X-ray crystallographic structures of the type II enzymes, where the active site Asp residue has displaced the catalytically essential water molecule. The active, R-state of the enzyme is assumed to closely resemble that of the type I β-CAs. The T-state of the enzyme is favored at low pH, when the metal-bound hydroxide is protonated to form a metal-bound water, which is more easily displaced than hydroxide by the side chain of the active site Asp residue. Figure 4.6 summarizes the principal elements of the T $\leftrightarrows$ R transition in HICA. Bicarbonate ion stabilizes the T-state by displacing the side chain of Val47 to interact with Trp39, Arg64, and Tyr181, as well as two water molecules and the carbonyl oxygen of Val47. Displacement of the polypeptide backbone from residues 44–48 disrupts the Asp44-Arg46 dyad, allowing Asp44 to coordinate to the metal ion. In the absence of HCO_3^-, Val47 can occupy its place and form a hydrophobic cluster with Trp39 and Tyr181, stabilizing the 44–48 loop in a way that allows for the formation of the Asp44-Arg46 dyad.

There is compelling evidence that type II β-CAs can adopt alternate conformations. The thiocyanate complex of Rv3588 crystallizes in an "R-state" conformation that resembles Type I β-CAs, whereas the uncomplexed enzyme crystallizes in a "T-state" conformation characteristic of Type II β-CAs. These structures suggest

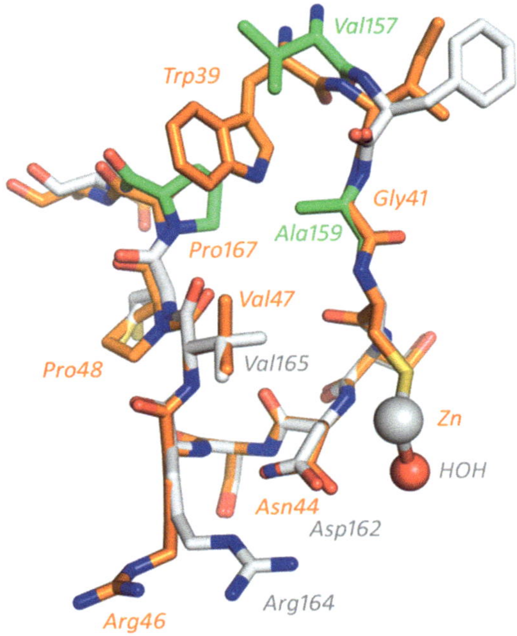

Fig. 4.7 Superposition of the allosteric loop region of HICA-D44N (PDB 3E1V, *orange backbone*) onto PSCA (PDB 1EKJ, *gray backbone*). HICA-D44N zinc ion and metal-bound water are depicted as *spheres*. *Green-backbone* residues indicate key structural differences in PSCA compared to HICA. HICA residues numbers are *orange*, PSCA *gray* or *green*

that Rv3588 can adopt two conformations in solutions that are distinguished by a "carboxylate shift" at the active site metal ion [39]. It is not known if HCO_3^- can bind to the noncatalytic site of Rv3588. Site-directed mutagenesis of HICA also supports the allosteric switching mechanism of Fig. 4.6. The allosteric site variant W39F retains a large fraction of the activity of the wild-type enzyme, but the apparent K_i for bicarbonate ion increases almost 5-fold for the variant enzyme [57]. This result is consistent with the loss of a stabilizing hydrogen bond for the binding of bicarbonate in the allosteric site.

The active site variant D44N, in which the active site Asp residue is replaced with an isosteric but non-nucleophilic Asn residue, is not active, but the critical 44–48 loop occupies precisely the same conformation (within 0.2 Å rmsd) as the analogous loop in the type I β-CA PSCA (Fig. 4.7) [57]. The observation of this structure in HICA-D44N suggests that (1) the enzyme can adopt more than one conformation about the active site, and (2) the stabilization of the "type I like" structure may be linked to the dissociation of the active site Asp from the metal ion. There are three key differences in the 39–50 loop (HICA numbering) between HICA and PSCA that may account for the former to adopt a bicarbonate-stabilized "T-state" while the latter cannot (Fig. 4.7). The first is the substitution of Ala for Gly41 in PSCA. This Ala residue pokes into the bicarbonate binding site, and would appear to preclude the binding of bicarbonate ion in the "T-state." The second is the substitution of Val for Trp39, which has been shown to significantly weaken bicarbonate binding to the allosteric site [57]. Finally, a Pro residue that occupies position 48 in HICA is shifted one residue later in the sequence in PSCA. This proline shift has the effect of pointing the proline ring directly into the bicarbonate binding site. The G41A variant of HICA is essentially identical to wild-type, and is observed to bind bicarbonate ion in the allosteric site despite the steric hindrance predicted by the protruding methyl group of Ala [71]. This result suggests that the proline shift may be a critical factor in distinguishing allosteric and non-allosteric β-CAs. In PSCA, the region that would correspond to the allosteric site in HICA forms what looks like a stable hydrophobic cluster with the ring of Pro167, Val157, Ala159, and Val165, stabilizing the R-state conformation.

Direct evidence pH-cooperativity and bicarbonate-induced changes at the active site comes from spectroscopic studies of Co-HICA [61]. The *d-d* transitions in Co-HICA change significantly in both molar absorptivity and the wavelength of maximum absorption in a highly cooperative manner with a $[H^+]^4$ dependence, and an apparent pK_a near 7.8, which corresponds closely to the pK_a observed in the pH-rate profile of k_{cat}/K_m for CO_2 hydration. The low-pH spectrum is presumed to be due to the T-state of the enzyme, whereas the high-pH spectrum is believed to be representative of the R-state. Near but just above the transition point (pH 8.0), the addition of 25 mM HCO_3^- causes Co-HICA to switch from the high-pH spectrum to that essentially identical to the low-pH spectrum. This suggests that HCO_3^- can stabilize the T-state of the HICA. Remarkably, heterogeneously substituted Co-HICA (for which the principal absorbing species are Co-Zn heterodimers) and homogeneously substituted Co-HICA (for which the principal absorbing species are Co-Co dimers), are different colors. Partially Co-substituted Co-HICA is green, whereas homogeneous Co-HICA is blue. This easily observable qualitative color difference is due to the shift of a ligand-to-metal charge transfer band from 410 nm in the partially substituted enzyme into the UV tail for homogeneous Co-HICA. This strange result suggests that the metal ion in one active site of the fundamental dimer is coupled to and can be affected by the metal present in the other half of the dimer: that is, the fundamental dimer metal sites are cooperatively coupled.

The likely ingress/egress route for allosteric bicarbonate ion was serendipitously discovered in bicarbonate complexes of the allosteric site HICA variants G41A and V47A. In these variants, bicarbonate ion is observed to bind to a crevice along the dimerization interface between two Arg64 and two Glu50 residues. This site is immediately adjacent to the allosteric binding site, and has been designated the bicarbonate "escort" site [71]. A bicarbonate ion could easily transition between the "escort" site and the allosteric site by a relatively simple movement of one of the Arg64 residues. Indeed in the V47A variant, bicarbonate ions are observed to be bound in both allosteric sites and the "escort" site in the fundamental dimer. Sulfate, an apparent activator of HICA [35], is also observed to bind at the "escort" site, suggesting that the mode of activation could be related to blocking access of bicarbonate to the allosteric site [71].

6 Conclusion

The β-CAs are diverse in distribution in nature, and are one of five convergently evolved classes of CA that have arrived at the same, efficient solution for catalysis: a metal-hydroxide mechanism that is rate-limited by proton transfer. Uniquely, the β-CAs harbor the only known allosteric CAs. The role of significance of allostery in β-CA is not presently known.

Acknowledgments The author wishes to acknowledge the support of the National Science Foundation (MCB-0741396, MCB-1157332, and CHE0819686) for support of β-CA research at Colgate University.

References

1. Burnell JN, Gibbs MJ, Mason JG (1990) Spinach chloroplastic carbonic-anhydrase—nucleotide-sequence analysis of cDNA. Plant Physiol 92:37–40
2. Neish AC (1939) Chloroplasts. II. Their chemical composition and the distribution of certain metabolites between the chloroplasts and the remainder of the leaf. Biochem J 33:300–308
3. Coleman JR, Luinenburg I, Majeau N, Provart N (1991) Sequence analysis and regulation of expression of a gene coding for carbonic anhydrase in *Chlamydomonas reinhardtii*. Can J Bot 69:1097–1102
4. Roeske CA, Ogren WL (1990) Nucleotide sequence of pea cDNA-encoding chloroplast carbonic anhydrase. Nucleic Acids Res 18:3413
5. Moroney JV, Bartlett SG, Samuelsson G (2001) Carbonic anhydrases in plants and algae. Plant Cell Environ 24:141–153
6. Smith KS, Ferry JG (2000) Prokaryotic carbonic anhydrases. FEMS Microbiol Rev 24:335–366
7. Smith KS, Jakubzick C, Whittam TS, Ferry JG (1999) Carbonic anhydrase is an ancient enzyme widespread in prokaryotes. Proc Natl Acad Sci U S A 96:15184–15189
8. Rowlett RS (2010) Structure and catalytic mechanism of the beta-carbonic anhydrases. Biochim Biophys Acta Proteins Proteomics 1804:362–373
9. Nishimori I, Minakuchi T, Kohsaki T, Onishi S, Takeuchi H, Vullo D, Scozzafava A, Supuran CT (2007) Carbonic anhydrase inhibitors: the beta-carbonic anhydrase from *Helicobacter pylori* is a new target for sulfonamide and sulfamate inhibitors. Bioorg Med Chem Lett 17:3585–3594
10. Nishimori I, Minakuchi T, Vullo D, Scozzafava A, Supuran CT (2011) Inhibition studies of the β-carbonic anhydrases from the bacterial pathogen *Salmonella enterica* serovar Typhimurium with sulfonamides and sulfamates. Bioorg Med Chem 19:5023–5030
11. Joseph P, Ouahrani-Bettache S, Montero JL, Nishimori I, Minakuchi T, Vullo D, Scozzafava A, Winum JY, Kohler S, Supuran CT (2011) A new beta-carbonic anhydrase from *Brucella suis*, its cloning, characterization, and inhibition with sulfonamides and sulfamates, leading to impaired pathogen growth. Bioorg Med Chem 19:1172–1178
12. Burghout P, Vullo D, Scozzafava A, Hermans PWM, Supuran CT (2011) Inhibition of the beta-carbonic anhydrase from *Streptococcus pneumoniae* by inorganic anions and small molecules: toward innovative drug design of antiinfectives? Bioorg Med Chem 19:243–248
13. Smith KS, Ferry JG (1999) A plant-type (beta-class) carbonic anhydrase in the thermophilic methanoarchaeon *Methanobacterium thermoautotrophicum*. J Bacteriol 181:6247–6253

14. Amoroso G, Morell-Avrahov L, Muller D, Klug K, Sultemeyer D (2005) The gene NCE103 (YNL036w) from *Saccharomyces cerevisiae* encodes a functional carbonic anhydrase and its transcription is regulated by the concentration of inorganic carbon in the medium. Mol Microbiol 56:549–558
15. So AKC, Espie GS (1998) Cloning, characterization and expression of carbonic anhydrase from the cyanobacterium Synechocystis PCC6803. Plant Mol Biol 37:205–215
16. Sawaya MR, Cannon GC, Heinhorst S, Tanaka S, Williams EB, Yeates TO, Kerfeld CA (2006) The structure of beta-carbonic anhydrase from the carboxysomal shell reveals a distinct subclass with one active site for the price of two. J Biol Chem 281:7546–7555
17. Eriksson M, Karlsson J, Ramazanov Z, Gardestrom P, Samuelsson G (1996) Discovery of an algal mitochondrial carbonic anhydrase: molecular cloning and characterization of a low-CO_2-induced polypeptide in *Chlamydomonas reinhardtii*. Proc Natl Acad Sci U S A 93:12031–12034
18. Mitsuhashi S, Mizushima T, Yamashita E, Yamamoto M, Kumasaka T, Moriyama H, Ueki T, Miyachi S, Tsukihara T (2000) X-ray structure of beta-carbonic anhydrase from the red alga, *Porphyridium purpureum*, reveals a novel catalytic site for CO_2 hydration. J Biol Chem 275:5521–5526
19. Syrjanen L, Tolvanen M, Hilvo M, Olatubosun A, Innocenti A, Scozzafava A, Leppiniemi J, Niederhauser B, Hytonen VP, Gorr TA, Parkkila S, Supuran CT (2010) Characterization of the first beta-class carbonic anhydrase from an arthropod (*Drosophila melanogaster*) and phylogenetic analysis of beta-class carbonic anhydrases in invertebrates. BMC Biochem 11, No pp given
20. Fasseas MK, Tsikou D, Flemetakis E, Katinakis P (2010) Molecular and biochemical analysis of the β class carbonic anhydrases in *Caenorhabditis elegans*. Mol Biol Rep 37:2941–2950
21. Majeau N, Coleman JR (1996) Effect of CO_2 concentration on carbonic anhydrase and ribulose-1,5-bisphosphate carboxylase/oxygenase expression in pea. Plant Physiol 112:569–574
22. Majeau N, Arnoldo M, Coleman JR (1994) Modification of carbonic anhydrase activity by antisense and over-expression constructs in transgenic tobacco. Plant Mol Biol 25:377–385
23. Fukuzawa H, Suzuki E, Komukai Y, Miyachi S (1992) A gene homologous to chloroplast carbonic anhydrase (icfA) is essential to photosynthetic carbon dioxide fixation by Synechococcus PCC7942. Proc Natl Acad Sci U S A 89:4437–4441
24. Merlin C, Masters M, McAteer S, Coulson A (2003) Why is carbonic anhydrase essential to *Escherichia coli*? J Bacteriol 185:6415–6424
25. Mitsuhashi S, Ohnishi J, Hayashi M, Ikeda M (2004) A gene homologous to beta-type carbonic anhydrase is essential for the growth of *Corynebacterium glutamicum* under atmospheric conditions. Appl Microbiol Biotechnol 63:592–601
26. Gotz R, Gnann A, Zimmermann FK (1999) Deletion of the carbonic anhydrase-like gene NCE103 of the yeast *Saccharomyces cerevisiae* causes an oxygen-sensitive growth defect. Yeast 15:855–864
27. Clark D, Rowlett RS, Coleman JR, Klessig DF (2004) Complementation of the yeast deletion mutant Delta NCE103 by members of the beta class of carbonic anhydrases is dependent on carbonic anhydrase activity rather than on antioxidant activity. Biochem J 379:609–615
28. Burghout P, Cron LE, Gradstedt H, Quintero B, Simonetti E, Bijlsma JJE, Bootsma HJ, Hermans PWM (2010) Carbonic anhydrase is essential for *Streptococcus pneumoniae* growth in environmental ambient air. J Bacteriol 192:4054–4062
29. Bury-Mone S, Mendz GL, Ball GE, Thibonnier M, Stingl K, Ecobichon C, Ave P, Huerre M, Labigne A, Thiberge J-M, De RH (2008) Roles of α and β carbonic anhydrases of *Helicobacter pylori* in the urease-dependent response to acidity and in colonization of the murine gastric mucosa. Infect Immun 76:497–509
30. Smeulders MJ, Barends TRM, Pol A, Scherer A, Zandvoort MH, Udvarhelyi A, Khadem AF, Menzel A, Hermans J, Shoeman RL, Wessels HJCT, van den Heuvel LP, Russ L, Schlichting I, Jetten MSM, den Camp HJMO (2011) Evolution of a new enzyme for carbon disulphide conversion by an acidothermophilic archaeon. Nature (London, UK) 478:412–416

31. Kimber MS, Pai EF (2000) The active site architecture of *Pisum sativum* beta-carbonic anhydrase is a mirror image of that of alpha-carbonic anhydrases. EMBO J 19:1407–1418
32. Cronk JD, Endrizzi JA, Cronk MR, O'Neill JW, Zhang KYJ (2001) Crystal structure of *E-coli* beta-carbonic anhydrase, an enzyme with an unusual pH-dependent activity. Protein Sci 10:911–922
33. Krissinel E, Henrick K (2007) Inference of macromolecular assemblies from crystalline state. J Mol Biol 372:774–797
34. So AK, Espie GS, Williams EB, Shively JM, Heinhorst S, Cannon GC (2004) A novel evolutionary lineage of carbonic anhydrase (epsilon class) is a component of the carboxysome shell. J Bacteriol 186:623–630
35. Cronk JD, Rowlett RS, Zhang KYJ, Tu CK, Endrizzi JA, Lee J, Gareiss PC, Preiss JR (2006) Identification of a novel noncatalytic bicarbonate binding site in eubacterial beta-carbonic anhydrase. Biochemistry 45:4351–4361
36. Strop P, Smith KS, Iverson TM, Ferry JG, Rees DC (2001) Crystal structure of the "cab"-type beta class carbonic anhydrase from the archaeon *Methanobacterium thermoautotrophicum*. J Biol Chem 276:10299–10305
37. Smith KS, Cosper NJ, Stalhandske C, Scott RA, Ferry JG (2000) Structural and kinetic characterization of an archaeal beta-class carbonic anhydrase. J Bacteriol 182:6605–6613
38. Covarrubias AS, Larsson AM, Hogbom M, Lindberg J, Bergfors T, Bjorkelid C, Mowbray SL, Unge T, Jones TA (2005) Structure and function of carbonic anhydrases from *Mycobacterium tuberculosis*. J Biol Chem 280:18782–18789
39. Covarrubias AS, Bergfors T, Jones TA, Hoegbom M (2006) Structural mechanics of the pH-dependent activity of beta-carbonic anhydrase from *Mycobacterium tuberculosis*. J Biol Chem 281:4993–4999
40. Pocker Y, Ng JSY (1973) Plant carbonic anhydrase. Properties and carbon dioxide hydration kinetics. Biochemistry 12:5127–5134
41. Rowlett RS, Tu C, McKay MM, Preiss JR, Loomis RJ, Hicks KA, Marchione RJ, Strong JA, Donovan GS, Chamberlin JE (2002) Kinetic characterization of wild-type and proton transfer-impaired variants of beta-carbonic anhydrase from *Arabidopsis thaliana*. Arch Biochem Biophys 404:197–209
42. Majeau N, Coleman JR (1992) Nucleotide sequence of a complementary DNA encoding tobacco chloroplastic carbonic anhydrase. Plant Physiol 100:1077–1078
43. Raines CA, Horsnell PR, Holder C, Lloyd JC (1992) *Arabidopsis thaliana* carbonic anhydrase: cDNA sequence and effect of carbon dioxide on mRNA levels. Plant Mol Biol 20:1143–1148
44. Hoang CV, Wessler HG, Local A, Turley RB, Benjamin RC, Chapman KD (1999) Identification and expression of cotton (Gossypium hirsutum L.) plastidial carbonic anhydrase. Plant Cell Physiol 40:1262–1270
45. Johansson IM, Forsman C (1992) Processing of the chloroplast transit peptide of pea carbonic anhydrase in chloroplasts and in *Escherichia coli*: identification of 2 cleavage sites. FEBS Lett 314:232–236
46. Forsman C, Pilon M (1995) Chloroplast import and sequential maturation of pea carbonic anhydrase: the roles of various parts of the transit peptide. FEBS Lett 358:39–42
47. Rowlett RS, Chance MR, Wirt MD, Sidelinger DE, Royal JR, Woodroffe M, Wang Y-FA, Saha RP, Lam MG (1994) Kinetic and structural characterization of spinach carbonic anhydrase. Biochemistry 33:13967–13976
48. Johansson IM, Forsman C (1993) Kinetic studies of pea carbonic anhydrase. Eur J Biochem 218:439–446
49. Johansson IM, Forsman C (1994) Solvent hydrogen isotope effects and anion inhibition of CO_2 hydration catalysed by carbonic anhydrase from *Pisum sativum*. Eur J Biochem 224:901–907
50. Smith KS, Ingram-Smith C, Ferry JG (2002) Roles of the conserved aspartate and arginine in the catalytic mechanism of an archaeal beta-class carbonic anhydrase. J Bacteriol 184:4240–4245

51. Innocenti A, Hall RA, Schlicker C, Muhlschlegel FA, Supuran CT (2009) Carbonic anhydrase inhibitors. Inhibition of the beta-class enzymes from the fungal pathogens *Candida albicans* and *Cryptococcus neoformans* with aliphatic and aromatic carboxylates. Bioorg Med Chem 17:2654–2657
52. Domsic JF, Avvaru BS, Kim CU, Gruner SM, Agbandje-McKenna M, Silverman DN, McKenna R (2008) Entrapment of carbon dioxide in the active site of carbonic anhydrase II. J Biol Chem 283:30766–30771
53. Bjorkbacka H, Johansson IM, Forsman C (1999) Possible roles for His 208 in the active-site region of chloroplast carbonic anhydrase from *Pisum sativum*. Arch Biochem Biophys 361:17–24
54. Rowlett RS, Tu C, Murray PS, Chamberlin JE (2004) Examination of the role of Gln-158 in the mechanism of CO_2 hydration catalyzed by beta-carbonic anhydrase from *Arabidopsis thaliana*. Arch Biochem Biophys 425:25–32
55. Hakkim V, Rajapandian V, Subramanian V (2011) Density functional theory investigations of the catalytic mechanism of β-carbonic anhydrase. Indian J Chem A Inorg Bio-inorg Phys Theor Anal Chem 50A:503–510
56. Provart NJ, Majeau N, Coleman JR (1993) Characterization of pea chloroplastic carbonic anhydrase—expression in *Escherichia coli* and site-directed mutagenesis. Plant Mol Biol 22:937–943
57. Rowlett RS, Tu C, Lee J, Herman AG, Chapnick DA, Shah SH, Gareiss PC (2009) Allosteric site variants of *Haemophilus influenzae* beta-carbonic anhydrase. Biochemistry 48:6146–6156
58. Silverman DN, Lindskog S (1988) The catalytic mechanism of carbonic anhydrase: implications of a rate-limiting protolysis of water. Acc Chem Res 21:30–36
59. Tripp BC, Ferry JG (2000) A structure-function study of a proton transport pathway in the gamma-class carbonic anhydrase from *Methanosarcina thermophila*. Biochemistry 39:9232–9240
60. Tu CK, Rowlett RS, Tripp BC, Ferry JG, Silverman DN (2002) Chemical rescue of proton transfer in catalysis by carbonic anhydrases in the beta- and gamma-class. Biochemistry 41:15429–15435
61. Hoffmann KM, Samardzic D, Heever K, Rowlett RS (2011) Co(II)-substituted *Haemophilus influenzae* beta-carbonic anhydrase: spectral evidence for allosteric regulation by pH and bicarbonate ion. Arch Biochem Biophys 511:80–87
62. Pocker Y, Ng JSY (1974) Plant carbonic anhydrase. Hydrase activity and its reversible inhibition. Biochemistry 13:5116–5120
63. Zimmerman S, Innocenti A, Casini A, Ferry JG, Scozzafava A, Supuran CT (2004) Carbonic anhydrase inhibitors. Inhibition of the prokaryotic beta- and gamma-class enzymes from Archaea with sulfonamides. Bioorg Med Chem Lett 14:6001–6006
64. Zimmerman SA, Ferry JG, Supuran CT (2007) Inhibition of the archaeal beta-class (Cab) and gamma-class (Cam) carbonic anhydrases. Curr Top Med Chem 7:901–908
65. Isik S, Kockar F, Arslan O, Guler OO, Innocenti A, Supuran CT (2008) Carbonic anhydrase inhibitors. Inhibition of the beta-class enzyme from the yeast *Saccharomyces cerevisiae* with anions. Bioorg Med Chem Lett 18:6327–6331
66. Isik S, Kockar F, Aydin M, Arslan O, Guler OO, Innocenti A, Scozzafava A, Supuran CT (2009) Carbonic anhydrase inhibitors: inhibition of the beta-class enzyme from the yeast *Saccharomyces cerevisiae* with sulfonamides and sulfamates. Bioorg Med Chem 17:1158–1163
67. Morishita S, Nishimori I, Minakuchi T, Onishi S, Takeuchi H, Sugiura T, Vullo D, Scozzafava A, Supuran CT (2008) Cloning, polymorphism, and inhibition of beta-carbonic anhydrase of *Helicobacter pylori*. J Gastroenterol 43:849–857
68. Innocenti A, Muhlschlegel FA, Hall RA, Steegborn C, Scozzafava A, Supuran CT (2008) Carbonic anhydrase inhibitors: inhibition of the beta-class enzymes from the fungal pathogens *Candida albicans* and *Cryptococcus neoformans* with simple anions. Bioorg Med Chem Lett 18:5066–5070

69. Lindskog S (1969) Mechanism of sulfonamide inhibition of carbonic anhydrase. Chalmers Institute and University of Technology, Goteborg, Sweden, pp 157–165
70. Innocenti A, Zimmerman S, Ferry JG, Scozzafava A, Supuran CT (2004) Carbonic anhydrase inhibitors. Inhibition of the beta-class enzyme from the methanoarchaeon *Methanobacterium thermoautotrophicum* (Cab) with anions. Bioorg Med Chem Lett 14:4563–4567
71. Rowlett RS, Hoffmann KM, Failing H, Mysliwiec MM, Samardzic D (2010) Evidence for a bicarbonate "escort" site in *Haemophilus influenzae* beta-carbonic anhydrase. Biochemistry 49:3640–3647

Chapter 5
Prokaryotic Carbonic Anhydrases of Earth's Environment

R. Siva Sai Kumar and James G. Ferry

Abstract Carbonic anhydrase is a metalloenzyme catalyzing the reversible hydration of carbon dioxide to bicarbonate. Five independently evolved classes have been described for which one or more are found in nearly every cell type underscoring the general importance of this ubiquitous enzyme in Nature. The bulk of research to date has centered on the enzymes from mammals and plants with less emphasis on prokaryotes. Prokaryotic carbonic anhydrases play important roles in the ecology of Earth's biosphere including acquisition of CO_2 for photosynthesis and the physiology of aerobic and anaerobic prokaryotes decomposing the photosynthate back to CO_2 thereby closing the global carbon cycle. This review focuses on the physiology and biochemistry of carbonic anhydrases from prokaryotes belonging to the domains *Bacteria* and *Archaea* that play key roles in the ecology of Earth's biosphere.

Keywords Anaerobe • Aerobe • Food chains • Biosphere • Decomposition • Prokaryotes • Carbon cycle

1 Introduction

There are five independently-evolved classes ($\alpha, \beta, \gamma, \delta, \zeta$) of carbonic anhydrases (CAs), with no structural or sequence similarity, that catalyze the reversible hydration of CO_2 ($CO_2 + H_2O \rightleftarrows HCO_3^- + H^+$). The α class is restricted

Susan C. Frost and Robert McKenna (eds.). Carbonic Anhydrase: Mechanism, Regulation, Links to Disease, and Industrial Applications

R.S.S. Kumar • J.G. Ferry (✉)
Department of Biochemistry and Molecular Biology, Ebery College of Science,
The Pennsylvania State University, University Park, PA, USA
e-mail: ssr14@psu.edu; jgf3@psu.edu

S.C. Frost and R. McKenna (eds.), *Carbonic Anhydrase: Mechanism, Regulation, Links to Disease, and Industrial Applications*, Subcellular Biochemistry 75,
DOI 10.1007/978-94-007-7359-2_5,

Fig. 5.1 The global carbon cycle

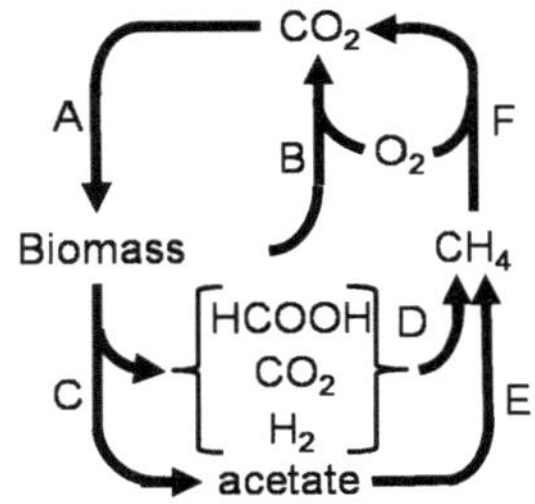

mostly to mammals and pathogenic prokaryotes, although a few have been described from non-pathogenic species [1]. The δ [2, 3] and ζ [3, 4] classes are found in marine diatoms [4]. A search of the Comprehensive Microbial Resource [5] of 656 completed prokaryotic genomes with the query "carbonic anhydrase" retrieved 1,300 sequences underscoring the widespread occurrence of this enzyme in diverse prokaryotes inhabiting Earth's biosphere. The great majority of prokaryotes harbor the β and γ class enzymes that play important roles in the global carbon cycle (Fig. 5.1). In the cycle, plants, algae and photosynthetic prokaryotes fix CO_2 into biomass (step A) that is decomposed back to CO_2 by O_2-requiring prokaryotes in aerobic environments (step B). Significant amounts of biomass enter diverse O_2-free environments where strictly anaerobic microbes decompose the biomass to methane and CO_2 (steps C,D and E). The remainder diffuses into aerobic environments where O_2-requiring prokaryotes oxidize it to CO_2 (Step F) closing the carbon cycle. CAs are a focus of investigation for developing technologies that capture and sequester CO_2 at point sources for reducing emissions that contribute to the greenhouse effect and global warming [6]. This review focuses on recent advances in understanding the physiology and biochemistry of CAs from prokaryotic microbes important to the ecology of Earth's biosphere.

2 Aerobic Species

Aerobes are prokaryotes that require O_2 to metabolize growth substrates. Cyanobacteria are aerobes and the most abundant photosynthetic organisms on Earth [11] that, together with plants, are responsible for the great majority of global CO_2 fixation into biomass (Step A, Fig. 5.1). Cyanobacteria utilize β and γ class CAs associated with an organelle called a carboxysome to enhance photosynthesis via a mechanism that concentrates and supplies CO_2 to ribulose-1,5-bisphosphate carboxylase [7]. Given that all cyanobacteria contain carboxysomes, and that cyanobacteria are considered the most abundant photosynthetic organisms on Earth [8], a substantial portion of global CO_2 fixation is dependent on CAs. The γ class CA (CcmM) from the aerobic cyanobacterium *Thermosynechococcus elongatus* strain BP-1 has been characterized [9]. The crystal structure closely resembles the archetype of the γ class (Cam) from *Methanosarcina thermophila* of the domain *Archaea* [10]. Cam is

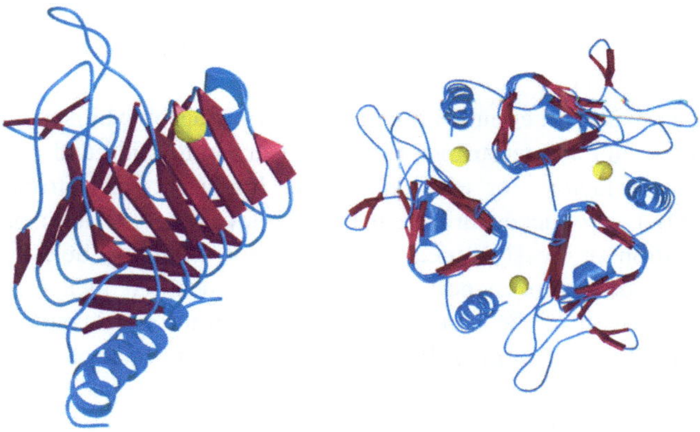

Fig. 5.2 Ribbon diagram of Cam, the archetype γ class CA. *Left panel*, a monomer. *Right panel*, view of the homotrimer along the axis of symmetry. The *yellow sphere* indicates the active site metal coordinated by two histidine residues from one of the three subunits and a third histidine ligand contributed by an adjacent subunit

a homotrimer with an overall left-handed β-helical fold (Fig. 5.2) [11]. However, CcmM differs from Cam by the addition of residues 198–207 that form a third α-helix stabilized by an essential cysteine 194-cysteine 200 disulfide bond. Deletion of this domain yields an inactive protein, the structure of which shows disordering of the N- and C-termini, and reorganization of the trimeric interface and active site. Under reducing conditions, CcmM is similarly partially disordered and inactive which underscores the importance of the disulfide bond. The crystal structure of the β class CA from the carboxysome of the aerobic species *Halothiobacillus neapolitanus* (CsoSCA) identifies it as one of three structural subclasses of the β class [12]. The active sites of β class enzymes from *Escherichia coli* [13], *Pisum sativum* [14], and the strict anaerobe *Methanothermobacter thermautotrophicus* [15] are organized as identical pairs through homodimerization. The enzyme from *Porphyridium purpureum*, a red alga, evolved an intermediate degree of symmetry wherein two domains are fused and organized with 2-fold pseudo-symmetry. The C-terminal domain of CsoSCA from *H. neapolitanus* diverged such that it lost the ligands for the catalytic zinc, including a loop on which one of the cysteine ligands would normally reside. Thus, the two domains in CsoSCA evolved such that only one in the pair retains a viable active site. It is proposed that this structure evolved to supply a new function specific to the carboxysome [12]. Nonetheless, the structural similarity among active site regions of the β class subclasses suggests a common catalytic mechanism.

A few non-photosynthetic chemolithotrophic aerobes contribute to CO_2 fixation such as the thermophiles *Sulfurihydrogenibium yellowstonense* and *Sulfurihydrogenibium azorense* [16]. These species, isolated from terrestrial hot springs, grow chemolithotrophically with CO_2 as the carbon source and H_2 as the energy source suggesting a potential role for CA in acquisition of CO_2 for cell carbon [17]. Characterization of the α class CAs from these aerobes has revealed the fastest CA known with a k_{cat} of 4.40×10^6 s^{-1} and a k_{cat}/K_m of 3.5×10^8 M^{-1} s^{-1} [1].

The decomposition of complex biomass in Earth's aerobic biosphere is mediated by prokaryotes called heterotrophs (Step B). Carbon dioxide and bicarbonate are essential growth factors for environmentally important heterotrophs. Although sufficient CO_2 is produced during catabolism, levels below atmospheric lead to growth inhibition or possibly death [18–20]. For example, deletion of a gene from *Ralstonia eutropha* encoding a β class CA resulted in a requirement for heterotrophic growth of the mutant at CO_2 concentrations elevated compared to wild-type [21]. The requirement for CO_2 is attributed to fixation in anaplerotic or other biosynthetic reactions. A recent bioinformatics analysis indicates that CA-deficient prokaryotes evolved to only occupy high-CO_2 niches [22]. Comparisons between CA-retaining and CA-deficient genomes indicated that CA was lost in some species during the course of evolution as a consequence of dispensability. On the other hand, pathogenic prokaryotes seem to be adapted to high CO_2 concentrations in their host environment, as they usually require 5–10 % (vol/vol) CO_2 for growth (9, 45, 60). Although genomes of aerobic heterotrophs isolated from the environment include CAs [23], none have been biochemically and structurally characterized in detail.

Characterized CAs from heterotrophic aerobes are not limited to the upper regions of Earth's surface. *Geobacillus kaustophilus* strain HTA426 is a heterotrophic thermophile isolated from deep sea sediment at 10,897 m [24]. The organism contains a γ class CA for which the crystal structure and documentation of CA activity was recently published albeit with no further characterization [25]. Characterization of CAs from methane-oxidizing prokaryotes that close the carbon cycle (Fig. 5.1, Step F) have not been reported although the genomic sequences of methylotrophic species have annotations for several CAs [23].

3 Anaerobic Species

Anaerobes are microbes that metabolize growth compounds in the absence of O_2. They inhabit environments such as freshwater and marine sediments, rice paddy soils, and the hind gut of termites to name a few. The metabolism of complex biomass is mediated by a diversity of anaerobes which include fermentatives, syntrophs, sulfate reducers and nitrate reducers.

3.1 Anaerobes Metabolizing Complex Biomass

Fermentative plus syntrophic species metabolize complex biomass to products (Fig. 5.1, Step C) that are substrates for methane-producing anaerobes

(Fig. 5.1, Steps D and E). CAs have been documented in several anaerobes producing acetate as the only product of their energy metabolism [26, 27]. These species reduce CO_2 to the methyl level for incorporation into the methyl group of acetate which places a high demand on acquisition of CO_2 from the environment; thus, a postulated role for CA is to facilitate a bicarbonate/acetate antiporter that transports acetate out of the cell and bicarbonate in. A similar physiological role has been postulated for the γ class CA encoded in the genome of *Pelobacter carbinolicus* whereby the enzyme functions in conversion of cytosolic bicarbonate to CO_2 for reduction to formate that is excreted by exchange with extracellular bicarbonate [28]. The Type I β class CA from *Clostridium perfringens* strain 13 is the only β class enzyme characterized from a strictly anaerobic species of the domain *Bacteria* [29]. Analyses of a deletion mutant showed the CA is strictly required for growth when cultured in semi-defined medium and an atmosphere without CO_2 suggesting a role in anaplerotic CO_2 fixation reactions by supplying bicarbonate to carboxylases.

Although the genomes of anaerobic prokaryotes are richly annotated with CAs [23], few have been characterized. The sequence and crystal structure of YrdA from *Escherichia coli* [30], a facultative anaerobe of the domain *Bacteria*, identifies YrdA as a homolog of the archetype γ class CA (Cam) from *Methanosarcina thermophila* belonging to the domain *Archaea* [10]. However, YrdA has no significant CA activity and does not contain residues corresponding to Glu62, Glu84 and Asn202 important for CA activity of Cam [31]. Interestingly, an acidic loop around the putative catalytic site of YrdA shows alternative conformations leading the authors to suggest that the enzyme accommodates substrates larger than carbonate. The only other characterized Cam homolog from an anaerobe is from *Pyrococcus horikoshii* which also belongs to the domain *Archaea* [32]. The crystal structure of the enzyme over-produced in *E. coli* and purified in the presence of air shows the overall fold similar to Cam with zinc in the active site, although activity was not reported and the enzyme was not further characterized. Interestingly, calcium was also shown to be present in the *P. horikoshii* structure.

3.2 Anaerobes Producing Methane

There are two major pathways (Fig. 5.1, steps D and E) producing methane from the products of fermentative and syntrophic species (Fig. 5.1, step C) which constitutes an anaerobic microbial food chain converting complex biomass to methane. Both methanogenic pathways involve CO_2 as either a reactant or product suggesting a role for CA in either CO_2 acquisition or removal. The CO_2 reduction pathway involves the reduction of CO_2 to methane with electrons derived from either H_2 or formate (Fig. 5.1 step D), whereas the aceticlastic pathway produces equimolar amounts of methane and CO_2 (Fig. 5.1, step E). Clearly there is a high demand for CO_2 in the CO_2 reduction pathway where CA could play a role in acquisition and retention of CO_2 by conversion to membrane impermeable bicarbonate inside the

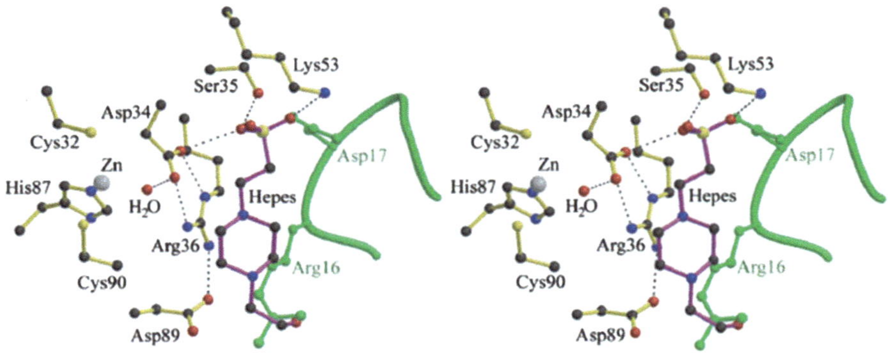

Fig. 5.3 Stereo view of the active-site of Cab from *Methanothermobacter thermautotrophicus* ΔH

cell. Indeed, a β class CA (Cab) has been characterized from *Methanothermobacter thermoautotrophicus* (f. *Methanobacterium thermoautotrophicum* strain ΔH) [33] which reduces CO_2 to methane with H_2 and synthesizes all cell carbon starting with CO_2 which places an additional demand on acquisition and retention of CO_2. Enzymes of biosynthetic pathways require either CO_2 or bicarbonate suggesting additional roles for CAs in supplying the required form of inorganic carbon.

Cab is the only β class CA characterized, both biochemically and structurally, from a methane-producing species. Although the overall fold is similar to that of other β-class CAs that are tetrameric and octameric, Cab is a dimer in the crystal structure [15]. The crystal structure shows the active site metal at the interface of the two monomers, although the two cysteines and one histidine coordinating the metal originate from the same monomer. The catalytic water is the fourth ligand (Fig. 5.3) which hydrogen bonds to an aspartate residue conserved in all β class CAs. The aspartate is in turn hydrogen bonded to an arginine also conserved in the active sites of β class CAs. Further, the crystal structure of Cab shows the active-site cavity open to bulk solvent such that a HEPES buffer molecule is present 8 Å from the catalytic metal (Fig. 5.3) distinct from the *P. sativum* enzyme. The kinetic analysis of Cab is consistent with a two-step metal-hydroxide mechanism proposed for all CAs. The conserved aspartate residue in the active site of Cab when replaced with alanine significantly reduces the k_{cat}/K_m value [34]. When the arginine interacting with aspartate is replaced with alanine, a still larger decrease in k_{cat}/K_m is observed. When the interaction between aspartate and arginine is disrupted, the aspartate may swing toward the active site and ligate the catalytic metal in a dead-end inactive conformation. Imidazole rescues the activity of both the replacement variants which suggests a role for the conserved residues in the proton transfer step. Thus, an exchange of ligands to aspartate during catalysis may be necessary if the pK_a of aspartate were decreased when interacting with arginine such that the aspartate residue is unable to abstract a proton from the zinc-bound water. In the aceticlastic pathway, the methyl group of acetate is reduced to methane with electrons derived from oxidation of the carbonyl group of acetate to CO_2.

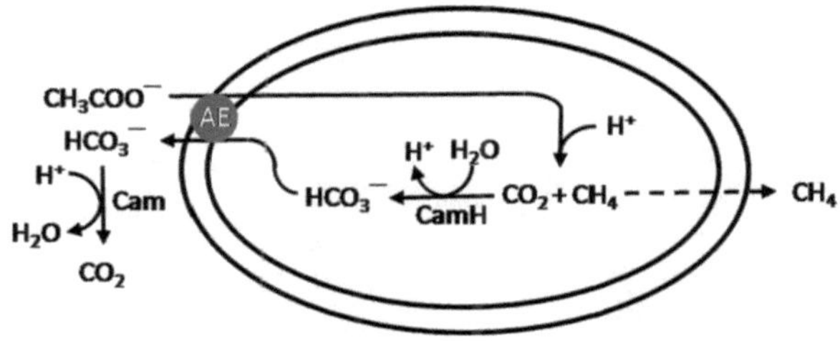

Fig. 5.4 Postulated roles for the γ class CAs Cam and CamH in the pathway of acetate conversion to methane and carbon dioxide in *Methanosarcina* species. The *double line* represents the cell membrane. *AE* anion exchanger

Methanosarcina thermophila contains two γ class CAs, Cam and CamH, postulated to facilitate transport of acetate into the cell and CO_2 out via an acetate/bicarbonate anion exchanger (Fig. 5.4) in analogy to the mammalian chloride/bicarbonate exchange system [35]. Cam was first investigated with the enzyme over produced in *Escherichia coli*. When purified from *E. coli* in the presence of air, Cam contains zinc in the active-site. However, when the zinc is exchanged with ferrous iron in vitro, the catalytic efficiency increases 3-fold over the zinc form [36]. Ferrous iron is present in the active-site when purified from *E. coli* in an atmosphere void of O_2. When exposed to air, ferrous is oxidized to ferric with rapid loss of CA activity. Cam from *M. thermophila* over-produced in the close relative *Methanosarcina acetivorans* contains stoichiometric amounts of ferrous iron when purified in the absence of O_2 establishing ferrous iron as the physiologically relevant metal in vivo [37]. When purified from *E. coli* in the presence of air, ferric iron is exchanged with contaminating amounts of zinc present in untreated buffers. A crystal structure is unavailable for Fe-Cam, although it is proposed to be similar to in vitro cobalt-substituted Cam with a coordination sphere containing three waters in addition to the metal ligands [11].

Cam is the most extensively investigated of the few γ class CAs characterized. The kinetic mechanism for a the reversible hydration of CO_2 by Cam is a two-step ping pong mechanism similar to all characterized CAs as shown in the following equations where E represents enzyme residues, M is metal and B is buffer.

$$\text{E-M}^{2+}\text{-OH}^- + \text{CO}_2 = \text{E-M}^{2+}\text{-HCO}_3{}^- \tag{5.1a}$$

$$\text{E-M}^{2+}\text{-HCO}_3{}^- + \text{H}_2\text{O} = \text{E-M}^{2+}\text{-H}_2\text{O} + \text{HCO}_3{}^- \tag{5.1b}$$

$$\text{E-M}^{2+}\text{-H}_2\text{O} = \text{H}^+\text{-E-M}^{2+}\text{-OH}^- \tag{5.2a}$$

$$\text{H}^+\text{-E-M}^{2+}\text{-OH}^- + \text{B} = \text{E-M}^{2+}\text{-OH}^- + \text{BH}^+ \tag{5.2b}$$

In step 5.1a, a lone pair of electrons on the metal-bound hydroxide attack carbon dioxide producing metal-bound bicarbonate that is displaced by water in step 5.1b. In step 5.2a, a proton is extracted from the metal-bound water and then transferred to buffer in step 5.2b. Cam exhibits a $k_{cat} > 10^4\ s^{-1}$ which is faster than the fastest

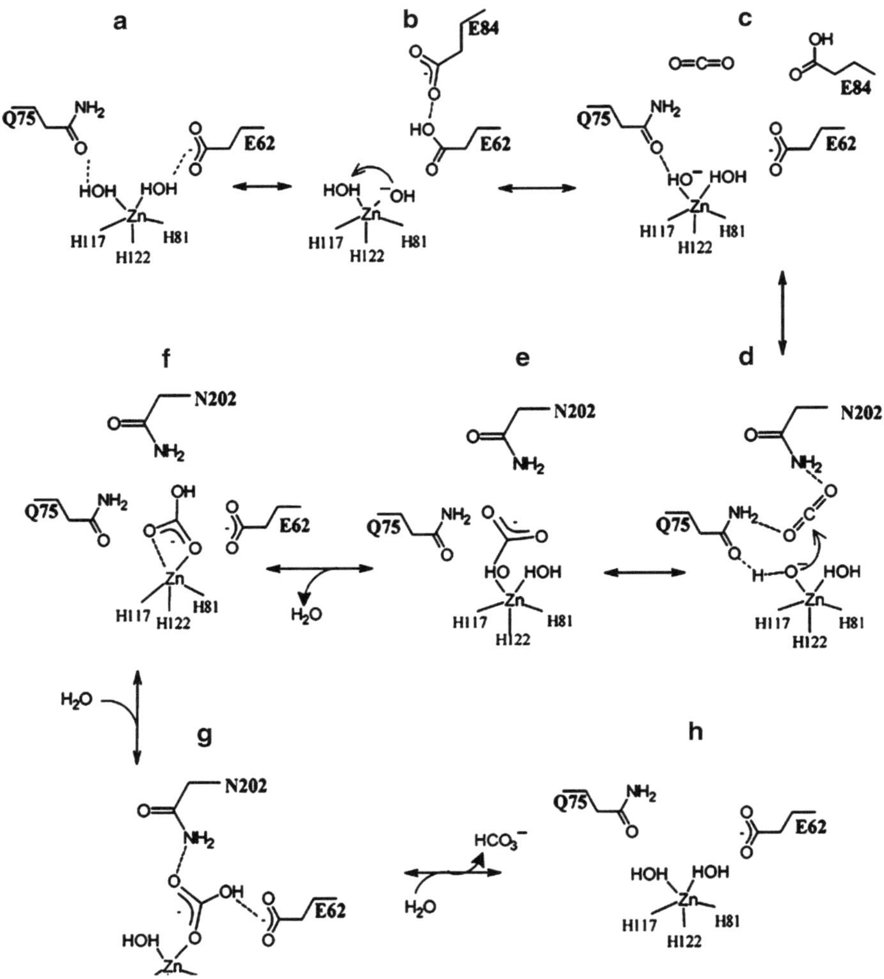

Fig. 5.5 The reaction mechanism proposed for the archetype γ class CA from *Methanosarcina thermophila*

rate at which protons can transfer from the zinc-bound water with a pK_a of 7 to water. Thus, Cam with $k_{cat} > 10^4$ s^{-1} must transfer the proton from the zinc-bound water to an intermediate proton shuttle residue (H^+-E in steps 5.2a and 5.2b) and then to the external buffer molecule. Replacement of active-site residues via site-directed mutagenesis, and kinetic analyses of the variant proteins, has identified several residues either essential or important for catalysis leading to the proposed mechanism for catalysis shown in Fig. 5.5 [31]. In the mechanism, glutamate 62 extracts a proton from either of two metal-bound waters (A) producing a hydroxyl group which extracts a proton from the adjacent water producing the metal-bound hydroxyl Zn-OH^- that attacks the carbon of CO_2 producing metal-bound HCO_3^- (B-F). Glutamine 75 and asparagine 73 participate in a hydrogen bond network to orient and increase the nucleophilicity of the metal-bound hydroxide. The bound HCO_3^- undergoes a bidentate transition state wherein the proton either rotates or transfers to the nonmetal-bound oxygen of HCO_3^- (G). Hydrogen bonding with glutamate 62 destablizes the HCO_3^- which is further destabilized by an incoming water that replaces one of the metal-bound oxygens (H). A second incoming water displaces HCO_3^- with product removal and regeneration of the water-bound active-site metal. Not shown in Fig. 5.5, tryptophan 19 and tyrosine 200 are second shell residues distant from the catalytic metal that fine tune catalysis. Tryptophan 19 is important for both CO_2/bicarbonate interconversion and the proton transfer step [38].

Inhibitor studies of Cab and Cam show differences in inhibition response to sulfonamides and metal-complexing anions, when compared to α class CAs [39–42]. In addition, inhibition between Cab and Cam differ. These inhibition patterns are consistent with the idea that although, α-, β-, and γ-class CAs participate in the same two-step mechanism, diverse active site architecture among these classes predicts variations on the catalytic mechanism.

CamH of *M. thermophila* when over produced in *E. coli* contains 0.15 iron per monomer and only a trace amount of zinc when purified in the absence of O_2 [43]. The enzyme losses CA activity when exposed to air which suggests ferrous iron has a role in the active site. Databases queried with the Cam sequence reveal the great majority of homologs comprising a subclass homologous to CamH for which there is conservation of most residues essential for the archetype Cam except an acidic loop often missing the proton shuttle residue glutamate 84 of Cam [43]. Interestingly, the γ class CAs isolated from *P. horikoshii* and *E. coli* belong to the CamH subclass for which neither is reported to have CA activity. Clearly, more research is warranted on CAs from this subclass.

References

1. Luca VD, Vullo D, Scozzafava A, Carginale V, Rossi M, Supuran CT, Capasso C (2012) An alpha-carbonic anhydrase from the thermophilic bacterium *Sulphurihydrogenibium azorense* is the fastest enzyme known for the CO_2 hydration reaction. Bioorg Med Chem 21:1465–1469
2. Cox EH, McLendon GL, Morel FM, Lane TW, Prince RC, Pickering IJ, George GN (2000) The active site structure of *Thalassiosira weissflogii* carbonic anhydrase 1. Biochemistry 39:12128–12130
3. Tripp BC, Smith K, Ferry JG (2001) Carbonic anhydrase: new insights for an ancient enzyme. J Biol Chem 276:48615–48618
4. Lane TW, Saito MA, George GN, Pickering IJ, Prince RC, Morel FM (2005) Biochemistry: a cadmium enzyme from a marine diatom. Nature 435:42
5. Anonymous (2013) The comprehensive microbial resource. J Craig Venter Inst. http://cmr.tigr.org/tigr-scripts/CMR/CmrHomePage.cgi
6. Savile CK, Lalonde JJ (2011) Biotechnology for the acceleration of carbon dioxide capture and sequestration. Curr Opin Biotechnol 22(6):818–823

7. Kinney JN, Salmeen A, Cai F, Kerfeld CA (2012) Elucidating essential role of conserved carboxysomal protein CcmN reveals common feature of bacterial microcompartment assembly. J Biol Chem 287:17729–17736
8. Partensky F, Hess WR, Vaulot D (1999) Prochlorococcus, a marine photosynthetic prokaryote of global significance. Microbiol Mol Biol Rev 63:106–127
9. Pena KL, Castel SE, de Araujo C, Espie GS, Kimber MS (2010) Structural basis of the oxidative activation of the carboxysomal γ-carbonic anhydrase, CcmM. Proc Natl Acad Sci U S A 107:2455–2460
10. Alber BE, Ferry JG (1994) A carbonic anhydrase from the archaeon *Methanosarcina thermophila*. Proc Natl Acad Sci U S A 91:6909–6913
11. Iverson TM, Alber BE, Kisker C, Ferry JG, Rees DC (2000) A closer look at the active site of γ-carbonic anhydrases: high resolution crystallographic studies of the carbonic anhydrase from *Methanosarcina thermophila*. Biochemistry 39:9222–9231
12. Sawaya MR, Cannon GC, Heinhorst S, Tanaka S, Williams EB, Yeates TO, Kerfeld CA (2006) The structure of β-carbonic anhydrase from the carboxysomal shell reveals a distinct subclass with one active site for the price of two. J Biol Chem 281:7546–7555
13. Cronk JD, Endrizzi JA, Cronk MR, O'Neill JW, Zhang KY (2001) Crystal structure of *E. coli* β-carbonic anhydrase, an enzyme with an unusual pH-dependent activity. Protein Sci 10:911–922
14. Kimber MS, Pai EF (2000) The active site architecture of *Pisum sativum* β-carbonic anhydrase is a mirror image of that of α-carbonic anhydrases. EMBO J 19:1407–1418
15. Strop P, Smith KS, Iverson TM, Ferry JG, Rees DC (2001) Crystal structure of the "cab"-type β-class carbonic anhydrase from the archaeon *Methanobacterium thermoautotrophicum*. J Biol Chem 276:10299–10305
16. Nakagawa S, Shtaih Z, Banta A, Beveridge TJ, Sako Y, Reysenbach AL (2005) *Sulfurihydrogenibium yellowstonense* sp. nov., an extremely thermophilic, facultatively heterotrophic, sulfur-oxidizing bacterium from Yellowstone National Park, and emended descriptions of the genus *Sulfurihydrogenibium*, *Sulfurihydrogenibium subterraneum* and *Sulfurihydrogenibium azorense*. Int J Syst Evol Microbiol 55:2263–2268
17. Aguiar P, Beveridge TJ, Reysenbach AL (2004) *Sulfurihydrogenibium azorense*, sp. nov., a thermophilic hydrogen-oxidizing microaerophile from terrestrial hot springs in the Azores. Int J Syst Evol Microbiol 54:33–39
18. Brown OR, Howitt HF (1969) Growth inhibition and death of *Escherichia coli* from CO2 deprivation. Microbios 3:241–246
19. Dehority BA (1971) Carbon dioxide requirement of various species of rumen bacteria. J Bacteriol 105:70–76
20. Gladstone GP, Fildes P, Richardson GM (1935) Carbon dioxide as an essential factor in the growth of bacteria. Br J Exp Pathol 16:335–348
21. Kusian B, Sultemeyer D, Bowien B (2002) Carbonic anhydrase is essential for growth of *Ralstonia eutropha* at ambient CO_2 concentrations. J Bacteriol 184:5018–5026
22. Ueda K, Nishida H, Beppu T (2012) Dispensabilities of carbonic anhydrase in proteobacteria. Int J Evol Biol 2012:324549
23. Anonomous (2013) The comprehensive microbial resource. J Craig Venter Inst. http://cmr.tigr.org/tigr-scripts/CMR/CmrHomePage.cgi
24. Takami H, Nishi S, Lu J, Shimamura S, Takaki Y (2004) Genomic characterization of thermophilic *Geobacillus* species isolated from the deepest sea mud of the Mariana Trench. Extremophiles 8:351–356
25. Ragunathan P, Raghunath G, Kuramitsu S, Yokoyama S, Kumarevel T, Ponnuraj K (2013) Crystallization, characterization and preliminary X-ray crystallographic analysis of GK2848, a putative carbonic anhydrase of *Geobacillus kaustophilus*. Acta Crystallogr Sect F Struct Biol Cryst Commun 69:162–164
26. Braus-Stromeyer SA, Schnappauf G, Braus GH, Gossner AS, Drake HL (1997) Carbonic anhydrase in *Acetobacterium woodii* and other acetogenic bacteria. J Bacteriol 179:7197–7200
27. Pitriuk AV, Pusheva MA, Kupriianova EV (2006) Carboanhydrase activity in halophilic anaerobes from soda lakes. Mikrobiologiia 75:18–21

28. Aklujkar M, Haveman SA, Didonato R Jr, Chertkov O, Han CS, Land ML, Brown P, Lovley DR (2012) The genome of *Pelobacter carbinolicus* reveals surprising metabolic capabilities and physiological features. BMC Genomics 13:690
29. Kumar RS, Hendrick W, Correll JB, Patterson AD, Melville SB, Ferry JG (2013) Biochemistry and physiology of the β class carbonic anhydrase (Cpb) from *Clostridium perfringens* strain 13. J Bacteriol 195(10):2262–2269
30. Park HM, Park JH, Choi JW, Lee J, Kim BY, Jung CH, Kim JS (2012) Structures of the gamma-class carbonic anhydrase homologue YrdA suggest a possible allosteric switch. Acta Crystallogr D Biol Crystallogr 68:920–926
31. Zimmerman SA, Ferry JG (2006) Proposal for a hydrogen bond network in the active site of the prototypic γ-class carbonic anhydrase. Biochemistry 45:5149–5157
32. Jeyakanthan J, Rangarajan S, Mridula P, Kanaujia SP, Shiro Y, Kuramitsu S, Yokoyama S, Sekar K (2008) Observation of a calcium-binding site in the γ-class carbonic anhydrase from *Pyrococcus horikoshii*. Acta Crystallogr D Biol Crystallogr 64:1012–1019
33. Smith KS, Ferry JG (1999) A plant-type (beta-class) carbonic anhydrase in the thermophilic methanoarchaeon *Methanobacterium thermoautotrophicum*. J Bacteriol 181:6247–6253
34. Smith KS, Ingram-Smith C, Ferry JG (2002) Roles of the conserved aspartate and arginine in the catalytic mechanism of an archaeal β-class carbonic anhydrase. J Bacteriol 184:4240–4245
35. Sterling D, Alvarez BV, Casey JR (2002) The extracellular component of a transport metabolon. Extracellular loop 4 of the human AE1 Cl/HCO_3 exchanger binds carbonic anhydrase IV. J Biol Chem 277:25239–25246
36. Tripp BC, Bell CB, Cruz F, Krebs C, Ferry JG (2004) A role for iron in an ancient carbonic anhydrase. J Biol Chem 279:6683–6687
37. MacAuley SR, Zimmerman SA, Apolinario EE, Evilia C, Hou Y, Ferry JG, Sowers KR (2009) The archetype γ-class carbonic anhydrase (Cam) contains iron when synthesized in vivo. Biochemistry 48:817–819
38. Zimmerman S, Domsic JF, Tu C, Robbins AH, McKenna R, Silverman DN, Ferry JG (2013) Role of Trp19 and Tyr200 in catalysis by the gamma-class carbonic anhydrase from *Methanosarcina thermophila*. Arch Biochem Biophys 529:11–17
39. Zimmerman SA, Ferry JG, Supuran CT (2007) Inhibition of the archaeal β-class (Cab) and γ-class (Cam) carbonic anhydrases. Curr Top Med Chem 7:901–908
40. Zimmerman S, Innocenti A, Casini A, Ferry JG, Scozzafava A, Supuran CT (2004) Carbonic anhydrase inhibitors. Inhibition of the prokaryotic β and γ-class enzymes from Archaea with sulfonamides. Bioorg Med Chem Lett 14:6001–6006
41. Innocenti A, Zimmerman S, Ferry JG, Scozzafava A, Supuran CT (2004) Carbonic anhydrase inhibitors. Inhibition of the beta-class enzyme from the methanoarchaeon *Methanobacterium thermoautotrophicum* (Cab) with anions. Bioorg Med Chem Lett 14:4563–4567
42. Innocenti A, Zimmerman S, Ferry JG, Scozzafava A, Supuran CT (2004) Carbonic anhydrase inhibitors. Inhibition of the zinc and cobalt γ-class enzyme from the archaeon *Methanosarcina thermophila* with anions. Bioorg Med Chem Lett 14:3327–3331
43. Zimmerman SA, Tomb JF, Ferry JG (2010) Characterization of CamH from *Methanosarcina thermophila*, founding member of a subclass of the γ class of carbonic anhydrases. J Bacteriol 192:1353–1360

Chapter 6
Carboxysomal Carbonic Anhydrases

Matthew S. Kimber

Abstract Cyanobacteria and some chemoautotrophic bacteria enhance their carbon fixation efficiency by actively concentrating bicarbonate within their cytosol. However, converting bicarbonate into carbon dioxide – the form required by RubisCO – would result in its rapid escape through cellular membranes. These organisms resolve this dilemma by sequestering RubisCO behind a semi-permeable protein shell; the resulting large insoluble bodies are known as carboxysomes. For the carbon concentrating mechanism to function, there is an absolute requirement for carbonic anhydrase activity within the carboxysome to convert the bicarbonate to cabon dioxide, and a simultaneous requirement that minimal carbonic anhydrase activity be found within the cystol. Carboxysomal carbomic anhydrases therefore contain additional motifs and domains that generally mediate protein-protein interactions, or encapsulation dependent activation mechanisms. Carboxysomes are found in two deeply divergent varieties. Alpha-Carboxysomes contain a β-carbonic anhydrase, CsoSCA, which is so divergent from canonical β-carbonic anhydrases that it was originally thought to be the founding member of a new class. Beta carboxysomes have CcmM whose N-terminal domain is an active γ-carbonic ahydrase in some strains, but in others has lost all activity and functions primarily as a protein complex assembly scaffold; in addition, a subset of β-carboxysomes also contain the β-carbonic anhydrase CcaA – either in addition to, or instead of, an active CcmM. Here we explore the structures, activities and interactions mediated by the three known carboxysomal carbonic anhydrases, and discuss the mechanisms by which they are recruited to the carboxysome.

Susan C. Frost and Robert McKenna (eds.). Carbonic Anhydrase: Mechanism, Regulation, Links to Disease, and Industrial Applications

M.S. Kimber (✉)
Department of Molecular and Cellular Biology, University of Guelph, Guelph, ON, Canada
e-mail: mkimber@uoguelph.ca

S.C. Frost and R. McKenna (eds.), *Carbonic Anhydrase: Mechanism, Regulation, Links to Disease, and Industrial Applications*, Subcellular Biochemistry 75,
DOI 10.1007/978-94-007-7359-2_6,

Keywords Carbonic anhydrase • Carboxysome • Carbon fixation • Bacterial microcompartment • β-carbonic anhydrases • Cyanobacteria

1 Introduction

Cyanobacteria are ancient organisms that originally evolved in an atmosphere far richer in CO_2 than its modern equivalent; as autotrophs, this represents a challenge, as ribulose 1,5-bisphosphate carboxylase/oxygenase (RubisCO) is an enzyme that operates most efficiently at elevated CO_2 levels [1]. Cyanobacteria have adapted to the challenges of fixing carbon in a carbon-depleted world by recreating the ancient high CO_2 atmosphere internally. Their inner membranes have a variety of ATP and NADH dependent CO_2 and HCO_3^- pumps [2] which raise the intracellular inorganic carbon concentration to as high as the millimolar range (Fig. 6.1). However, this inorganic carbon can only be retained in the form of HCO_3^-; CO_2 is lipid soluble, and readily escapes the cytosol through cellular membranes. Instead, CO_2 is only evolved in particles known as carboxysomes, where RubisCO is sequestered behind a CO_2 impermeable protein shell. RubisCO reacts the CO_2 with ribulose 1,5-bisphosphate (RuBP) to form two molecules of 3-phosphoglycerate (3PGA); 3PGA is then converted back to RuBP by the enzymes of the Calvin-Benson-Bassham cycle, located in the cytosol, with the additional reduced carbon atom being diverted into general metabolism. Carbonic anhydrases are absolutely required within carboxysomes in order to equilibrate the HCO_3^-/CO_2 pool within the enclosed environment of the shell.

Carboxysomes are can be thought of as specialized variants of a more general class of bacterial inclusions, known as bacterial microcompartments. Microcompartments all share a generalized architecture where enzymes are encapsulated behind a thin shell built from multiple members of two protein families [3]. The enzymes generally comprise part of a pathway where the intermediates are either highly volatile or toxic. The ethanolamine and propanediol degradation microcompartments are well studied [4], while ethanol [5] and choline degrading microcompartments [6] have also been partially characterized; in all, bioinformatics comparison of gene complements indicates that about ten different varieties can be discerned in known genomes [7]. Carboxysomes themselves are found in two distinct varieties. Alpha-carboxyomes have type 1a RubisCO, and have genes designated *cso*; α-carboxysomes are found in α-cyanobacteria (comprised of the pelagic *Prochlorococcus* and *Synechococcus* strains that fix over 25 % of the planet's CO_2) and are also found in chemoautotrophic bacteria. Beta-carboxysomes have type 1b RubisCO, and their genes are designated *ccm* for carboxysome structural proteins, and *rbs* for RuBisCO. Beta-carboxysomes are found in β-cyanobacteria, which occupy an wide variety of environments. Surprisingly, the homologous proteins in α- and β-carboxysomes are quite distantly related, and may result from convergent evolution.

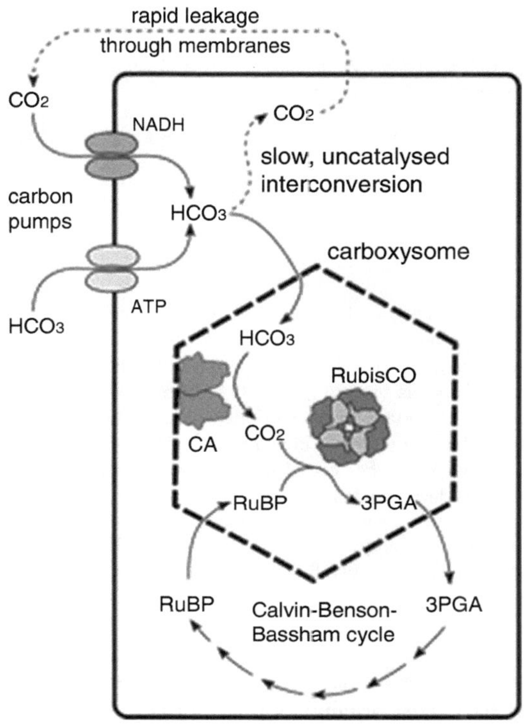

Fig. 6.1 The role of the carboxysome in carbon fixation. Carboxysomes function as an integral part of the carbon concentrating mechanism in cyanobacteria and chemoautotrophs. Inorganic carbon, in the form of HCO_3^-, is actively pumped into the cytosol by a series of ATP and NAD dependent HCO_3^- and CO_2 pumps. Because it is highly polar, HCO_3^- can accumulate in the cytosol to high concentrations, but any CO_2 generated by dehydration of HCO_3^- (which does occur spontaneously at a slow rate) is rapidly lost by diffusion through cellular membranes. It is therefore important that active carbonic anhydrases not accumulate in the cytosol and accelerate this process. Accumulated HCO_3^- can enter the carboxysome through small pores, where it is converted to CO_2 by carboxysomal carbonic anhydrases; the carboxysome shell acts as a barrier to CO_2 escape, allowing it to accumulate to high concentrations. Ribulose 1,5-bisphosphate can also enter the carboxysome, where encapsulated RubisCO combines RuBP with CO_2 to make two molecules of 3-phosphoglycerate (3PGA). 3PGA can then exit the carboxysome through pores in the shell, and be converted back to RuBP (with one excess carbon atom being diverted to general metabolism) by cytosolic enzymes in the Calvin-Benson-Bassham cycle

Carboxysomes are relatively large (90–400 nm) and at least approximately icosahedral [8–10]. The shells are relatively thin (about 4 nm for α-carboxyomes, and 6 nm for β-carboxysomes) and are built up from proteins from two protein families. Numerically dominant are the bacterial microcompartment proteins (BMC), which are small (~90 a.a.) and organized around a thioredoxin-like fold [11]. This protein forms hexamers with a regular hexagonal outline. In crystal structures,

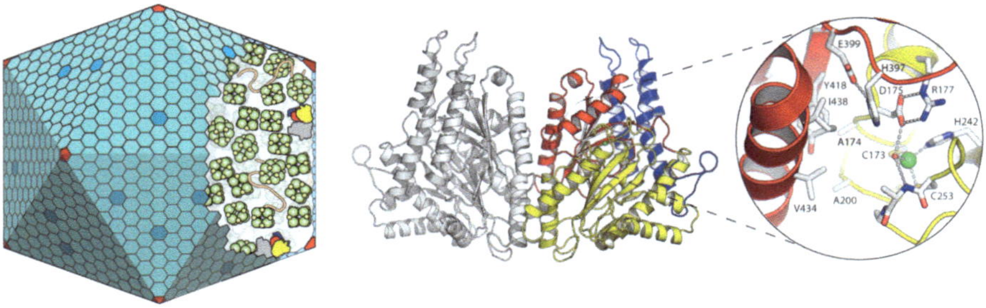

Fig. 6.2 Carbonic anhydrases in α-carboxysomes. CsoSCA is the only known carbonic anhydrase in α-carboxysomes. This protein associates tightly with the shell, but is not an essential structural component. CsoSCA is a dimeric protein, with each protomer built from three distinct domains. The N-terminal domain (*blue*) is unique to CsoSCA, forms an α-helical bundle. Its role is not understood, but is possibly important for interactions with the shell. The catalytic domain (*yellow*) contains the majority of the catalytic site, including all of the zinc binding residues (inset). The C-terminal domain (*red*) provides the second half of the catalytic site, contributing His397 and a wall of hydrophobic residues. Interestingly, the role of this domain is fulfilled by a second catalytic domain in most other β-carbonic anhydrases, and the C-terminal domain in CsoSCA shows appreciable structural homology to the catalytic domain. This domain therefore most likely arose by an N-to-C terminal fusion of two β-CA domains, followed by gradual devolvement of the second catalytic site

BMC proteins are typically found to tile into extended flat sheets, with no gaps (see Figs. 6.2 and 6.3). This organization is thought to underpin the overall form of the facet [11–13]. In β-carboxysomes, this basic BMC protein organization is generally represented by four distinct paralogs, named CcmK1, CcmK2, CcmK3 and CcmK4; of these, only CcmK2 seems indispensible. Beta-carboxysomes also have a second high abundance essential BMC protein, CcmO, that is organized as a head-to-tail fusion of two BMC domains [14]. In α-carboxysomes, the CsoS1A-C proteins closely resemble the CcmK proteins in structure and organization. Alpha-carboxysomes also have a CsoS1D protein that is built as a head-to-tail fusion of two circularly permuted BMC domains. This protein is organized as a back-to-back dimer of pseudo-hexameric rings; interestingly, this complex is asymmetric, opening on one side to a chamber; it has been suggested that this protein may passage larger metabolites by alternatingly opening the two halves [15]. A related protein, provisionally called CcmP, is found in β-cyanobacteria. The final shell protein is the pfam03319 protein, called CcmL in β-carboxysomes and CsoS4A/CsoS4B in α-carboxysomes. The structure of this protein is pentameric, and it has been argued to form the vertex of the shell, where five facets meet [16].

The passage of metabolites into and out of the carboxysome is controlled by these shell proteins. The carboxyome's function requires, at minimum, HCO_3^-, ribulose 1-5-bisphosphate be able to enter, and 3-phosphoglycerate be able to exit. Optimal functioning would also entail that CO_2 be selectively retained, and O_2 selectively excluded (to minimize RubisCO's competing oxygenase reaction). The structures of the CcmK1, CcmK2 and CcmK4 and CsoS1A and CsoS1B proteins (which form the bulk of their respective shells) all show a small (~6 nm) central pore surrounded by hydrophilic residues [11–13]. These pores are, in principle, large enough to passage any of these metabolites. However, their simple six-fold symmetry implies weak discrimination. In addition, the pore lining residues appear somewhat flexible, and able to adapt in shape to irregularly shaped metabolites [17]. The CcmO structure is not known, but studies of similar fusion proteins suggest that it might possibly possess asymmetric off-centered pores which are more specific for an individual metabolite. Direct measurements comparing intact and disrupted α-carboxysomes suggest that both HCO_3^- and RuBP can pass through the shell, but both are somewhat impeded [18].

The specialized roles of carboxysomal carbonic anhydrases impose constraints on their functioning over and above simple catalytic activity. First and foremost, they are required to be properly targeted to the interior of the carboxysome through a suitable protein-protein interaction network. This need for proper targeting extends beyond the requirement that sufficient carbonic anhydrase activity be localized to the carboxysome; the carbon concentrating mechanism breaks down if carbonic anhydrase activity is present within the cytosol at appreciable levels, as it promotes the escape of cytosolic CO_2 from the cell. This idea has been directly demonstrated by experiments where human α-CA was expressed in the cytosol of *Synechococcus* sp. PCC7942; the cells were unable to fix CO_2 and grow at atmospheric CO_2 concentrations [19]. This implies that carboxysomal carbonic anhydrases require either rapid turnover (or other related mechanisms) to prevent accumulation of significant quantities in the cytosol, or, alternatively, post-translational regulatory mechanisms that maintain the carbonic anhydrase in an inactive form until encapsulated. These additional requirements are partly reflected in the fact that all three varieties of carboxysomal CAs possess motifs and domains beyond those required for carbonic anhydrase functioning alone; as a result, all carboxysomal carbonic anhydrases are significantly larger than related non-carboxysomal enzymes, and display structural motifs and biochemical behaviors not seen in related enzymes. Here we explore the structures, activities, interactions and additional roles of these unusual carbonic anhydrases.

2 α-Carboxysomes Contain a Deeply Divergent β-Carbonic Anhydrase, CsoSCA

The genomes of α-carboxysome containing α-cyanobacteria and chemoautotrophs do not generally contain readily detectable homologs of canonical α-, β- and γ-carbonic anhydrases. The identity of the carbonic anhydrase in these organisms was therefore something of a puzzle for several years. However, testing of *Halothiobacillus neapolitanus* carboxysome extracts showed the presence of unambiguous CA activity [20]. The search for the appropriate candidate enzyme was greatly facilitated by the tight association of all α-carboxysome containing genes (all known

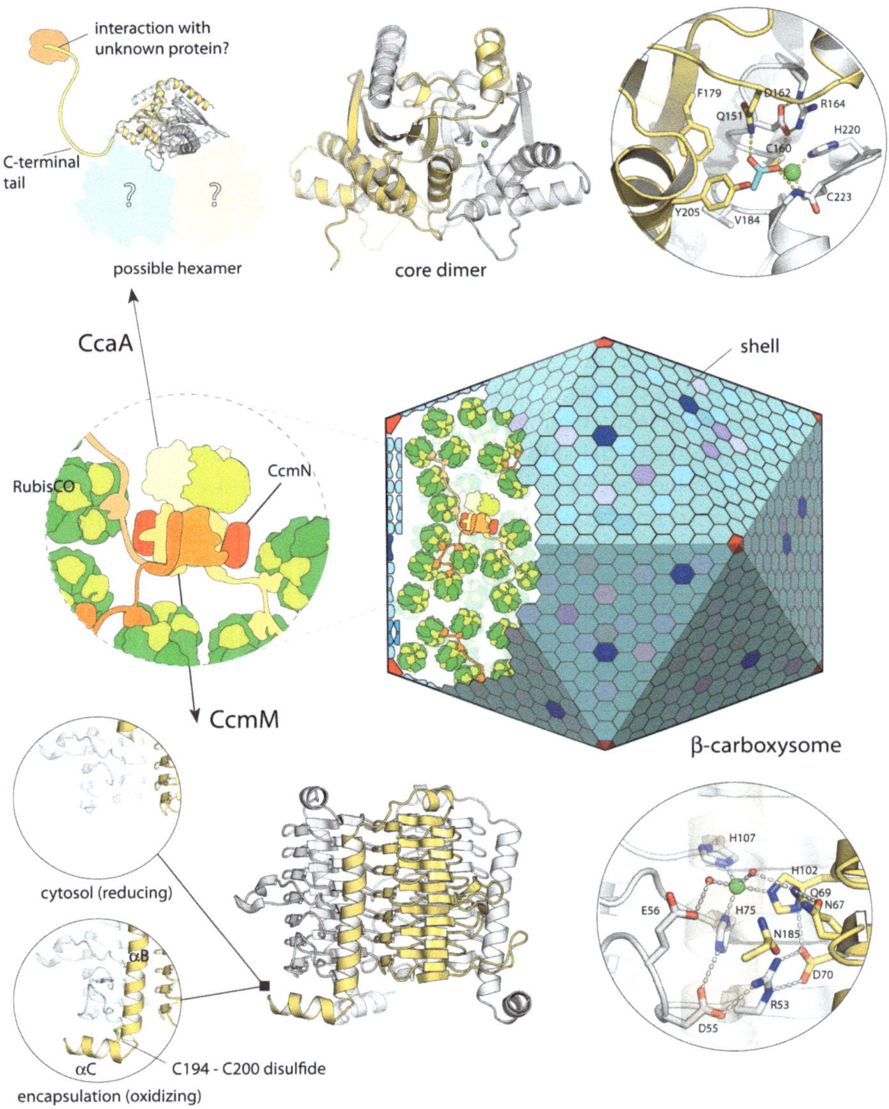

Fig. 6.3 Carbonic anhydrases in β-carboxysomes. The β-carboxysome is built of approximately 11 distinct polypeptides (depending on strain). These are organized into polyhedral bodies with a distinct shell and interior. The carbonic anhydrases are sequestered to the interior, where the (potentially) catalytic N-terminal domain of CcmM interacts with CcaA (where present), as well as CcmN. This domain may also potentially interact directly with the shell, though the evidence for this is equivocal. CcmM also has an extended C-terminal region which interacts with RubisCO through its RbcS-like domains. This region of the protein is also separately translated at a higher level from an independent ribosome entry site. CcaA (*top*) is a β-carbonic anhydrase, with a well-conserved structural core dimer (shown is the 1ekj dimer), but also with a unique extended C-terminal tail. The proximal part of this tail appears to be important for oligomerization and activity, while the distal 17 amino acids are conserved and most likely mediate an interaction with

genes form a single operon, with the exception of CsoS1D, which is immediately 5′ to this operon, and transcribed divergently [15]. Testing each open reading frame in the *Halothiobacillus neapolitanus cso* operon showed that a previously functionally uncharacterized gene *csoS3* encodes a 57 kDa protein with carbonic anhydrase activity when expressed in *E. coli*; this gene was therefore renamed *csoSCA* [15, 20]. Similar activity was shown in the equivalent CsoSCA proteins of the cyanobacteria *Procholococcus marinus* MED4 and MIT9313, as well as *Synechococcus* WH8102. Consistent with an essential role in carbon fixation, a strain where this gene was inactivated by replacement with a kanamycin cassette does not grow at ambient CO_2 concentrations, but can be rescued by growth at 5 % CO_2 [21]. Intriguingly, all detectable homologs of CsoSCA are associated with *cso* gene clusters, and no sequence similarity to the established CA classes was apparent in CsoSCA; it was therefore proposed that these enzymes were represent a new CA fold class, the ε-CAs [20, 22]. However, the determination of the structure of *Halothiobacillus neapolitanus* CsoSCA revealed a central domain whose structural core of five β-strands and three α-helices is shared with canonical β-CAs, and also possesses the same organization of the metal binding site and the Asp/Arg catalytic motif (Fig. 6.2). CsoSCA is therefore a β-CA, albeit a highly divergent example.

Canonical β-CAs function as a dimer, with the second half of the catalytic pocket formed by a symmetry related copy of the basic protomer. In CsoSCA, in contrast, the second half of the catalytic pocket is formed by a separate C-terminal domain. This domain shows very weak structural similarity to β-CAs, and likely arose by a head to tail gene fusion event, followed by gradual loss of non-essential functional motifs. This second β-CA domain does not contain an independent zinc binding or catalytic site, and contributes only a few motifs to the catalytic site formed by the catalytic domain [23]. CsoSCA also possesses a unique N-terminal domain built as a bundle of four α-helices. The role of this domain is not clear, but may mediate protein-protein interactions.

The catalytic site of CsoSCA is a unique variant on the canonical β-CA catalytic site (Fig. 6.2). The catalytic zinc ion is bound by Cys173, His242(Nε2), and Cys253. Two other key catalytic residues – Asp175 and Arg177 – are also conserved relative to other β-CAs [23]. A water molecule occupying the fourth zinc site, and is

Fig. 6.3 (continued) an as yet unknown protein. The 50 or so amino acids in between these regions are poorly conserved and hydrophobic, and are proposed to form an unstructured linker. The catalytic site is shown on the *right* (shown is the pea β-CA structure 1EKJ, which differs only from CcaA homologs in some minor changes in the hydrophobic residues along the *left* hand side of the pocket). The CcmM N-terminal domain (*bottom*) acts as a functional γ-CA in a subset of β-cyanobacteria, and is organized as a trimer. CcmM undergoes a highly unusual encapsulation dependent maturation step. The protein is expressed in the cytosol, where the reducing conditions keep the Cys194 and Cys200 residues reduced. As a consequence, αC, the C-terminal end of αB, and the N-terminal most 20 amino acids are all disordered, and the catalytic site is also disorganized. Encapsulation shifts the environment to be oxidizing, resulting in the formation of a Cys194-Cys200 disulfide bond, stabilization of the N- and C-terminus, and reorganization of the catalytic site into a catalytically active conformation (*right inset*)

stabilized by hydrogen bonds with Asp175 and Ala254N, and is likely the catalytic water. His397 presents its Nε2 into the catalytic pocket, in a position analogous to Gln151 in the plant β-CA [24]. The hydrogen bond Nδ1 of this residue forms with Glu399 likely stabilizes this residue in the protonated form, allowing it to act as a hydrogen bond donor. The back of the pocket forms a hydrophobic surface, comprised of the side chains of residues Ala174, Ala200, Tyr418, Val434 and Ile438.

For *Halothiobacillus* CsoSCA, the rate of hydration of CO_2 increases with pH; at pH 8.0 k_{cat} is 8.9×10^4 s^{-1} and K_M is 3.2 mM, and the calculated catalytic efficiency is (k_{cat}/K_M) was calculated to be 2.8×10^7 M^{-1} s^{-1}. For the dehydration of bicarbonate at pH 7.0 k_{cat} is 4.6×10^4 s^{-1}, K_M is 9.3 mM, and k_{cat}/K_M is 4.9×10^6 M^{-1} s^{-1}. Intriguingly, ethoxzolamide is a fairly poor inhibitor, with an IC_{50} of 1.9 mM, while for the common laboratory reducing agent dithiothreitol, IC_{50} is 240 μM. Cyanide, at 2 μM, is the most effective inhibitor characterized to date. Chelex treated protein is reactivated by the addition of zinc, but not by Co^{2+} or Mg^{2+} [18].

Studies of the proteomic complement of *Prochlorococcus* MED4 carboxysomes show that CsoSCA is relatively rare, with an estimated 58 copies per particle, versus 1216 copies of the CbbL, the large subunit of RubisCO [25]. CsoSCA is associated with the α-carboxysome shell, as both the protein and the CA activity partitions with the shell fractions after carboxysomes are ruptured by freeze-thaw [20]. This idea is supported by the observation that CsoSCA is associated with the periphery of the carboxysome in immuno-gold labeling studies [26]. The catalytic rate of CsoSCA associated with the α-carboxysome shell is similar to the rate of an equivalent quantity of recombinant protein, implying that the protein is not significantly activated by the interaction [18]. CsoSCA does not appear to play an important structural role in shell formation, as the carboxysomes formed in *csoSCA* knockout strains remain wild type in their morphology [21]. These carboxysomes still support CO_2 fixation in vitro, but require approximately three times as much HCO_3^- to saturate [18].

At this time, there are no concrete models proposed as to how CsoSCA interacts with the shell. One potential complication is the apparent symmetry mismatch: CsoSCA is organized as a dimer, while the underlying symmetry of the shell should be dominated by the six-fold symmetry of the CsoS1 shell proteins. In addition, the footprint of CsoSCA is larger than the individual shell protein oligomers, meaning that its interaction site is likely only created once the shell is partially assembled. One possibility is that CsoSCA binds along the two-fold axis where two adjacent CsoS1 hexamers meet, with each CsoSCA protomer interacting with one hexamer. This idea is consistent with the observation that CsoSCA remains tightly associated with the shell unless released with high concentrations of urea or guanadinium – the same conditions that results in dissolution of the shell. Despite this close association with the shell, CsoSCA is not required to form the shell, as carboxysomes from CsoSCA knockout strains are morphologically indistinguishable from those of wild type *Halothiobacillus* cells [18]. An intriguing possibility is that CsoSCA binds along the icosahedral edges, where two adjacent facets meet; the number of carbonic anhydrase binding sites would then naturally be limited by to a few sites per edge. Such an association geometry of CsoSCA could perhaps explain the more limited range in sizes of α-carboxysomes – as the number of CA recruitment sites would increase more slower than the number of RubisCO molecules as carboxysomal size increased.

3 CcmM: The γ-Carbonic Anhydrase in β-Carboxysomes

CcmM is the largest protein in the β-carboxysome, and is present in all β-cyanobacterial species sequenced to date. CcmM was originally discovered as an open reading frame, located within a larger operon encoding carboxysomal genes in *Synechococcus* sp. PCC 7942; it was noted that the C-terminal region contained three repeats 24 % identical to RubisCO's small subunit (in other species up to five copies are present), but the affinities of the N-terminal region were obscure [27]. However, Alber and Ferry noted a year later that the N-terminal domain of CcmM has 35 % sequence identity to the newly discovery γ-CA's in the archaeon *Methanosarcina thermophila* [28]. Ongoing research in the Ferry lab has highlighted several critical catalytic residues [29], all of which have proven to be conserved in CcmM. This affinity is therefore highly suggestive, but direct testing of recombinant CcmM from *Synechococcus* sp. PCC 7942 and *Synechocystis* sp. PCC6803 (the standard model cyanobacterial strains) showed no detectable carbonic anhydrase activity [30]. It was only very recently, when CcmM from *Thermosynechococcus elongatus* (a species that lacks CcaA) was tested, that the N-terminal domain of CcmM was shown to be capable of carbonic anhydrase activity [31]. While the kinetics of CcmM were not characterized in detail, the electrometric assay in the hydration direction shows activity at about 13 % of that seen in bovine α-CAII. The reaction is weakly inhibited by ethoxzolamide [31].

Like the well characterized γ-CA from *Methanosarcina* [32], the structure of CcmM is organized as a trimer of seven turn left handed β-helices, with a long α-helix (αB) that fits into the extended groove between protomers and covers the catalytic site [31] (Fig. 6.3). CcmM differs from other γ-CAs in having an additional C-terminal α-helix (αC) that is stabilized by a disulfide bond between residues Cys194 and Cys200. This disulfide bonded motif in unique to CcmM, and appears to be essential for the enzyme's carbonic anhydrase function. Either deleting the residues C-terminal to residue 193, or mutating the cysteine residues, results in an inactive construct. In addition, reducing conditions were found to inactivate the enzyme. The structure of the CcmM193 deletion construct showed that both the N-terminal helix, and an N-terminal loop close to the catalytic site become disordered. This leads to a shift in the subunits, and a reorganization of the catalytic site, inactivating it (Fig. 6.3, inset). This concerted structural change also occurs upon reduction of the disulfide, as revealed by changes in tryptophan fluorescence. Intriguingly, the motifs at the N- and C-termini which coordinate this activation are precisely those that diverge in sequence in *Synechocytis* sp. PCC6803 and *Synechococcus* sp. PCC7942. Analysis shows that these motifs are lost only in strains that contain CcaA, though many CcaA containing strains do also retain the relevant functional motifs. CcmM therefore appears to be active as a carbonic anhydrase in many β-cyanobacterial strains, but is inactivated in other strains for reasons that are presently obscure [31].

The presence of activation/inactivation mechanism associated with the Cys194-Cys200 disulfide bond is interesting, and would appear to be functionally relevant. Significantly, CcmM expressed in the bacterial cytosol is inactive. It is only upon oxidation (e.g. as routinely occurs during purification in an oxygen atmosphere) that the enzyme becomes active. In vivo, one would expect that CcmM would remain inactive as long as it remains in the cytosol. However, encapsulation in the carboxysome would presumably be required to somehow activate the carbonic anhydrase activity. This implies that the shell of the carboxysome prevents the free passage of cellular reducing agents such as thioredoxin, and the interior carboxysome is then oxidizing [31].

4 The Role of CcmM in Organizing the Carboxysome

CcmM is indispensible for forming carboxysomes, even in strains such as *Synechococcus* sp. PCC7942 where it is not active as a carbonic anhydrase [27], implying that CcmM's has a critical role in the carboxysome that is not dependent upon its CA activity. Unusually for a bacterial protein, CcmM is produced in two variants: a 58 kDa full length variant (encompassing both the CA and RubisCO-like domains and linkers), and a smaller 35 kDa variant (which excludes the CA domain) which is translated from an internal ribosome entry site at Val226 [22]. The longer variant is only produced at a level of a quarter of the shorter variant, but both are abundant and required to produce viable carboxysomes [33].

The key role of CcmM appears to be as an organizing center for other carboxyomal proteins. It interacts with CcaA (where present – details discussed below), the large subunit of RubisCO, CcmN, and possibly also various shell proteins. The C-terminal RubisCO-like domains of CcmM interact with the large subunit of RubisCO in a yeast two hybrid experiment [34]. When histidine tagged CcmM was used in IMAC pull down experiments in *Synechococcus*, RubisCO was found to bind strongly to both the 58 and 35 kDa isoforms, confirming that the C-terminal domains are the principal interaction determinant. These RubisCO/CcmM complexes contain less RbcS than bulk RubisCO, consistent with the idea that the small subunits of CcmM displace RcbS in some sites in the $RbcL_8RbcS_8$ complex [35]. CcmM also binds CcmN, a ~23 kDa protein found in all β-carboxysome gene clusters [34, 36]. CcmN has an N-terminal domain with clear sequence similarity to left-handed β-helical proteins, including CcmM. However, this domain lacks all of CcmM's known catalytic residues. This N-terminal domain binds CcmM in a 1:1 stoichiometric complex [36]. The C-terminal domain appears to be comprised of a long stretch of unconserved, hydrophilic residues that appear prone to intrinsic disorder, with a short stretch of approximately 17 conserved residues near the very C-terminus. CcmN is essential for forming carboxysomes [36, 37], suggesting a role primarily in organizing the compartment, but current evidence does not preclude the idea that the protein also somehow modulates the carbonic anhydrase activity of CcmM or CcaA.

The means by which CcmM is recruited to the carboxysome is not fully understood. Yeast two-hybrid data suggests that the N-terminal domain of CcmM may interact directly with CcmK2 and CcmL in the shell [34]. However, the growth of yeast in these experiments was weak, and the experiment was only completed in one configuration; in the absence of any confirmatory experiments, these results are inconclusive. More recently, constructs encompassing the C-terminal domain of CcmN have been shown to interact with CcmK2 via pull-downs [36]. This is an interesting finding given that enzymes in the propanediol utilization microcompartment are recruited via helical peptides at their N-terminus [38]; however the interaction mediated by CcmN appears to be fairly weak, and in vivo relevance has not yet been demonstrated. An alternative model, so far not discussed in detail in the literature, is that RubisCO itself recruits CcmM to the carboxysome. This idea is supported by the fact that the majority of RubisCO is recruited to the carboxysome without forming a complex with CcmM [35]. Presumably recruitment depends upon RubisCO-RubisCO interactions, as none of the other protein components are large enough to bridge the ~100 nm gap between the shell and the center of the carboxysome. Indeed, the paracrystalline organization of RubisCO in β-carboxysomes [10] suggests that RubisCO-RubisCO interactions may be one of the foundations on which carboxysome assembly is built. In *Synechococcus* sp. PCC7942 cells that express the N- and C-terminal domains of CcmM as separate proteins, CcaA is not recruited to the carboxysome; this implies that the C-terminal domains of CcmM are the determinant for carboxysome recruitment, while the N-terminal domain mediate interactions with CcmN (and CcaA where present) only, and cannot independently bind the structural components of the carboxysome [35].

5 CcaA –a β-Carboxysomal β-Carbonic Anhydrase

CcaA was the first carboxysomal carbonic anhydrase identified. It was originally discovered as DNA fragment that complemented a mutants that cause a high CO_2 requiring phenotype in *Synechoccoccus* sp. PCC 7942; the only open reading frame in the DNA fragment proved to be homologous to bacterial and plant β-carbonic anhydrases [39, 40]. The carbonic anhydrase activity in this strain was shown to be associated with the carboxysomal fraction in purifications, an activity absent in a CcaA mutant strain, supporting the idea that this is a carboxysomal carbonic anhydrase [41]. CcaA from *Synechococcus* is inhibited by ethoxzolamide with an IC_{50} of 4 μM [41]. Surprisingly, while CcaA has been demonstrated to be active in recombinant form [42], the specific activity of the purified enzyme has never been reported. Therefore, despite being discovered 15 years before either other carboxysomal carbonic anhydrase, from the perspective of both activity and structure, CcaA remains the least well characterized.

CcaA's closest non-cyanobacterial homologs are found in the γ-proteobacteria; however, members of this group of enzymes are presently unstudied, and the plant chloroplast enzymes are the most closely related β-CA homologs that are well characterized. In contrast to CcmM and CsoSCA, no structure in known for CcaA, and the closest homolog with known structure is the pea enzyme, with 32 % identity (to *Synechocystis* sp. PCC6803 CcaA) [24]. CcaA is characterized by a unique C-terminal extension about 80 amino acids relative to the plant sequences. This region is largely hydrophilic and weakly conserved, but does have two regions with at least hints of defined function. First, the 20 amino acids immediately C-terminal to the predicted β5 (195–214 in *Synechocystis* PCC6803) are conserved, and required for enzymatic activity and the ability to form oligomers [42]; this suggests that this motif is involved in interactions required to stabilize the core β-CA dimer structure. Secondly, an approximately 17 amino acid motif at the very C-terminus is fairly conserved in many strains [42], although it does appear to be missing in others (e.g. *Synechococcus* sp. PCC7002). It has been suggested that this motif might mediate interactions with a carboxysomal shell protein, but there is no direct evidence for this idea [36]. Certainly, deleting the C-terminal most 60 residues, including this motif, does not appreciably affect enzymatic activity of recombinantly expressed protein [42].

Aside from self-interactions required for oligomerization, CcaA has only been identified in yeast two-hybrid assays to interact with CcmM [34]. Supporting this idea, His-tagged CcmM pulls down CcaA from carboxysomal preparations [35]. This interaction does not require the C-terminal most 60 residues of CcaA, and only requires CcmM's N-terminal domain [34]. CcmM from cyanobacterial strains that do not contain CcaA nevertheless are able to bind this protein in vitro [31]; this suggests that CcaA may share a CcmM binding site with another protein, or, alternatively, binds a motif conserved for other reasons. Expressing the full length CcmM from a plasmid in *Synechococcus* sp. PCC7942 resulted in CcaA levels in the carboxysome tightly tracking those of full length CcmM [43]; this suggests that the CcmM-CcaA interaction may be required to stabilize CcaA. Possibly CcaA activity is prevented from accumulating in the cytosol by rapid turnover in the absence of formation of a CcmM complex.

The exact nature of the complex between CcaA and CcmM is currently unknown. The structural determinants of this complex appear to be limited to the core carbonic anhydrase domains, but it is not clear how such a complex would be formed. CcmM has three-fold symmetry, while β-carbonic anhydrase oligomers are built as multimers of a fundamental dimeric unit. The most obvious way to form a complex would be if CcaA forms a trimer-of-dimers hexamer, and binds to the three-fold symmetric site at either the N- or C-terminal end of CcmM's β-helix. However, while examples of tetrameric and octomeric β-CA's are known, no examples of hexameric β-CA's are known to date.

6 Challenges and Future Research Directions

While the broad role of carbonic anhydrases in carboxysomes are now understood, at least in outline, there are several areas where further research is needed. For CsoSCA, the motifs that recruit it to the shell, and the protein(s) it interacts with, remain unknown. It is also not clear whether some mechanism acts to attenuate activity in the cytosol, prior to encapsulation. CcaA remains comparably little characterized. The catalytic parameters in particular are important for constraining models of overall carboxysome functioning, but the structure, means of binding CcmM, role of the C-terminal tail, and the basis of co-regulation with CcmM expression levels are all important questions. CcmM's kinetic characteristics are also unknown, and the structural basis of, and rationale for its rich interaction behavior remains a major area of research. Finally, the complexity of the complement carbonic anhydrase activities in different β-carboxysomes remains a major puzzle. Many species appear to have both CcaA (sometimes present in two copies) and CcmM as active carbonic anhydrases, while others have either one or the other. In principle, any one of these enzymes should have more than sufficient activity to support carbon fixation, and there is no apparent reason why one enzyme should not be functionally equivalent to another. Understanding this problem will likely require expanding the set of model β-cyanobacterial strains studied to include examples of these different carbonic anhydrase complements.

Acknowledgments This work was funded by a Discovery Grant from the National Science and Engineering Research Council of Canada (# 327280).

References

1. Andersson I, Backlund A (2008) Structure and function of Rubisco. Plant Physiol Biochem 46:275–291
2. Price GD (2011) Inorganic carbon transporters of the cyanobacterial CO_2 concentrating mechanism. Photosynth Res 109:47–57
3. Kerfeld CA, Heinhorst S, Cannon GC (2010) Bacterial microcompartments. Annu Rev Microbiol 64:391–408
4. Yeates TO, Crowley CS, Tanaka S (2010) Bacterial microcompartment organelles: protein shell structure and evolution. Annu Rev Biophys 39:185–205
5. Seedorf H, Fricke WF, Veith B, Bruggemann H, Liesegang H, Strittmatter A, Miethke M, Buckel W, Hinderberger J, Li F, Hagemeier C, Thauer RK, Gottschalk G (2008) The genome of *Clostridium kluyveri*, a strict anaerobe with unique metabolic features. Proc Natl Acad Sci U S A 105:2128–2133
6. Craciun S, Balskus EP (2012) Microbial conversion of choline to trimethylamine requires a glycyl radical enzyme. Proc Natl Acad Sci U S A 109:21307–21312
7. Jorda J, Lopez D, Wheatley NM, Yeates TO (2013) Using comparative genomics to uncover new kinds of protein-based metabolic organelles in bacteria. Protein Sci 22:179–195
8. Iancu CV, Ding HJ, Morris DM, Dias DP, Gonzales AD, Martino A, Jensen GJ (2007) The structure of isolated Synechococcus strain WH8102 carboxysomes as revealed by electron cryotomography. J Mol Biol 372:764–773

9. Schmid MF, Paredes AM, Khant HA, Soyer F, Aldrich HC, Chiu W, Shively JM (2006) Structure of *Halothiobacillus neapolitanus* carboxysomes by cryo-electron tomography. J Mol Biol 364:526–535
10. Kaneko Y, Danev R, Nagayama K, Nakamoto H (2006) Intact carboxysomes in a cyanobacterial cell visualized by hilbert differential contrast transmission electron microscopy. J Bacteriol 188:805–808
11. Kerfeld CA, Sawaya MR, Tanaka S, Nguyen CV, Phillips M, Beeby M, Yeates TO (2005) Protein structures forming the shell of primitive bacterial organelles. Science 309:936–938
12. Tanaka S, Sawaya MR, Phillips M, Yeates TO (2009) Insights from multiple structures of the shell proteins from the beta-carboxysome. Protein Sci 18:108–120
13. Tsai Y, Sawaya MR, Cannon GC, Cai F, Williams EB, Heinhorst S, Kerfeld CA, Yeates TO (2007) Structural analysis of CsoS1A and the protein shell of the *Halothiobacillus neapolitanus* carboxysome. PLoS Biol 5:e144
14. Rae BD, Long BM, Badger MR, Price GD (2012) Structural determinants of the outer shell of β-carboxysomes in *Synechococcus elongatus* PCC 7942: roles for Ccm, K2, K3–K4, CcmO, and CcmL. PLoS One 7:e43871
15. Klein MG, Zwart P, Bagby SC, Cai F, Chisholm SW, Heinhorst S, Cannon GC, Kerfeld CA (2009) Identification and structural analysis of a novel carboxysome shell protein with implications for metabolite transport. J Mol Biol 392:319–333
16. Tanaka S, Sawaya MR, Yeates TO (2010) Structure and mechanisms of a protein-based organelle in *Escherichia coli*. Science 327:81–84
17. Samborska B, Kimber MS (2012) A dodecameric CcmK2 structure suggests β-carboxysomal shell facets have a double-layered organization. Structure 20:1353–1362
18. Heinhorst S, Williams EB, Cai F, Murin CD, Shively JM, Cannon GC (2006) Characterization of the carboxysomal carbonic anhydrase CsoSCA from *Halothiobacillus neapolitanus*. J Bacteriol 188:8087–8094
19. Price GD, Badger MR (1989) Expression of human carbonic anhydrase in the cyanobacterium Synechococcus PCC7942 creates a high CO(2)-requiring phenotype: evidence for a central role for carboxysomes in the CO(2) concentrating mechanism. Plant Physiol 91:505–513
20. So AKC, Espie GS, Williams EB, Shively JM, Heinhorst S, Cannon GC (2004) A novel evolutionary lineage of carbonic anhydrase (epsilon class) is a component of the carboxysome shell. J Bacteriol 186:623–630
21. Dou Z, Heinhorst S, Williams EB, Murin CD, Shively JM, Cannon GC (2008) CO2 fixation kinetics of *Halothiobacillus neapolitanus* mutant carboxysomes lacking carbonic anhydrase suggest the shell acts as a diffusional barrier for CO_2. J Biol Chem 283:10377–10384
22. Long BM, Price GD, Badger MR (2005) Proteomic assessment of an established technique for carboxysome enrichment from Synechococcus PCC7942. Can J Bot 83:746–757
23. Sawaya MR, Cannon GC, Heinhorst S, Tanaka S, Williams EB, Yeates TO, Kerfeld CA (2006) The structure of beta-carbonic anhydrase from the carboxysomal shell reveals a distinct subclass with one active site for the price of two. J Biol Chem 281:7546–7555
24. Kimber MS, Pai EF (2000) The active site architecture of *Pisum sativum* beta-carbonic anhydrase is a mirror image of that of alpha-carbonic anhydrases. EMBO J 19:1407–1418
25. Roberts EW, Cai F, Kerfeld CA, Cannon GC, Heinhorst S (2012) Isolation and characterization of the Prochlorococcus carboxysome reveal the presence of the novel shell protein CsoS1D. J Bacteriol 194:787–795
26. Baker SH, Williams DS, Aldrich HC, Gambrell AC, Shively JM (2000) Identification and localization of the carboxysome peptide Csos3 and its corresponding gene in *Thiobacillus neapolitanus*. Arch Microbiol 173:278–283
27. Price GD, Howitt SM, Harrison K, Badger MR (1993) Analysis of a genomic DNA region from the cyanobacterium Synechococcus sp. strain PCC7942 involved in carboxysome assembly and function. J Bacteriol 175:2871–2879
28. Alber BE, Ferry JG (1994) A carbonic anhydrase from the archaeon *Methanosarcina thermophila*. Proc Natl Acad Sci U S A 91:6909–6913

29. Ferry JG (2010) The γ class of carbonic anhydrases. Biochim Biophys Acta Proteins Proteomic 1804:374–381
30. Espie GS, So AK-C (2005) Cyanobacterial carbonic anhydrases. Can J Bot 83:721–734
31. Peña KL, Castel SE, de Araujo C, Espie GS, Kimber MS (2010) Structural basis of the oxidative activation of the carboxysomal gamma-carbonic anhydrase, CcmM. Proc Natl Acad Sci U S A 107:2455–2460
32. Kisker C, Schindelin H, Alber BE, Ferry JG, Rees DC (1996) A left-hand beta-helix revealed by the crystal structure of a carbonic anhydrase from the archaeon *Methanosarcina thermophila*. EMBO J 15:2323–2330
33. Long BM, Tucker L, Badger MR, Price GD (2010) Functional cyanobacterial beta-carboxysomes have an absolute requirement for both long and short forms of the CcmM protein. Plant Physiol 153:285–293
34. Cot SS-W, So AKC, Espie GS (2008) A multiprotein bicarbonate dehydration complex essential to carboxysome function in cyanobacteria. J Bacteriol 190:936–945
35. Long BM, Badger MR, Whitney SM, Price GD (2007) Analysis of carboxysomes from Synechococcus PCC7942 reveals multiple Rubisco complexes with carboxysomal proteins CcmM and CcaA. J Biol Chem 282:29323–29335
36. Kinney JN, Salmeen A, Cai F, Kerfeld CA (2012) Elucidating essential role of conserved carboxysomal protein CcmN reveals common feature of bacterial microcompartment assembly. J Biol Chem 287:17729–17736
37. Friedberg D, Kaplan A, Ariel R, Kessel M, Seijffers J (1989) The 5′-flanking region of the gene encoding the large subunit of ribulose-1, 5-bisphosphate carboxylase/oxygenase is crucial for growth of the cyanobacterium Synechococcus sp. strain PCC 7942 at the level of CO_2 in air. J Bacteriol 171:6069–6076
38. Fan C, Cheng S, Sinha S, Bobik TA (2012) Interactions between the termini of lumen enzymes and shell proteins mediate enzyme encapsulation into bacterial microcompartments. Proc Natl Acad Sci U S A 109:14995–15000
39. Fukuzawa H, Suzuki E, Komukai Y, Miyachi S (1992) A gene homologous to chloroplast carbonic anhydrase (icfA) is essential to photosynthetic carbon dioxide fixation by *Synechococcus* PCC7942. Proc Natl Acad Sci U S A 89:4437–4441
40. Yu JW, Price GD, Song L, Badger MR (1992) Isolation of a putative carboxysomal carbonic anhydrase gene from the cyanobacterium *Synechococcus* PCC7942. Plant Physiol 100:794–800
41. Price GD, Coleman JR, Badger MR (1992) Association of carbonic anhydrase activity with carboxysomes isolated from the cyanobacterium *Synechococcus* PCC7942. Plant Physiol 100:784–793
42. So AK-C, Cot SS-W, Espie GS (2002) Characterization of the C-terminal extension of carboxysomal carbonic anhydrase from Synechocystis sp. PCC6803. Funct Plant Biol 29:183–194
43. Long BM, Rae BD, Badger MR, Price GD (2011) Over-expression of the β-carboxysomal CcmM protein in Synechococcus PCC7942 reveals a tight co-regulation of carboxysomal carbonic anhydrase (CcaA) and M58 content. Photosynth Res 109:33–45

Chapter 7
Carbonic Anhydrases and Their Interplay with Acid/Base-Coupled Membrane Transporters

Holger M. Becker, Michael Klier, and Joachim W. Deitmer

Abstract Carbonic anhydrases (CAs) have not only been identified as ubiquitous enzymes catalyzing the fast reversible hydration of carbon dioxide to generate or consume protons and bicarbonate, but also as intra- and extracellular proteins, which facilitate transport function of many acid/base transporting membrane proteins, coined 'transport metabolon'. Functional interaction between CAs and acid/base transporters, such as chloride/bicarbonate exchanger (AE), sodium-bicarbonate cotransporter (NBC) and sodium/hydrogen exchanger (NHE) has been shown to require both catalytic CA activity as well as direct binding of the enzyme to specific sites on the transporter. In contrast, functional interaction between different CA isoforms and lactate-proton-cotransporting monocarboxylate transporters (MCT) has been found to be isoform-specific and independent of CA catalytic activity, but seems to require an intramolecular proton shuttle within the enzyme. In this chapter, we review the various types of interactions between acid/base-coupled membrane

---

Susan C. Frost and Robert McKenna (eds.). Carbonic Anhydrase: Mechanism, Regulation, Links to Disease, and Industrial Applications

H.M. Becker (✉)
Division of Zoology/Membrane Transport, Department of Biology,
University of Kaiserslautern, Kaiserslautern, Germany
e-mail: h.becker@biologie.uni-kl.de

M. Klier
Division of Zoology/Membrane Transport and Division of General Zoology, Department of Biology, University of Kaiserslautern, Kaiserslautern, Germany
e-mail: michael.klier@biologie.uni-kl.de

J.W. Deitmer
Division of General Zoology, Department of Biology, University of Kaiserslautern, Kaiserslautern, Germany
e-mail: deitmer@biologie.uni-kl.de

S.C. Frost and R. McKenna (eds.), *Carbonic Anhydrase: Mechanism, Regulation, Links to Disease, and Industrial Applications*, Subcellular Biochemistry 75,
DOI 10.1007/978-94-007-7359-2_7,

carriers and different CA isoforms, as studied in vitro, in intact *Xenopus* oocytes, and in various mammalian cell types. Furthermore, we discuss recent findings that indicate the significance of these 'transport metabolons' for normal cell functions.

Keywords Protein-protein interaction • Transport metabolon • Proton • Bicarbonate • Monocarboxylates

1 Acid/Base Transporters

Virtually all biologically active molecules are equipped with chemical groups acting as weak acid or base, which is why already minor changes in pH can have significant physiological impact, e.g. by influencing a molecule's surface charge, conformation and finally its biological activity. Thus, H^+ ions are not only chemically most active, but also potent modulators of most biological processes [1, 2]. Consequently, a wide variety of biological processes are dependent, or modulated by pH. To guarantee functionality of proteins and membranes and hence of metabolism, intra- and extracellular pH must be tightly regulated. This regulation of intra- and extracellular pH depends on the sophisticated interplay of various acid/base transporting proteins: Besides active pH-regulation by H^+-ATPases and H^+/K^+-ATPases, regulation of H^+ is mainly attributable to Na^+ or Cl^- dependent H^+/HCO_3^- carriers, which can recover intracellular and extracellular pH from acidification or alkalinisation. These transporters include (see Table 7.1): the SLC4 family of HCO_3^- transporters, which is divided in two groups of three Cl^-/HCO_3^- exchangers (AE1-3) and five Na^+-coupled HCO_3^- transporters (NBCe1, NBCe2, NBCn1, NDCBE, NCBE); six members of the SLC26 family of Cl^-/HCO_3^- exchangers (SLC26A2, SLC26A3 (DRA), SLC26A4 (Pendrin), SLC26A6 (PAT-1), SLC26A7, SLC26A9); and the SLC9 family of Na^+/H^+ exchangers (NHEs), consisting of nine known isoforms, NHE1-NHE9. Furthermore, acid/base-coupled metabolite carriers like the H^+-linked monocarboxylate transporters (MCT1-4) or H^+-coupled amino acid transporters like the excitatory amino acid transporters (EAATs, SLC1) or glutamine transporters (SNAT3, SLC38) can participate in acid/base regulation.

2 The Concept of the Transport Metabolon

The majority of these transporters had been demonstrated to physically and/or functionally interact with isoforms of carbonic anhydrase (CA) to form a 'transport metabolon' (Fig. 7.1). A metabolon has been defined as a 'temporary, structural-functional, supramolecular complex of sequential metabolic enzymes and cellular structural elements, in which metabolites are passed from one active site to another without complete equilibration with the bulk cellular fluids' [3–5]. This substrate channeling should decrease transit time of intermediates, prevent loss of

Table 7.1 Overview of the known interactions between acid/base transporters and carbonic anhydrases. The type of interaction is grouped into functional interactions, as measured by physiological methods and physical interactions, as determined by binding studies

Transporter	Interacts with CA isoform	Type of interaction	Reference
AE1 (SLC4A1)	CAII, CAIV, CAIX	Functional	[15–17, 56]
	CAII, CAIV, CAIX	Physical	[7, 11, 12, 15, 16, 56, 57]
AE2 (SLC4A2)	CAII, CAIV, CAIX	Functional	[15, 16, 56, 61]
	CAII, CAIX	Physical	[12, 16]
AE3 (SLC4A3)	CAII, CAIV, CAIX, CAXIV	Functional	[15, 16, 56, 58]
	CAIX, CAXIV	Physical	[16, 58]
SLC26A3	CAII	Functional	[25]
	—	Physical	—
SLC26A6	CAII	Functional	[26]
	CAII	Physical	[26]
SLC26A7	CAII	Functional	[16]
	—	Physical	—
NBCe1 (SLC4A4)	CAI, CAII, CAIII, CAIV, CAIX	Functional	[48–53, 60, 61]
	CAII, CAIV, CAIV	Physical	[47–49, 60]
NBCn1 (SLC4A7)	CAII	Functional	[52]
	CAII	Physical	[52]
NHE1 (SLC9A1)	CAII, CAIV	Functional	[53, 80–82]
	CAII	Physical	[81, 82]
SNAT1 (SLC38A3)	CAI, CAII, CAIII, CAIV	Functional	[83, 84]
	—	Physical	—
MCT1 (SLC16A1)	CAII	Functional	[112, 113, 116, 119]
	CAII	Physical	[116]
MCT2 (SLC16A7)	CAIV	Functional	[115]
	—	Physical	—
MCT4 (SLC16A3)	CAII	Functional	[114, 119]
	—	Physical	—

intermediates by diffusion, protect labile intermediates from solvent, and prevent entrance of intermediates into competing metabolic pathways [5, 6]. Such metabolons have been found in many metabolic pathways like glycolysis and tricarboxylic acid cycle, as well as in the biosynthesis of DNA, RNA and proteins [4].

2.1 The Bicarbonate Transport Metabolon

First evidence for a transport metabolon, formed between CA and an acid/base transporter, has been presented in 1993 by Kifor et al. [7] for CA isoform II

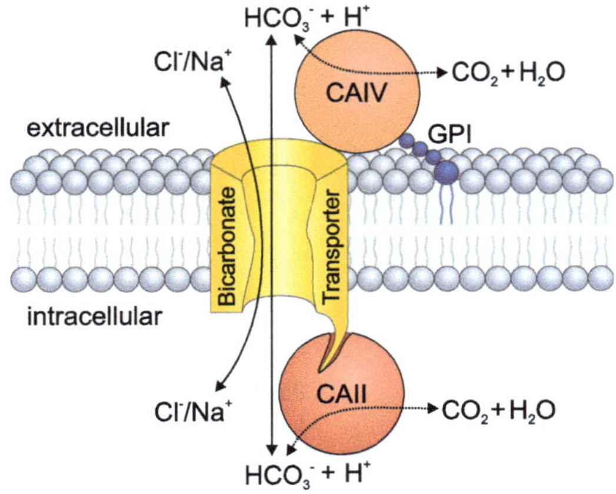

Fig. 7.1 Hypothetical model of the interaction between bicarbonate transporters and CAs. Cytosolic CAII (*red circle*) binds to the C-terminal tail of various bicarbonate transporters (*yellow structure*), while extracellular CAIV (*orange circle*), anchored to the extracellular side of the plasma membrane by GPI (*dark blue structure*), would bind to an extracellular loop of the transporter. By catalyzing the reversible conversion of CO_2 to HCO_3^- and H^+ in the direct vicinity of the transporter, CAs suppress depletion of HCO_3^- at the *cis* side, and suppress accumulation of HCO_3^- at the *trans* side of the transporter to facilitate transport function

(CAII) and the Cl^-/HCO_3^- exchanger AE1 (anion exchanger 1; Band3). AE1 belongs to the anion exchanger (AE) family which is composed of the three isoforms AE1 (SLC4A1), AE2 (SLC4A2), and AE3 (SLC4A3) sharing about 65 % amino acid sequence identity in their transmembrane region. They are ubiquitously distributed in vertebrate tissues and physiologically most relevant; AEs mediate the electroneutral counter-transport of Cl^- and HCO_3^- across the plasma membrane with a stoichiometry of 1:1 and a turnover rate of 5×10^4/s [8]. AE-mediatetd electroneutral Cl^-/HCO_3^- exchange contributes to the regulation of intracellular pH, cell volume, and membrane potential. Besides Cl^-/HCO_3^- exchange AEs have also been demonstrated to transport several other anions including bromide, fluoride, iodide, phosphate, and sulphate [9, 10].

Kifor et al. [7] could demonstrate that binding of the stilben-disulfonate inhibitor 4,4′-diisothiocyano-2,2′-stilbenedisulfonic acid (DIDS) to AE1 alters the binding kinetics of the CA inhibitor dansylsulfonamide to CAII in human red cells, indicating that AE1 and CAII may form a physical complex. Evidence for direct interaction between red cell AE1 and CAII was then shown by Reithmeier and coworkers [11–13]: The authors examined the distribution of AE1 and CAII in intact red cell ghost membranes by immunofluorescent staining, following treatment of the membrane with tomato lectin. Tomato lectin specifically binds to the repeating polylactosaminoglycan structure on AE1 [14] and causes their clustering within the membrane. Treatment with tomato lectin lead to clustering of CAII and AE1 which suggests that AE1 is physically tethered to CAII in intact red cell membranes [11]. Furthermore, CAII could be co-immunoprecipitated with AE1 when antiserum against the N-terminal of AE1 was used, while immunoprecipitation of AE1 with serum directed against the C-terminal of the transporter failed to co-immunoprecipitate CAII [11]. These data suggest that CAII physically binds to the C-terminal tail of AE1. These data were confirmed by affinity blotting and a solid phase binding assay with CAII and a GST fusion protein of the AE1 C-terminal, which showed saturable binding of CAII with a $K_{1/2}$ of 20 nM. The binding rate was found to be low ($t_{1/2} = 12$ h) at physiological pH and ionic strength, but increased with decreasing concentration of NaCl and KCl, respectively, and decreasing pH [11]. Single site mutations identified the acidic cluster L^{886}DADD in the C-terminal tail of AE1 as binding site for CAII with at least two out of the three acidic amino acids being crucial for binding [12, 13]. These data suggest a hydrophobic amino acid followed by at least two acidic amino acids as the CAII binding motif. In line with these findings direct binding of CAII to a GST fusion protein of the C-terminal tail of AE2, which has a corresponding cluster LDANE, could be demonstrated with a solid phase binding assay [12].

In contrast to CAII, no physical interaction between CAI and AE1 could be identified [13]. CAII contains a cluster of five histidine residues within its N-terminal (His3, 4, 10, 15 and 17 in human CAII), which is highly conserved within CA isoform II in different species, but is absent in CAI. Truncation of the first 17 amino acids in CAII caused a loss of binding, as did the mutation of those five histidine residues (and one lysine) to the analogous residues in CAI (H3P, H4D, K9D, H10K, H15Q and H17S; Ref. [13]), suggesting that binding of CAII to the C-terminal of AE1 is mediated via the histidine-rich cluster in the N-terminal of CAII.

The amino acid cluster L^{886}DADD in the C-terminal of AE1 is located close to the last transmembrane domain. Therefore, binding of CAII to this cluster would position the enzyme close to the transporter pore of AE1 near the inner cell surface. This location may ideally position CAII to hydrate incoming CO_2 and directly supply the AE1 transporter with a localized substrate pool [12].

The CA catalytic activity requirement for maximum AE1 transport was first shown by Sterling et al. [15] by demonstrating that inhibition of endogenous CA with acetazolamide decreases transport activity of AE1, heterologously expressed in HEK293 cells, by 50–60 %. Furthermore, coexpression of AE1, AE2 or cardiac AE3c with the catalytically inactive mutant CAII-V143Y in HEK293 cells inhibited AE transport function by up to 60 % [15, 16]. From these experiments the authors concluded, that heterologously expressed CAII-V143Y which can bind to AE1, as confirmed by a microtiter plate binding assay, displaces endogenous CAII-WT from the transporter. However, due to the lack of CAII-V143Y catalytic activity, the bound mutant cannot supply HCO_3^- to the transporter, which results in an overall reduction in AE transport activity. The need of a physical interaction between AE1 and CAII for maximum AE1 transport function was further confirmed by measuring the transport activity of AE1 mutants, which had been shown to be unable to bind CAII [12] in HEK293 cells [15]. Mutation of the putative CAII binding site L^{886}DADD to either L^{886}AAAA or L^{886}NANN decreased AE1 transport activity to approx. 10 % of wild-type activity, while the expression level of the mutants did not differ from the expression level of AE1-WT. These results indicated that

CAII enhances transport activity of AE1 by substrate channeling. However, in order to investigate the hypothesis that CAII might enhance AE1 activity by allosteric interaction, Dahl and coworkers [17] compared Cl^-/Cl^- and Cl^-/HCO_3^- exchange activities of AE1 polypeptides with truncation and missense mutations in the C-terminal tail. Expression of AE1 mutants in *Xenopus* oocytes showed, that the two truncation mutants AE1-A891X and AE1-D896X, which both still contain the CAII binding site L^{886}DADD, as well as the missense mutant AE1-L^{886}DAAA and AE1- A^{886}DADD were inactive as Cl^-/HCO_3^- exchangers, but exhibited normal Cl^-/Cl^- exchange activities, while the truncation mutant AE1-R901X, associated with distal renal tubular acidosis, exhibited both Cl^-/Cl^- and Cl^-/HCO_3^- exchange activities [17]. While coexpression of CAII could enhance Cl^-/HCO_3^-, but not Cl^-/Cl^- exchange activity of AE1-WT in *Xenopus* oocytes, no effect of CAII on either Cl^-/Cl^- or Cl^-/HCO_3^- exchange activity of either one of the mutants could be detected. These findings do not favour an allosteric effect of CAII binding on AE1 transport function, since binding of CAII cannot enhance transport of a non-bicarbonate ion. Furthermore, coexpression of either one of the mutants hAE1-SAO (Δ400-408; associated with Southeast Asian ovalocytosis) or mAE1-E699Q (which both retain the CAII binding site, but are practically deficient of Cl^-/HCO_3^- exchange activity) with the missense mutant AE1-LDAAA (lacking the CAII binding site), could rescue AE1-LDAAA-mediated HCO_3^- transport activity [17]. Since AE1 proteins have been shown to form dimers within the plasma membrane [18] the authors concluded, that AE1-SAO/AE1-E699Q and AE1-LDAAA can form a heterodimer with CAII bound to the C-terminal of AE1-SAO/AE1-E699Q, which locates the enzyme close enough to the transporter pore of AE1-LDAAA to facilitate HCO_3^- transport activity.

Interestingly, not only CAII enhances transport activity of AE1, but it has also been shown that di-/tri- and especially tetrapeptides incorporating the amino acid sequence DADD strongly activate human CAII [19]. This suggests that within the AE1-CAII transport metabolon, CAII and AE1 can mutually increase activity of each other.

2.1.1 Interactions Between SLC26A and CAII

The SLC26 gene family encodes 11 members of anion exchangers which are capable of transporting chloride, iodide, sulfate, bicarbonate, hydroxyl, oxalate, and formate anions with different affinities and anion specificity profiles [20], except SLC26A10 which is a pseudogene [21]. Six isoforms of this protein family, namely SLC26A2, SLC26A3 (DRA), SLC26A4, SLC26A6, SLC26A7, and SLC26A9 have meanwhile been found to mediate electrogenic or electroneutral Cl^-/HCO_3^- or Cl^-/OH^- exchange [22–24], with the inward Cl^- gradient providing the driving force for net HCO_3^- efflux through these anion exchangers [23]. The transporters encoded by this gene family have 10–14 transmembrane domains, with the N- and C-terminal tails facing the intracellular side [21].

Out of the six Cl^-/HCO_3^- or Cl^-/OH^- transporting SLC26A proteins, three isoforms have been demonstrated to show some kind of interaction with CAII: For SLC26A3 and SLC26A7 it could be shown, that acetazolamide (AZA) reduces transporter-mediated Cl^-/HCO_3^- exchange activity both in SLC26A3- and SLC216A7-transfected HEK cells (SLC26A3: Ref. [25]; SLC26A7: Ref. [16]). Coexpression of CAII-V143Y with SLC26A3 and SLC26A7, respectively, in HEK cells did not change Cl^-/HCO_3^- exchange activity (SLC26A3: Ref. [25]; SLC26A7: Ref. [16]), suggesting no direct interaction between SLC26A3/A7 and CAII. Since SLC26A3 and SLC26A7 are the only bicarbonate transport proteins, whose C-terminal lacks a putative CAII binding motif, it seems that functional coupling between SLC26A3/A7 and CAII follows a different mechanism than observed for the AEs.

In contrast to SLC26A3 and SLC26A7, SLC26A6 was found to bind to CAII as shown by co-immunoprecipitation and micro titer plate binding assays with a GST fusion protein of the SLC26A6 C-terminal tail [26]. Truncation experiments identified the acidic cluster D^{546}VDF within the C-terminal of SLC26A6 as CAII binding site [26]. Furthermore, both mutation of the binding site to N^{546}NVNF and coexpression of SLC26A6-WT with the catalytically inactive mutant CAII-V143Y, significantly reduced Cl^-/HCO_3^- exchange activity in HEK cells to values observed after inhibition of endogenous CAII with AZA [26]. In addition, activation of protein kinase C (PKC) via the angiotensin II receptor (AT_{1a}), significantly decreased Cl^-/HCO_3^- exchange activity of coexpressed SLC26A6. Activation of PKC reduced SLC26A6/CAII association in immunoprecipitates and displaced CAII from the plasma membrane, as monitored by immunofluorescence. Finally, mutation of a PKC site by exchange of serin at position 553 to alanin rendered the SLC26A6 unresponsive to PKC. It was concluded, that phosphorylation of S^{553} by PKC reduces CAII/SLC26A6 interaction by displacing CAII from its binding site within the STAS (Sulphate Transporter and AntiSigma factor antagonist) domain, thereby reducing bicarbonate transport rate. This mechanism for acute regulation of membrane transport was termed "metabolon disruption".

2.1.2 Interactions Between Na^+-HCO_3^- Cotransporters and CAII

The sodium-bicarbonate cotransporters (NBCs), which also belong to the SCL4A gene family, consist of six members mediating the electrogenic or electroneutral transport of bicarbonate: NBCe1 (SLC4A4), NBCe2 (SLC4A5), NBCn1 (SLC4A7), NDCBE (Na^+-driven Cl^-/HCO_3^- exchanger, SLC4A8), NBCn2 (NCBE (Na^+-driven Cl^-/HCO_3^- exchanger), SLC4A10), and AE4 (SLC4A9). Physiologically, NBCe1 and NBCe2 mediate the electrogenic symport of HCO_3^- and Na^+, and directly influence the membrane potential [27, 28], whereas NBCn1, NDCBE, and NCBE mediate an electroneutral transport of HCO_3^- and Na^+. In case of NDCBE and NCBE this cotransport is additionally coupled to an antiport of Cl^- [27].

The *NBCe1* gene encodes the three splice variants NBCe1-A to C, which differ in their N- and/or C-termini [29–32]. NBCe1-A normally operates with a 1:3 stoichiometry (1 Na^+ : 3 HCO_3^-; Ref. [33]), while NBCe1-B and NBCe1-C function with a 1:2 stoichiometry [34–39]. It has been reported that protein kinase A (PKA) -dependent phosphorylation within the cytoplasmic C-terminal of the A and B variant, changes transporter stoichiometry from 1:3 to 1:2 Na^+:HCO_3^- [39, 40]. Part of this Na^+-dependent clade is also AE4, but varying Na^+ dependence of its function has been reported [41–43].

Reports on a functional interaction between NBCs and carbonic anhydrase go back to the finding that application of the CA inhibitor AZA inhibits the transport of bicarbonate across the basolateral cell membrane in renal proximal tubule of rabbits [44–46]. First evidence for a direct interaction between NBCe1 and CAII was presented by Gross et al. [47] using isothermal titration calorimetry. CAII was found to bind to the C-terminal peptide of kNBCe1 (amino acids 915–1035) with a K_D of 160 ± 10 nM [47]. In analogy to the acidic CAII-binding cluster D^{887}ADD found in AE1 [12], the authors suggested the cluster D^{986}NDD within the C-terminal of NBCe1 as the putative CAII binding site. This assumption was supported by the finding that CAII increases the short-circuit current of the transporter, heterologously expressed in the mouse proximal convoluted tubule cell line mPCT, in an AZA-sensitive manner, when the NBCe1 was operating in the 3 HCO_3^- : 1 Na^+ mode, but not in the 2:1 mode. It has been shown previously that phosphorylation of S^{982} by PKA shifts the transport mode of NBCe1 from 3:1 to 2:1 [39]. The increased negative charge on phosphorylated S^{982} might induce a local conformational change within the C-terminal of NBCe1, which favors an electrostatic interaction between the nearby D^{986} and D^{988} with a bicarbonate-binding site [47]. Since D^{986} and D^{988} are also involved in binding of CAII, phosphorylation of S^{982} would drive CAII away from the C-terminal of NBCe1 in a similar fashion as observed for CAII and SLC26A6 [26, 47].

A follow-up study by Pushkin et al. [48], applying a GST-CAII pull-down assay with various mutants of a His-tagged NBCe1-C-terminal, revealed that both the acidic clusters L^{958}DDV and D^{986}NDD are involved in binding CAII. Physiological measurements in mPTC cells showed, that binding of CAII to both clusters is crucial for full activation of NBCe1, since the binding affinity of CAII to various mutants of the two binding clusters, as measured by pull-down assay, showed a linear correlation with the AZA-inhibitable acid/base flux measured with the corresponding NBCe1-mutants [48]. The finding that binding of catalytically active CAII is crucial for enhancement of NBCe1 activity is also supported by the result showing that the rate of NBCe1-mediated pH_i recovery after acid load was inhibited by 40 % when the transporter was cotransfected with catalytically inactive CAII-V143Y in HEK293 cells [49]. In analogy to AE1, CAII-V143Y might compete with endogenous CAII for interaction with NBCe1 at the inner surface of the plasma membrane, which indicates that physical interaction between NBCe1 and CAII is required for full transport activity.

Functional interaction between NBCe1 and CAII was further investigated by heterologous protein expression in *Xenopus* oocytes [50]. Both injection and coexpression of CAII increased NBCe1-mediated membrane current and membrane conductance during application of CO_2/HCO_3^--buffered solution in an ethoxzolamide (EZA)-sensitive manner. Measurements of intracellular Na^+ concentrations with ion-selective microelectrodes showed an increase in the rate of NBCe1-mediated Na^+-flux by two to fourfold when CAII was injected or coexpressed. CAII-mediated increase in NBCe1 activity as determined by changes in membrane current and conductance was dependent on the CAII concentration with a half-maximum enhancement in NBCe1-activity at 20–30 ng CAII/oocyte [50]. In contrast to the measurements in mPTC cells, CAII increased NBCe1 activity in oocytes when the transporter was operating at a stoichiometry of 2 HCO_3^- : 1 Na^+ which indicates, that either no direct interaction between NBCe1 and CAII is required for an increase in NBCe1 transport activity in oocytes, or physical binding of CAII to NBCe1 follows another mechanism in the frog oocytes than it does in mPTC cells.

While expression of the catalytically inactive mutant CAII-V143Y could not enhance activity of NBCe1 in *Xenopus* oocytes, coexpression of the N-terminal CAII mutant CAII-HEX (H3P, H4D, K9D, H10K, H15Q, H17S), which had been shown to be unable to interact with AE1 [13], lead to an enhancement of NBCe1 activity as did CAII-WT [50]. Indeed, not only the CAII-mutant mimicking the N-terminal of CA isoform I (CAI) enhanced NBCe1 transport activity, also CAI itself as well as CA isoform III (CAIII) were able to augment NBCe1 activity when coexpressed in *Xenopus* oocytes [51].

Both physical and functional interaction could also be demonstrated for CAII and NBCn1 (NBC3): A micro titer plate binding assay with immobilized CAII and GST fusion proteins of NBCn1-C-terminal (AA 1127–1214 of human NBCn1) showed physical binding of CAII to NBCn1 with a K_D of 101 nM [52]. The binding strength increased with decreasing pH value. Antibody staining showed that expression of NBCn1 recruited overexpressed CAII to the plasma membrane of HEK293 cells, indicating binding of CAII to NBCn1 also under physiological conditions. Physical interaction of catalytically active CAII with NBCn1 for full transport activity was needed as demonstrated by a significant reduction in NBCn1 activity, when the transporter was coexpressed with the catalytically inactive mutant CAII-V143Y in HEK293 cells [52]. Furthermore, mutation of the two aspartic acid residues D^{1135} and D^{1136} in the C-terminal of NBCn1 reduced both binding of CAII as well as NBC-mediated pH recovery from acid loads in HEK293 cells. These data suggest that CAII binds to the amino acids D^{1135} and D^{1136} within the C-terminal of NBCn1 to locate the enzyme close enough to the transporter pore to supply HCO_3^- to the carrier and enhance transport activity.

A physiological role for a transport metabolon between NBC and CAII could be demonstrated in rodent oligodentrocytes [53]. In mature oligodentrocytes, intracellular pH has been found to be predominantly regulated by activities of Na^+/H^+ exchanger (NHE), CAII and NBC [54, 55]. NBC exports HCO_3^- and Na^+ in a suggested 3:1 stoichiometry [54], thereby acting as an acid loader, while NHE exports H^+ from the cell, leading to an intracellular alkalization. The two transporters have differential subcellular distribution in oligodentrocytes: NBC showed punctate distribution in distal dendrites, while NHE is localized in a punctate distribution in the perikaryon and proximal processes. Fluorescence spectroscopy correlation analysis of CAII microinjected into oligodentrocytes reveals freely diffusing protein throughout the cell as well as protein associated with NHE predominantly in the perikaryon and with NBC predominantly in the dendrites.

By fluorimetric measurements of intracellular pH the authors could show that spatial non-uniformity in the differential subcellular distribution of the two transporters, which both are colocalized with CAII, generates spatial non-uniformity of intracellular pH [53]. However, inhibition of CA by EZA reduced the NBC-mediated acidification of dendrites. From these data the authors concluded, that transport metabolons consisting of CAII with NHE and/or NBC could generate alkaline and acidic microdomains within one cell, which may provide a mechanism for differential regulation of pH-sensitive enzyme activities in different subcellular compartments.

2.1.3 Interactions Between HCO_3^- Transporters and Extracellular CAs

Functional interaction between the chloride/bicarbonate exchangers AE1-3 and extracellular CAIV was first demonstrated in HEK293 cells [56]. Expression of CAIV could restore transport activity of AE1, AE2 and AE3 when the carriers were cotransfected with CAII-V143Y. Gel overlay assays indicated a specific interaction between AE1, AE2, AE3 and CAIV [56]. Physical interaction between AE1 and CAIV was further investigated by GST pull-down assays: CAIV was selectively pulled down from a lysate produced of CAIV-transfected HEK293 cells with a GST fusion protein of the 4th extracellular loop of AE1, while no physical interaction between CAIV and a GST fusion protein of the AE1's 3rd extracellular loop could be detected. Mutation studies of the AE1 suggested that the amino acid cluster S^{643}-L^{655} within the 4th extracellular loop has a folded structure that was inaccessible to hydrophilic reagents, whereas the R^{656}-I^{661} region had an open structure with maximum accessibility at R^{656} [57]. From this it was concluded that CAIV might interact with AE1 somewhere within the R^{656}-I^{661} cluster. Interestingly, this region has also been suggested to form an outer vestibule that funnels anions to and from the transport site [57]. It was concluded that localization of CAIV to the 4th extracellular loop of AE1 would place the enzyme close to the extracellular domain of the anion transport site to form the extracellular component of a bicarbonate transport metabolon, which then increases the rate of AE-mediated bicarbonate transport [56].

Both physical and functional interaction between AE1, AE2, AE3 and transmembrane carbonic anhydrase CAIX were investigated in transfected HEK293 cells [16]. Coexpression of CAIX increased HCO_3^- flux via AE1, AE2 and AE3, respectively. Furthermore, CAIX was co-immunoprecipitated with the coexpressed AE1, AE2, and AE3. GST pull-down assays with several domain-deleted forms of CAIX revealed that the catalytic domain of CAIX mediated the interaction with AE2. CAIX can apparently bind to these Cl^-/HCO_3^- exchangers to form a bicarbonate transport metabolon.

In addition to the interaction with CAII, CAIV and CAIX, AE3 also seems to interact both physically and functionally with CAXIV [58]. In mouse brain and retinal lysates, CAXIV immunoprecipitated with anti-AE3 antibody, and both of the two AE3 isoforms, "full-length" AE3fl and "cardiac" AE3c, were immunoprecipitated using anti-CAXIV antibody, indicating CAXIV and AE3 interaction in the CNS and heart. Functional interaction between AE3fl and CAXIV was also demonstrated by pH imaging in HEK293 cells, where overexpression of CAXIV increased the rate of AE3fl-mediated HCO_3^- transport in an AZA-sensitive manner [58].

A functional interaction between AE3 and the extracellular carbonic anhydrases CAIV and CAXIV could also be shown by Svichar et al. [59] in mouse hippocampal neurons. Enhancement of the NH_4^+-induced alkalinization became apparent when extracellular CAs were inhibited by the poorly membrane-permeable CA blocker benzolamide (BZA), or by inhibitory antibodies specific for either CAIV or CAXIV. Quantitative PCR in cultured and isolated neurons showed that AE3 was the predominant AE isoform in these cells. Furthermore, NH_4^+-induced alkalinization was found to be much larger in hippocampal neurons from AE3-null mice, and BZA had no effect on the NH_4^+-induced alkalinization in these cells. It was suggested that CAIV and CAXIV both play important roles in the regulation of intracellular pH in hippocampal neurons by facilitating AE3-mediated Cl^-/HCO_3^- exchange.

Physical and functional interaction with extracellular CAIV could also be demonstrated for the Na^+/HCO_3^- cotransporter NBCe1 [49]: Transport activity of NBCe1 was determined by fluorometric pH measurements in NBCe1-transfected HEK293 cells subjected to acid loads. Cotransfection of NBCe1 with CAIV significantly increased the rate NBCe1-mediated pH_i recovery. In contrast, CAIV did not increase activity of the NBCe1-mutant $G^{767}T$ (positioned in the 4th extracellular loop). In line with the physiological findings, pull-down assays demonstrated physical binding between CAIV and a GST fusion protein of the NBCe1's 4th extracellular loop, but neither to a GST fusion protein of the 4th extracellular loop in which G^{767} was mutated to T, nor to a GST fusion protein of the transporter's 3rd extracellular loop. These data indicate that CAIV can bind to the 4th extracellular loop of NBCe1 (as it binds to the 4th extracellular loop of AE1 [56]) to form the extracellular part of a CAIV-NBCe1 transport metabolon [49].

Recently, both physical and functional interaction between NBCe1 and CAIX could be demonstrated in native rat cardiac ventricular myocytes [60]. NBCe1 co-immunoprecipitated with anti-CAIX antibody in myocyte lysate, indicating physical interaction between NBCe1 and CAIX in the myocardium. In analogy to the interaction between NBCe1 and CAIV, a GST pull-down assay demonstrated binding of CAIX to the fourth extracellular loop of the transporter, but not to the third [60]. Antibody staining of NBCe1 and CAIX further indicated colocalization of the two proteins in isolated rat ventricular myocytes. Functional interaction between NBCe1 and CAIX in cardiomyocytes was examined by monitoring NBCe1 transport activity via pH imaging using BCECF after membrane depolarization with high extracellular K^+. This depolarization protocol produced an NBCe1-mediated intracellular alkalinization which could be blocked by application of the poorly membrane-permeable CA inhibitor BZA [60]. These data indicate that extracellular CA activity is essential for NBCe1 transport function in cardiomyocytes. The specific interaction between NBCe1 and CAIX was confirmed by a parallel set of experiments in HEK293 cells: Co-immunoprecipitation in HEK cells showed that anti-CAIX antibody immunoprecipitated NBCe1 only when the cells were transfected with CAIX. Furthermore, cotransfection of CAIX increased the rate in NBCe1-mediated pH_i recovery from an acid load by about fourfold in HEK293 cells in a BZA-sensitive manner [60].

The significance of transport metabolons involving extracellular CAIX for cell function has recently been demonstrated by Svatova and coworkers in various tumor cell lines [61]. The authors investigated the role of CAIX in cell migration, which promotes the metastatic cascade in cancer cells. Cell migration itself depended on the development of polarized pH gradients with lower pericellular pH value at the cell front than at the rear end of migrating cells and a reversed intracellular pH gradient, which is presumably generated by different acid/base transporters including NBCe1 and AE2 [62]. Ectopically expressed CAIX increased migration of MDCK cells, while an inactive CAIX variant lacking the catalytic domain failed to do so. Correspondingly, hypoxic HeLa cells exhibited diminished migration upon inactivation of the endogenous CAIX either by overexpression of the catalytically inactive CAIX or by treatment with AZA [61]. Furthermore, immuno double labeling revealed that CAIX colocalizes with NBCe1 and AE2 at the leading edge of lamellipodia of migrating tumor cells. It was concluded that CAIX can form a transport metabolon with NBCe1 and AE2 and actively contributes to cell migration via its ability to facilitate ion transport and pH control at protruding processes of moving cells.

2.1.4 Controversies on Bicarbonate Transport Metabolons

While a considerable amount of data indicates a physical and functional interaction between various bicarbonate transporters and carbonic anhydrases, several studies questioned such transport metabolons. Lu et al. [63] measured the membrane conductance in NBCe1-expressing *Xenopus* oocytes, before and 1 day after injection of 300 ng CAII. The authors could not observe an increase in membrane conductance after injection of CAII compared to the same oocytes measured 1 day before injection. Furthermore, when measuring the same oocytes 1 day later, after 3 h incubation in 400 μM EZA, no reduction on membrane conductance could be observed. Even when fusing CAII to the C-terminal of NBCe1, no change in membrane conductance could be observed. In line with these experiments, Yamada et al. [64] could find no increase in the membrane current during application of CO_2/HCO_3^- when coexpressing wild-type NBCe1A or the mutant NBCe1-Δ65bp (lacking the putative CAII binding site D^{986}NDD) with CAII.

These results are in contrast to the study of Becker & Deitmer [50] and Schüler et al. [51] who could detect a robust, EZA-sensitive increase in membrane current, membrane conductance and Na^+ flux during application of CO_2/HCO_3^- in

NBCe1-expressing oocytes when catalytically active CA was either injected or co-expressed. Since the four studies had been carried out under quite similar condition in *Xenopus* oocytes, the diverging results are difficult to explain. The striking difference between the study of Lu et al. and Becker & Deitmer is that Lu et al. measured G_m on two consecutive days, were they observed a slight decrease in conductance both in the CAII-Tris- and just Tris-injected oocytes, while Becker & Deitmer compared oocytes either injected with CAII or H_2O on the same day. This might explain, why Becker & Deitmer could observe a small increase in G_m which was missing in the study of Lu et al. Furthermore Lu et al. injected 300 ng CAII/oocyte, a value 100 ng above the maximum amount of CAII Becker & Deitmer were able to inject to retain intact oocytes. In addition, Becker & Deitmer [50] not only measured conductance, but also CO_2/HCO_3^--induced membrane current, intracellular pH and Na^+ changes. In particular the changes in membrane current and intracellular Na^+ unambiguously indicated an increased NBCe1 activity, when co-expressed with CAII.

While Lu et al. only measured G_m but not I_m, Yamada et al. [64] could find no CAII-mediated increase in membrane current as did Becker & Deitmer. Yamada et al. did not inject CAII as mature protein, but coexpressed NBCe1 and CAII by injecting 5 ng of cRNA for each protein (while Becker & Deitmer used 14 ng of NBCe1 cRNA and 12 ng of CAII cRNA). Since the expression level of NBCe1 was not controlled in the oocytes, it cannot be excluded, that coexpression of CAII leads to a reduction on the expression level of NBCe1 and thereby to a reduction in CO_2/HCO_3^--induced membrane current.

The concept of a physical interaction between HCO_3^- transporters and CAII has also been challenged by a binding study of Piermarini and colleagues [65]. The authors were able to reproduce the findings of other groups [11–13, 15, 47, 48, 52] by showing that GST fusion proteins of the C-terminal tails of NBCe1, AE1 and NDCBE (SLC4A8) can bind to immobilized CAII in a micro titer plate binding assay. However, when reversing the assay, no increased binding of CAII to the immobilized GST fusion proteins could be detected as compared to the binging between CAII and GST alone [65]. Furthermore, when using pure peptides of the C-termini neither in the micro titer plate binding assay, nor by surface plasmon resonance spectroscopy an interaction with CAII could be observed [65]. It was concluded that a bicarbonate transport metabolon may exist, but that CAII might not directly bind to the transporters. These findings are in contrast to the results of several other groups which detected a direct interaction between CAII and these bicarbonate transporters, as described above [7, 11–15, 47–49, 52].

2.2 *Catalytic Interactions Between H^+ Transporters and Carbonic Anhydrase*

The Na^+/H^+ exchangers (NHEs) are integral membrane transporters that catalyze the electroneutral exchange of Na^+ (or K^+) with H^+ down their respective concentration gradients and are crucial for numerous physiological processes, such as e.g. cellular pH regulation, cell growth and proliferation, cell volume regulation and cell metabolism [66–69]. The mammalian NHEs belong to the SLC9 gene family, and nine distinct isoforms (NHE1-9) have been identified [66–68]. The isoforms share approx. 25–70 % amino acid identity and a common secondary structure of 12 transmembrane domains, a cytoplasmic N-terminal, and a more divergent cytoplasmic C-terminal domain. Under normal physiological conditions, NHEs extrude excess cytosolic acid and altered NHE activity has been linked to several diseases, including essential hypertension, congenital secretory diarrhea, diabetes, or tissue damage caused by ischemia/reperfusion, and cancer [66–68, 70].

NHE1, the only NHE isoform for which an interaction with CA has been demonstrated so far, is the 'housekeeping' isoform of the NHEs. It is ubiquitously expressed in the plasma membrane of virtually all tissues, and the only plasma membrane isoform expressed in the myocardium [68, 71–73]. NHE1 exhibits a simple Michaelis-Menten dependence on extracellular Na^+, with a K_m of 5–50 mM [74]. Additionally, NHE1 activity is highly regulated by pH, several protein interactions, phosphorylation, or signaling molecules [67, 70, 75, 76]. Under normal physiological conditions NHE1 activity is negligible, but is rapidly activated with decreasing pH_i [68, 76]. NHE1 function is fundamental for the maintenance of pH_i and cell volume thereby affecting cell growth, proliferation, migration and apoptosis [70, 77]. NHE1's pathological roles are in particular linked to kidney and heart disease, and cancer [78, 79].

Functional interaction between the Na^+/H^+ exchanger NHE1 and extracellular CAIV had first been investigated in a cell line derived from rabbit non-pigmented ephithelium [80]. By measuring intracellular pH with BCECF, the authors could demonstrate that inhibition of extracellular CA activity by a dextran-bound CA inhibitor (DBI) or AZA significantly decreased the magnitude of pH_i reduction caused by amiloride. Since these observations are consistent with a reduction in NHE activity in the presence of DBI or AZA, the authors suggested that CAIV might catalyze the rapid equilibration between H^+/HCO_3^- and CO_2 in an unstirred layer outside the plasma membrane to prevent local accumulation of H^+ and drive proton efflux via the NHE [80].

Evidence for an interaction between NHE1 and intracellular CAII was obtained by measuring the recovery from CO_2-induced acid load in the Chinese hamster ovary cell line AP1, transfected with NHE1 [81]. Cotransfection of NHE1 with CAII almost doubled the rate of pH recovery as compared to cells expressing NHE1 alone. The CAII-induced increase in NHE1 activity could be blocked by inhibition of CA catalytic activity with AZA. In contrast to CAII-WT, cotransfection with the catalytically inactive mutant CAII-V143Y even decreased the rate of pH recovery, indicating a physical interaction between NHE1 and catalytically active CAII. Physical interaction between the two proteins in vivo was demonstrated by co-immunoprecipitation of heterologously expressed NHE1 and CAII in AP1 cells [81]. Using a micro titer plate binding assay with a GST fusion protein of the NHE1 C-terminal tail, it was shown that the C-terminal region of NHE1 binds CAII, and that binding was stimulated by low pH and blocked by antibodies against the C-terminal [81]. Affinity blotting further revealed that phosphorylation of the

C-terminal of NHE1 greatly increased the binding of CAII, which lead to the suggestion that regulation of NHE1 transport activity by phosphorylation could involve modulation of CAII binding [81].

The binding mechanism was further investigated by the same group in a later study [82]. A micro titer plate binding assay with a GST fusion protein of the NHE1 C-terminal tail revealed that CAII did not bind to acidic sequences E^{753}EDEDDD of NHE1 that were similar to the CAII binding site of bicarbonate transporters, but to the penultimate group of 13 amino acids of the cytoplasmic tail (R^{790}IQRCL**SD**PGPHP), with the amino acids S^{796} and D^{797} playing an essential role in binding. Phosphorylation of the last 26 amino acids within the C-terminal of NHE1 did not alter CAII binding to this region, while phosphorylation greatly increased binding of CAII to a protein containing the last 182 amino acids of NHE1-C-terminal. It was concluded that an upstream region of the cytoplasmic tail might act as an inhibitor of CAII binding to the penultimate group of 13 amino acids [82].

2.2.1 Interaction Between SNAT3 and CAs

Another type of interaction between CAII and a proton transporter was found for the glutamine transporter SNAT3 [83]. SNAT3-mediated transport of glutamine and asparagine displays a non-stoichiometric conductance, when the transporter is heterologously expressed in *Xenopus* oocytes. This glutamine-induced membrane conductance could be suppressed by coexpression of SNAT3 with CAI, CAII, CAIII and CAIV, respectively, while no influence of CA on the asparagine-induced conductance could be observed [83, 84]. The reduction of the glutamine-induced membrane conductance by CA was dependent on the presence of CO_2/HCO_3^-, and could be reversed by blocking the CA catalytic activity with EZA. Furthermore, the catalytically inactive mutant CAII-V143Y failed to suppress the membrane conductance. Although the mechanism of the substrate specificity of the CA effect on the SNAT3-associated membrane conductance remains unresolved at this point, these findings demonstrate another important role of carbonic anhydrases for the modulation of acid/base-coupled metabolite transporters.

2.3 Non-catalytic Transport Metabolons

2.3.1 Interaction Between Different Isoforms of MCT and CA

Lactate, pyruvate, and ketone bodies are transported into and out of cells via monocarboxylate transporters (MCT, SLC16), of which 14 isoforms have been described. The first four of these 14 isoforms (MCT1-4) have been shown to transport monocarboxylates together with H^+ in a 1:1 stoichiometry [85–87]. MCT1 is the ubiquitous isoform that is found in nearly all tissues, where it operates either as a lactate/pyruvate importer or as an exporter, and has an intermediate K_m

value of 3–5 mM for L-lactate [86, 87]. MCT2 is primarily found in liver, kidney, testis, and in the brain [88, 89], and has the highest affinity for L-lactate among all MCTs with a K_m value of about 0.7 mM [90]. In liver and kidney, MCT2 facilitates the uptake of lactate, which is then partly used for glyconeogenesis [91]. In the rodent brain, MCT2 is exclusively expressed in neurons, where it facilitates the import of lactate, which is exported from astrocytes and vascular endothelial cells via MCT1 and MCT4 [92–94]. MCT3, the expression of which is restricted to retinal pigment epithelium and choroid plexus epithelia [95, 96], transports L-lactate with a K_m of about 6 mM [97]. MCT4 is a low-affinity, high-capacity carrier with a K_m value of 20–35 mM for L-lactate [98], and primarily acts as a lactate exporter in glycolytic cells and tissues like astrocytes, skeletal muscle and tumor cells [91, 99–102]. All MCTs have a classical 12 transmembrane-helix structure, with both the C- and N-terminal located intracellularly [91, 103]. Trafficking, but also regulation of transport activity of MCT1-4 is mediated by the ancillary proteins basigin (CD147) or embigin (gp70), which bind to the transporter [104, 105].

First evidence for a facilitation of lactate transport by carbonic anhydrases was presented by Wetzel et al. [106]. By measuring changes in intracellular and surface pH in rat extensor digitorum longus and soleus fibers using ion-selective microelectrodes, it was shown that inhibition of extracellular carbonic anhydrase with benzolamide and acetazolamide, respectively, lead to a significant reduction in the lactate-induced acidification, while additional inhibition of intracellular CA isoforms with chlorzolamide and ethoxzolamide, respectively, had no additional effect on lactate transport. It was concluded that extracellular sarcolemmal CA plays an important role in muscular lactate transport, while intracellular CA does not seem to have an effect on the transporter. These results have later been confirmed on mouse muscle by Hallerdei et al. [107], who could additionally show that knock-out of the extracellular CA isoforms IV, IX and XIV all lead to a reduction in lactate in- and efflux. Furthermore, immunohistochemical studies could show a co-localization of MCT4 with CAIV and CAIX in mouse skeletal muscle [107, 108]. In the mouse heart, co-localization of MCT1 with CAIV, and to a lesser extent CAXIV, could be detected [109]. However, direct binding between MCTs and extracellular CAs has not been investigated so far.

Augmentation of MCT activity by extracellular CAs has also been found in the brain: By inhibition of extracellular CA activity with benzolamide and an antiserum against CAIV, respectively, Svichar et al. [110, 111] could show a significant reduction in lactate-induced intracellular acidification in rat hippocampal pyramidal neurons and cultured astrocytes.

A potential direct interaction between MCTs and intracellular CAs has first been studied by heterologous expression of rat MCT1 in *Xenopus* oocytes injected with bovine CAII [112]. Like in the studies on muscle and brain, transport activity of MCT was determined by measuring the rate of lactate-induced acidification with ion-selective microelectrodes. The experiments revealed that injection of 50 ng CAII into the oocyte increased MCT1 transport rate by more than two-fold. However, in contrast to the observations on MCTs and extracellular CAs and the other transport metabolons described above, the CAII-induced augmentation of MCT activity persisted in the absence of CO_2/HCO_3^-, and was insensitive to inhibition of CAII catalytic activity with EZA [112]. Furthermore, co-expression of MCT1 with the catalytically inactive mutant CAII-V143Y resulted in the same augmentation of transport activity than did the co-expression of CAII-WT [113], suggesting that the augmentation of MCT activity does not depend on the reversible conversion of CO_2 and HCO_3^-/H^+ by CAII. In a later study, the same effects could also be observed for CAII and rat MCT4 [114]: Injection of CAII into MCT4 expressing *Xenopus* oocytes increased transport activity of MCT4, even in the absence of CO_2/HCO_3^- and in the presence of EZA, and after co-expression of MCT4 with CAII-V143Y. In contrast to MCT1 and MCT4, no interaction between CAII and rat MCT2 could be detected, when the enzyme was injected into oocytes co-expressing MCT2 together with its trafficking protein embigin [115].

2.3.2 Mechanism of MCT1/4 : CAII Interaction

First evidence that the non-catalytic interaction between MCT and CAII depends on a direct interaction between the two proteins was shown by injection of CAII that was bound to an antibody prior to the injection. In this experiment CAII was not able to enhance transport activity of MCT1 in *Xenopus* oocytes, suggesting a sterical suppression of the interaction by the antibody [112]. In the same study, truncation of the MCT1 C-terminal tail (MCT1-D56) led to loss of interaction between MCT1 and CAII in *Xenopus* oocytes. A pull-down assay with CAII and MCT1 showed binding between CAII and both MCT1-WT and MCT1-D56, respectively [112]. For the chloride/bicarbonate exchanger AE1, Vince and Reithmeier [12] could show that binding between CAII and the AE1 C-terminal tail is mediated by the acidic cluster D^{887}ADD in AE1. By introduction of single site mutations in the C-terminal of MCT1 and subsequent expression of these mutants in CAII-injected *Xenopus* oocytes, the two glutamate residues E^{489} and E^{491} flanking the acidic cluster E^{489}EE within the MCT1 C-terminal tail could be identified to be crucial for the functional interaction with CAII [116]. In the same study, a direct binding between CAII and the MCT1 C-terminal tail could be shown by co-immunoprecipitation when the acidic cluster E^{489}EE was intact, while mutation of E^{489} and/or E^{491} suppressed the binding between MCT1-CT and CAII. This suggests that cytosolic CAII can bind to the C-terminal tail of MCT1, which presumably positions the enzyme close enough to the pore of the transporter for efficient H^+ shuttling. The binding of CAII to a glutamic acid cluster within the MCT C-terminal may also explain the isoform specificity of the interaction between MCTs and CAII, since rat MCT4, but not MCT2, possesses a similar cluster of three glutamate residues (E^{431}EE) (Fig. 7.2c).

It can only be speculated at this point, whether there is a second MCT binding site for CAII, as suggested by Becker et al. [112] using a truncated MCT1. However, these results were not entirely conclusive, because the assay was performed with a transporter produced in an in vitro translation. Since the transporter was produced in a cell-free system in the absence of chaperones needed for proper expression, the transporter was presumably incorrectly folded. Furthermore, the transmembrane

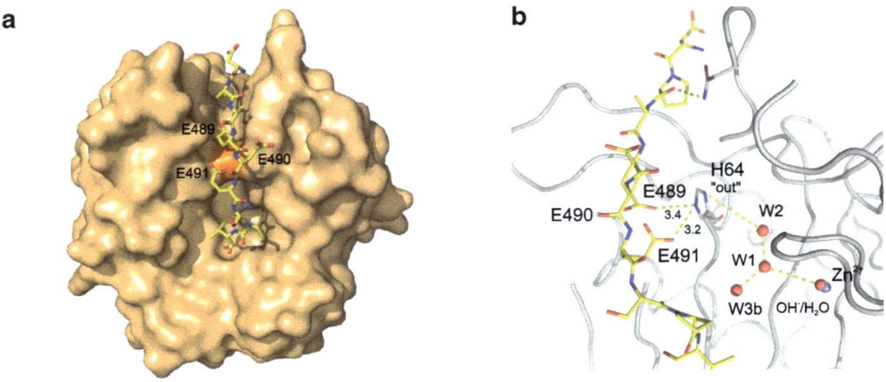

c

MCT1 -**nyrlvakeqkaeekkrdgkedetstdvdekpkktmketqspaplqnssgdpaeeespv**

MCT2 -**nyrllekerkrekarrkksasqaskemealsrskqddvtvkvsnthnppsdrdkessi**

MCT4 -**cigkrkrpevtkpeevaseeeklhkppvdvrvdsrevehflkaepekngevvhtpetsv**

Fig. 7.2 Structural model of the interaction between MCT1 and CAII. (**a**) Structural model of CAII (ochre surface representation) and the last nine amino acids from the C-terminal of MCT1 (*yellow ribbon*). (**b**) Close-up of the model structure shown in *A*. The model shows hydrogen bonds between the residues E489 and E491 of MCT1 with the residue H64 of CAII in the 'out' confirmation. (From Stridh et al. [116]) (**c**) Protein sequence of the C-termini of rat MCT1, MCT2 and MCT4. The CAII binding site in MCT1 is marked in *green*. A potential CAII binding site in MCT4 is depicted in *blue*. In MCT2, no analogous binding site can be identified

domains, which normally remain hidden within the plasma membrane, are exposed under these conditions. Therefore, such an assay is very prone to show false positive binding, which would not occur under physiological conditions. Hence, a second binding site for CAII outside of the MCT C-terminal tail cannot be excluded.

The amino acids within CAII involved in binding to the C-terminal of MCT1/4 have not yet been identified. The structural model presented by Stridh et al. [116] suggests that the last 10 amino acids of the MCT1 C-terminal tail reside within the surface groove of CAII with MCT1-E489 and E491 forming hydrogen bonds with H64 in the catalytic pocket of CAII (Fig. 7.2a, b). Another study suggests, that the N-terminal of CAII might be involved in the functional interaction, and therefore in the binding to MCT1: Vince et al. [13] showed that binding between the acidic cluster D[887]ADD in AE1 and CAII is mediated by a histidine-rich cluster within the N-terminal of CAII. The authors could show that mutation of the amino acids (H3P, H4D, K9D, H10K, H15Q, H7S) within this cluster to the analogous residues in CAI, which does not interact with AE1, suppressed binding of the CA mutant to AE1. Co-expression of this mutant with MCT1 in *Xenopus* oocytes led to no augmentation in MCT1 transport activity [113]. The same study also demonstrated that injection of up to 200 ng of CAI does not enhance transport activity of MCT1 in oocytes.

Thus, it was suggested that the histidine-rich cluster within the N-terminal of CAII may also mediate the binding to MCT1. However, although the study provides good evidence on the functional level, the exact role of this cluster for physical binding between MCT1 and CAII has not yet been elucidated. Therefore, it remains speculative whether the histidine-rich cluster in CAII mediates physical binding to the MCT1 (as it does for AE1), or whether it is involved in another mechanism underlying the CAII-induced augmentation of MCT1 transport activity.

It has been demonstrated that the enhancing effect of CAII on H^+/lactate influx via MCT1 and MCT4 increased with increasing extracellular proton concentration, but decreased with extracellular lactate concentration [113, 114, 117]. This gave rise to the hypothesis that CAII-induced augmentation of MCT transport activity is linked to the H^+ gradient across the cell membrane. Mathematical modeling of the transport mechanism suggested that CAII increases the rate constant for the binding and the release of H^+ at the transporter by providing additional H^+ binding sites, and thereby speeds up lactate/proton co-transport [117]. The question then comes up how CAII provides additional H^+ binding sites to the MCT. CAII facilitates H^+ transfer between the zinc-bound water and the solvent surrounding the enzyme by the side chain of H64, which shuttles H^+ between the bulk solvent and a network of well-ordered hydrogen-bonded water molecules in the enzyme's active-site cavity [118]. Co-expression of the mutant CAII-H64A, lacking the intramolecular H^+ shuttle, with MCT1 or MCT4 in *Xenopus* oocytes resulted in no increase in transport activity [119]. This led to the assumption that CAII may provide or subtract protons to or from the transporter, respectively, via its intramolecular H^+ shuttle. This is supported by the finding that injection of 4-methylimidazole (4-MI), which acts as a H^+ donor/acceptor, can restore the ability of the CAII-H64A mutant to enhance transport activity of MCT1/4 [119]. It is hypothezised that 4-MI binds at various positions within the active site cavity of CAII, which would then rescue proton shuttling in the enzyme [120]. In contrast to H64, the intramolecular water wire in CAII does not seem to participate in the shuttle, since various mutations, which induced changes of the water wire (CAII-Y7F, CAII-N62L, CAII-N67L), had no effect on the CAII-induced augmentation of MCT transport activity [119]. In line with this is the finding that CAIII—an isoform which lacks the intramolecular H^+ shuttle—was not able to enhance transport activity when co-expressed with MCT1 or MCT4, in *Xenopus* oocytes [119]. However, neither restoring the intramolecular H^+ shuttle in CAIII (CAIII-K64H), nor injection of 4-MI into MCT1/4 and CAIII co-expressing oocytes could enable CAIII to enhance lactate transport [119]. Therefore it has been speculated that CAIII, which also lacks the histidine-rich cluster in the N-terminal, may not bind MCT, and therefore cannot interact. Direct binding of CAIII to MCT1 or MCT4 has, however, not been investigated so far.

2.3.3 Can CAs Act as "Proton-collecting Antennae"?

It has been proposed by Martinez et al. [121] that membrane transporters, whose substrates are available at very low concentrations, need to receive their substrate from an additional source at the membrane, since these transporters would extract molecules from the cytoplasm at rates well above the capacity for simple diffusion to replenish the substrate in the immediate vicinity. In the case of MCT, which has a turnover number of 85 s^{-1} [122], this means that lactate, which has been calculated to diffuse at a rate of $5.6*10^5$ s^{-1} to the transporter [121], would be maintained in sufficient quantity to maintain maximal transport activity. For protons, however, which are highly buffered and as free H^+ only present in nanomolar concentrations, a supply rate of 5.7 s^{-1} (a rate well below the turnover number of MCT1) was calculated [121]. Based on this scenario, CAII might supply additional protons to the MCT by subtracting protons from other protonatable residues near the plasma membrane and provide them to the transporter (Fig. 7.3 a). It has indeed been shown that protonatable residues that are up to 1–1.5 nm apart from each other could form proton-attractive domains and could share the proton at a very fast rate, exceeding the upper limit of diffusion-controlled reactions [123]. When these residues are located in proteins or lipid head groups at the plasma membrane, they can collect protons from the solution and "funnel" them to the entrance of a proton-transfer pathway of a membrane-anchored protein, a process coined "H^+-collecting antenna" [124, 125]. The residues within the plasma membrane that might be involved in this proton transfer between the MCT1-CAII complex and its environment have not been investigated so far; however, first evidence for a long range proton shuttle mediated by CAII has been shown by proton imaging in *Xenopus* oocytes [113]. By measuring the change in proton concentration at the inner face of the plasma membrane at defined areas of interest during focal application of lactate, it could be demonstrated that protons travel a longer distance along the inner face of the oocytes plasma membrane when CAII is present in the cell. The idea of CAII being involved in a long range proton shuttle is also supported by the finding that augmentation of MCT1/4 transport activity is increased with rising CAII concentrations [113, 117]. This could be explained by at least two alternative mechanisms: In the first scenario, CAII in the vicinity of the membrane surface reversibly binds to MCT1, the fraction of MCT1-CAII complexes being dependent on the concentrations of both MCT1 and CAII. However, calculations by Almquist et al. [117] resulted in a half maximum saturation of MCT1 with CAII at a concentration of approximately 200 ng. For the 50 ng of CAII used in most of the experiments described so far, 15 % of the MCT1 would be estimated to be bound to CAII. Since the other 85 % would not operate in the transport enhancing mode, this would require an enormously large transport enhancement, mediated by only a small fraction of MCT1 proteins. This seems hardly compatible with the mathematical model of the CAII-induced augmentation of MCT1 transport activity, as the effect of CAII on a single MCT1 molecule saturates before reaching the amplification needed to explain the observed transport rates [117]. In a second scenario, all MCT1 binding sides would already be saturated with CAII at a concentration of 50 ng CAII/oocyte. The amount of MCT1 expressed in a single oocyte has been estimated to be 10^{10} molecules/oocyte [126], while the amount of 50 ng CAII amounts to 10^{12} molecules/oocyte. A potential mechanism by which transport enhancement could

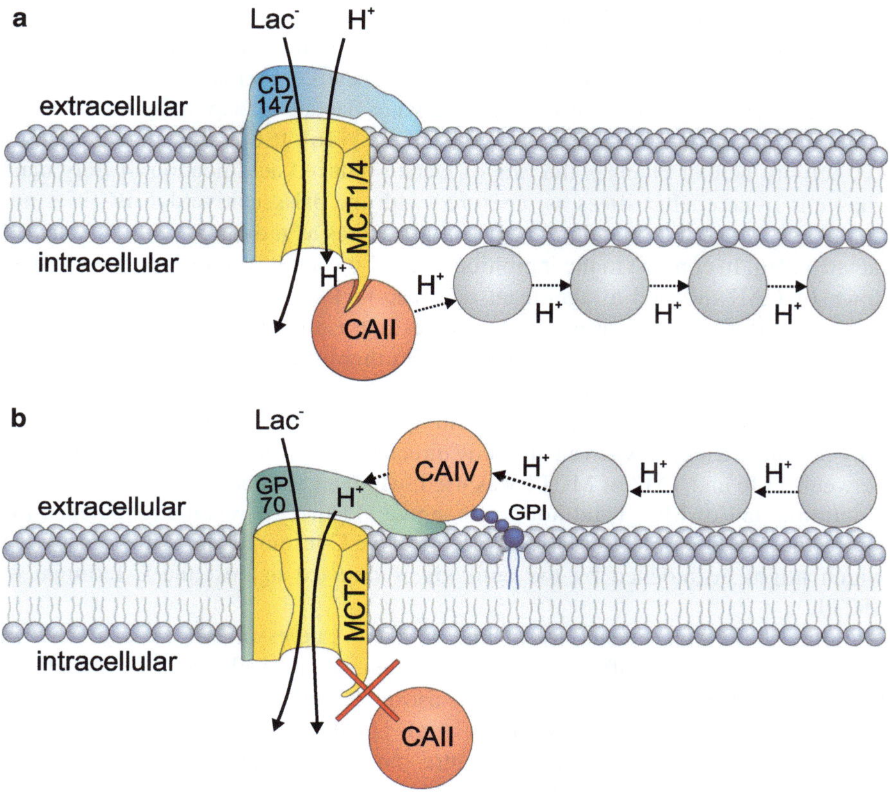

Fig. 7.3 Hypothetical model of isoform-specific interaction between MCTs and CAs. (**a**) Model of the interaction between MCT1/4 and CAII: CAII (*red circle*) binds to the C-terminal tail of MCT1/4 (*yellow structure*), which is associated to the chaperone CD147 (*green structure*). In this position, CAII can apply its intramolecular H^+ shuttle to withdraw protons from MCT1/4 and pass them to other protonatable residues near the plasma membrane (*gray circles*). (**b**) Model of the interaction between MCT2 and CAIV: CAIV (*orange circle*), anchored to the extracellular side of the plasma membrane by GPI (*dark blue structure*) is located close to the MCT2 (*yellow structure*) by direct interaction with GP70 (*green*). In this position, CAIV can apply its intramolecular H^+ shuttle to subtract H^+ from protonatable residues near the plasma membrane (*gray circle*) and pass them to the MCT2. In contrast, CAII (*red circle*) does not interact with MCT2, possibly due to a lack of a CAII-binding site in the C-terminal of MCT2, which prevents CAII from getting close enough to the transporter to establish an efficient H^+ shuttle

still depend on the CAII concentration would be that CAII molecules themselves form clusters or chains. Indeed, it has been shown that CAII can associate with the inner leaflet of the plasma membrane of human cancerous pancreatic duct cells [127], which might thus provide a high concentration of CAII molecules near the membrane. Higher concentrations of CAII would result in larger complexes of CAII connected to MCT1, which in turn also would result in more effective proton antennae [113, 117]. If such a network of CAII molecules is not restricted to the oocyte overexpression system, but would also be found within other cells, has to be explored. However, data on mouse cortical and cerebellar astrocytes suggest a non-catalytic augmentation of MCT-mediated lactate transport by CAII in the brain, in cells in culture as well as in cells *in situ* [116]. It was shown that blocking CA activity neither changes the rate of lactate-induced acidification in Bergmann glia cells in cerebellar slices as measured with the pH-sensitive dye BCECF, nor modulates lactate flux in cultured cortical astrocytes as measured with ^{14}C labeled lactate [116]. In contrast to the inhibition of catalytic activity, knock-out and knock-down of CAII induced a significant decrease in lactate flux in cerebellar slices and cultured astrocytes, respectively, suggesting a non-catalytic augmentation of lactate transport by CAII in mouse brain. Furthermore, co-localization of MCT1 and CAII, with a maximum distance of 40 nm between the two proteins, was shown in astrocyte cultures using an *in situ* proximity ligation assay. Interestingly, the assay only produced a signal when an antibody directed against the large intracellular loop between TM6 and TM7 of MCT1 was used, while an antibody directed against the C-terminal tail of MCT1 did not give a signal [116]. This result strengthened the notion that binding of CAII to MCT1 occurs at the C-terminal tail of the transporter, which might be blocked by the antibody if directed against the C-terminal of MCT1.

While these experiments are in line with the studies on MCT1 and CAII in *Xenopus* oocytes, they differ from the experiments on the influence of extracellular CAs on lactate flux, carried out in rat astrocytes by Svichar & Chesler [110]. The experiments performed by Stridh et al. [116] could not show any effect of inhibiting catalytic activity of either intra- and extracellular CAs with EZA on lactate flux, whereas Svichar & Chesler [110] found a complete block of lactate-induced acidification when extracellular CAs were inhibited with BZA. The reason for these differences remains obscure, since the two studies have been performed under quite similar condition, with the exception that the study of Svichar & Chesler [110] was performed on astrocytes from rats instead of mice. Therefore, species-dependent differences both in protein expression levels and in the microenvironment at the cell surface, including the H^+ buffer capacity, that may influence H^+ depletion at the membrane, cannot be excluded. Furthermore, Svichar & Chesler [110] used a concentration of 20 mM lactate in their experiments, which would produce higher lactate/H^+ fluxes, than the fluxes produced by application of 5 and 10 mM, respectively, used in the experiments by Stridh et al. [116]. The higher overall transport activity would therefore tend to deplete surface H^+ faster than in the case of lower lactate concentrations. The facilitation of lactate influx by a surface carbonic anhydrase is expected to depend critically on the rate of this surface H^+ depletion. A slow enough rate of influx would therefore not be influenced measurably by surface carbonic anhydrase, owing to local diffusion and replenishment of buffer, local buffer capacity, and the uncatalyzed hydration of CO_2. Hence, an influence of extracellular CA activity on lactate flux may have not been detected in the study of Stridh et al. [116], possibly due to higher intra- and/or extracellular H^+ buffer capacity.

2.3.4 Non-catalytic Interaction Between MCT2 and Extracellular CAIV

While injection of CAII does not increase transport activity of MCT2, heterologously expressed in *Xenopus* oocytes, co-expression of the extracellular isoform CAIV led to a doubling in the rate of lactate-induced acidification [115]. As already observed for MCT1/4 and CAII, the interaction between MCT2 and CAIV persisted in the nominal absence of CO_2/HCO_3^- and was insensitive to inhibition of the enzyme's catalytic activity with EZA. The non-catalytic nature of the interaction was confirmed by co-expression of MCT2 with the catalytically inactive mutant CAIV-V165Y, which increased MCT2 activity as did CAIV-WT. Interestingly, removing the intramolecular H^+-shuttle (CAIV-H88A), analogous to the proton shuttle in CAII, which had been shown to be crucial for the interaction of CAII with MCT1/4, led to a significantly smaller increase in MCT2 activity, than did CAIV-WT, but did not fully abolish the interaction with MCT2 [115]. It has been suggested that this residual enhancement could be due to a second, not yet identified, proton shuttle in CAIV [115, 128]. Furthermore, CAIV could only increase activity of MCT2 when the transporter was co-expressed with its trafficking protein embigin (gp70). From this it was concluded that the interaction between MCT2 and CAIV may not depend on a direct binding between MCT2 and CAIV, as suggested for MCT1/4 and CAII, but might be mediated by binding of CAIV to embigin, which could act as a mediator for CAIV-induced transport enhancement of MCT2 [115]. As both N- and C-terminal tails of MCT2 are located intracellularly, and the transporters' short extracellular loops might not provide enough space for binding of CAIV, the two extracellular globular domains of embigin might provide a binding site for CAIV (Fig. 7.3b). An intriguing hypothesis is that embigin mediates the functional interaction between MCT2 and CAIV, although this still needs experimental proof.

3 Conclusions

Transport metabolons between carbonic anhydrases and acid/base-coupled membrane transporters display a great versatility. Not only that most of the H^+ and HCO_3^- carrying transporters interact with one or several isoforms of carbonic anhydrase, the mode of interaction also shows some unexpected variability. Membrane transporters may interact with intracellular and extracellular carbonic anhydrases, and this interaction may be direct including physical binding, or may be mediated by chaperones associated with the transport proteins. In addition, while most of the interactions require the catalytic activity of the carbonic anhydrases, the interplay between all monocarboxylate transporters and carbonic anhydrase isoforms so far tested, is *in*dependent of the enzymatic activity. More detailed studies on the interaction between the two protein groups, transporters and carbonic anhydrase family, should help to unravel the molecular mechanisms of this important functional metabolon system.

References

1. Whitten ST, García-Moreno EB, Hilser VJ (2005) Local conformational fluctuations can modulate the coupling between proton binding and global structural transitions in proteins. Proc Natl Acad Sci U S A 102:4282–4287
2. Hulikova A, Harris AL, Vaughan-Jones RD, Swietach P (2012) Acid-extrusion from tissue: the interplay between membrane transporters and pH buffers. Curr Pharm Des 18:1331–1337
3. Srere P (1985) The metabolon. Trends Biochem Sci 14:313–314
4. Srere PA (1987) Complexes of sequential metabolic enzymes. Annu Rev Biochem 56:89–124
5. Miles EW, Rhee S, Davies DR (1999) The molecular basis of substrate channeling. J Biol Chem 274:12193–12196
6. Ovádi J (1991) Physiological significance of metabolic channelling. J Theor Biol 152:1–22
7. Kifor G, Toon MR, Janoshazi A, Solomon AK (1993) Interaction between red cell membrane band 3 and cytosolic carbonic anhydrase. J Membr Biol 134:169–179
8. Sterling D, Casey JR (2002) Bicarbonate transport proteins. Biochem Cell Biol 80:483–497
9. Knauf PA, Breuer W, McCulloch L, Rothstein A (1978) N-(4-azido-2-nitrophenyl)-2-aminoethylsulfonate (NAP-taurine) as a photoaffinity probe for identifying membrane components containing the modifier site of the human red blood cell anion exchange system. J Gen Physiol 72:631–649
10. Jennings ML (1989) Structure and function of the red blood cell anion transport protein. Annu Rev Biophys Biophys Chem 18:397–430
11. Vince JW, Reithmeier RA (1998) Carbonic anhydrase II binds to the carboxyl terminus of human band 3, the erythrocyte Cl^-/HCO_3^- exchanger. J Biol Chem 273:28430–28437
12. Vince JW, Reithmeier RA (2000) Identification of the carbonic anhydrase II binding site in the Cl(−)/HCO(3)(−) anion exchanger AE1. Biochemistry 39:5527–5533
13. Vince JW, Carlsson U, Reithmeier RA (2000) Localization of the Cl^-/HCO_3^- anion exchanger binding site to the amino-terminal region of carbonic anhydrase II. Biochemistry 39:13344–13349
14. Casey JR, Pirraglia C, Reithmeier R (1992) Enzymatic deglycosylation of human band 3, the anion transport protein of the erythrocyte membrane. Effect on protein structure and transport properties. J Biol Chem 267:11940–11948
15. Sterling D, Reithmeier RA, Casey JR (2001) A transport metabolon. Functional interaction of carbonic anhydrase II and chloride/bicarbonate exchangers. J Biol Chem 276:47886–47894
16. Morgan PE, Pastoreková S, Stuart-Tilley AK, Alper SL, Casey JR (2007) Interactions of transmembrane carbonic anhydrase, CAIX, with bicarbonate transporters. Am J Physiol Cell Physiol 293:738–748
17. Dahl NK, Jiang L, Chernova MN, Stuart-Tilley AK, Shmukler BE, Alper SL (2003) Deficient HCO_3^- transport in an AE1 mutant with normal Cl- transport can be rescued by carbonic anhydrase II presented on an adjacent AE1 protomer. J Biol Chem 278:44949–44958
18. Jennings ML (1984) Oligomeric structure and the anion transport function of human erythrocyte band 3 protein. J Membr Biol 80:105–117
19. Scozzafava A, Supuran CT (2002) Carbonic anhydrase activators: human isozyme II is strongly activated by oligopeptides incorporating the carboxyterminal sequence of the bicarbonate anion exchanger AE1. Bioorg Med Chem Lett 12:1177–1180
20. Kere J (2006) Overview of the SLC26 family and associated diseases. Novartis Found Symp 273:2–11
21. Dorwart MR, Shcheynikov N, Wang Y, Stippec S, Muallem S (2007) SLC26A9 is a Cl(−) channel regulated by the WNK kinases. J Physiol 584:333–345
22. Mount DB, Romero MF (2004) The SLC26 gene family of multifunctional anion exchangers. Pflügers Archiv Eur J Physiol 447:710–721
23. Casey JR, Grinstein S, Orlowski J (2010) Sensors and regulators of intracellular pH. Nat Rev Mol Cell Biol 11:50–61

24. Ohana E, Shcheynikov N, Park M, Muallem S (2012) Solute carrier family 26 member a2 (Slc26a2) protein functions as an electroneutral $SO_4^{2-}/OH^-/Cl^-$ exchanger regulated by extracellular Cl^-. J Biol Chem 287:5122–5132
25. Sterling D, Brown NJD, Supuran CT, Casey JR (2002) The functional and physical relationship between the DRA bicarbonate transporter and carbonic anhydrase II. Am J Physiol Cell Physiol 283:1522–1529
26. Alvarez BV, Vilas GL, Casey JR (2005) Metabolon disruption: a mechanism that regulates bicarbonate transport. EMBO J 24:2499–2511
27. Soleimani M, Burnham CE (2001) Na^+:HCO($_3{}^-$) cotransporters (NBC): cloning and characterization. J Membr Biol 183:71–84
28. Romero MF, Fulton CM, Boron WF (2004) The SLC4 family of HCO_3^- transporters. Pflügers Archiv Eur J Physiol 447:495–509
29. Abuladze N, Lee I, Newman D (1998) Molecular cloning, chromosomal localization, tissue distribution, and functional expression of the human pancreatic sodium bicarbonate cotransporter. J Biol Chem 273:17689–17695
30. Choi I, Romero MF, Khandoudi N, Bril A, Boron WF (1999) Cloning and characterization of a human electrogenic Na^+-$HCO^-{}_3$ cotransporter isoform (hhNBC). Am J Physiol 276:576–584
31. Abuladze N, Song M, Pushkin A, Newman D, Lee I, Nicholas S, Kurtz I (2000) Structural organization of the human NBC1 gene: kNBC1 is transcribed from an alternative promoter in intron 3. Gene 251:109–122
32. McAlear SD, Liu X, Williams JB, McNicholas-Bevensee CM, Bevensee MO (2006) Electrogenic Na/HCO3 cotransporter (NBCe1) variants expressed in *Xenopus* oocytes: functional comparison and roles of the amino and carboxy termini. J Gen Physiol 127:639–658
33. Soleimani M, Grassi S, Aronson P (1987) Stoichiometry of Na^+-$HCO^-{}_3$ cotransport in basolateral membrane vesicles isolated from rabbit renal cortex. J Clin Invest 79:1276–1280
34. Deitmer JW, Schlue W (1989) An inwardly directed electrogenic sodium-bicarbonate cotransport in leech glial cells. J Physiol 411:179–194
35. Bevensee MO, Weed RA, Boron WF (1997) Intracellular pH regulation in cultured astrocytes from rat hippocampus. I. Role of HCO_3^-. J Gen Physiol 110:453–465
36. Bevensee MO, Apkon M, Boron WF (1997) Intracellular pH regulation in cultured astrocytes from rat hippocampus. II. Electrogenic Na/HCO_3 cotransport. J Gen Physiol 110:467–483
37. Aiello EA, Petroff MG, Mattiazzi AR, Cingolani HE (1998) Evidence for an electrogenic Na^+-HCO_3^- symport in rat cardiac myocytes. J Physiol 512:137–148
38. Gross E, Hawkins K, Abuladze N, Pushkin A, Cotton CU, Hopfer U, Kurtz I (2001) The stoichiometry of the electrogenic sodium bicarbonate cotransporter NBC1 is cell-type dependent. J Physiol 531:597–603
39. Gross E, Hawkins K, Pushkin A, Sassani P, Dukkipati R, Abuladze N, Hopfer U, Kurtz I (2001) Phosphorylation of Ser(982) in the sodium bicarbonate cotransporter kNBC1 shifts the HCO(3)(−) : Na(+) stoichiometry from 3 : 1 to 2 : 1 in murine proximal tubule cells. J Physiol 537:659–665
40. Gross E, Fedotoff O, Pushkin A, Abuladze N, Newman D, Kurtz I (2003) Phosphorylation-induced modulation of pNBC1 function: distinct roles for the amino- and carboxy-termini. J Physiol 549:673–682
41. Ko SBH, Luo X, Hager H, Rojek A, Choi JY, Licht C, Suzuki M, Muallem S, Nielsen S, Ishibashi K (2002) AE4 is a DIDS-sensitive Cl(−)/HCO(−)(3) exchanger in the basolateral membrane of the renal CCD and the SMG duct. Am J Physiol Cell Physiol 283:1206–1218
42. Pushkin A, Kurtz I (2006) SLC4 base (HCO_3^-, CO_3^{2-}) transporters: classification, function, structure, genetic diseases, and knockout models. Am J Physiol Renal Physiol 290:580–599
43. Alper SL (2009) Molecular physiology and genetics of Na^+-independent SLC4 anion exchangers. J Exp Biol 212:1672–1683
44. Burg M, Green N (1977) Bicarbonate transport by isolated perfused rabbit proximal convoluted tubules. Am J Physiol 233:307–314

45. Sasaki S, Marumo F (1989) Effects of carbonic anhydrase inhibitors on basolateral base transport of rabbit proximal straight tubule. Am J Physiol 257:947–952
46. Seki G, Frömter E (1992) Acetazolamide inhibition of basolateral base exit in rabbit renal proximal tubule S2 segment. Pflügers Archiv Eur J Physiol 422:60–65
47. Gross E, Pushkin A, Abuladze N, Fedotoff O, Kurtz I (2002) Regulation of the sodium bicarbonate cotransporter kNBC1 function: role of Asp986, Asp988 and kNBC1-carbonic anhydrase II binding. J Physiol 544:679–685
48. Pushkin A, Abuladze N, Gross E, Newman D, Tatishchev S, Lee I, Fedotoff O, Bondar G, Azimov R, Ngyuen M, Kurtz I (2004) Molecular mechanism of kNBC1-carbonic anhydrase II interaction in proximal tubule cells. J Physiol 559:55–65
49. Alvarez BV, Loiselle FB, Supuran CT, Schwartz GJ, Casey JR (2003) Direct extracellular interaction between carbonic anhydrase IV and the human NBC1 sodium/bicarbonate co-transporter. Biochemistry 42:12321–12329
50. Becker HM, Deitmer JW (2007) Carbonic anhydrase II increases the activity of the human electrogenic Na^+/HCO_3^- cotransporter. J Biol Chem 282:13508–13521
51. Schueler C, Becker HM, McKenna R, Deitmer JW (2011) Transport activity of the sodium bicarbonate cotransporter NBCe1 is enhanced by different isoforms of carbonic anhydrase. PLoS One 6:e27167
52. Loiselle FB, Morgan PE, Alvarez BV, Casey JR (2004) Regulation of the human NBC3 Na^+/HCO_3^- cotransporter by carbonic anhydrase II and PKA. Am J Physiol Cell Physiol 286:1423–1433
53. Ro H, Carson JH (2004) pH microdomains in oligodendrocytes. J Biol Chem 279:37115–37123
54. Boussouf A, Lambert R, Gaillard S (1997) Voltage-dependent Na^+/HCO_3^- cotransporter and Na^+/H^+ exchanger are involved in intracellular pH regulation of cultured mature rat cerebellar oligodendrocytes. Glia 19:74–84
55. Boussouf A, Gaillard S (2000) Intracellular pH changes during oligodendrocyte differentiation in primary culture. J Neurosci Res 59:731–739
56. Sterling D, Alvarez BV, Casey JR (2002) The extracellular component of a transport metabolon. Extracellular loop 4 of the human AE1 Cl^-/HCO_3^- exchanger binds carbonic anhydrase IV. J Biol Chem 277:25239–25246
57. Tang XB, Fujinaga J, Kopito R, Casey JR (1998) Topology of the region surrounding Glu681 of human AE1 protein, the erythrocyte anion exchanger. J Biol Chem 273:22545–22553
58. Casey JR, Sly WS, Shah GN, Alvarez BV (2009) Bicarbonate homeostasis in excitable tissues: role of AE3 Cl^-/HCO_3^- exchanger and carbonic anhydrase XIV interaction. Am J Physiol Cell Physiol 297:1091–1102
59. Svichar N, Waheed A, Sly WS, Hennings JC, Hübner CA, Chesler M (2009) Carbonic anhydrases CA4 and CA14 both enhance AE3-mediated Cl^--HCO_3- exchange in hippocampal neurons. J Neurosci 29:3252–3258
60. Orlowski A, De Giusti VC, Morgan PE, Aiello EA, Alvarez BV (2012) Binding of carbonic anhydrase IX to extracellular loop 4 of the NBCe1 Na^+/HCO_3^- cotransporter enhances NBCe1-mediated HCO_3^- influx in the rat heart. Am J Physiol Cell Physiol 303:69–80
61. Svastova E, Witarski W, Csaderova L, Kosik I, Skvarkova L, Hulikova A, Zatovicova M, Barathova M, Kopacek J, Pastorek J, Pastorekova S (2012) Carbonic anhydrase IX interacts with bicarbonate transporters in lamellipodia and increases cell migration via its catalytic domain. J Biol Chem 287:3392–33402
62. Stock C, Schwab A (2009) Protons make tumor cells move like clockwork. Pflügers Archiv Eur J Physiol 458:981–992
63. Lu J, Daly CM, Parker MD, Gill HS, Piermarini PM, Pelletier MF, Boron WF (2006) Effect of human carbonic anhydrase II on the activity of the human electrogenic Na/HCO_3 cotransporter NBCe1-A in *Xenopus* oocytes. J Biol Chem 281:19241–19250
64. Yamada H, Horita S, Suzuki M, Fujita T, Seki G (2011) Functional role of a putative carbonic anhydrase II-binding domain in the electrogenic Na^+-HCO_3^- cotransporter NBCe1 expressed in *Xenopus* oocytes. Channels 5:106–109

65. Piermarini PM, Kim EY, Boron WF (2007) Evidence against a direct interaction between intracellular carbonic anhydrase II and pure C-terminal domains of SLC4 bicarbonate transporters. J Biol Chem 282:1409–1421
66. Malo ME, Fliegel L (2006) Physiological role and regulation of the Na^+/H^+ exchanger. Can J Physiol Pharmacol 84:1081–1095
67. Slepkov ER, Rainey JK, Sykes BD, Fliegel L (2007) Structural and functional analysis of the Na^+/H^+ exchanger. Biochem J 401:623–633
68. Fliegel L (2009) Regulation of the Na(+)/H(+) exchanger in the healthy and diseased myocardium. Expert Opin Ther Targets 13:55–68
69. Ohgaki R, Van IJzendoorn SCD, Matsushita M, Hoekstra D, Kanazawa H (2011) Organellar Na^+/H^+ exchangers: novel players in organelle pH regulation and their emerging functions. Biochemistry 50:443–450
70. Orlowski J, Grinstein S (2004) Diversity of the mammalian sodium/proton exchanger SLC9 gene family. Pflügers Archiv Eur J Physiol 447:549–565
71. Fliegel L, Dyck JR, Wang H, Fong C, Haworth RS (1993) Cloning and analysis of the human myocardial Na^+/H^+ exchanger. Mol Cell Biochem 125:137–143
72. Fliegel L, Sardet C, Pouyssegur J, Barr A (1991) Identification of the protein and cDNA of the cardiac Na^+/H^+ exchanger. FEBS Lett 279:25–29
73. Kemp G, Young H, Fliegel L (2008) Structure and function of the human Na^+/H^+ exchanger isoform 1. Channels 2:329–336
74. Pedersen SF, O'Donnell ME, Anderson SE, Cala PM (2006) Physiology and pathophysiology of Na^+/H^+ exchange and Na^+ -K^+ -$2Cl^-$ cotransport in the heart, brain, and blood. Am J Physiol Regul Integr Comp Physiol 291:1–25
75. Karmazyn M, Moffat MP (1993) Role of Na^+/H^+ exchange in cardiac physiology and pathophysiology: mediation of myocardial reperfusion injury by the pH paradox. Cardiovasc Res 27:915–924
76. Wallert MA, Fröhlich O (1989) Na^+-H^+ exchange in isolated myocytes from adult rat heart. Am J Physiol 257:207–213
77. Denker SP, Huang DC, Orlowski J, Furthmayr H, Barber DL (2000) Direct binding of the Na–H exchanger NHE1 to ERM proteins regulates the cortical cytoskeleton and cell shape independently of H(+) translocation. Mol Cell 6:1425–1436
78. Avkiran M, Marber MS (2002) Na(+)/H(+) exchange inhibitors for cardioprotective therapy: progress, problems and prospects. J Am Coll Cardiol 39:747–753
79. Cardone RA, Casavola V, Reshkin SJ (2005) The role of disturbed pH dynamics and the Na^+/H^+ exchanger in metastasis. Nat Rev Cancer 5:786–795
80. Wu Q, Pierce WM, Delamere NA (1998) Cytoplasmic pH responses to carbonic anhydrase inhibitors in cultured rabbit nonpigmented ciliary epithelium. J Membr Biol 162:31–38
81. Li X, Alvarez BV, Casey JR, Reithmeier RAF, Fliegel L (2002) Carbonic anhydrase II binds to and enhances activity of the Na^+/H^+ exchanger. J Biol Chem 277:36085–36091
82. Li X, Liu Y, Alvarez BV, Casey JR, Fliegel L (2006) A novel carbonic anhydrase II binding site regulates NHE1 activity. Biochemistry 45:2414–2424
83. Weise A, Becker HM, Deitmer JW (2007) Enzymatic suppression of the membrane conductance associated with the glutamine transporter SNAT3 expressed in *Xenopus* oocytes by carbonic anhydrase II. J Gen Physiol 130:203–215
84. Weise A, Schneider H, McKenna R, Deitmer JW (2011) Substrate-dependent interference of carbonic anhydrases with the glutamine transporter SNAT3-induced conductance. Cell Physiol Biochem 27:79–90
85. Carpenter L, Halestrap AP (1994) The kinetics, substrate and inhibitor specificity of the lactate transporter of Ehrlich-Lettre tumour cells studied with the intracellular pH indicator BCECF. Biochem J 304:751–760
86. Bröer S, Rahman B, Pellegri G, Pellerin L, Martin JL, Verleysdonk S, Hamprecht B, Magistretti PJ (1997) Comparison of lactate transport in astroglial cells and monocarboxylate transporter 1 (MCT 1) expressing *Xenopus laevis* oocytes. Expression of two different monocarboxylate transporters in astroglial cells and neurons. J Biol Chem 272:30096–30102

87. Bröer S, Schneider HP, Bröer A, Deitmer JW (1998) Characterization of the monocarboxylate transporter 1 expressed in *Xenopus laevis* oocytes by changes in cytosolic pH. Biochem J 174:167–174
88. Garcia CK, Brown MS, Pathak RK, Goldstein JL (1995) cDNA cloning of MCT2, a second monocarboxylate transporter expressed in different cells than MCT1. J Biol Chem 270:1843–1849
89. Jackson VN, Price NT, Carpenter L, Halestrap AP (1997) Cloning of the monocarboxylate transporter isoform MCT2 from rat testis provides evidence that expression in tissues is species-specific and may involve post-transcriptional regulation. Biochem J 324:447–453
90. Bröer S, Bröer A, Schneider HP, Stegen C, Halestrap AP, Deitmer JW (1999) Characterization of the high-affinity monocarboxylate transporter MCT2 in *Xenopus laevis* oocytes. Biochem J 535:529–535
91. Halestrap AP, Price NT (1999) The proton-linked monocarboxylate transporter (MCT) family: structure, function and regulation. Biochem J 343:281–299
92. Pellerin L, Pellegri G, Bittar PG, Charnay Y, Bouras C, Martin JL, Stella N, Magistretti PJ (1998) Evidence supporting the existence of an activity-dependent astrocyte-neuron lactate shuttle. Dev Neurosci 20:291–299
93. Bergersen L, Wærhaug O, Helm J, Thomas M, Laake P, Davies AJ, Wilson MC, Halestrap AP, Ottersen OP (2001) A novel postsynaptic density protein: the monocarboxylate transporter MCT2 is co-localized with δ-glutamate receptors in postsynaptic densities of parallel fiber-Purkinje cell synapses. Exp Brain Res 136:523–534
94. Pierre K, Pellerin L (2005) Monocarboxylate transporters in the central nervous system: distribution, regulation and function. J Neurochem 94:1–14
95. Yoon H, Fanelli A, Grollman EF, Philp NJ (1997) Identification of a unique monocarboxylate transporter (MCT3) in retinal pigment epithelium. Biochem Biophys Res Commun 234:90–94
96. Philp NJ, Yoon H, Lombardi L (2001) Mouse MCT3 gene is expressed preferentially in retinal pigment and choroid plexus epithelia. Am J Physiol Cell Physiol 280:1319–1326
97. Grollman EF, Philp NJ, McPhie P, Ward RD, Sauer B (2000) Determination of transport kinetics of chick MCT3 monocarboxylate transporter from retinal pigment epithelium by expression in genetically modified yeast. Biochemistry 39:9351–9357
98. Dimmer KS, Friedrich B, Lang F, Deitmer JW, Bröer S (2000) The low-affinity monocarboxylate transporter MCT4 is adapted to the export of lactate in highly glycolytic cells. Biochem J 350:219–227
99. Wilson MC, Jackson VN, Heddle C, Price NT, Pilegaard H, Juel C, Bonen A, Montgomery I, Hutter OF, Halestrap AP (1998) Lactic acid efflux from white skeletal muscle is catalyzed by the monocarboxylate transporter isoform MCT3. J Biol Chem 273:15920–15926
100. Pinheiro C, Reis RM, Ricardo S, Longatto-Filho A, Schmitt F, Baltazar F (2010) Expression of monocarboxylate transporters 1, 2, and 4 in human tumours and their association with CD147 and CD44. J Biomed Biotechnol 2010:427694
101. Halestrap AP, Wilson MC (2012) The monocarboxylate transporter family–role and regulation. IUBMB Life 64:109–119
102. Pinheiro C, Longatto-Filho A, Azevedo-Silva J, Casal M, Schmitt FC, Baltazar F (2012) Role of monocarboxylate transporters in human cancers: state of the art. J Bioenerg Biomembr 44:127–139
103. Halestrap AP (2012) The monocarboxylate transporter family–structure and functional characterization. IUBMB Life 64:1–9
104. Wilson MC, Meredith D, Fox JEM, Manoharan C, Davies AJ, Halestrap AP (2005) Basigin (CD147) is the target for organomercurial inhibition of monocarboxylate transporter isoforms 1 and 4: the ancillary protein for the insensitive MCT2 is EMBIGIN (gp70). J Biol Chem 280:27213–27221
105. Wilson MC, Meredith D, Bunnun C, Sessions RB, Halestrap AP (2009) Studies on the DIDS-binding site of monocarboxylate transporter 1 suggest a homology model of the open conformation and a plausible translocation cycle. J Biol Chem 284:20011–20021

106. Wetzel P, Hasse A, Papadopoulos S, Voipio J, Kaila K, Gros G (2001) Extracellular carbonic anhydrase activity facilitates lactic acid transport in rat skeletal muscle fibres. J Physiol 531:743–756
107. Hallerdei J, Scheibe RJ, Parkkila S, Waheed A, Sly WS, Gros G, Wetzel P, Endeward V (2010) T tubules and surface membranes provide equally effective pathways of carbonic anhydrase-facilitated lactic acid transport in skeletal muscle. PLoS One 5:e15137
108. Scheibe RJ, Mundhenk K, Becker T, Hallerdei J, Waheed A, Shah GN, Sly WS, Gros G, Wetzel P (2008) Carbonic anhydrases IV and IX: subcellular localization and functional role in mouse skeletal muscle. Am J Physiol Cell Physiol 294:402–412
109. Scheibe RJ, Gros G, Parkkila S, Waheed A, Grubb JH, Shah GN, Sly WS, Wetzel P (2006) Expression of membrane-bound carbonic anhydrases IV, IX, and XIV in the mouse heart. J Histochem Cytochem 54:1379–1391
110. Svichar N, Chesler M (2003) Surface carbonic anhydrase activity on astrocytes and neurons facilitates lactate transport. Glia 41:415–419
111. Svichar N, Esquenazi S, Waheed A, Sly WS, Chesler M (2006) Functional demonstration of surface carbonic anhydrase IV activity on rat astrocytes. Glia 53:241–247
112. Becker HM, Hirnet D, Fecher-Trost C, Sültemeyer D, Deitmer JW (2005) Transport activity of MCT1 expressed in *Xenopus* oocytes is increased by interaction with carbonic anhydrase. J Biol Chem 280:39882–39889
113. Becker HM, Deitmer JW (2008) Nonenzymatic proton handling by carbonic anhydrase II during H^+-lactate cotransport via monocarboxylate transporter 1. J Biol Chem 283:21655–21667
114. Becker HM, Klier M, Deitmer JW (2010) Nonenzymatic augmentation of lactate transport via monocarboxylate transporter isoform 4 by carbonic anhydrase II. J Membr Biol 234:125–135
115. Klier M, Schüler C, Halestrap AP, Sly WS, Deitmer JW, Becker HM (2011) Transport activity of the high-affinity monocarboxylate transporter MCT2 is enhanced by extracellular carbonic anhydrase IV but not by intracellular carbonic anhydrase II. J Biol Chem 286:27781–27791
116. Stridh MH, Alt MD, Wittmann S, Heidtmann H, Aggarwal M, Riederer B, Seidler U, Wennemuth G, McKenna R, Deitmer JW, Becker HM (2012) Lactate flux in astrocytes is enhanced by a non-catalytic action of carbonic anhydrase II. J Physiol 590:2333–2351
117. Almquist J, Lang P, Prätzel-Wolters D, Deitmer JW, Jirstrand M, Becker HM (2010) A Kinetic Model of the Monocarboxylate Transporter MCT1 and its Interaction with Carbonic Anhydrase II. J Comput Sci Syst Biol 3:107–116
118. Fisher SZ, Maupin CM, Budayova-Spano M, Govindasamy L, Tu C, Agbandje-McKenna M, Silverman DN, Voth GA, McKenna R (2007) Atomic crystal and molecular dynamics simulation structures of human carbonic anhydrase II: insights into the proton transfer mechanism. Biochemistry 46:2930–2937
119. Becker HM, Klier M, Schüler C, McKenna R, Deitmer JW (2011) Intramolecular proton shuttle supports not only catalytic but also noncatalytic function of carbonic anhydrase II. Proc Natl Acad Sci U S A 108:3071–3076
120. Duda D, Tu C, Qian M, Laipis P, Agbandje-McKenna M, Silverman DN, McKenna R (2001) Structural and kinetic analysis of the chemical rescue of the proton transfer function of carbonic anhydrase II. Biochemistry 40:1741–1748
121. Martínez C, Kalise D, Barros LF (2010) General requirement for harvesting antennae at ca and h channels and transporters. Front Neuroenerg 2:1–8
122. Ovens MJ, Davies AJ, Wilson MC, Murray CM, Halestrap AP (2010) AR-C155858 is a potent inhibitor of monocarboxylate transporters MCT1 and MCT2 that binds to an intracellular site involving transmembrane helices 7–10. Biochem J 425:523–530
123. Gutman M, Nachliel E, Friedman R (2006) The dynamics of proton transfer between adjacent sites. Photochem Photobiol Sci 5:531–537
124. Adelroth P, Brzezinski P (2004) Surface-mediated proton-transfer reactions in membrane-bound proteins. Biochim Biophys Acta 1655:102–115
125. Brändén M, Sandén T, Brzezinski P, Widengren J (2006) Localized proton microcircuits at the biological membrane-water interface. Proc Natl Acad Sci U S A 103:19766–19770

126. Zampighi GA, Kreman M, Boorer KJ, Loo DD, Bezanilla F, Chandy G, Hall JE, Wright EM (1995) A method for determining the unitary functional capacity of cloned channels and transporters expressed in *Xenopus laevis* oocytes. J Membr Biol 148:65–78
127. Alvarez L, Fanjul M, Carter N, Hollande E (2001) Carbonic anhydrase II associated with plasma membrane in a human pancreatic duct cell line (CAPAN-1). J Histochem Cytochem 49:1045–1053
128. Hurt JD, Tu C, Laipis PJ, Silverman DN (1997) Catalytic properties of murine carbonic anhydrase IV. J Biol Chem 272:13512–13518

Chapter 8
Carbonic Anhydrase Related Proteins: Molecular Biology and Evolution

Ashok Aspatwar, Martti E.E. Tolvanen, Csaba Ortutay, and Seppo Parkkila

Abstract The catalytically inactive isoforms of α-carbonic anhydrases are known as carbonic anhydrase related proteins (CARPs). The CARPs occur independently or as domains of other proteins in animals (both vertebrates and invertebrates) and viruses. The catalytic inactivity of CARPs is due to the lack of histidine residues required for the coordination of the zinc atom. The phylogenetic analysis shows that these proteins are highly conserved across the species. The three CARPs in vertebrates are known as CARP VIII, X and XI. CARPs orthologous to CARP VIII are found in deuterostome invertebrates, whereas protostomes only possess orthologs of CARP X. The CA-like domains of receptor-type protein tyrosine phosphatases (PTPR) are found only in PTPRG and PTPRZ. Most of these CARPs are predominantly expressed in central nervous system. Among the three vertebrate CA isoforms, CARP VIII is functionally associated with motor coordination in human, mouse and zebrafish and certain types of cancers in humans. Vertebrate expression studies show that CARP X is exclusively expressed in the brain. CARP XI is only found in tetrapods and is highly expressed in the central nervous system (CNS) of humans and mice and is also associated with several cancers. CARP VIII, PTPRZ and PTPRG have been shown to coordinate the function of other proteins by protein-protein interaction, and viral CARPs participate in attachment to host

---

Susan C. Frost and Robert McKenna (eds.). Carbonic Anhydrase: Mechanism, Regulation, Links to Disease, and Industrial Applications

A. Aspatwar (✉) • S. Parkkila
Institute of Biomedical Technology and School of Medicine, University of Tampere and BioMediTech, Tampere, Finland
e-mail: ashok.aspatwar@uta.fi; seppo.parkkila@uta.fi

M.E.E. Tolvanen • C. Ortutay
Institute of Biomedical Technology, University of Tampere and BioMediTech, Tampere, Finland
e-mail: martti.tolvanen@uta.fi; csaba.ortutay@uta.fi

S.C. Frost and R. McKenna (eds.), *Carbonic Anhydrase: Mechanism, Regulation, Links to Disease, and Industrial Applications*, Subcellular Biochemistry 75,
DOI 10.1007/978-94-007-7359-2_8,

cells, but the precise biological function of CARPs X and XI is still unknown. The findings so far suggest many novel functions for the CARP subfamily, most likely related to binding to other proteins.

Keywords Expression • Cerebellar-ataxia • Phylogeny • Cancer • Bioinformatics • Cell adhesion

1 Introduction

The α-carbonic anhydrases (CAs), EC 4.2.1.1, are zinc containing metalloenzymes that catalyze the reversible hydration of CO_2, which is relevant in many important biological functions, such as photosynthesis, respiration, renal tubular acidification and bone resorption [1–4]. With the progress in genome sequencing of different organisms the number of CA and CA like proteins has been increasing constantly. Recently, we have reported several duplications in zebrafish CAs, and consequently the number of α-CAs has increased up to 17 [5]. The CA isozymes are categorized into different groups depending on the subcellular localization and enzymatic activity.

Apart from the enzymatically active CAs, there are CA isoforms which are evolutionarily well conserved with a sequence and structural similarity to active CAs, but which lack the classical enzymatic activity of active CAs. The inactive CA isoforms either occur independently as carbonic anhydrase related proteins (CARPs) or as domains of other proteins. The catalytic inactivity is due to lack of one or more of the three histidine residues which are necessary for co-ordination of zinc metal ion in the active site. The enzymatic activity of mammalian CARP isoforms (CARP VIII, X, and XI) can be regained by restoration of these histidines [6, 7]. In the family of protein tyrosine phosphatases (PTPs), there are two receptor-type protein tyrosine phosphatases, PTPR zeta (ζ) and PTPR gamma (γ), that contain an N-terminal CA-like domain [8, 9], termed "CARP XVI domain" [5, 10].

In addition to the above acatalytic CAs and CA-like domains found in vertebrates, CARP X like sequences have been identified in invertebrates, including groups as primitive and as distinct as insects and nematodes [10]. Furthermore, invertebrate genomes contain other, unrelated seemingly acatalytic CAs (Ortutay C and Tolvanen MEE, unpublished), and the D8 surface antigen in vaccinia viruses has been known to contain a CA-like domain since long time [11, 12].

The biological functions of CARPs and CA domains have started emerging recently. However, the precise physiological roles of these proteins are poorly understood till date. In the present chapter, we chronicle recent data on CARPs and CARP-like domains.

2 Carbonic Anhydrase Related Protein VIII

The CARP VIII is sequentially and structurally highly similar to catalytically active CAs. The catalytic inactivity of the CARP VIII is due to substitution of the first of the three histidine residues required for co-ordination of zinc atom in the active site (by Arg in all vertebrates). CARP VIII seems universal in vertebrates, and is also found in a number of deuterostome invertebrate species, and the sequence is highly conserved across the species [13]. A recent report [14] describes a CARP VIII sequence in pearl oyster, *Pinctada fucata*, which is a protostome, indicating that the evolutionary origin of CARP VIII might be much earlier than previously thought.

Among the catalytically inactive proteins of α-CA gene family, CARP VIII was the first to be reported based on its expression pattern in mouse brain in 1990s by Kato [15]. Subsequently, several studies have been done on CARP VIII, which include expression pattern during development and in adult tissues both in human and mouse and in pathological conditions [16–20] The structure of CARP VIII has been resolved and the basis of its inactivity has been studied [13, 21]. Its association with ataxia and neurodegeneration both in human and mouse has started emerging recently from several independent studies [22–25]. The previously reported interaction of CARP VIII with ITPR1 [26] provides a plausible mechanism for these cerebellar disorders. CARP VIII is known to be upregulated particularly in lung cancer and several others cancers [27–29]. However, the precise mechanism of how the regulation of ITPR1 leads to the biological effects is still not known. Very recently, we have developed an ataxic zebrafish model by knocking down *CA8* gene using antisense morpholino oligonucleotides which may help in elucidating the function of CARP VIII [30].

2.1 *Molecular Properties and Bioinformatics of CARP VIII*

Human CARP VIII is encoded by nine exons, and the exon boundaries in *CA8* have been maintained in the same locations through deuterostome evolution . The gene product consists of 290 amino acids with about 40 % of amino acid identity with other mammalian CA paralogs [31]. The N terminus of CARP VIII contains a unique feature among CAs, an acidic region. In mammals, the acidic region is considerably longer than in the orthologs in non-mammalian vertebrates. With the exclusion of glutamic acid stretch in mammalian *CA8*, the amino acid identity between human and zebrafish CARP VIII is 84 %. Such a high degree conservation between CARP VIII sequences of distant vertebrate species speaks for a conserved, essential function. An exhaustive review on CARP VIII has been published recently, describing the data related to molecular biology, expression in healthy tissues and disease conditions and role in human diseases [32]. The data related to basis of inactivity and sequence and phylogenetic analysis has also been published in the recent past [13, 33].

Interaction of CARP VIII with inositol 1,4,5-trisphosphate receptor type 1 (ITPR1), an ion channel protein that regulates internal Ca^{2+} ion release is well known [26]. The binding of CARP VIII to ITPR1 is believed to reduce the sensitivity of ITPR1 to inositol trisphosphate (IP3) and control the release of Ca^{2+} ions from the internal stores of the endoplasmic reticulum [34]. Deletion mutagenesis experiments in a yeast two-hybrid system showed that almost the entire CA domain of CARP VIII is required for ITPR1 binding [14]. We studied the functional association of ITPR1 and CARP VIII throughout the vertebrate evolution; two coevolution analyses of genes encoding these two proteins from 31 species were performed. In fish species ITPR1 gene has been duplicated into genes which are named *itpr1a* and *itpr1b* in zebrafish. The analysis of *itpr1a* orthologs show more sites that appear to coevolve with CARP VIII compared to the analysis of *itpr1b* [30]. This indicates that the gene product of itpr1a is more likely to have retained the interaction with CARP VIII in fish.

2.2 *Catalytic Activity and Inhibition Studies*

Replacement of Arg117 to His and Glu115 to Gln in CARP VIII restored its catalytic activity, which was significantly higher than that of the mammalian CA III isozyme [7]. The inhibition studies on this protein showed its activity could be strongly inhibited by acetazolamide (5-acetamido-1,3,4-thiadiazole-2-sulfonamide). The replacement of Gln92 to Glu in human CA II has only slight effect on zinc affinity suggesting that the single mutation of Arg117 to His is enough to change inactive CARP VIII to an active CA [7, 35, 36]. Indeed, later studies by the same group showed that replacement of single arginine residue with histidine was able to restore the catalytically inactive mouse CARP VIII to active carbonic anhydrase with CO_2 hydration turnover number of $1.2 \times 104\ s^{-1}$. Further mutations in CARP VIII of residues His94, Gln92, Val121, Val143 and Thr200 (numbering system of human CA I) greatly increases CA activity similar to CA I with a kcat value of 4×104 s-1 [36].

Recently, another site-directed mutagenesis study was performed to restore the catalytic activity of CARP VIII [6], resulting to a kcat/km value which was 67.3 % of the value for human CA II, one of the most highly active enzymes in nature [3]. The mutant CARP VIII was effectively inhibited by acetazolamide, with inhibition constants in the range of 12–35 nM (the KI of this sulfonamide against human CA I is of 250 nM). Interestingly, the catalytic activity of human CA VIII mutant was higher than was reported for the mouse CA VIII mutant reported earlier, although there is only difference of three amino acid residues between these two mutant enzymes [6]. The inhibition study done with inorganic anions showed much weaker inhibition compared to CA X and CA XI. However, inhibitory effect of thiocyanate to restored CA VIII was stronger than to other restored CARPs.

2.3 *Ataxic Zebrafish Model Knocked Down for CA8*

Many types of congenital ataxia and mental retardation are caused by alterations in genes that affect the development of the brain. Studies done in the past in mouse showed that spontaneously occurring mutation in *CA8* gene leads to ataxia and lifelong gait disorder [24]. Similarly, recent reports from human studies also show that mutations in *CA8* gene cause mental retardation and ataxia with a reduction in the cerebellar volume [22, 23]. Recently, we have developed a zebrafish model by knocking down the *CA8* gene using antisense morpholinos against *CA8*, which imitated the human disease [30].

2.3.1 Expression of *CA8* in Zebrafish

The developmental expression studies using RT-qPCR showed that *CA8* mRNA is expressed throughout the embryonic development period. The expression of *CA8* was particularly high at 0 hpf, 24 hpf and 120 hpf. The presence of *CA8* at 0 hpf suggests that *CA8* is of maternal origin and that it is required early in the development. The expression analysis of *CA8* in a panel of tissues showed that *CA8* is expressed in almost all the tissues of zebrafish at mRNA level. The *CA8* mRNA was predominantly high in the brain, similar to the expression in mouse and human [33, 37]. The high level of expression was also seen in kidney, eye, gills and heart. The overall expression pattern zebrafish tissues was similar to the expression pattern in mouse tissues [33]. Immunohistochemistry of cerebellar region showed intense signal in cerebellar Purkinje cells, which is similar to the expression in mouse and human brain [33, 37]. Similarly, developmental expression analysis of *itpr1a* showed an expression pattern similar to *CA8* mRNA during the early embryonic development, suggesting the need of their interaction during development.

2.3.2 Morphant Zebrafish with Ataxia

The zebrafish larvae injected with 125 μM translation-blocking antisense morpholinos (MOs) against *CA8* completely suppressed its expression, which was confirmed by western blotting at 3 dpf in zebrafish larvae. So, this fairly low concentration mimics a null mutation. The larvae lacking the *CA8* gene product showed a variety of phenotypic defects, including a curved and fragile body, pericardial edema, absence of otolith sacs, and incomplete utilization of yolk. Swim pattern of the

5 dpf morphant larvae injected with 2 μM and 6 μM *CA8*-MOs, which had an otherwise normal phenotype, showed difficulty in balancing the body and increased turning angle, suggesting an ataxic phenotype similar to mice and humans with a mutation in *CA8* gene [22–24]. Histological examination showed reduction in cerebellar volume similar to the human patients from Saudi Arabian family [23]. Similarly, electron microscopy and TUNEL assay showed increased neuronal cell death. The results obtained from zebrafish model suggest an important role for CARP VIII during embryonic development. Lack of CARP VIII leads to defects in motor and coordination functions, mimicking the ataxic human phenotype. Zebrafish knockdown work also reveals an evolutionarily conserved function of CARP VIII in brain development. The novel zebrafish model will be helpful in investigating the mechanisms of CARP VIII-related ataxia and mental retardation in humans.

2.4 Distribution of CARP VIII in Neuronal Cells Expressing Mutant Ataxin-3

In a recent study, changes in the gene expression profile of human neuroblastoma cells expressing a repeat expansion mutant ataxin 3 were studied using microarray technology [38]. This microarray study showed a nearly 9-fold increase in *CA8* expression in the presence of mutant ataxin 3 compared to cells expressing normal ataxin 3. Further analysis of distribution of CARP VIII in these cells showed significant increase in the expression of CARP VIII in the neuroblastoma cells (SK-N-SH) expressing mutant ataxin 3 [39]. Perinuclear and cytoplasmic localization study using monoclonal antibodies showed intense and even distribution of CARP VIII. The expression of CARP VIII in the perinuclear region was consistent with its co-localization with ITPR1, which is expressed on endoplasmic reticulum membrane [26]. It was shown earlier that defective ataxin 3 disturbs neuronal calcium signaling by binding to ITPR1 [40]. In vitro and cultured cell studies showed that the sensitivity of ITPR1 to IP3 is increased when it is bound to mutant ataxin 3. The increased expression and altered localization of CARP VIII in the cells expressing mutant ataxin 3 might be linked to the disruption of ITPR1-mediated calcium signaling in neurons by mutant ataxin-3 [39].

3 Carbonic Anhydrase X-like Sequences

CARP X-like sequences seem to be found in all bilaterian animal genomes, both in deuterostomes and protostomes (Tolvanen et al., unpublished results), unlike CARP VIII orthologs, which have been observed only in some deuterostome invertebrates

[33] and in a single protostome species [14], in addition to their ubiquitous presence in vertebrates. In protostomes many species harbor several copies of CARP X-like genes. For example, in the nematode *C. elegans* the α-CA genes *cah-1* and *cah-2* code for CARP X-like proteins, and in most insects we see a system of two CARPs which are one-to-one orthologs of *D. melanogaster* genes *CG1402* and *CG32698*, for which we have used the names CARP-A and CARP-B [10].

The data related to mammalian CARP X and XI expression studies and role in diseases has been reviewed recently [32]. In this chapter, we review recent data related to catalytic activity of mutated human CARP X and XI. We also discuss the information related to distribution of CARP XI in human brain with spinocerebellar ataxia 3 (SCA3) and in a transgenic mouse model with Machado-Joseph disease (MJD). Finally, we present evolutionary data related to CARP X-like sequences.

3.1 Catalytic and Inhibition Studies of Engineered CARP X and CARP XI

The CARP X is catalytically inactive due to the absence of two out of the three histidine residues necessary for coordination of the zinc atom in the active site. Similarly, mammalian CARP XI lacks all the three histidine residues and is enzymatically inactive [33]. In the past, studies on catalytic activity of mutated CARP VIII have been described, but no information was available on restoring the catalytic activity of CARP X and CARP XI till very recently [6]. The mutated CARP X and CARP XI demonstrated high catalytic activity. Human CARP X had the highest activity among the mutated CARPs, showing 92 % catalytic activity compared to human CA II, which is considered as one of the fastest enzymes in nature [3]. The catalytic activity of mutated CARP XI was only slightly lower than that of CARP X, at 82.6 % of the activity of human CA II.

These enzymes were effectively inhibited by acetazolamide, with inhibition constants ranging from 12 to 35 nM [6]. Inorganic anions represent another well-known group of CA inhibitors. They bind metal ions in solution or they bind within the active site of the metalloenzyme [41]. The inorganic anions showed higher inhibition of these enzymes than of CARP VIII. The anion thiocyanate which inhibits CARP VIII strongly was less effective against CARP X. Nitrate, nitrite, hydrogen sulfide, cyanide, and azide were effective CARP X inhibitors. The best CARP X anion inhibitors were sulfamide, sulfamic acid, phenylboronic acid, and phenylarsonic acid.

The inhibition rate of mutated CARP XI was intermediate between CARP VIII and CARP X, similar to its catalytic activity. The effective inhibitors against CARP XI were cyanide, hydrogen sulfide, sulfamide, sulfamate, phenylboronic acid and phenylarsonic acid, with K_I values in the range of 5.1–88 μM [6].

3.2 Altered Expression of Mammalian CARP X and XI

Expansion mutations in CAG trinucleotide repeats of ataxin 3 lead to spinocerebellar ataxia 3/Machado-Joseph disease (SCA3/MJD), a neurodegenerative disorder in humans and mice [42]. Microarray studies on mRNAs isolated from human neuroblastoma cells expressing mutant ataxin-3 demonstrated that *CA11* is highly upregulated and *CA10* slightly upregulated in these cells [39]. As a follow-up of these results, the authors studied the cellular distribution of CARP X and CARP XI in cultured neuronal cells, in the brain tissues of transgenic MJD mice and in the brain tissues of a human patient with SCA3 and of a normal human. There was no significant change in the cellular localization of CARP X in the cultured MJD-positive or -negative cells. However, localization of CARP XI was changed in the cells expressing mutant ataxin-3 compared to the cells expressing normal ataxin-3. Namely, cellular localization studies in normal and MJD84.2 mouse and in human SCA3 patients showed that the cerebellar Purkinje cells in MJD mouse stained strongly especially in the nucleus and cytoplasm compared to the normal mouse. The neural cells in the cortex also showed a stronger staining in MJD mouse compared to normal mouse. The immunostaining of cerebellar tissue of SCA3 mutation-positive human patients confirmed the results of the staining obtained for MJD mouse. The cerebellar white matter in the cytoplasmic region and Purkinje cells showed increased immunoreactivity compared to the staining of age-matched control brains. The results on the distribution of CARP XI suggest that in the presence of mutant ataxin-3, CARP XI changes its expression level and cellular re-localization to nuclei of neuronal cells. It remains an interesting question whether this plays a role in the progression of the disease and dysfunction of nervous system.

3.3 Evolutionary Aspects of CARP X-like Sequences

The expression of mammalian CARP X and CARP XI is most abundant in the central nervous system (CNS). In Drosophila, *CG1402* (CARP-A) is similar in being most highly expressed in larval CNS and in adult fly brain (http://flybase.org/reports/FBgn0029962.html). In *C. elegans*, both of the CARP X-like genes *cah-1* and *cah-2* are expressed in several types of neurons (http://gbrowse.raetschlab.org/cgi-bin/gbrowse/ce200/) [43]. So, neural expression of CARP X-like genes seems to have been conserved since the earliest bilaterian animals. We can find orthologs of CARP X even in flatworms (Tolvanen et al., unpublished), but there is no data on their expression pattern.

It is remarkable how well the overall structure of CARP X and CARP XI has withstood the changes during its history. Despite having been acatalytic since very early metazoan evolution, these proteins can still reach nearly the same reaction rates as the very highly active CA II when the critical zinc-binding histidines are restored in their sequence [6].

4 Receptor-Type Protein Tyrosine Phosphatases (PTPR) Gamma and Zeta

The phosphorylation of tyrosine residues in proteins is a basic regulatory mechanism which controls proliferation, differentiation, communication, and adhesion of cells. Receptor-type protein tyrosine phosphatases (PTPRs) are a family of integral cell surface proteins that possess intracellular protein tyrosine phosphatase (PTP) activity, and contain extracellular domains that have sequence homology to cell adhesion molecules. The function of PTPRs is less well known compared to receptor-type protein tyrosine kinases (RPTKs), especially as regards to their substrate specificities, regulatory mechanisms, biological functions, and their roles in human diseases. The PTPRs are divided into eight subfamilies (R1/R6; R2A; R2B; R3; R4; R5; R7; and R8) based on structure of their extracellular domains [44, 45]. In the context of the present chapter, PTPR-zeta (ζ) which is also known as PTPR-beta (β), (henceforth, it will be referred as PTPRZ) and PTPR-gamma (γ) (PTPRG) are of particular interest. PTPRZ and PTPRG contain carbonic anhydrase (CA) domains and belong to the R5 sub-family.

The PTPRZ/G proteins are encoded by the genes *PTPRZ1/PTPRG*, and possess an N-terminal CA-like domain and a single Fibronectin type III domain in their extracellular domains (ECD), and an intracellular domain with tandem PTP domains [8, 46]. The CA domains of PTPRs lack the CA enzymatic activity due to absence of two of the possible three histidine residue required for co-ordination of zinc atom at active site. The CA like domain at N-terminal end is comprised of seven exons and six introns, and the positions of introns are similar to the introns found in mammalian *CA1, 2, 3, 5 and 7* genes [47]. The CA domains of these proteins show up to 50 % similarity with the other CA-related domains and active CAs [8, 9]. The two of the histidine residues in human PTPRZ and PTPRG are replaced by threonine and glutamine and therefore, these proteins are catalytically inactive [6, 13, 14]. However, the CA domains in these proteins can serve as a hydrophobic binding pocket for contactin molecules [48]. In this chapter, we mainly present the data on structure of CA domains of PTPRZ and PTPRG and their interaction with contactin molecules emerged recently and the role of these proteins in diseases.

4.1 Molecular Biology and Bioinformatics

The CA domain of PTPRZ binds to contactin 1 (CNTN-1) and promotes cell adhesion and neurite outgrowth of primary neurons [48]. The other ligands that can bind to ECD are tenascin, N-CAM, pleitrophin and midkine [49, 50]. The PTPRZ is expressed by astrocytes and oligodendrocytes in the developing and adult nervous system and it binds to contactin-1 (CNTN1) which is expressed on the surface of oligodendrocyte precursor cells and is involved in proliferation and differentiation of these cells [50–52]. Similarly, PTPRZ is also involved in oligodendrogenesis, suggesting that the interaction between PTPRZ and CNTN1 is required for maturation of oligodendrocytes.

A full length PTPRZ contains, an N-terminal carbonic anhydrase-like (CA-like) domain followed by a fibronectin type III (FNIII) repeat, a spacer region, a large insert with attachment sites for chondroitin sulfate proteoglycan and two tyrosine phosphatase domains (D1 and D2) in the cytoplasmic region [53]. It is expressed in soluble form, phosphacan, which is one of the most abundant proteoglycans in the brain and is expressed during myelination [51]. CNTN1 is a glycophosphatidyl-anchored protein composed of six Ig repeats and four FNIII domains and is the first identified member of the CNTN family of neural cell adhesion molecules, which are involved in the construction of neural networks [54]. It is shown that Ig repeats two and three of CNTN1 interact with the CA-domain of PTPRZ [55]. Recently, the structural basis of interaction between PTPRZ and CNTN1 has been shown to demonstrate that this interaction regulates the proliferation and differentiation of oligodendrocyte precursor cells OPCs [56].

4.2 *Crystal Structures of PTPRZ-CNTN-1 Complex and of CA Domains of PTPRs*

The crystal structure of CNTN1 fragment Ig2-Ig3 bound to CA-domain of PTPRZ shows that the Ig2-Ig3 repeats in CNTN1 adopt a horseshoe-like conformation in which apolar contacts between Ig2 and Ig3 bury 1,355 $Å^2$ with a shape complementarity coefficient of 0.77 [57]. The interface between the CA domain and Ig2-Ig3 repeats of CNTN1 occludes 1,658 $Å^2$ with a shape complementarity coefficient of 0.68, both of which compare favorably to those of known biological interfaces and are similar to the values obtained for the complex between PTPRG and CNTN4 [55]. The crystal structure complex shows that the interface between PTPRZ and CNTN1 consists of a β-hairpin loop contacting repeats of Ig2 and Ig3 domains of CNTN1 (PTPRZ residues 265–280) and a short loop (PTPRZ residues 205–208) that contacts the Ig3 domain of CNTN1 exclusively.

The PTPRZ and CNTN1 are known to be involved in development of oligodendrocytes [52, 58]. Studies on the interaction of PTPRZ and CNTN1 at cell surface of oligodendrocytes showed that PTPRZ co-localizes with CNTN1 [56]. Similarly, binding studies between different CNTN molecules and PTPRZ confirmed that PTPRZ binds specifically to CNTN1 and not other CNTN molecules [55]. Further examination of proliferation of oligodendrocyte precursor cells (OPC) in $PTPRZ^{-/-}$ mice during early postnatal development showed that the lack of PTPRZ increases the proliferation of OPCs [56]. Taken together these results suggest that formation of PTPRZ and CNTN1 complex represses proliferation of OPCs and promotes oligodendrocyte maturation.

Recently, crystal structure of CNTN4-Ig1-4 and PTPRG-CA complex has also been resolved and structural basis of interaction of CNTN4-Ig1-4 with PTPRG-CA was studied [55]. In PTPRG-CA, the flexible β-hairpin (residues 288–301) accounts for almost 80 % of the interaction with CNTN4; rest of the contacts are mediated by residues 225–229 and the contacts between CNTN4 and PTPRG 4 are restricted to Ig domains 2 and 3. The two strands of the β-hairpin in PTPRG-CA complement the 3-strand antiparallel β-sheet in CNTN4 Ig 2 to form a 5-strand antiparallel β-sheet with the main chain atoms of residues 295–299 of PTPRG, forming hydrogen bonds with the main chain atoms of residues 139–143 in CNTN4. All the CNTN4 residues that interact with PTPRG are conserved in CNTN3, 5, and 6, thus explaining why PTPRG binds specifically to these four CNTN family members. Similarly, In-vitro affinity-isolation assay showed binding of PTPRG-CA to CNTN3,4,5, and 6, and further binding studies of CNTN3,4,5 and 6 lacking different Ig domains showed that the PTPRG binds to 2 and 3 Ig domains of CNTN3,4,5, and 6 forming 1:1 complex [55].

Similarly, the crystal structures of mouse and human PTPRG-CA domain and CA domain of human PTPRZ have been resolved at a resolution of 1.7 Å. The CA domain of all three receptors are similar and have central 8 strand antiparallel β-sheet surrounded by three to four α-helices and extensive loop regions [55]. Mouse and human PTPRG-CA domains superimpose with a rmsd of 0.5 Å for 257 C-α positions, whereas mouse PTPRG-CA and human PTPRZ-CA superimpose with a rmsd of 1.4 Å for 257 C-α positions. However, unlike PTPRG-CA, PTPRZ-CA includes an additional disulfide bond between C133 and C264, which is conserved in all known orthologs of PTPRZ. The presence of this additional disulfide bridge has little effect on the structure of PTPRZ-CA when compared to PTPRG-CA and its biological significance is currently unknown [55].

Structural studies show that the CA domains of PTPRG and PTPRZ resemble α-CAs. Mouse PTPRG-CA, human PTPRG-CA, and human PTPRZ-CA superimpose with mouse CA-XIV with rmsd of 1.5–1.7 Å for 253–255 C-α positions and most of the differences reside in the orientation of an extended β-hairpin loop in type V PTPRs [59]. This hairpin loop is disordered in human PTPRG-CA and in three of the four molecules found in the asymmetric unit of mouse PTPRG-CA-crystals, indicating that this region is flexible.

4.3 Role in Diseases

The PTPRZ plays critical role in oligodendrocyte survival, in recovery from demyelinating disease, and in memory formation [58, 60]. The up-regulation of pleiotrophin is associated with the repair of injured nervous system, which suggests that pleiotrophin may have protective effects in Parkinson's disease [61]. Binding of pleiotrophin to PTPRZ modulates the PTP activity by increasing β-catenine, Fyn and β-adducin tyrosine phosphorylation by inhibiting PTP activity of PTPRZ [49, 62]. The activation of these substrates is needed for proliferation of dopaminergic progenitors and survival and differentiation of dopaminergic neurons [63]. The inhibition of phosphatase activity of PTPRZ using specific inhibitor may be useful in treatment of Parkinson's disease.

PTPRZ plays a role in pathogenesis of gastric ulcers by acting as a cell surface receptor for VacA, a protein toxin produced by *H. pylori* [64]. The VacA protein toxin inhibits the PTP activity of PTPRZ and induces the detachment of gastric epithelial cells by increasing tyrosine phosphorylation of its substrate Git-1 a protein involved in cell adhesion and cytoskeleton [65]. Git-1 tyrosine phosphorylation accompanies epithelial detachment, leading to gastric ulcers with direct action of gastric acid. The defect in signaling of PTPRZ is a basic mechanism of gastric ulcers caused by *H. pylori*. The mice which are knocked out for PTPRZ are resistant for gastric ulcers caused by *H. pylori* also support the above conclusions [66].

The neurogulin-1 (NGR) is a signaling molecule and functions through the interaction of epidermal growth factor receptor (ErbB4) receptor. Interestingly, PTPRZ is known to inhibit its role in signaling [67]. The expression studies showed that the gene encoding PTPRZ is up-regulated in the brain of patients with schizophrenia [68]. Similarly, over expression of PTPRZ in transgenic mouse model reduced the NRG1 signaling, leading to molecular and cellular changes such as, altered glutamatergic, GABAergic and dopaminergic activity and delayed oligodendrocyte developments, which occur during the pathogenesis of schizophrenia. These mice also showed reduced sensory motor gating, hyperactivity and working memory deficit which are typical symptoms of schizophrenia [68].

The PTPRZ is also implicated in remyelination, as it is upregulated in human oligodendrocytes during the repair of lesions to myelin sheath [58]. The experimental allergic encephalomyelitis-induced PTPRZ$^{-/-}$ mice have an impairment of repair of demylinating lesions. These findings suggest that the formation of PTPRZ/CNTN1 complex is necessary for the differentiation of OPCs leading to remyelination. Involvement of PTPRG in different diseases is less well known. However, it has been implicated in renal and lung cancers as it was found to be frequently deleted in these cancers in humans [69]. The PTPRG is also associated with motor coordination function, and the mice deficient in this protein exhibit impaired motor coordination during rod walking and string tests [70].

The PTPRZ and RTPRG play different roles in cellular processes and can be potential drug targets for treating certain diseases. More importantly, the discovery of secondary substrate-binding sites in these proteins provides an opportunity for development of highly specific inhibitors. Similarly, small molecules with high affinity targeting ECDs can up-regulate the activity of these proteins. The deeper understanding of physiological function and regulatory mechanisms of these proteins will help us to design specific inhibitory molecules for balancing of activity of these proteins.

5 Carbonic Anhydrase Related Proteins in Poxviruses

The vaccinia virus transmembrane protein D8 [11, 12, 71] was reported to contain a CA-like N-terminal domain in 1990 [11, 12, 71]. The D8 protein was observed to participate in attachment to the host cell [11, 12, 71], and later the ligand was discovered to be chondroitin sulfate [72] Now we know that altogether four proteins mediate attachment of vaccinia virus, including A27 and H3, which bind heparin, and laminin-binding A26 [73].

5.1 Structure

The vaccinia virus *D8L* gene codes for a 304-residue protein. Figure 8.1 presents a multiple sequence alignment of 14 viral sequences, detailed in Table 8.1, together with selected human CA and CARP sequences. In the N-terminal ectodomain the first 236 residues are clearly homologous with the catalytic domain of carbonic anhydrases. There is no counterpart for the first 24 residues of CA II in the viral sequences. The truncated CA domain is followed by a spacer of almost 40 residues, and residues 275–294 form the predicted transmembrane domain (indicated in Fig. 8.1), leaving just 11 C-terminal residues on the other side of the membrane.

The three-dimensional structure of vaccinia D8 binding to an antibody has been determined using 225-residue and 269-residue constructs of the ectodomain [74]. In the longer construct, the last 60 residues following the CA domain are not visible. Figure 8.2a shows that despite missing the first 24 residues of the CA fold, the backbone of vaccinia D8 follows very closely that of other CAs and CARPs. Of the 236 visible residues, 214 are superimposable (within 5.0 Å) on all of the other four structures in Fig. 8.2a with RMSD values ranging from 1.16 to 1.53 Å. This is similar to the RMSD values between all other structures shown: 1.10–1.54 Å, except for PTPRG to PTPRZ which gives an RMSD of 0.86 Å. Figure 8.2b–d show the surfaces of D8 (2B), CARP VIII (2C), and PTPRZ (2D), demonstrating that the missing 24 residues leave a large dent at the upper right.

The binding site for the negatively charged chondroitin sulfate chain is predicted to be at the bottom of the cavity created by the N-terminal deletion (arrow in Fig. 8.2b), lined with positively charged residues [74]. The sequence alignment of Fig. 8.1 shows all Lys and Arg residues on grey background. Residues K41, R44, K108, and R220, on black background in the bottom sequence, are the residues implicated in binding in a docking experiment with a short chondroitin sulfate fragment [74]. The large amount of additional conserved Lys and Arg residues (Figs. 8.1 and 8.2b) indicates that even a longer CS chain could bind efficiently.

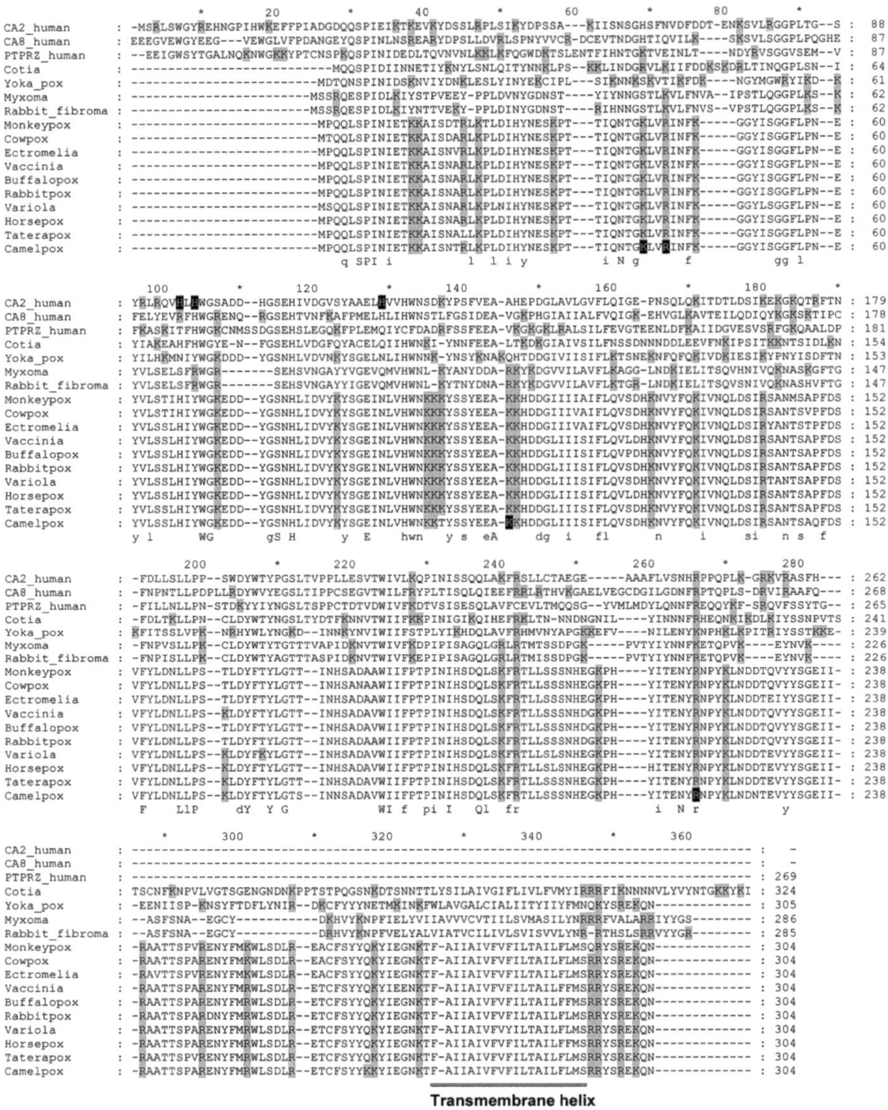

Fig. 8.1 Multiple sequence alignment of viral CARPs with selected human CAs. CA8_human includes only the visible part of structure 2W2J, and PTPRZ_human the visible part of 3JXF. Alignment with Clustal Omega iterated four times and manually edited to match a structure-based alignment near the end of the CA domain. Grey highlighting for all K and R residues. Black highlighting in the bottom row for residues K41, R44, K108, and R220 in the predicted chondroitin sulfate binding site. Black highlighting in the top row for for the three Zn-binding residues in CA II

Table 8.1 List of virus proteins homologous to carbonic anhydrases included in the alignment of Fig. 1

Probable chondroitin sulfate binding 32 kDa proteins	Name in GenBank	GenBank protein accession number
Buffalopox virus	immunogenic envelope protein	ADW84807.1
Camelpox virus	CMLV111	NP_570501.1
Cowpox virus	CPXV125 protein	NP_619909.1
Ectromelia virus	EVM097	NP_671615.1
Horsepox virus	HSPV114	ABH08220.1
Monkeypox virus Zaire-96-I-16	E8L	NP_536532.1
Rabbitpox virus	RPXV102	AAS49815.1
Taterapox virus	IMV membrane protein	YP_717422.1
Vaccinia virus	IMV membrane protein	YP_232995.1
Variola virus	hypothetical protein VARVgp098	NP_042142.1
Other CA-like viral proteins		
Cotia virus SPAn232	carbonic anhydrase-like protein	YP_005296290.1
Myxoma virus	m83L	NP_051797.1
Rabbit fibroma virus	gp083L	NP_051972.1
Yoka poxvirus	IMV membrane protein	YP_004821447.1

5.2 Evolution and Function of Viral CARP Sequences

We have constructed a phylogenetic tree of selected viral CA-like sequences for this review, together with all human CA isoforms (Fig. 8.3). The tree shows that the viral CA-like proteins are most likely to have been derived from cytoplasmic CA isozyme genes of the group CA1/CA2/CA3/CA13. Even if the tree indicates human CA13 as the most closely related gene, it remains uncertain, since we do not know from which host animal the gene was originally captured in a poxvirus.

The Orthopoxvirus sequences (bottom 10 sequences in Fig. 8.1, and the tightly linked group indicated in Fig. 8.3) are highly similar to each other, with protein sequence identities of at least 93 % to the prototype vaccinia D8 sequence. It is highly probable that they all retain the same function of CS binding. The other four virus sequences are less than 40 % identical to the vaccinia sequence, so they may or may not share this function. In case of myxoma and rabbit fibroma viruses, some of the assumed CS-binding site residues are conserved, so they might retain the function, but for Yoka pox virus and Cotia virus, this seems less likely. However, this is highly speculative since the binding site has been defined only by a docking simulation.

The zinc-binding His residues of CA II are shown with black highlight in Fig. 8.1. In all of the viral sequences some of these His residues are missing, and they are predicted to be devoid of CA enzymatic activity. It was verified that the vaccinia D8

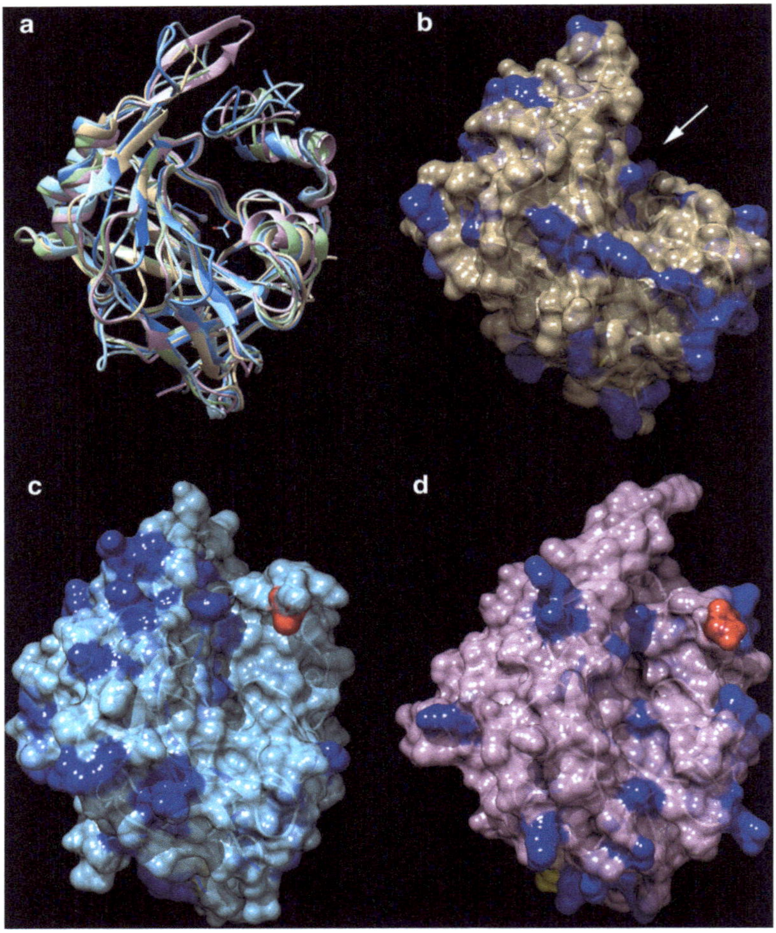

Fig. 8.2 Structures of CARPs. (**a**): four CARPs and CA II superimposed. Protein backbones of vaccinia D8 ectodomain (4E9O, *beige*), CARP VIII (2W2J, *cyan*), CA-like domain of human PTPRZ (3JXF, *lilac*), CA-like domain of human PTPRG (3JXH, *light green*), and human CA II (2D0N, *darker blue*). Note the small beige stump at 3 o'clock which is the N-terminus of 4E9O, residue 2 of the viral sequence. (**b**): Vaccinia D8 ectodomain (4E9O). (**c**): CARP VIII (2W2J). (**d**): CA-like domain of human PTPRZ (3JXF). In all of 2B to 2D, blue shows positively charged Lys and Arg residues on the surface, *yellow* (*at the bottom*) indicates the C-terminus, and *red* the N-terminus of each protein

protein does not have catalytic activity, but when the two histidines missing from the catalytic site are returned, the activity and acetazolamide inhibition sensitivity is restored [75]. This proves the surprising robustness of the CA fold—despite the lack of the N-terminal 24 residues, the reconstituted enzyme worked at the

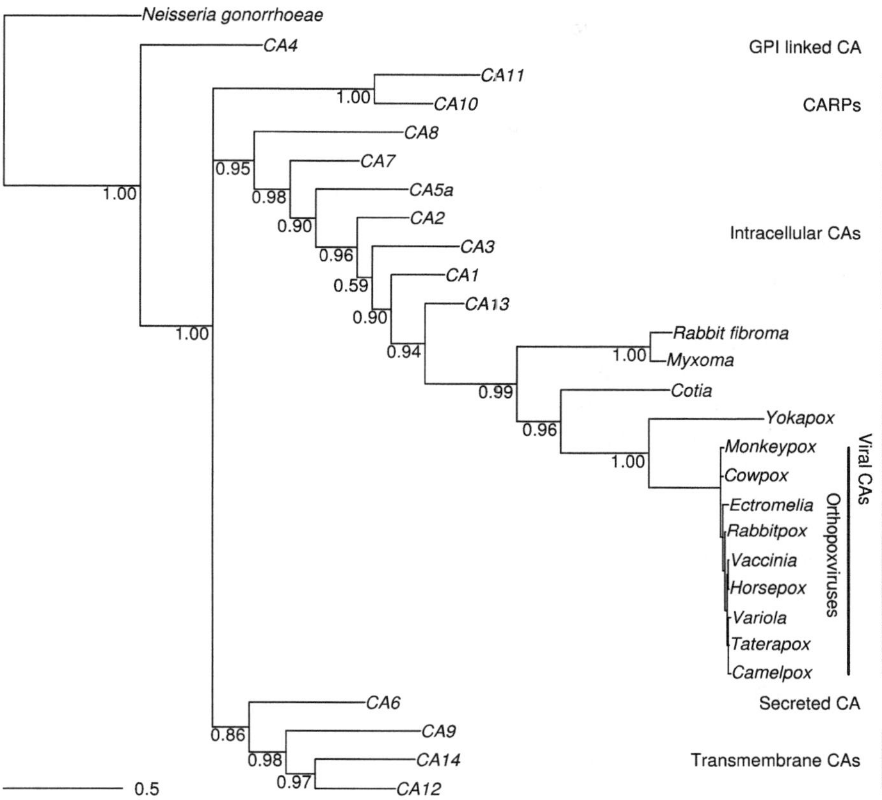

Fig. 8.3 Phylogenetic tree of viral CA-like sequences together with human CA isoforms. Abbreviated virus species names are indicated, for full names and strains and database accession numbers, see Table 8.1. Protein sequences were aligned with Clustal Omega, then replaced with their corresponding cDNA sequences using PAL2NALweb service [78]. This alignment was used with MrBayes v 3.2 [79] to estimate the phylogeny of the sequences using Bayesian inference. Bayesian estimation was run for 40,000 generations, with flat *a priori* distribution of base frequencies, substitution rates, proportion of invariable sites, and gamma shape parameter. The arithmetic mean of the estimated marginal likelihoods for runs sampled was −12386.95. The 50 % majority rule consensus tree was saved and visualized using the APE R package [80]. The tree was rooted using the *Neisseria gonorrhoeae* alpha CA gene as outgroup

rate of medium-active human CA isoforms. Incidentally, the phylogenetic analysis presented in the above paper suggested the closest relatedness with CARP X and XI, but we believe that the methods used in our analysis are more robust, especially if we take into account the approximately two folds substitution rates observed in pox viruses compared to the humans [76, 77]. Therefore, we suggest that viral CARPs are related more closely to intracellular CAs than to genomic CARPs.

6 Conclusions

We have reviewed above the major groups of acatalytic α-CAs, a.k.a CARPs, namely CARP VIII; CARP X-like, including CARP X, CARP XI, and their invertebrate orthologs; CARP XVI domains in PTPRG and PTPRZ; and viral CARPs. With the exception of CARP X and XI, we now know ligands for all of these CARPs, and in some cases even the consequences of binding. The study of CARPs has made it evident that the CA fold can serve as a scaffold for developing various binding functions instead of the original catalytic activity. This can be taken as evidence for the robustness of the fold, a robustness which is also highlighted by the experiments which have restored the catalytic activity of CARP VIII, CARP X, CARP XI, and a viral CARP, still after tens or hundreds of millions of years of evolution in acatalytic functions.

CARP X and XI are the isoforms which remain the most mysterious, without any published information on their actual functions. The very high degree of sequence conservation of the CARP X-like proteins and their presence in neural tissues in all animals which have a distinct brain (triploblasts) would open much room for speculation of very essential functions, but we need to wait for more experimental data before any conclusions can be made.

References

1. Tashian RE (1989) The carbonic anhydrases: widening perspectives on their evolution, expression and function. Bioessays 10:186–192
2. Sly WS, Hu PY (1995) Human carbonic anhydrases and carbonic anhydrase deficiencies. Annu Rev Biochem 64:375–401
3. Supuran CT (2008) Carbonic anhydrases: novel therapeutic applications for inhibitors and activators. Nat Rev Drug Discov 7:168–181
4. Tashian RE (1992) Genetics of the mammalian carbonic anhydrases. Adv Genet 30:321–356
5. Tolvanen ME, Ortutay C, Barker HR, Aspatwar A, Patrikainen M, Parkkila S (2012) Analysis of evolution of carbonic anhydrases IV and XV reveals a rich history of gene duplications and a new group of isozymes. Bioorg Med Chem 21:1503–1510
6. Nishimori I, Vullo D, Minakuchi T, Scozzafava A, Clemente Capasso C, Supuran CT (2012) Restoring catalytic activity to the human carbonic anhydrase (CA) related proteins VIII, X and XI affords isoforms with high catalytic efficiency and susceptibility to anion inhibition. Bioorg Med Chem Lett 23:256–260
7. Sjoblom B, Elleby B, Wallgren K, Jonsson BH, Lindskog S (1996) Two point mutations convert a catalytically inactive carbonic anhydrase-related protein (CARP) to an active enzyme. FEBS Lett 398:322–325
8. Barnea G, Silvennoinen O, Shaanan B, Honegger AM, Canoll PD, D'Eustachio P, Morse B, Levy JB, Laforgia S, Huebner K (1993) Identification of a carbonic anhydrase-like domain in the extracellular region of RPTP gamma defines a new subfamily of receptor tyrosine phosphatases. Mol Cell Biol 13:1497–1506
9. Levy JB, Canoll PD, Silvennoinen O, Barnea G, Morse B, Honegger AM, Huang JT, Cannizzaro LA, Park SH, Druck T, Huebner K, Sap J, Ehrlich M, Musacchio JM, Schlessinger J (1993) The cloning of a receptor-type protein tyrosine phosphatase expressed in the central nervous system. J Biol Chem 268:10573–10581

10. Ortutay C, Olatubosun A, Parkkila S, Vihinen M, Tolvanen ME (2010) An evolutionary analysis of insect carbonic anydrases. In: Berhardt LV (ed) Advances in medicine and biology, vol 7. Nova Science Publishers, Hauppauge, pp 145–168
11. Maa JS, Rodriguez JF, Esteban M (1990) Structural and functional characterization of a cell surface binding protein of vaccinia virus. J Biol Chem 265:1569–1577
12. Niles EG, Condit RC, Caro P, Davidson K, Matusick L, Seto J (1986) Nucleotide sequence and genetic map of the 16-kb vaccinia virus HindIII D fragment. Virology 153:96–112
13. Aspatwar A, Tolvanen ME, Ortutay C, Parkkila S (2010) Carbonic anhydrase related protein VIII and its role in neurodegeneration and cancer. Curr Pharm Des 16:3264–3276
14. Miyamoto H (2012) Sequence of the pearl oyster carbonic anhydrase-related protein and its evolutionary implications. Biochem Genet 50:269–276
15. Kato K (1990) Sequence of a novel carbonic anhydrase-related polypeptide and its exclusive presence in Purkinje cells. FEBS Lett 271:137–140
16. Taniuchi K, Nishimori I, Takeuchi T, Ohtsuki Y, Onishi S (2002) cDNA cloning and developmental expression of murine carbonic anhydrase-related proteins VIII, X, and XI. Brain Res Mol Brain Res 109:207–215
17. Lakkis MM, Bergenhem NC, O'Shea KS, Tashian RE (1997) Expression of the acatalytic carbonic anhydrase VIII gene, *Car8*, during mouse embryonic development. Histochem J 29:135–141
18. Taniuchi K, Nishimori I, Takeuchi T, Fujikawa-Adachi K, Ohtsuki Y, Onishi S (2002) Developmental expression of carbonic anhydrase-related proteins VIII, X, and XI in the human brain. Neuroscience 112:93–99
19. Miyaji E, Nishimori I, Taniuchi K, Takeuchi T, Ohtsuki Y, Onishi S (2003) Overexpression of carbonic anhydrase-related protein VIII in human colorectal cancer. J Pathol 201:37–45
20. Nishikata M, Nishimori I, Taniuchi K, Takeuchi T, Minakuchi T, Kohsaki T, Adachi Y, Ohtsuki Y, Onishi S (2007) Carbonic anhydrase-related protein VIII promotes colon cancer cell growth. Mol Carcinog 46:208–214
21. Picaud SS, Muniz JR, Kramm A, Pilka ES, Kochan G, Oppermann U, Yue WW (2009) Crystal structure of human carbonic anhydrase-related protein VIII reveals the basis for catalytic silencing. Proteins 76:507–511
22. Turkmen S, Guo G, Garshasbi M, Hoffmann K, Alshalah AJ, Mischung C, Kuss A, Humphrey N, Mundlos S, Robinson PN (2009) CA8 mutations cause a novel syndrome characterized by ataxia and mild mental retardation with predisposition to quadrupedal gait. PLoS Genet 5:e1000487
23. Kaya N, Aldhalaan H, Al-Younes B, Colak D, Shuaib T, Al-Mohaileb F, Al-Sugair A, Nester M, Al-Yamani S, Al-Bakheet A, Al-Hashmi N, Al-Sayed M, Meyer B, Jungbluth H, Al-Owain M (2011) Phenotypical spectrum of cerebellar ataxia associated with a novel mutation in the CA8 gene, encoding carbonic anhydrase (CA) VIII. Am J Med Genet B Neuropsychiatr Genet 156:826–834
24. Jiao Y, Yan J, Zhao Y, Donahue LR, Beamer WG, Li X, Roe BA, Ledoux MS, Gu W (2005) Carbonic anhydrase-related protein VIII deficiency is associated with a distinctive lifelong gait disorder in waddles mice. Genetics 171:1239–1246
25. Yan J, Jiao Y, Jiao F, Stuart J, Donahue LR, Beamer WG, Li X, Roe BA, LeDoux MS, Gu W (2007) Effects of carbonic anhydrase VIII deficiency on cerebellar gene expression profiles in the wdl mouse. Neurosci Lett 413:196–201
26. Hirota J, Ando H, Hamada K, Mikoshiba K (2003) Carbonic anhydrase-related protein is a novel binding protein for inositol 1,4,5-trisphosphate receptor type 1. Biochem J 372:435–441
27. Ishihara T, Takeuchi T, Nishimori I, Adachi Y, Minakuchi T, Fujita J, Sonobe H, Ohtsuki Y, Onishi S (2006) Carbonic anhydrase-related protein VIII increases invasiveness of non-small cell lung adenocarcinoma. Virchows Arch 448:830–837
28. Lu SH, Takeuchi T, Fujita J, Ishida T, Akisawa Y, Nishimori I, Kohsaki T, Onishi S, Sonobe H, Ohtsuki Y (2004) Effect of carbonic anhydrase-related protein VIII expression on lung adenocarcinoma cell growth. Lung Cancer 44:273–280

29. Kilpinen S, Autio R, Ojala K, Iljin K, Bucher E, Sara H, Pisto T, Saarela M, Skotheim RI, Bjorkman M, Mpindi J-P, Haapa-Paananen S, Vainio P, Edgren H, Wolf M, Astola J, Nees M, Hautaniemi S, Kallioniem O (2008) Systematic bioinformatic analysis of expression levels of 17,330 human genes across 9,783 samples from 175 types of healthy and pathological tissues. Genome Biol 9:R139
30. Aspatwar A, Tolvanen ME, Jokitalo E, Parikka M, Ortutay C, Harjula SK, Ramet M, Vihinen M, Parkkila S (2013) Abnormal cerebellar development and ataxia in CARP VIII morphant zebrafish. Hum Mol Genet 22:417–432
31. Skaggs LA, Bergenhem NC, Venta PJ, Tashian RE (1993) The deduced amino acid sequence of human carbonic anhydrase-related protein (CARP) is 98% identical to the mouse homologue. Gene 126:291–292
32. Aspatwar A, Tolvanen ME, Parkkila S (2013) An update on carbonic anhydrase-related proteins VIII, X and XI. J Enzyme Inhib Med Chem 2013:1–14
33. Aspatwar A, Tolvanen ME, Parkkila S (2010) Phylogeny and expression of carbonic anhydrase-related proteins. BMC Mol Biol 11:25
34. Mikoshiba K (2007) IP3 receptor/Ca2+ channel: from discovery to new signaling concepts. J Neurochem 102:1426–1446
35. Kiefer LL, Paterno SA, Fierke CA (1995) Hydrogen bond network in the metal binding site of carbonic anhydrase enhances zinc affinity and catalytic efficiency. J Am Chem Soc 117:6831–45837
36. Elleby B, Sjoblom B, Tu C, Silverman DN, Lindskog S (2000) Enhancement of catalytic efficiency by the combination of site-specific mutations in a carbonic anhydrase-related protein. Eur J Biochem 267:5908–5915
37. Okamoto N, Fujikawa-Adachi K, Nishimori I, Taniuchi K, Onishi S (2001) cDNA sequence of human carbonic anhydrase-related protein, CA-RP X: mRNA expressions of CA-RP X and XI in human brain. Biochim Biophys Acta 1518:311–316
38. Wen FC, Li YH, Tsai HF, Lin CH, Li C, Liu CS, Lii CK, Nukina N, Hsieh M (2003) Down-regulation of heat shock protein 27 in neuronal cells and non-neuronal cells expressing mutant ataxin-3. FEBS Lett 546:307–314
39. Hsieh M, Chang WH, Hsu CF, Nishimori I, Kuo CL, Minakuchi T (2012) Altered expression of carbonic anhydrase-related protein XI in neuronal cells expressing mutant ataxin-3. Cerebellum 12:338–349
40. Chen X, Tang TS, Tu H, Nelson O, Pook M, Hammer R, Nukina N, Bezprozvanny I (2008) Deranged calcium signaling and neurodegeneration in spinocerebellar ataxia type 3. J Neurosci 28:12713–12724
41. De Simone G, Supuran CT (2011) (In)organic anions as carbonic anhydrase inhibitors. J Inorg Biochem 111:117–129
42. Kawaguchi Y, Okamoto T, Taniwaki M, Aizawa M, Inoue M, Katayama S, Kawakami H, Nakamura S, Nishimura M, Akiguchi I et al (1994) CAG expansions in a novel gene for Machado-Joseph disease at chromosome 14q32.1. Nat Genet 8:221–228
43. Spencer WC, Zeller G, Watson JD, Henz SR, Watkins KL, McWhirter RD, Petersen S, Sreedharan VT, Widmer C, Jo J et al (2011) A spatial and temporal map of C. elegans gene expression. Genome Res 21:325–341
44. Andersen JN, Jansen PG, Echwald SM, Mortensen OH, Fukada T, Del Vecchio R, Tonks NK, Moller NP (2004) A genomic perspective on protein tyrosine phosphatases: gene structure, pseudogenes, and genetic disease linkage. FASEB J 18:8–30
45. Brady-Kalnay SM, Tonks NK (1995) Protein tyrosine phosphatases as adhesion receptors. Curr Opin Cell Biol 7:650–657
46. Krueger NX, Saito H (1992) A human transmembrane protein-tyrosine-phosphatase, PTP zeta, is expressed in brain and has an N-terminal receptor domain homologous to carbonic anhydrases. Proc Natl Acad Sci USA 89:7417–7421
47. Kastury K, Ohta M, Lasota J, Moir D, Dorman T, LaForgia S, Druck T, Huebner K (1996) Structure of the human receptor tyrosine phosphatase gamma gene (PTPRG) and relation to the familial RCC t(3;8) chromosome translocation. Genomics 32:225–235

48. Peles E, Nativ M, Campbell PL, Sakurai T, Martinez R, Lev S, Clary DO, Schilling J, Barnea G, Plowman GD, Grumet M, Schlessinger J (1995) The carbonic anhydrase domain of receptor tyrosine phosphatase beta is a functional ligand for the axonal cell recognition molecule contactin. Cell 82:251–260
49. Maeda N, Ichihara-Tanaka K, Kimura T, Kadomatsu K, Muramatsu T, Noda M (1999) A receptor-like protein-tyrosine phosphatase PTPzeta/RPTPbeta binds a heparin-binding growth factor midkine. Involvement of arginine 78 of midkine in the high affinity binding to PTPzeta. J Biol Chem 274:12474–12479
50. Peles E, Schlessinger J, Grumet M (1998) Multi-ligand interactions with receptor-like protein tyrosine phosphatase beta: implications for intercellular signaling. Trends Biochem Sci 23:121–124
51. Faissner A, Heck N, Dobbertin A, Garwood J (2006) DSD-1-Proteoglycan/Phosphacan and receptor protein tyrosine phosphatase-beta isoforms during development and regeneration of neural tissues. Adv Exp Med Biol 557:25–53
52. Czopka T, von Holst A, Ffrench-Constant C, Faissner A (2010) Regulatory mechanisms that mediate tenascin C-dependent inhibition of oligodendrocyte precursor differentiation. J Neurosci 30:12310–12322
53. Barnea G, Grumet M, Milev P, Silvennoinen O, Levy JB, Sap J, Schlessinger J (1994) Receptor tyrosine phosphatase beta is expressed in the form of proteoglycan and binds to the extracellular matrix protein tenascin. J Biol Chem 269:14349–14352
54. Shimoda Y, Watanabe K (2009) Contactins: emerging key roles in the development and function of the nervous system. Cell Adh Migr 3:64–70
55. Bouyain S, Watkins DJ (2010) The protein tyrosine phosphatases PTPRZ and PTPRG bind to distinct members of the contactin family of neural recognition molecules. Proc Natl Acad Sci USA 107:2443–2448
56. Lamprianou S, Chatzopoulou E, Thomas JL, Bouyain S, Harroch S (2011) A complex between contactin-1 and the protein tyrosine phosphatase PTPRZ controls the development of oligodendrocyte precursor cells. Proc Natl Acad Sci USA 108:17498–17503
57. Lawrence MC, Colman PM (1993) Shape complementarity at protein/protein interfaces. J Mol Biol 234:946–950
58. Harroch S, Furtado GC, Brueck W, Rosenbluth J, Lafaille J, Chao M, Buxbaum JD, Schlessinger J (2002) A critical role for the protein tyrosine phosphatase receptor type Z in functional recovery from demyelinating lesions. Nat Genet 32:411–414
59. Whittington DA, Grubb JH, Waheed A, Shah GN, Sly WS, Christianson DW (2004) Expression, assay, and structure of the extracellular domain of murine carbonic anhydrase XIV: implications for selective inhibition of membrane-associated isozymes. J Biol Chem 279:7223–7228
60. Tamura H, Fukada M, Fujikawa A, Noda M (2006) Protein tyrosine phosphatase receptor type Z is involved in hippocampus-dependent memory formation through dephosphorylation at Y1105 on p190 RhoGAP. Neurosci Lett 399:33–38
61. Blondet B, Carpentier G, Lafdil F, Courty J (2005) Pleiotrophin cellular localization in nerve regeneration after peripheral nerve injury. J Histochem Cytochem 53:971–977
62. Meng K, Rodriguez-Pena A, Dimitrov T, Chen W, Yamin M, Noda M, Deuel TF (2000) Pleiotrophin signals increased tyrosine phosphorylation of beta beta-catenin through inactivation of the intrinsic catalytic activity of the receptor-type protein tyrosine phosphatase beta/zeta. Proc Natl Acad Sci USA 97:2603–2608
63. Herradon G, Ezquerra L (2009) Blocking receptor protein tyrosine phosphatase beta/zeta: a potential therapeutic strategy for Parkinson's disease. Curr Med Chem 16:3322–3329
64. Padilla PI, Wada A, Yahiro K, Kimura M, Niidome T, Aoyagi H, Kumatori A, Anami M, Hayashi T, Fujisawa J, Saito H, Moss J, Hirayama T (2000) Morphologic differentiation of HL-60 cells is associated with appearance of RPTPbeta and induction of Helicobacter pylori VacA sensitivity. J Biol Chem 275:15200–15206

65. Manabe R, Kovalenko M, Webb DJ, Horwitz AR (2002) GIT1 functions in a motile, multi-molecular signaling complex that regulates protrusive activity and cell migration. J Cell Sci 115:1497–1510
66. Fujikawa A, Shirasaka D, Yamamoto S, Ota H, Yahiro K, Fukada M, Shintani T, Wada A, Aoyama N, Hirayama T, Fukamachi H, Noda M (2003) Mice deficient in protein tyrosine phosphatase receptor type Z are resistant to gastric ulcer induction by VacA of Helicobacter pylori. Nat Genet 33:375–381
67. Buxbaum JD, Georgieva L, Young JJ, Plescia C, Kajiwara Y, Jiang Y, Moskvina V, Norton N, Peirce T, Williams H, Craddock NJ, Carroll L, Corfas G, Davis KL, Owen MJ, Harroch S, Sakurai T, O'Donovan MC (2008) Molecular dissection of NRG1-ERBB4 signaling implicates PTPRZ1 as a potential schizophrenia susceptibility gene. Mol Psychiatry 13:162–172
68. Takahashi N, Sakurai T, Bozdagi-Gunal O, Dorr NP, Moy J, Krug L, Gama-Sosa M, Elder GA, Koch RJ, Walker RH, Hof PR, Davis KL, Buxbaum JD (2011) Increased expression of receptor phosphotyrosine phosphatase-beta/zeta is associated with molecular, cellular, behavioral and cognitive schizophrenia phenotypes. Transl Psychiatry 1:e8
69. LaForgia S, Morse B, Levy J, Barnea G, Cannizzaro LA, Li F, Nowell PC, Boghosian-Sell L, Glick J, Weston A (1991) Receptor protein-tyrosine phosphatase gamma is a candidate tumor suppressor gene at human chromosome region 3p21. Proc Natl Acad Sci USA 88:5036–5040
70. Lamprianou S, Vacaresse N, Suzuki Y, Meziane H, Buxbaum JD, Schlessinger J, Harroch S (2006) Receptor protein tyrosine phosphatase gamma is a marker for pyramidal cells and sensory neurons in the nervous system and is not necessary for normal development. Mol Cell Biol 26:5106–5119
71. Strayer DS, Jerng HH (1992) Sequence and analysis of the BamHI "D" fragment of Shope fibroma virus: comparison with similar regions of related poxviruses. Virus Res 25:117–132
72. Hsiao JC, Chung CS, Chang W (1999) Vaccinia virus envelope D8L protein binds to cell surface chondroitin sulfate and mediates the adsorption of intracellular mature virions to cells. J Virol 73:8750–8761
73. Moss B (2012) Poxvirus cell entry: how many proteins does it take? Viruses 4:688–707
74. Matho MH, Maybeno M, Benhnia MR, Becker D, Meng X, Xiang Y, Crotty S, Peters B, Zajonc DM (2012) Structural and biochemical characterization of the vaccinia virus envelope protein D8 and its recognition by the antibody LA5. J Virol 86:8050–8058
75. Ohradanova A, Vullo D, Kopacek J, Temperini C, Betakova T, Pastorekova S, Pastorek J, Supuran CT (2007) Reconstitution of carbonic anhydrase activity of the cell-surface-binding protein of vaccinia virus. Biochem J 407:61–67
76. Babkin IV, Babkina IN (2011) Molecular dating in the evolution of vertebrate poxviruses. Intervirology 54:253–260
77. Keightley PD (2012) Rates and fitness consequences of new mutations in humans. Genetics 190:295–304
78. Suyama M, Torrents D, Bork P (2006) PAL2NAL: robust conversion of protein sequence alignments into the corresponding codon alignments. Nucleic Acids Res 34:W609–W612
79. Ronquist F, Teslenko M, van der Mark P, Ayres DL, Darling A, Hohna S, Larget B, Liu L, Suchard MA, Huelsenbeck JP (2012) MrBayes 3.2: efficient bayesian phylogenetic inference and model choice across a large model space. Syst Biol 61:539–542
80. Paradis E, Claude J, Strimmer K (2004) APE: analyses of phylogenetics and evolution in R language. Bioinformatics 20:289–290

Chapter 9
Membrane Associated Carbonic Anhydrase IV (CA IV): A Personal and Historical Perspective

Abdul Waheed and William S. Sly

Abstract Carbonic anhydrase IV is one of 12 active human isozymes and one of four expressed on the extracellular surfaces of certain endothelial and epithelial cells. It is unique in being attached to the plasma membrane by a glycosyl-phosphatiydyl-inositol (GPI) anchor rather than by a membrane-spanning domain. It is also uniquely resistant to high concentrations of sodium dodecyl sulfate (SDS), which allows purification from tissues by inhibitor affinity chromatography without contamination by other isozymes. This unique resistance to SDS and recovery following denaturation is explained by the two disulfide bonds. The 35-kDa human CA IV is a "high activity" isozyme in CO_2 hydration activity, like CA II, and has higher activity than other isozymes in catalyzing the dehydration of HCO_3^-. Human CA IV is also unique in that it contains no oligosaccharide chains, where all other mammalian CA IVs are glycoproteins with one to several oligosaccharide side chains.

Although CA IV has been shown to be active in mediating CO_2 and HCO_3^- transport in many important tissues like kidney and lung, and in isolated cells from brain and muscle, the gene for CA IV appears not to be essential. The CA IV knockout mouse produced by targeted mutagenesis, though slightly smaller and produced in lower than expected numbers, is viable and has no obvious mutant phenotype. Conversely, several dominant negative mutations in humans are associated with one form of reitinitis pigmentosa (RP-17), which we attribute to unfolded protein accumulation in the choreocapillaris, leading to apoptosis of cells in the overlying retina.

A. Waheed • W.S. Sly (✉)
Edward A. Doisy Department of Biochemistry and Molecular Biology,
Saint Louis University, School of Medicine, St. Louis, MO, USA
e-mail: waheeda@slu.edu; slyws@slu.edu

S.C. Frost and R. McKenna (eds.), *Carbonic Anhydrase: Mechanism, Regulation, Links to Disease, and Industrial Applications*, Subcellular Biochemistry 75,
DOI 10.1007/978-94-007-7359-2_9,

Keywords Carbonic anhydrase IV • Purification • Characterization • Catalytic properties • Expression • Structure • Localization • Mutation • Disease • Taste perception

1 Introduction

Carbonic anhydrases are zinc metalloenzymes and are the products of a gene family that encodes 16 different functionally active proteins. There are three carbonic anhydrase related proteins (CARPs) that are coded by distinct genes and are functionally inactive. Based on sub-cellular locations of different carbonic anhydrases, they are divided into four subgroups: cytosolic, secretory, membrane associated and mitochondrial. These 16 mammalian enzymes vary significantly in their enzyme kinetics, sensitivities towards carbonic anhydrase inhibitors, and tissue distributions [1–6].

The members of carbonic anhydrase family catalyze the reversible reaction of hydration and dehydration of CO_2 and HCO_3^-. Therefore, carbonic anhydrases play important roles in homeostasis of CO_2, HCO_3^-, and pH in different tissues to regulate different physiological processes. Physiological functions that are regulated by CA consist of removal of HCO_3^- in lung by respiration, reutilization of HCO_3^- in the kidney, production of aqueous humor in eyes, cerebrospinal fluids in brain, gastric juice production in stomach, pancreatic juice, and bone resorption by osteoclasts. Carbonic anhydrase family members also play important roles in metabolic processes that include ureagenesis, gluconeogenesis, and lipogenesis [7–12].

The clinical significance of the carbonic anhydrase family of proteins became apparent when CA II deficiency was discovered 29 years ago in patients with a syndrome of osteopetrosis, renal tubular acidosis, and brain calcification [13, 14]. Overexpression of CA IX and CA XII associated with different cancers and tumors was discovered 20 years ago in cancer patients [15, 16]. Carbonic anhydrase XIV is highly expressed on a unique membrane domain of glial cells and retinal pigment epithelium (RPE), where CA XIV regulates CO_2 removal, pH, and subretinal fluid absorption to modulate photoreceptor function [17, 18]. CA XVI, a GPI- linked enzyme is not present in humans, but was expressed in mouse tissues; therefore, it may have a physiological role in mouse kidney [7]. Around 10 years ago, a mutation in the *CA IV* gene was linked to a retinal degeneration disease, retinitis pigmentosa 17 (RP-17) in families of different ethnicity [19, 20]. Biochemical, molecular, and cellular studies in vitro and in an RP-17 disease model of mice suggested that RP-17 is a disease of protein folding that could be treated with chemical chaperones [19, 21–23].

Carbonic anhydrase IV is a very unique carbonic anhydrase enzyme that was discovered very early on as a membrane-associated enzyme, which was purified from SDS extracts of bovine lung membranes. The bovine enzyme was characterized as an SDS-resistant disulfide bond-containing glycoprotein of 52 kDa. CA IV purified from different tissues of different mammals are N-linked glycoproteins with a molecular weight of 39–52 kDa. Upon treatment with endoglycosidase F, the polypeptide mass is reduced to 35–36 kDa. However, CA IV purified from human lung or kidney membranes is a non-glycoprotein of 35 kDa and is anchored to the plasma membrane via GPI-anchor [24–27].

2 History of the Purification and Characterization of the Membrane Associated CA IV Enzyme Between 1982 and 1990

The presence of membrane-associated carbonic anhydrase was suspected from CA activity based on histochemical staining of membranes of different tissues, including lung, kidney, and skeletal muscle [28, 29]. However, the first membrane-associated CA was purified to homogeneity from bovine lung membrane extracts in high concentrations of SDS (1–5 %) using affinity chromatography by Whitney and Briggle [24]. The enzyme was named CA IV to distinguish it from the cytoplasmic enzymes, CA I, CA II, and CA III. Bovine CA IV was characterized as a 52-kDa glycoprotein with disulfide bonds. The enzyme activity was stable for several hours in 1–5 % SDS, which allowed purification by affinity chromatography. However, prolonged storage under these conditions resulted an inactivation of the CA IV activity. The poor recovery of the protein prevented an extensive characterization of kinetic and physical chemical properties of the enzyme [24]. Two more purification procedures for CA IV from human kidney were described in 1984 and 1989, respectively [25]. The main difference was the use of Triton X-100 instead of SDS detergent for enzyme extraction from the membrane and microsomes of human kidney. The purified membrane-associated carbonic anhydrase was functionally inactive and showed two different polypeptides of apparent molecular mass of 68 and 34.4 kDa. It was confusing when antibodies raised against the 34.4-kDa polypeptides reacted to a polypeptide with an apparent molecular mass of 55 kDa in different tissues [30]. This discrepancy led Zhu and Sly to develop a new purification protocol for isolation of homogeneous human CA IV with functionally active enzyme from lung and kidney membranes. In their report in 1990, 5 % SDS extracts of human lung or kidney membranes were allowed to bind to pAMBS-affinity resins at room temperature for an hour, similar to Whitney and Briggle [24]. However, after this step, Zhu and Sly [27] made a fundamental change in the procedure to maintain the enzyme activity and improve enzyme recovery. In brief, after the enzyme was bound to the affinity resin in the presence of 5 % SDS, the resin slurries were washed step-wise, first with buffer containing 1 % SDS followed by a buffer containing 0.1 % Brij 35 or 0.1 % Triton X-100 to remove the deleterious detergent, SDS, before eluting the enzyme from the affinity resin at 4 °C. The CA IV recovered from this procedure was stable at 4 °C indefinitely, was fully active, and was homogenous on SDS-PAGE. Using this modified and improved protocol for CA IV purification, CA IVs were purified from several different mammalian tissues, including lung from cows, rats, and mice, and kidney from rabbit [26]. These CA IV enzyme preparations were used to raise antibodies in rabbits and the serum has been found to be very useful in biochemical characterization and demonstration of cellular localizations [22, 26, 31, 32].

2.1 Molecular Characterization of Human CA IV

Full-length cDNA for human CA IV was isolated and characterized from a human kidney library [19–23, 33]. The deduced amino acid sequence contains an 18-amino acid signal sequence at the N-terminal of the protein for endoplasmic reticulum (ER) translocation and a 260-amino acid "CA domain" containing active site amino acid residues that shows 30–36 % homology with cytoplasmic CAs. Later, the R14W mutation in the leader sequence, and the R65S and R219S mutations in the CA domain, were found to be associated with retinitis pigmentosa in RP-17 patients [21, 22]. At the C-terminal, an additional 27 amino acid residues containing the hydrophobic sequences of 21 amino acids sufficient to span the membrane are preceeded by the 6-amino acid signal sequence for GPI-anchoring [34]. The amino acid residue, Ser 266, was identified as the site for the attachment of the GPI anchor. The missense mutation, S266F, prevented the removal of the C-terminal domain, GPI-anchoring, and cell surface expression and resulted in an ER retained, inactive enzyme. The G267S mutation has no effect on the activity, GPI-anchoring, and the cell surface expression of the enzyme. However, the G267X C-terminal deletion mutation resulted in secretion of the soluble, secretory form of fully active enzyme found in amounts several fold higher than amounts found in cell lysates transfected with wild-type cDNA. From these results, we concluded that removal of the C-terminal hydrophobic domain is necessary for GPI-anchoring, cell surface expression, and realization of the enzyme activity [34].

Based on amino acid sequences deduced from the nucleotide sequence, human CA IV contains no classical consensus sites (Asn-Xxx-Ser/Thr) for N-glycosylation. These results suggested that N-linked oligosaccharides are not necessary for enzyme activity [26]. However, human CA IV is an exception in this respect. All other mammalian CA IVs contain one or more N- glycosylation site and are glycoproteins. The precursor form of human CA IV contains five cysteine residues at positions 6, 18, 28, 211, and 286. During maturation of the precursor enzyme into mature enzyme in endoplasmic reticulum (ER), the C-terminal cysteine at position 286 is removed [33]. The remaining cysteines form two pairs of disulfide bonds. Using biochemical procedures, microsequencing of amino acids and metabolic labeling with ^{35}S cysteine, we established that the first disulfide bond was between Cys 6 and Cys 28, and the second disulfide bond between Cys 28 and Cys 211. Both disulfide bonds contribute to the remarkable stability of human CA IV to high concentrations of SDS and to elevated temperature [35]. Contribution of individual disulfide bonds in biosynthesis, cell surface expression, and structural stability will be discussed in a separate section.

Comparison of the amino acid sequence of human CA IV with that of other human carbonic anhydrases showed 30–36 % similarity. Human CA IV has 16 of the 17 highly conserved amino acid residues that are at the active sites of most other CAs. One exception is a highly conserved proline residue at position 202 (with respect to the numbering system of CA I), which is replaced by a threonine residue [33]. We characterized five different mammalian CA IVs, including human CA IV and CA IV from rat, murine, rabbit, and bovine tissues. They are highly homologous with a few interesting differences. All five CA IVs contain conserved histidine residues. Histidine residue 64 (His 64) is implicated as a proton shuttle in the carbonic anhydrase mechanism. After closer examination of the amino acid sequences around the conserved histidines, we found the amino acid immediately upstream of conserved His 64 I (Glu 63) was different in murine and rat CA IV from the highly conserved Gly 63 found in bovine, rabbit, and human CA IVs [36]. This difference was an important clue to the cause for the differences between the high activity CA IV enzymes (bovine, rabbit, and human) and the low activity enzymes of rat and mouse. To test this hypothesis, we changed Gln 63 to Gly 63 in mouse and rat CA IVs, and Gly 63 to Gln 63 in bovine and rabbit CA IVs, and compared activities after transient transfection of COS-7 cells with the different cDNAs. We found that replacing Gln 63 for Gly 63 in bovine and rabbit enzymes resulted in reuction of the CA activities of both enzymes by 37–42 %, respectively. When Gly 63 was replaced for Gln 63 in mouse CA IV, the enzyme activities were 3-fold higher than the native rodent enzymes. These results were confirmed using affinity purified secretory forms of the mutant CA IVs. These results led us to conclude that the decrease in the enzyme activity in the rodent CA IV is due to the presence of the bulkier Gln 63 at the N-terminal of His 64, which impairs its' proton transfer from His 64 to water. Subsequently, Hurt et al. [37] measured the k_{cat} for CO_2 hydration by the murine enzyme and compared it with human CA IV. They found that murine CA IV is 10-fold slower than the human enzyme due to a slower rate of proton transfer. The structural interpretation for the slower rate of proton transfer was confirmed after we solved the three-dimensional structure of human CA IV. The bulkier Gln side chain causes a conformational change around His 64, which reduces the rate constant for proton transfer [36, 37].

2.2 Genomic Organization and Chromosomal Localization of Human CA IV

The human *CA IV* gene is 9.5 kb and localized to chromosome 17q23. The human *CA IV* gene has 1–7 exons, like other human carbonic anhydrases. However, exon 1 is split. Exon 1a encodes the ER signal sequence. Exons 1b through 7 encode the remaining coding sequences and the 3′ end untranslated region [38]. The genomic organization and chromosomal localization of other human CA genes are summarized in an earlier review [2].

2.3 Catalytic Properties of Human CA IV

Human CA IV is a "high activity" enzyme, like human CA II in hydrating CO_2. However, CA IV is more active than CA II in HCO_3^- dehydration, as is evident from the 3-fold increase in k_{cat}/K_m. The esterase activity of CA IV is decreased 150-fold compared to that of CA II. In fact, it is hard to measure any esterase activity at room temperature with prolonged incubation time because of non-enzymatic hydrolysis of the substrate, 4-nitrophenylacetate [39]. The very low esterase activity of CA IV is very similar to the esterase activity of CA III. The catalysed hydration and dehydration rates at low concentrations of buffer are increased. Further analyses of activities of CA IV indicate a ping-pong kinetic mechanism for CA IV similar to the kinetic mechanism of CA II [39].

Human CA IV activity is inhibited by sulfate ions. The results of pH dependent inhibition of CA IV catalyzed dehydration of bicarbonate by sulfate ions suggested sulfate ions binding to zinc-bound water. When sulfate inhibition of the CA IV catalyzed reaction was compared with that of CA II [39], it was found that CA IV is much more sensitive to inhibition by sulfate than CA II. Other anions, such as chloride, bromide, iodide, formate, acetate, and phosphate, were found to be inhibitory as well. These anions were also able to increase the ratio of k_{cat}/K_m when the concentration of anions was less than 20 mM [39].

The kinetic properties of mouse CA IV are similar to human CA IV with the exception that the CO_2 hydration rate catalyzed by mouse CA IV is 10-fold lower than human CA IV at pH 7. As discussed above, these results led us to conclude that the proton transfer shuttle is partially blocked [37].

Sulfonamide Inhibitors. Sulfonamides are known to be clinically useful inhibitors of carbonic anhydrases. They are effective inhibitors of CA IV by directly binding to zinc ion at the active site. Dissociation constants of different sulfonamides, dansylamide,acetazolamide, ethoxzolamide, sulfanilamide, and benzenesulfonamide, have been measured and compared with another high activity enzyme, CA II [8, 39]. The affinity constant for dansylamide is comparable with that of CA II. However, binding of other sulfonamides to CA IV is 3–65-fold lower than to CA II. The side chain of amino acids in the active site pocket of CA IV might be responsible for reduced affinity of sulfonamide. This structure could also explain why CA IV shows such poor esterase activity compared to CA II when a phenyl derivative of ester is used as substrate [39].

2.4 Structure-Function Relationship and Structural Stability of Human CA IV

Among all CAs, only CA IV is rapidly renatured on dilution or dialysis after exposure to high concentrations of SDS and urea [35]. CA IV is also renatured on ice after exposure to high heat. It is this property that allowed us and others to purify CA IV from mammalian tissue extracts using CA-affinity resin and high concentrations of the ionic detergent, SDS, in which all other CAs are inactive. However, CA IVs are no less sensitive to denaturation by SDS than other CAs under reducing conditions [35]. This result suggested that the structure and function of CA IV are stabilized by disulfide bonds [35]. As mentioned earlier, the cDNA sequence of mature CA IV predicts four cysteine residues at amino acid position 6, 18, 28, and 211. Using metabolic labeling of affinity purified CA IV from secretion of CHO clones expressing the secretory form of the enzyme, we identified the arrangement of disulfide bonds chemically [35]. These results identified the two disulfide bonds between Cys 6 to Cys 18 and Cys 28 to Cys 211 in native enzyme. Reduction of disulfide bonds or reduction followed by alkylation of reduced sulfhydryl groups results in 70 % inactivation, and the reduced enzyme is susceptible to complete inactivation by denaturants. The native enzyme with intact disulfide bonds is remarkably more resistant to denaturants, and the loss in activity is rapidly recovered after removal of denaturants. Using fluorescence anisotropy, the structure of CA IV in the presence and absence of disulfide bonds was determined. The results suggested that the enzyme retains significant elements of structure in the presence of denaturants. We concluded that disulfide bonds play significant roles in retaining the native structure and some enzyme activity in the presence of denaturants and account for the rapid recovery of full enzyme activity after removal of denaturants [35].

CA IV is anchored to the plasma membranes of epithelial and endothelial cells of different organs by a glycosylphosphatidylinositol (GPI) tail [26, 27, 33, 34]. Based on nucleotide sequence of human CA IV cDNA, the human enzyme is synthesized as a 310-amino acid precursor in the endoplasmic reticulum with an amino terminal signal sequence (residues −18 to −1) that is removed co-translationally, and a C-terminal hydrophobic domain that is also cleaved and removed for preassembled GPI anchor attachment [33]. Using biochemical and molecular methods, we identified Ser 266 as the site for GPI-anchor attachment. We also observed that the S266F missense mutation prevented removal of the C-terminal hydrophobic domain, resulting in ER retention and accelerated degradation of the inactive enzyme, neither GPI-anchored nor expressed at the cell surface. However, the G267F mutant was fully active without any loss in GPI-anchoring and expression at the cell surface [34]. The C-terminal deletion mutant, G267X, produced a secretory form of the fully active enzyme and was made in amounts several fold higher than the enzyme found in cell lysates transfected with wild-type cDNA. From these results, we concluded that removal of C-terminal hydrophobic domain is required for both GPI-anchoring and realization of the enzyme activity [34]. Consistently, bypassing GPI-anchoring by deletion of the C-terminal hydrophobic domain resulted in secretion of fully active enzyme in amounts far greater (6–10 fold) than those seen when the wild-type enzyme accumulates at the cell surface as a GPI-anchored protein [34].

Additionally, we observed that the truncated, secretory form of the enzyme had similar catalytic properties to those of wild-type, GPI-anchored enzyme purified from human tissues [26, 27]. The three dimensional structure of the secretory form of human CA IV was determined by x-ray crystallographic methods to a resolution of 2.8 Å [40]. Structural integrity near the active site pocket, including Zn binding and substrate binding sites of CA IV, are similar to other mammalian CA isozymes. However, unique structural differences are present elsewhere in the active site. In particular, CA IV contains two disulfide bonds, Cys 6 to Cys 11 and Cys 23 to Cys 203, which stabilize the conformation of the N-terminal domain. The second disulfide bond, Cys 23 to Cys 203, additionally stabilizes an active loop containing cis peptide linkage between Pro 201 and Thr 202. This loop contains Thr 199, a catalytic residue [40]. The second disulfide bond is present in several membrane associated CAs. On the opposite side of the active site, the Val 131-Asp 136 segment forms an extended loop structure instead of an alpha-helix structure found in other isozymes [40]. The C-terminus of the enzyme is on the opposite side of the globular protein from the canonical active site cleft, and the mode and orientation of the enzyme association with membrane is optimal for the biological function of the CA IV. The active site of CA IV is oriented towards the lumen of microcapillaries for optimal enzyme activity [40]. The C-terminus is surrounded by many electropositive residues, which have been proposed to mediate the interaction of cell surface CA IV protein with negatively charged phosphate groups of the phospolipid membrane. This interaction orients the active site of the enzyme at membrane for efficient catalysis of CO_2 hydration and HCO_3^- dehydration in the lumen of capillary endothelial cells and renal tubular epithelial cells [40].

2.5 Biosynthesis and Cell Specific Localization of CA IV in Mammalian Tissues

Biosynthesis of CA IV was studied in COS-7 cells transfected with wild-type and mutant cDNAs. Metabolic labeling of the polypeptide backbone with radioactive methionine and the GPI-anchor with radioactive ethanolamine for 30 min followed by immunoprecipitation of CA IV protein, SDS-PAGE under reducing conditions, and fluorography, suggested that CA IV is synthesized as a35-kDa polypeptides containing radioactive ethanolamine [34]. However, radioactive ethanolamine was not incorporated in the S267F mutant polypeptide and the protein was rapidly degraded [34]. Immunohistochemical staining of COS-7 cells expressing wild-type enzyme showed expression of the wild-type enzyme at the cell surface with some intracellular accumulation in punctated vesicles and in the ER around the nucleous of the cell. When S266F mutant CA IV was expressed, there was no expression at the cell surface. The majority of the residual protein accumulated inside the ER and was functionally inactive [34]. From these results, we concluded that the wild-type human CA IV is synthesized as a GPI-anchored protein and the active enzyme is expressed at the cell surface of COS-7 cells [34].

A membrane-associated carbonic anhydrase had been detected in several tissues by activity based histochemical staining that showed no positive staining with antibodies against soluble CAs. Once CA IV enzyme was purified from human lung and kidney to homogeneity, antibodies against CA IV were raised. Using the CA IV specific antibodies, expression of a membrane-associated CA IV could be studied in different human tissues [26, 27].

2.5.1 Human Eye

In human eye, CA IV immunolocalization was different from human CA II, a soluble CA. An intense immunostaining for CA IV was associated with endothelial cells of choroidal capillaries, the choriocapillaries [41]. At higher magnification, labeling with CA IV antibodies was observed at both the luminal and abluminal surfaces of the endothelial cells [41]. CA IV was also detected in epithelial and fiber cells of the lens, but and was not detected in the neuroretina, the cilliary process, and the cornea where CA II staining was intense. However, another report suggested that a 54-kDa protein is present in human retina, which they called CA IV [42]. This misidentification created considerable confusion in the field. Human CA IV is a 35-kDa nonglycosylated GPI-anchored protein. Human CA XIV is a 54-kDa protein later shown to be highly expressed in neural retina [17, 18]. Therefore, it seems likely that the authors' antibody reacted with CA XIV and they misidentified CA XIV in retina as CA IV. Another membrane associated carbonic anhydrase, CA XII, is also not present in retina. We repeat: CA XIV is the major membrane associated enzyme in mouse retina. CA IV is absent in the retina but highly expressed in choriocappillaris. The physiological function of CA IV in the choriocapillaries was suggested to regulate transport of ions and CO_2 between vasculature and retinal pigmented epithelium (RPE). Surprisingly, CA IV null mice showed no functional or morphological differences compared with normal littermates. However, CA IV/CA XIV double knockout mice showed a greater deficit in light response than that seen in the CA XIV null retina. From these results, it was concluded that CA IV also plays an important role in pH regulation, at least in the absence of CA XIV [43].

2.5.2 Mouse and Rat Kidney

Mouse and rat kidney CA IV has been immunolocalized in the apical plasma membranes of the proximal convoluted tubules and thick ascending limb of Henle [44–46]. Both segments are involved in bicarbonate resorption. CA IV has been also localized in the basolateral membranes of proximal tubules and epithelial cells of thick-ascending limb. CA IV is not detected in intercalated cells of the collecting ducts, but they are rich in cytosolic CA II. These studies suggested that CA IV plays important role in bicarbonate transport across the basolateral plasma membrane of the rat nepharon.

Human patients with retinitis pigmentosa-17 due to CA IV mutation showed no deficit in renal function. Thus, the renal epithelial cells are less sensitive to the protein folding defect and unfolded protein response than R14W CA IV expression cells in the eye [20]. Also, mice deficient in CA IV showed no defect in bicarbonate reabsorption unless CA XIV, which overlaps in distribution with CA IV in kidney, was also absent [47].

2.5.3 Pulmonary Expression

Pulmonary expression of CA IV was found to be developmentally regulated and was immunolocalized to the luminal side of alveolar capillary endothelial cells [24, 27, 43, 48, 49]. The large pulmonary vessels do not express CA IV. Selective expression of CA IV in the alveolar capillaries places CA IV in an important location to catalyze the dehydration of plasma bicarbonate to carbon dioxide, which can then diffuse across the capillary endothelial surface and be released during respiration. Although the physiological significance of membrane associated CA IV in lung function has been studied using selective CA inhibitors, results from studies were not conclusive [9]. It has been also suggested that membrane associated CA IV in the lung may be involved in regulatory control of ventilation. The mouse model deficient in CA IV has been developed, but study of respiratory function and acidosis in this mouse has yet to be done.

2.5.4 Skeletal and Cardiac Muscle

There was physiological and histochemical evidence suggesting membrane associated carbonic anhydrase in sarcolemmal membranes more than 50 years ago [50]. Only in 1992 [51], using immunological, biochemical, and functional procedures, was the membrane associated CA identified as CA IV, both in sarcolemmal and sarcoplasmic vesicle preparations of rat skeletal muscle. In 1994, very strong immunostaining with anti-rat CA IV or anti-human CA IV were reported in endothelial cells in the lumen of muscle capillary [52]. These results identified CA IV as the CA producing histochemical evidence of enzyme activity on capillary endothelial cells that had been reported earlier as a membrane associated CA IV. From these results, it was concluded that the majority of the CA IV in skeletal muscle is the GPI- anchored protein expressed on the surface of endothelial cells of capillaries. Only a relatively small amount of CA IV was also expressed in the sarcoplasmic and sarcolemmal membranes [51, 52]. More recently, CA IV was found to be located on the plasma membranes and on the sarcoplasmic reticulum membrane [53], using mouse specific antibodies to CA IV and confocal laser scanning microscopy. CA IV staining was homogeneously distributed on the plasma membrane, but more concentrated at the opening of T tubules [54, 55]. The physiological function of the membrane associated CAs of muscle fibers has been determined using single knockout or double and triple knockout animals, which are described in a separate section.

Mouse hearts do not express cytosolic CA, but the gene transcript for CA 13 has been demonstrated [56, 57]. In 1992, histochemical staining using the dansylsulfonamide technique stained a sarcolemmal CA activity in the rabbit heart that was not further identified. Using a modified Hansson technique, sarcolemmal associated CA was confirmed, though this assay could not distinguish among different CA isoforms [29]. In 1998, using electron microscopy and staining of ultrathin section with anti-rat or anti-human CA IV-immunogold, a weak intracellular staining was observed, suggesting CA IV association with sarcoplasmic reticulum of heart [58]. Finally, in 2006, several membrane associated carbonic anhydrase were identified using CA isoform specific antibodies and heart from CA IV, CA IX, and CA XIV knockout mice. Carbonic anhydrases IV, IX, and XIV are expressed in the heart but CA XII was not detected in the tissue [53]. These isoforms are located to unique subcellular locations, suggesting different functions of the different isozymes in excitation-contraction coupling [53].

2.5.5 Central Nervous System

The majority of brain cytosolic carbonic anhydrase is present in oligodendrocytes and astrocytes. The principle isoform of CA is CA II [59]. During synaptic transmission, a rapid neutralization in extracellular space of the brain takes place. Pharmacological inhibition of CA activity by the CA inhibitor, acetazolamide, resulted in amplification of alkaline pH transients in rat cerebellum [60]. Later, studies using brain slices suggested that CA-inhibitors impermeable to membranes also cause similar amplification in pH transients [60]. From these results, we concluded that an extracellular CA was the target for impermeant CA-inhibitors. This conclusion was supported by later studies where CA II deficient mice were used and evidence shown that different isoforms of CA may be responsible for extracellular buffering. In 1992, using mice and rat CA IV specific antibodies, both for immunostaining and electron microscopy, the membrane associated CA in brain tissues was identified as CA IV [61]. CA IV is located on the luminal surface of cerebral capillaries and associated with the blood–brain barrier. Thus, CA IV is one of the targets to explain the cortical effects of CA inhibitors on brain function [61].

In 2000, CA IV in mouse brain was further characterized by treatment of hippocampal slices with phosphatidylinositol-specific phospholipase C (PI-PLC), which cleaves the CA IV link from the surface of plasma membranes [62]. Before and after PI-PLC treatment, evoked alkaline pH shifts were recorded. These results demonstrated that the ability to buffer a rapid alkaline load of PI-PLC treated brain slices was reduced in comparison to control, untreated brain slices [62]. From these results, it was concluded that the interstitial CA activity of brain is due to CA IV. However, some residual CA activity remained after removal of CA IV from the brain slices. In 2001, the remaining CA activity was identified as membrane associated CA XIV by immunostaining using CA XIV specific antibodies. Expression of CA XIV in both human and mice brain was seen on neuronal membranes and axons [63]. Using carbonic anhydrase IV and XIV double knockout mice in 2005, it was observed that both membrane associated CAs contribute to extracellular buffering in the central nervous system. CA IV seems to be the more important extracellular CA in the hippocampus [47, 64].

2.5.6 Gastrointestinal Tract

Histochemical studies of rat alimentary tract have demonstrated cytosolic and membrane associated CA-activity in several regions of gut [31]. In 1995, a review on carbonic anhydrase in the alimentary tract nicely summarized the tissue-specific distribution of different CAs and the roles of CA in the physiological process of the gut [65]. The membrane-associated CA was identified as CA IV by Northern blot, RT-PCR analysis, and immunohistochemistry using anti-rat CA IV antibodies. Taking advantage of the resistance of the CA IV enzyme to denaturation by SDS and loss in the activities of other CAs, the contribution of CA IV activity to total CA activity was determined in different regions of the rat gut [31]. The CA IV activities are spread along the gut, but maximum contribution of CA IV activity in total CA activity was observed in the ileum. Immunohistochemically, CA IV was localized to the apical (luminal) side of the epithelial plasma membranes [31]. Therefore, it has been proposed that functional loss of CA IV in the ileum and colon may affect the water and salt reabsorption contributing to pathogenesis of diarrheal condition [31].

2.5.7 Liver, Gall Bladder, and Pancreas

In 1991, the plasma membrane associated CA activity in hepatocytes of regenerating liver was demonstrated by biochemical method [66]. It was found that the CA activity in the plasma membrane was reduced by 55 % in regenerating liver and the activity returned to normal in 7 days. A year later, membrane associated CA was observed in the endoplasmic reticulum of male rat liver [28]. Since CA IV was the only known membrane associated CA at that time, the membrane associated CA was characterized as CA IV. However, in 1996 it was shown that human liver did not show any immunoreactive CA IV on Western blot [32]. Finally, in 2002, the membrane associated CA on hepatocyte plasma membranes was identified as CA XIV using mouse CA XIV specific antibodies. Therefore, it will be interesting to see the expression level of CA XIV in regenerating liver after partial hepatectomy [67].

In 1986, histochemical staining demonstrated membrane associated CA in the gallbladder epithelium [32]. Using total gallbladder homogenates from human and detergent phasing procedure to separate membrane associated GPI-anchored proteins, CA IV was identified by Western blot [32]. The gallbladder was also analyzed by immunohistochemistry using confocal laser scanning microscopy. The staining for CA IV was seen at the surface epithelial cells. The subepithelial capillary endothelium also showed expression of CA IV. From these results, it was concluded that typical GPI-anchored CA IV is expressed in human gallbladder where its function has not been clarified [67].

In 1990, a 55-kDa immunoreactive CA IV was identified in human pancreas [68]. Twelve years later, in the human pancreatic cell line, CFPAC-1, investigators found a 35-kDa CA IV. In 2004, using two different antibodies raised against the 18-amino acid peptide at the N-terminal of mature CA IV and 19-amino acid peptide at the C-terminal of precursor enzyme, which is removed in the endoplasmic reticulum during maturation and GPI-anchoring of CA IV, a 35-kDa peptide was observed in pancreatic homogenates [68]. The antibodies against the C-terminal peptide did not react with mature recombinant GPI-anchored human CA IV. The antibodies against the C-terminal peptide were also used for immunohistochemical staining of CA IV in normal human pancreas. From these results, it was concluded that the C-terminal antibodies recognize the precursor peptide of CA IV and it is expressed at the luminal plasma membranes or in the cytoplasm. The mode of anchoring of the CA IV in the plasma membranes of pancreatic tissues or in isolated human pancreatic duct cells was characterized as the hydrophobic C-terminal peptide present in the precursor CA IV. It is not clear yet whether this expression of CA IV was at the cell surface or in transit to its final destination at the cell surface. Whether the immune-reactive enzyme associated with the membrane via a hydrophobic peptide is functionally active in human pancreas remained to be answered [68]. These observations are puzzling since there is evidence that blocking the GPI-anchoring process results in intracellular accumulation of inactive CA IV in the ERT that is subject to rapid degradation [34].

2.5.8 Reproductive Organs

In 1976, histochemical staining of epididymis and testes showed membrane associated CA-mediated reaction [69]. In 1993, using human CA IV specific antibodies, the membrane associated staining was identified as human CA IV by immunohistochemistry and Western blot analysis [70]. In 2005, using RT-PCR and CA isozyme specific antibodies, several membrane-associated CAs, CA IV, CA XII, and CA XIV, were identified in specific cells of different regions of epididymis of rat. The physiological functions of the membrane associated CAs in the reproductive organs remain to be studied. However, single, double, and triple knockout mice have been produced and their fertility was not severely affected [47].

2.6 *Physiological Function of CA IV*

2.6.1 Renal Physiology

In cultured mammalian kidney cells, co-expression of CA IV and NBC1, a sodium bicarbonate co-transporter, results in association between these proteins and an increase in the transport of proton and bicarbonate from the cells [46, 71]. By analogy, in the epithelial cells of the proximal tubule, filtered bicarbonate and protons provided by NHE1, the sodium hydrogen exchanger form of carbonic acid that is dehydrated by CA IV and/or CA XIV into CO_2 at the brush border of proximal tubule for easy diffusion from the cells. At the basolateral surface of the epithelial cells, CA IV and/or CA XII may associate with sodium bicarbonate transporter, kNBC1, to remove bicarbonate at the basolatateral surface, affecting net bicarbonate reabsorption. Intercalated cells reabsorb the remaining 10–15 % of filtered bicarbonate by dehydration by apically expressed CA IV [71]. In the mouse model of CA IV knockout and in human patients with inactive mutant CA IV (RP-17), abnormal urinary pH was not observed, likely because of redundancy in membrane associated carbonic anhydrases in the nephron [20, 47].

2.6.2 Neurophysiology

Rapid shifts in interstitial pH are associated with neural discharge [60]. The endogenous pH change can influence neuronal function through ion-gated channels [62]. The level of influence depends on the size, speed, and spread of endogenous pH changes. The most important factor that controls these variables is the buffering capacity of the extracellular fluids [60]. Several lines of evidence suggested that the membrane-associated extracellular CA is responsible for the regulation of intracellular pH transients in brain [62]. In 2000, using biochemical and immunological techniques, it was demonstrated that the extracellular CA activity is mainly due to CA IV [62]. However, in 2005, both CA IV and CA XIV membrane associated enzymes were implicated in extracellular buffering of mouse brain [47]. The relative contributions of these enzymes were analyzed using CA IV and CA XIV single or double knockout animals. Electrophysiological studies on hippocampal slices suggested that either CA can catalyze the buffering after synaptic transmission. In the absence of both CA IV and CA XIV, complete inhibition in buffering after synaptic transmission was observed, as is seen when the CA-inhibitor, benzolamide, was present [47]. From these observations, it appears that both membrane associated CA IV and CA XIV contribute to extracellular buffering, although CA IV seems to be more important in the hippocampus [47].

Using a heterologous expression system, it has been also observed that CA IV plays an important role in the transport of protons or bicarbonate by associating with the chloride-bicarbonate exchangers, AE1-AE3 [64]. These observations became more relevant when AE3 knockout mice showed a lower threshold for the generation of seizures in mice because of the deficit in neural pH regulation in hippocampal neurons. To study the physiological role of extracellular CA-activated transport by AE3 in hippocampal neurons, the rise in intracellular pH due to short exposure to NH^{4+} was studied. The results from such studies suggested that acid loading to compensate for alkalosis is enhanced by extracellular CA. Using specific inhibitory antibodies against CA IV and CA XIV, it was again demonstrated that both extracellular CAs have roles in opposing intracellular alkaline load. Thus, one can conclude that both CA IV and CA XIV contribute to neuronal pH regulation [64].

2.6.3 Eye Physiology

A role for carbonic anhydrases in vision physiology has been appreciated for long time. CA II, a cytosolic enzyme, is highly expressed in the cilliary epithelium and is responsible for aqueous humour production and regulation of intraocular pressure in the eye [72]. The human eye disease, glaucoma, has been treated with CA-inhibitors to prevent retinal degeneration due to high ocular pressure. Human patients and mice with mutations in CA II show visual defects [72]. The molecular mechanism of the visual defect in humans is not well understood. Optic atrophy is seen in some patients but osteopetrosis, part of the CA II deficiency syndrome, may contribute by narrowing of the optic foramina, causing optic nerve compression [72].

Histochemical staining of the eye in CA II deficient mice showed membrane associated CA that was later identified as CA IV. However, it was discovered that CA IV in human and mouse eyes is not expressed in retina [17, 18, 41]. It is strongly expressed in endothelial cells of the choriocapillaris, a vascular layer supplying nutrients to retina. After it was discovered that a signal sequence mutation in the *CA IV* gene results an autosomal dominant form of retinitis pigmentosa (RP-17), interest in the role of CA IV in eye physiology has been renewed [19–21]. In 2005, another membrane associated enzyme, CA XIV was identified on retinal pigment epithelium (RPE), Mueller cells, and apical and basolateral membranes of RPE [17, 18]. Because of its strategic location, a physiological function for CA XIV in eye has been predicted.

Using single and double knockouts of CA IV and CA XIV mice, retinal function and morphology of the eye have been studied [43]. These results suggested that CA IV null mice show minimum defect in retinal function and morphology of the retina. On the other hand, CA XIV deficiency results in significant reduction in the amplitude of the electroretinogram (ERG) without any morphological defects. These results suggested that CA XIV plays an important role in normal retinal light response [43]. A larger functional deficit in retinal light response was found in the mice doubly deficient in both CA IV and CA XIV enzymes. From these results, it was concluded that CA IV also contributes to the light response in eye physiology, at least in the absence of CA XIV [43].

2.6.4 Skeletal and Cardiac Muscle Physiology

The results from several biochemical and immunological studies suggested that CA IV and CA XIV are both located at the plasma membranes and sarcoplasmic reticulum membranes [53–55]. CA IX is not expressed at the plasma membrane, but is seen in the t-tubular (TT) terminal SR membrane [55]. CA XII is not expressed in either sarcoplasmic or sarcolemmal membranes. It has been proposed that CA in muscle, associated with the sarcoplasmic reticulum, plays an important role in the kinetics of Ca^{2+} transport, resulting in control of muscle contractions [55]. Using single knockouts of CA IV, CA IX, or CA XIV mice, physiology of fast-twitch (EDL) and slow-twitch (Sol) muscle fibers was studied [73]. From such studies, it was concluded that the lack of any single CA in the muscle fibers does not affect the muscle contraction. However, lack of all three membrane associated CAs, CA IV, CA IX, and CA XIV, affects muscle contraction [72].

Like skeletal muscle fibers, microsomal membranes of mouse heart show expression of membrane associated CAs, CA IV, CA IX, and CA XIV [53]. Isozyme CA XII is not expressed in mouse heart. Before 2006, expression of CA IX was not appreciated in cardiac muscle [53]. Finally, CA IX was also identified in heart muscle using biochemical and immunological procedures, and its' absence shown in CA IX deficient mouse heart. There have been no direct measurements of cardiac muscle function. However, based on the unique subcellular locations of CA IV, CA IX, and CA XIV, it has been suggested that these membrane associated CAs might play roles in excitation-contraction coupling in heart muscle fibers [53].

2.7 CA IV in Metabolons

Metabolons are complexes of proteins involved in a metabolic pathway that allow rapid transport of substrates and metabolites from the active sites of one protein component to the next. The rapid transport of substrate and metabolites is accomplished by "substrate channeling" to maintain the optimum local concentration of substrates for metabolic processes. Direct or indirect physical association between protein components of metabolons improve the efficiency of metabolic processes. The first study of CAs in a metabolon in 1998 showed association of chloride-bicarbonate anion exchange proteins (AEs) with carbonic anhydrase II (CA II) [74]. It was also demonstrated that activity of AE proteins to transport of bicarbonate was enhanced after physical association with CA II [74].

The strategic location of membrane associated CA IV and AE proteins at the plasma membranes of endothelial and epithelial cells in different tissues raised the question of a similar physical and functional relationship between AE and CA IV. In 2002, GPI-anchored CA IV was reported to associate with AE and to increase the bicarbonate transport rate [75]. Using pull-down assays, it was observed that association between CA IV and AE1 takes place on the large fourth extracellular loop of AE1. It was also observed that cytosolic CA II interacts with cytoplasmic C-terminal tails of AE1, resulting in an increase in bicarbonate transport. Based on these results, a working model with a "push-pull" mechanism for bicarbonate transport has been proposed [75].

In 2003, physical and functional association of CA IV with the sodium/bicarbonate co-transporters (NBC) has been studied using transfection of heteralogous mammalian cells with NBC alone and/or together with CA IV [76]. It was found that CA IV associates with NBC1 and intracellular pH recovery is increased upon CA IV expression. These results suggested that CA IV also forms a metabolon with NBC1 and interaction with NBC1 is necessary for full activity of the cotransporter for bicarbonate transport. Direct interaction between NBC1 and CA IV was demonstrated using sucrose-density gradients and by gel overlay assay. However, CA IV was not found to facilitate the bicarbonate transport in bovine corneal endothelial cells [77]. The major difference between these studies is the nature of cell types used. A stable heterologous mammalian cell line expressing both

NBC1 and CA IV proteins was used whereas the bovine corneal endothelial cells were in primary cell cultures. Further, questions have been raised about the chloride-bicarbonate exchanger, AE1, for bicarbonate transport, and for NBC metabolons where CA does not bind directly to AE1 or other transporters. In addition, CA II, upon binding, does not appear to accelerate bicarbonate transport [78].

2.8 *Mutations in CA IV Resulting in Pathological Disease*

2.8.1 Eye Pathology

Retinal pigmentosa-17 (RP-17) is a progressive retinal degenerative disease of eye. Clinical manifestations of the disease include night blindness, gradual loss of visual fields, and, finally, complete blindness. Several missense mutations in the leader sequence of CA IV, R14W, or A12T, and in the body of the CA IV protein, R219S or R69H, are linked to RP-17 disease. RP-17 patients have been identified in several unrelated families [19–23, 79]. Yet, CA IV is not expressed in human retina. Thus, RP-17 is an eye disease caused by the mutation in a non-retinal gene, CA IV. There are other examples of mutations in a non-retinal gene causing retinal degeneration disease [19].

The molecular basis of RP-17 was studied using transiently transfected heterologous mammalian cells, COS-7 and HEK-293, with disease associated mutant proteins. Several biochemical and cellular properties were compared with the cells transfected with cDNA expressing wild-type protein. At the steady state, the cell lysates transfected with R14W and A12T mutant CA IV showed 70 and 71 %, respectively, of wild-type CA activity [21, 79]. R69H mutant enzyme has nearly full activity, whereas R219S mutant has lost most of the CA activity. From these results, it was concluded that the loss of CA IV activity itself is not the molecular basis of RP-17 pathogenesis. The results from several physiological studies on CA IV deficient mice suggested that CA IV is a non-essential enzyme. To answer the question of why R14W, A12T, R69H, and R219S mutations in CA IV results in a dominant RP disease, the effects of expression of these mutant CA IV in heterologous mammalian cells, COS-7 and HEK-293, were studied [21–23, 79]. All mutations associated with RP-17 produced misfolding of the CA IV enzyme, resulting in endoplasmic reticulum (ER) stress and activation of unfolded protein response (UPR). A sustained activation of UPR is linked to inflammatory cascade signaling and results in apoptotic cell death of endothelial cells of the choriocapillaris [21–23]. Therefore, RP-17 is primarily a disease of the choriocapillaris and the pathology in the retina is the result of ER stress and apoptotic death of endothelial cells of capillaries, which help to nourish the retina. Similarly, ER stress induced apoptosis of cells causing pathology has been found in many human diseases, including diabetes, Parkinson's, Alzheimer's, hypoparathyroidism, and one form of retinal pigmentosa caused by mutant rhodopsin [21, 22]. Since all CA IV mutants associated with RP-17 result in ER stress and cell death, we suggested a common mechanism for RP-17 disease and a single mutant allele that produces the dominant disease by a gain of function mechanism. The heterologous cells expressing mutant CA IV associated with RP-17 also show dominant negative effects on biosynthesis, secretion, and cell surface expression of other secretory and membrane associated proteins [19–23].

In 2005, an alternate hypothesis was proposed to explain the molecular basis of RP-17 disease. It ascribed the disease to haplo insufficiency of CA IV activity in CA IV mutants associated with retinitis pigmentosa-17 [20]. The major assumption of the hypothesis is the association of CA IV with the sodium/bicarbonate co-transporter 1 (NBC 1) forming a functional metabolon in endothelial cells of the choriocapillaries that regulates the pH homeostasis of retina. Therefore, loss of a functional metabolon complex due to missense mutations in CA IV associated with RP-17 would disrupt bicarbonate transport and pH homeostasis in endothelial cells of choriocapillaries, resulting indirectly in photoreceptor degeneration [20]. Building on this speculation, it has been hypothesized that long-term use of CA inhibitors for treatment of glaucoma should also result in functional deficiency of CA IV, loss in function of the CA IV-metabolon for bicarbonate transport and thereby result in retinal degeneration. This hypothesis fails to explain: (1) how loss of only 1 allele can cause a dominant disease of retinal degeneration; (2) lack of retinal abnormality and function in the CA IV knockout mice, suggesting that complete loss CA IV function does not cause RP-17 disease; and, (3) long-term use of CA-inhibitors for glaucoma has not been reported to cause retinal abnormality.

Several studies between 2007 and 2009 on RP-17 associated mutants in COS-7 and HEK-293 cells suggested that RP-17 is a protein folding disease. We reasoned that pharmacological "chemical chaperones" might rescue the misfolded proteins in cells expressing such mutant proteins [19–23]. Using specific CA-inhibitors and non-specific chaperones, phenyl butyric acid, a histone deacetylate (HDAC) inhibitor, we showed improvement in protein folding and processing, and reduction in cell death. We reversed the dominant negative effects on protein synthesis and trafficking of the mutant enzyme [19–23]. It should be noted that CA inhibitors have been proven effective in treatment of glaucoma and cystoids macular edema in RP. Thus, we suggested that CA-inhibitors may be an effective and safe pharmacological compound in the treatment of RP-17 [19, 21, 22].

An interesting observation was made in 2008 that RP-17 associated R14W mutation in CA IV is a benign polymorphism in a Swedish population [80]. This discovery raised the question of how the RP-17 causing mutation in patients of South Africa does not cause disease in the Swedish population. Although the genetic background of these patients is a possible reason, an environmental factor could also play an important role. Exposure to light has been known to show damaging effect on photoreceptors and choroidal endothelial cells [81]. We suggested that longer exposure to sunlight may act as a second hit for the retinal pathogenesis. Based on this hypothesis, we suggested that the protection of eyes from sunlight exposure in South African patients might be beneficial and will result in milder phenotype or prevent the pathology altogether [22].

2.8.2 Kidney Pathology

Patient with RP-17 show no abnormality in kidney function, although expression of CA IV in renal tubular epithelial cells were reported in 1990 [20]. Using transgenic mice expressing RP-17 associated human CA IV under control of the endogenous human CA IV promoter, slowly progressive renal injuries in kidney of male mice at 20 weeks of age were observed [23]. The renal injuries increased dramatically in the transgenic mice when RP-17 associated mutants were expressed together with a heterozygous deficiency for the ER, cochaperone p58IPK [23]. These results suggested that expression of RP-17 associated mutants may also affect tissues other than retina, at least when the unfolded protein response pathway is compromised. The in vivo toxicity of RP-17 associated mutant CA IV seems to be related to folding capacities of the ER of different cells of the tissues [23].

2.9 CA IV and the Taste of Carbonation

It had long been known that a side effect of oral CA inhibitors, like acetazolamide, was the loss of the taste for carbonated beverages. We recently showed that it was CA IV on sour cells in taste buds that explains this side effect of CA inhibitors [82]. By targeted ablation and silencing of synapses of defined populations of taste receptor cells in the mouse, we identified the sour receptor cells as mediators of the acidic taste of carbonated beverages. These cells, which required the ion channel PKD2-L1 for response to sour stimuli, including CO_2, were also found on microarray studies to express very high levels of transcripts for membrane attached CA IV. Immunolocalization of CA IV in these cells supported these findings. Next, CA IV null mice were shown to lack the electro-physiological response to CO_2 that was seen in wild-type mice. From these observations, we concluded that the CO_2 hydration reaction catalyzed by CA IV on the surface of the sour cells provides the protons to PKD2-LI, which mediates "the taste of CO_2." Thus, the sour cells both provide the GPI anchor that retains CA IV at the cell surface and positions CA IV to play a key role the cellular sensation of carbonation.

Acknowledgments The authors wish to thank Tracey Baird for excellent editorial assistance in the preparation of this manuscript. Research on carbonic anhydrases in the Sly lab was supported by National Institutes of Health grant DK040163 to W.S.S.

References

1. Tashian RE (1989) The carbonic anhydrases: widening perspectives on their evolution, expression and function. Bioessays 10:186–192
2. Sly WS, Hu PY (1995) Human carbonic anhydrases and carbonic anhydrase deficiencies. Annu Rev Biochem 64:375–401

3. Christianson DW, Fierke CA (1996) Carbonic anhydrase: evolution of the zinc binding site by nature and by design. Acc Chem Res 29:331–339
4. Hewett-Emmett D (2000) Evolution and distribution of the carbonic anhydrase gene families. In: Chegwidden WR, Carter ND, Edwards YH (eds) The carbonic anhydrases: new horizons. Birkhauser Verlag, Basel, pp 29–76
5. Tashian RE, Hewett-Emmett D, Carter ND, Bergenhem NC (2000) Carbonic anhydrase (CA)-related proteins (CA-RPs) and transmembrane proteins with CA or CA-RP domains. In: Chegwidden WR, Carter ND, Edwards YH (eds) The carbonic anhydrases: new horizons. Birkhouser Verlag, Basel, pp 105–120
6. Nishimori I (2004) Acatalytic CAs: carbonic anhydrase-related proteins. In: Supuran CT, Scozzafava A, Conway J (eds) Carbonic anhydrase: its inhibitors and activators. CRC Press, Boca Raton, pp 25–42
7. Hilvo M, Tolvanen M, Clark A, Shen B, Shah GN, Waheed A, Halmi P, Hanninen M, Hamalainen JM, Vihinen M, Sly WS, Parkkila S (2005) Characterization of CA XV, a new GPI-anchored form of carbonic anhydrase. Biochem J 392:83–92
8. Supuran CT (2008) Carbonic anhydrases: novel therapeutic applications for inhibitors and activators. Nat Rev Drug Discov 7:168–181
9. Esbaugh AJ, Tufts BL (2006) The structure and function of carbonic anhydrase isozymes in the respiratory system of vertebrates. Respir Physiol Neurobiol 154:185–198
10. Dodgson SJ, Forster RE 2nd, Schwed DA, Storey BT (1983) Contribution of matrix carbonic anhydrase to citrulline synthesis in isolated guinea pig liver mitochondria. J Biol Chem 258:7696–7701
11. Imtaiyaz Hassan M, Shajee B, Waheed A, Ahmad F, Sly WS (2013) Structure, function and applications of carbonic anhydrase isozymes. Bioorg Med Chem 21:1570–1582
12. Supuran CT (2008) Carbonic anhydrases–an overview. Curr Pharm Des 14:603–614
13. Sly WS, Lang R, Avioloi L, Haddad J, Lubowitz H, McAlister W (1972) Recessive osteopetrosis: new clinical phenotype. Am J Hum Genet 24:A34
14. Sly WS, Hewett-Emmett D, Whyte MP, Yu YS, Tashian RE (1983) Carbonic anhydrase II deficiency identified as the primary defect in the autosomal recessive syndrome of osteopetrosis with renal tubular acidosis and cerebral calcification. Proc Natl Acad Sci U S A 80:2752–2756
15. Wykoff CC, Beasley N, Watson PH, Campo L, Chia SK, English R, Pastorek J, Sly WS, Ratcliffe P, Harris AL (2001) Expression of the hypoxia-inducible and tumor-associated carbonic anhydrases in ductal carcinoma in situ of the breast. Am J Pathol 158:1011–1019
16. Kivela AJ, Parkkila S, Saarnio J, Karttunen TJ, Kivela J, Parkkila AK, Pastorekova S, Pastorek J, Waheed A, Sly WS, Rajaniemi H (2000) Expression of transmembrane carbonic anhydrase isoenzymes IX and XII in normal human pancreas and pancreatic tumours. Histochem Cell Biol 114:197–204
17. Nagelhus EA, Mathiisen TM, Bateman AC, Haug FM, Ottersen OP, Grubb JH, Waheed A, Sly WS (2005) Carbonic anhydrase XIV is enriched in specific membrane domains of retinal pigment epithelium, Muller cells, and astrocytes. Proc Natl Acad Sci U S A 102:8030–8035
18. Ochrietor JD, Clamp MF, Moroz TP, Grubb JH, Shah GN, Waheed A, Sly WS, Linser PJ (2005) Carbonic anhydrase XIV identified as the membrane CA in mouse retina: strong expression in Muller cells and the RPE. Exp Eye Res 81:492–500
19. Rebello G, Ramesar R, Vorster A, Roberts L, Ehrenreich L, Oppon E, Gama D, Bardien S, Greenberg J, Bonapace G, Waheed A, Shah GN, Sly WS (2004) Apoptosis-inducing signal sequence mutation in carbonic anhydrase IV identified in patients with the RP17 form of retinitis pigmentosa. Proc Natl Acad Sci U S A 101:6617–6622
20. Yang Z, Alvarez BV, Chakarova C, Jiang L, Karan G, Frederick JM, Zhao Y, Sauve Y, Li X, Zrenner E, Wissinger B, Hollander AI, Katz B, Baehr W, Cremers FP, Casey JR, Bhattacharya SS, Zhang K (2005) Mutant carbonic anhydrase 4 impairs pH regulation and causes retinal photoreceptor degeneration. Hum Mol Genet 14:255–265
21. Bonapace G, Waheed A, Shah GN, Sly WS (2004) Chemical chaperones protect from effects of apoptosis-inducing mutation in carbonic anhydrase IV identified in retinitis pigmentosa 17. Proc Natl Acad Sci U S A 101:12300–12305

22. Datta R, Waheed A, Bonapace G, Shah GN, Sly WS (2009) Pathogenesis of retinitis pigmentosa associated with apoptosis-inducing mutations in carbonic anhydrase IV. Proc Natl Acad Sci U S A 106:3437–3442
23. Datta R, Shah GN, Rubbelke TS, Waheed A, Rauchman M, Goodman AG, Katze MG, Sly WS (2010) Progressive renal injury from transgenic expression of human carbonic anhydrase IV folding mutants is enhanced by deficiency of p58IPK. Proc Natl Acad Sci U S A 107: 6448–6452
24. Whitney PL, Briggle TV (1982) Membrane-associated carbonic anhydrase purified from bovine lung. J Biol Chem 257:12056–12059
25. Wistrand PJ, Knuuttila KG (1989) Renal membrane-bound carbonic anhydrase. Purification and properties. Kidney Int 35:851–859
26. Waheed A, Zhu XL, Sly WS (1992) Membrane-associated carbonic anhydrase from rat lung. Purification, characterization, tissue distribution, and comparison with carbonic anhydrase IVs of other mammals. J Biol Chem 267:3308–3311
27. Zhu XL, Sly WS (1990) Carbonic anhydrase IV from human lung. Purification, characterization, and comparison with membrane carbonic anhydrase from human kidney. J Biol Chem 265:8795–8801
28. Ono Y, Ridderstrale Y, Forster RE 2nd, Chu ZG, Dodgson SJ (1992) Carbonic anhydrase in the membrane of the endoplasmic reticulum of male rat liver. Proc Natl Acad Sci U S A 89: 11721–11725
29. Bruns W, Gros G (1992) Membrane-bound carbonic anhydrase in the heart. Am J Physiol 262:H577–H584
30. Carter ND, Fryer A, Grant AG, Hume R, Strange RG, Wistrand PJ (1990) Membrane specific carbonic anhydrase (CAIV) expression in human tissues. Biochim Biophys Acta 1026: 113–116
31. Fleming RE, Parkkila S, Parkkila AK, Rajaniemi H, Waheed A, Sly WS (1995) Carbonic anhydrase IV expression in rat and human gastrointestinal tract regional, cellular, and subcellular localization. J Clin Invest 96:2907–2913
32. Parkkila S, Parkkila AK, Juvonen T, Waheed A, Sly WS, Saarnio J, Kaunisto K, Kellokumpu S, Rajaniemi H (1996) Membrane-bound carbonic anhydrase IV is expressed in the luminal plasma membrane of the human gallbladder epithelium. Hepatology 24:1104–1108
33. Okuyama T, Sato S, Zhu XL, Waheed A, Sly WS (1992) Human carbonic anhydrase IV: cDNA cloning, sequence comparison, and expression in COS cell membranes. Proc Natl Acad Sci U S A 89:1315–1319
34. Okuyama T, Waheed A, Kusumoto W, Zhu XL, Sly WS (1995) Carbonic anhydrase IV: role of removal of C-terminal domain in glycosylphosphatidylinositol anchoring and realization of enzyme activity. Arch Biochem Biophys 320:315–322
35. Waheed A, Okuyama T, Heyduk T, Sly WS (1996) Carbonic anhydrase IV: purification of a secretory form of the recombinant human enzyme and identification of the positions and importance of its disulfide bonds. Arch Biochem Biophys 333:432–438
36. Tamai S, Waheed A, Cody LB, Sly WS (1996) Gly-63– > Gln substitution adjacent to His-64 in rodent carbonic anhydrase IVs largely explains their reduced activity. Proc Natl Acad Sci U S A 93:13647–13652
37. Hurt JD, Tu C, Laipis PJ, Silverman DN (1997) Catalytic properties of murine carbonic anhydrase IV. J Biol Chem 272:13512–13518
38. Okuyama T, Batanian JR, Sly WS (1993) Genomic organization and localization of gene for human carbonic anhydrase IV to chromosome 17q. Genomics 16:678–684
39. Baird TT Jr, Waheed A, Okuyama T, Sly WS, Fierke CA (1997) Catalysis and inhibition of human carbonic anhydrase IV. Biochemistry 36:2669–2678
40. Stams T, Nair SK, Okuyama T, Waheed A, Sly WS, Christianson DW (1996) Crystal structure of the secretory form of membrane-associated human carbonic anhydrase IV at 2.8-A resolution. Proc Natl Acad Sci U S A 93:13589–13594
41. Hageman GS, Zhu XL, Waheed A, Sly WS (1991) Localization of carbonic anhydrase IV in a specific capillary bed of the human eye. Proc Natl Acad Sci U S A 88:2716–2720

42. Wolfensberger TJ, Mahieu I, Jarvis-Evans J, Boulton M, Carter ND, Nogradi A, Hollande E, Bird AC (1994) Membrane-bound carbonic anhydrase in human retinal pigment epithelium. Invest Ophthalmol Vis Sci 35:3401–3407
43. Ogilvie JM, Ohlemiller KK, Shah GN, Ulmasov B, Becker TA, Waheed A, Hennig AK, Lukasiewicz PD, Sly WS (2007) Carbonic anhydrase XIV deficiency produces a functional defect in the retinal light response. Proc Natl Acad Sci U S A 104:8514–8519
44. Brown D, Zhu XL, Sly WS (1990) Localization of membrane-associated carbonic anhydrase type IV in kidney epithelial cells. Proc Natl Acad Sci U S A 87:7457–7461
45. Kaunisto K, Parkkila S, Rajaniemi H, Waheed A, Grubb J, Sly WS (2002) Carbonic anhydrase XIV: luminal expression suggests key role in renal acidification. Kidney Int 61:2111–2118
46. Schwartz GJ (2002) Physiology and molecular biology of renal carbonic anhydrase. J Nephrol 15(Suppl 5):S61–S74
47. Shah GN, Ulmasov B, Waheed A, Becker T, Makani S, Svichar N, Chesler M, Sly WS (2005) Carbonic anhydrase IV and XIV knockout mice: roles of the respective carbonic anhydrases in buffering the extracellular space in brain. Proc Natl Acad Sci U S A 102:16771–16776
48. Effros RM, Chang RS, Silverman P (1978) Acceleration of plasma bicarbonate conversion to carbon dioxide by pulmonary carbonic anhydrase. Science 199:427–429
49. Fleming RE, Crouch EC, Ruzicka CA, Sly WS (1993) Pulmonary carbonic anhydrase IV: developmental regulation and cell-specific expression in the capillary endothelium. Am J Physiol 265:L627–L635
50. Ridderstrale Y (1979) Observations on the localization of carbonic anhydrase in muscle. Acta Physiol Scand 106:239–240
51. Waheed A, Zhu XL, Sly WS, Wetzel P, Gros G (1992) Rat skeletal muscle membrane associated carbonic anhydrase is 39-kDa, glycosylated, GPI-anchored CA IV. Arch Biochem Biophys 294:550–556
52. Sender S, Gros G, Waheed A, Hageman GS, Sly WS (1994) Immunohistochemical localization of carbonic anhydrase IV in capillaries of rat and human skeletal muscle. J Histochem Cytochem 42:1229–1236
53. Scheibe RJ, Gros G, Parkkila S, Waheed A, Grubb JH, Shah GN, Sly WS, Wetzel P (2006) Expression of membrane-bound carbonic anhydrases IV, IX, and XIV in the mouse heart. J Histochem Cytochem 54:1379–1391
54. Scheibe RJ, Mundhenk K, Becker T, Hallerdei J, Waheed A, Shah GN, Sly WS, Gros G, Wetzel P (2008) Carbonic anhydrases IV and IX: subcellular localization and functional role in mouse skeletal muscle. Am J Physiol Cell Physiol 294:C402–C412
55. Hallerdei J, Scheibe RJ, Parkkila S, Waheed A, Sly WS, Gros G, Wetzel P, Endeward V (2010) T tubules and surface membranes provide equally effective pathways of carbonic anhydrase-facilitated lactic acid transport in skeletal muscle. PLoS One 5:e15137
56. Geers C, Kruger D, Siffert W, Schmid A, Bruns W, Gro G (1992) Carbonic anhydrase in skeletal and cardiac muscle from rabbit and rat. Biochem J 282(Pt 1):165–171
57. Lehtonen J, Shen B, Vihinen M, Casini A, Scozzafava A, Supuran CT, Parkkila AK, Saarnio J, Kivela AJ, Waheed A, Sly WS, Parkkila S (2004) Characterization of CA XIII, a novel member of the carbonic anhydrase isozyme family. J Biol Chem 279:2719–2727
58. Sender S, Decker B, Fenske CD, Sly WS, Carter ND, Gros G (1998) Localization of carbonic anhydrase IV in rat and human heart muscle. J Histochem Cytochem 46:855–861
59. Cammer W, Tansey FA (1988) Carbonic anhydrase immunostaining in astrocytes in the rat cerebral cortex. J Neurochem 50:319–322
60. Chesler M (1990) The regulation and modulation of pH in the nervous system. Prog Neurobiol 34:401–427
61. Ghandour MS, Langley OK, Zhu XL, Waheed A, Sly WS (1992) Carbonic anhydrase IV on brain capillary endothelial cells: a marker associated with the blood–brain barrier. Proc Natl Acad Sci U S A 89:6823–6827
62. Tong CK, Brion LP, Suarez C, Chesler M (2000) Interstitial carbonic anhydrase (CA) activity in brain is attributable to membrane-bound CA type IV. J Neurosci 20:8247–8253

63. Parkkila S, Parkkila AK, Rajaniemi H, Shah GN, Grubb JH, Waheed A, Sly WS (2001) Expression of membrane-associated carbonic anhydrase XIV on neurons and axons in mouse and human brain. Proc Natl Acad Sci U S A 98:1918–1923
64. Svichar N, Waheed A, Sly WS, Hennings JC, Hubner CA, Chesler M (2009) Carbonic anhydrases CA4 and CA14 both enhance AE3-mediated Cl–HCO3- exchange in hippocampal neurons. J Neurosci 29:3252–3258
65. Parkkila S, Parkkila AK (1996) Carbonic anhydrase in the alimentary tract. Roles of the different isozymes and salivary factors in the maintenance of optimal conditions in the gastrointestinal canal. Scand J Gastroenterol 31:305–317
66. Garcia Marin JJ, Tabernero Urbieta A, Perez Barriocanal F, Rodriguez Barbero E, Eleno N (1991) Plasma membrane-bound carbonic anhydrase activity in the regenerating rat liver. Biochim Biophys Acta 1061:9–14
67. Parkkila S, Kivela AJ, Kaunisto K, Parkkila AK, Hakkola J, Rajaniemi H, Waheed A, Sly WS (2002) The plasma membrane carbonic anhydrase in murine hepatocytes identified as isozyme XIV. BMC Gastroenterol 2:13
68. Fanjul M, Alvarez L, Salvador C, Gmyr V, Kerr-Conte J, Pattou F, Carter N, Hollande E (2004) Evidence for a membrane carbonic anhydrase IV anchored by its C-terminal peptide in normal human pancreatic ductal cells. Histochem Cell Biol 121:91–99
69. Cohen JP, Hoffer AP, Rosen S (1976) Carbonic anhydrase localization in the epidimymis and testis of the rat: histochemical and biochemical analysis. Biol Reprod 14:339–346
70. Parkkila S, Parkkila AK, Kaunisto K, Waheed A, Sly WS, Rajaniemi H (1993) Location of a membrane-bound carbonic anhydrase isoenzyme (CA IV) in the human male reproductive tract. J Histochem Cytochem 41:751–757
71. Purkerson JM, Schwartz GJ (2007) The role of carbonic anhydrases in renal physiology. Kidney Int 71:103–115
72. Krupin T, Sly WS, Whyte MP, Dodgson SJ (1985) Failure of acetazolamide to decrease intraocular pressure in patients with carbonic anhydrase II deficiency. Am J Ophthalmol 99:396–399
73. Wetzel P, Scheibe RJ, Hellmann B, Hallerdei J, Shah GN, Waheed A, Gros G, Sly WS (2007) Carbonic anhydrase XIV in skeletal muscle: subcellular localization and function from wild-type and knockout mice. Am J Physiol Cell Physiol 293:C358–C366
74. Vince JW, Reithmeier RA (1998) Carbonic anhydrase II binds to the carboxyl terminus of human band 3, the erythrocyte C1-/HCO3- exchanger. J Biol Chem 273:28430–28437
75. Sterling D, Alvarez BV, Casey JR (2002) The extracellular component of a transport metabolon. Extracellular loop 4 of the human AE1 Cl-/HCO3- exchanger binds carbonic anhydrase IV. J Biol Chem 277:25239–25246
76. Alvarez BV, Loiselle FB, Supuran CT, Schwartz GJ, Casey JR (2003) Direct extracellular interaction between carbonic anhydrase IV and the human NBC1 sodium/bicarbonate co-transporter. Biochemistry 42:12321–12329
77. Sun XC, Li J, Cui M, Bonanno JA (2008) Role of carbonic anhydrase IV in corneal endothelial HCO3- transport. Invest Ophthalmol Vis Sci 49:1048–1055
78. Boron WF (2010) Evaluating the role of carbonic anhydrases in the transport of HCO3–related species. Biochim Biophys Acta 1804:410–421
79. Tian Y, Tang L, Cui J, Zhu X (2010) Screening for the carbonic anhydrase IV gene mutations in Chinese retinitis pigmentosa patients. Curr Eye Res 35:440–444
80. Kohn L, Burstedt MS, Jonsson F, Kadzhaev K, Haamer E, Sandgren O, Golovleva I (2008) Carrier of R14W in carbonic anhydrase IV presents Bothnia dystrophy phenotype caused by two allelic mutations in RLBP1. Invest Ophthalmol Vis Sci 49:3172–3177
81. Fain GL (2006) Why photoreceptors die (and why they don't). Bioessays 28:344–354
82. Chandrashekar J, Yarmolinsky D, von Buchholtz L, Oka Y, Sly W, Ryba NJ, Zuker CS (2009) The taste of carbonation. Science 326:443–445

Chapter 10
Carbonic Anhydrase Expression in Kidney and Renal Cancer: Implications for Diagnosis and Treatment

Egbert Oosterwijk

Abstract Four different carbonic anhydrases are expressed in the human nephron, the functional unit of the kidney. These are specifically expressed in different nephron segments, emphasizing the critical role carbonic anhydrases play in maintaining the homeostasis of this crucial organ.

Whereas the localization of carbonic anhydrases in the kidney has been long established, interest in carbonic anhydrases has increased dramatically for renal cancer, in particular for the clear cell variant of renal cell carcinoma (ccRCC) because carbonic anhydrase IX is specifically expressed in ccRCC. Therefore carbonic anhydrase IX is being studied as potential diagnostic and therapeutic target, despite carbonic anhydrase IX expression in non-renal tissues.

Keywords Kidney • Renal cancer • Diagnosis • Treatment • Clinical trial • Antibody • Carbonic anhydrase IV • Carbonic anhydrase IX • Carbonic anhydrase XII

1 Carbonic Anhydrases in Normal Human Kidney

In the kidney carbonic anhydrases (CA) facilitate renal acidification [1]. Of the 15 described CA, four are expressed in most segments of the nephron, the functional unit of the kidney, with CAII and CAIV being most prominent. Cytosolic CAII

Susan C. Frost and Robert McKenna (eds.). Carbonic Anhydrase: Mechanism, Regulation, Links to Disease, and Industrial Applications

E. Oosterwijk (✉)
Department of Urology, University Medical Center St Radboud, Nijmegen, The Netherlands
e-mail: e.oosterwijk@uro.umcn.nl

S.C. Frost and R. McKenna (eds.), *Carbonic Anhydrase: Mechanism, Regulation, Links to Disease, and Industrial Applications*, Subcellular Biochemistry 75,
DOI 10.1007/978-94-007-7359-2_10,

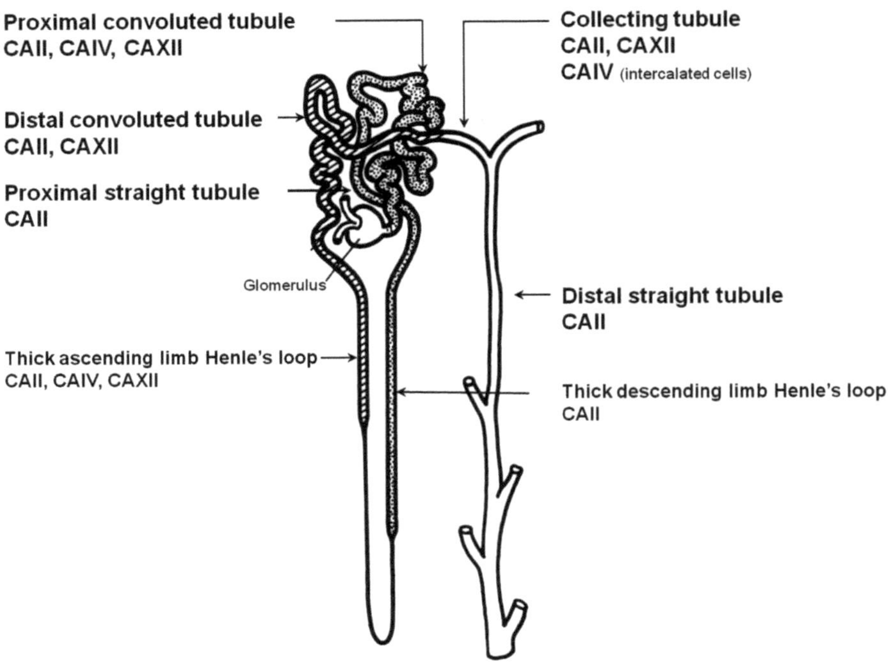

Fig. 10.1 Schematic representation of the nephron and distribution of carbonic anhydrases

comprises approximately 95 % of renal CA activity [2]. The other CA expressed in kidney tissue (CAIV [3] and CAXII [4]) are membrane-associated. Finally, CAXIV and CAXV expression in the kidney is detected in rodent species only.

CAII is expressed by a large number of nephron segments, including proximal convoluted and straight tubules, thin descending limbs of Henle's loop, thick ascending limbs of Henle's loop in some species, intercalated cells of the cortical and medullary collecting ducts, and weakly in principal cells and inner medullary collecting ducts of some species [2]. CAIV, a GPI-anchored protein, is expressed in the proximal tubule and in the distal nephron, including outer and inner medullary collecting ducts and intercalated cells of the cortical collecting duct [3]. CAXII is abundantly expressed in the basolateral membranes of the thick ascending limb of Henle's loop, distal convoluted tubules, and principal cells of the collecting duct. Moreover, weak basolateral expression in proximal convoluted tubules has been described [4], i.e., expression of CA in kidney tissue is quite diverse emphasizing that they are crucial in maintenance of the acid–base balance and other solute transport along the nephron. Figure 10.1 summarizes CA expression in the nephron.

2 Carbonic Anhydrases in Renal Cancer

Renal cancer accounts for approximately 3 % of all malignancies and ranks among the top 10 most common malignancies in both men and women [5]. In 2012, it is estimated that approximately 64,000 individuals will be diagnosed, with an estimated 13,570 dying from this disease in the United States [6]. It is now clear that at least 6 different renal cancers can be recognized, and molecular analysis of renal cancers has revealed that specific mutational events underlie these different renal malignancies [7, 8]. Renal Cell Carcinoma (RCC) accounts for about 90 % of renal cancers, and approximately 80 % of RCC are clear cell RCC (ccRCC) [9].

3 Renal Cancer, CAIX and Transcriptional Control

While the importance of CA for the kidney has long been recognized, interest in CA in relation to renal cancer has increased dramatically once it became apparent that CAIX is expressed in a high proportion of renal cancers, whereas expression in normal kidney tissue is absent. CAIX, a 54/58 kDa N-glycosylated single-pass transmembrane protein containing an N-terminal proteoglycan-like domain that functions in cell adhesion and can be shed by stimulus-dependent activation of metalloproteinase activity [10–12], is catalytically active and exhibits hydratase activity that is comparable to CAIV [13]. Although CAIX absent in normal kidney tissue, high CAIX expression can be observed in the GI system: gastric surface epithelium, gastric glands, pyloric/Brunner's glands, crypt cells of the small intestine, and the biliary trees, including the gallbladder all express CAIX [14].

CAIX expression is governed by a complex of transcription factors. The central regulatory element in the CAIX promoter is a hypoxia-responsive element (HRE), close to the CAIX initiation site while other elements (particularly SP1 binding sites) play lesser roles [15, 16]. Under normoxic conditions the HRE is not occupied by the hypoxic inducible transcription factor (HIF) because the activity of the HIFα–ARNT(HIF1β) heterodimer is determined by hypoxic stabilization of HIF [17]. Under normal oxygen conditions, HIFα is marked by prolyl hydroxylation for proteolytic degradation: hydroxylated HIFα binds to the Von Hippel Lindau protein (pVHL) leading to ubiquitination and subsequent degradation of HIFα [18]. Thus far three HIFα genes have been identified [19–21] and in general HIF-responsive genes can be activated by HIF1 and HIF2. Remarkably, CAIX gene expression is exclusively governed by HIF1 [16]. Under hypoxic conditions HIFα hydroxylation does not occur, which results in inhibition of binding of pVHL to HIFα, leading to accumulation of HIFα and subsequent association with HIFβ. The resulting HIFs cause the downstream transcription of a number of hypoxia-inducible genes, such as angiogenic genes and CAIX [22].

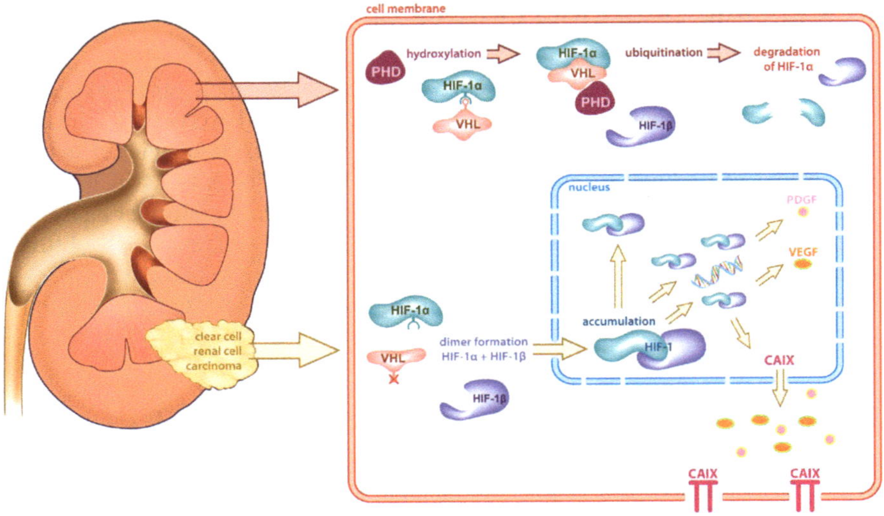

Fig. 10.2 Schematic representation of regulation CAIX expression in RCC. In normal kidney, hypoxia-inducible factor-1α (HIF-1α) is rapidly degraded via the S26 proteasome after polyubiquitylation after prolyl hydroxylation and binding to VHL. In ccRCC, VHL cannot bind to HIF-1α. Thus, HIF-1α escapes degradation, associates with the constitutively stable partner HIF-1β and forms the active heterodimeric HIF-1 transcription factor, which binds to hypoxia-responsive elements located in the promoter/enhancer regions of hypoxia-inducible genes, including pro-angiogenic factors (such as vascular endothelial growth factor and platelet-derived growth factor) and CAIX, which is ultimately expressed on the tumor cell surface

The link between VHL, HIF, CAIX expression and renal cancer became apparent once it became clear that the mutational defect in families suffering from a Von Hippel Lindau syndrome, a rare, autosomal dominant genetic condition that predisposes individuals to benign and malignant tumors, could be linked with sporadic clear cell renal cell cancer (ccRCC). After the gene responsible for the VHL syndrome was cloned [23] comparative analyses showed that mutations [24] and promotor silencing [25] in the VHL gene were also present in sporadic ccRCC. Thus, the ubiquitous CAIX expression in ccRCC could be explained by mutated VHL leading to HIF1 stabilization (see Fig. 10.2).

Although CAIX transcription is clearly HIF1 dependent, CAIX transcription is also regulated by extracellular acidosis [22, 26], which accounts for the CAIX expression in the GI tract. Moreover, CAIX promoter methylation influences CAIX expression. In addition to transcriptional regulation, control of CAIX expression may involve phosphatidyl inositol 3- kinase (PI3-kinase) activity or alternative splicing of the CAIX transcript [27, 28]. Nevertheless, in ccRCC and under hypoxic conditions, CAIX transcription is heavily dependent on HIF1.

Another VHL-dependent carbonic anhydrase tumor-associated antigen whose expression is induced by hypoxia is CAXII [29]. However, in contrast to CAIX,

CAXII is also expressed in normal kidney tissue, and only in approximately 10 % of patients with RCC higher expression of the CAXII transcript in the RCC than in surrounding normal kidney tissue was observed [4, 30]. Therefore, most research has focused on CAIX as potential target for therapy/diagnosis of RCC, with emphasis on ccRCC as CAIX expression in this subtype is almost invariably high and homogeneous [31], the consequence of the VHL defect in this tumor type.

4 CAIX Distribution in Renal Cancer

The distribution of CAIX in renal cancer was first described in 1986 in an immunohistochemical study describing the specificity analysis and tumor specificity of monoclonal antibody G250 (mAbG250), albeit that at the time of publication the authors had been unable to identify the molecule recognized by mAbG250 [32]. MAbG250-positivity was noted in a very restricted number of normal tissues. Almost all renal cancers investigated—no distinction between different renal cancer subtypes was described—were mAbG250-positive, whereas normal kidney tissue did not express the recognized molecule. The authors suggested that the recognized moiety could potentially serve as a target for therapy and/or diagnosis in RCC. Additionally, heterogeneous activity was observed in non-renal tumors, an observation for which no satisfactory hypothesis was formulated [32]. Cloning of the antigen recognized by mAbG250 demonstrated that the antibody recognized a conformational epitope in CAIX [11]. High CAIX expression and subsequent acidification is thought to aid tumors to adept to hypoxic conditions and hamper the uptake of weakly basic chemotherapeutic agents, leading to chemoresistance of CAIX-expressing tumors [33].

Numerous studies have confirmed the CAIX distribution normal tissues and malignancies. For renal cancer CAIX is almost homogeneously expressed in the ccRCC subtype [31]. In view of the favorable tissue distribution the potential of CAIX targeting of RCC for diagnosis or therapy has been studied extensively. Table 10.1 summarizes the clinical experience with CAIX targeting. Additionally, several studies have addressed whether it is of value to include CAIX expression in prognostic RCC nomograms.

5 CAIX Targeting in RCC for Diagnosis

Improved radiologic imaging methods have lead to a dramatic increase in the incidental detection of renal lesions and almost 50 % of all renal lesions are found incidentally [34]. However, these methods cannot adequately differentiate between benign and malignant renal lesions. Unfortunately, the sensitivity and specificity of renal biopsies is low and therefore less than 10 % of patients with suspected RCC

Table 10.1 Overview of clinical studies targeting CAIX in RCC

Ref	Agent/study	Patients (N)	Imaging/therapy	Outcome
Oosterwijk et al. [41]	131I-mG250	Primary RCC (15)	Imaging	12/12 imaged
Steffens et al. [43]	131I-cG250	Primary RCC (16)	Imaging	13/13
Brouwers et al. [58]	131I-cG250 111In-cG250	Metastatic RCC (5)	Imaging, 131I versus 111In-cG250	Imaging 111In-cG250 superior to 131I-cG250
Divgi et al. [44]	124I-cG250	Primary RCC (26)	Imaging	15/16 ccRCC identified
Divgi et al. [45]	124I-cG250	Primary RCC (202)	Imaging	124/143 ccRCC imaged (86,7 %)
Steffens et al. [78]	131I mG250 vs 131I-mG250	Primary RCC	Imaging	5/5 imaged
Bleumer et al. [51]	cG250, Phase 2, 50 mg/	Metastatic RCC (36)	Therapy	1 CR; 1 PR; 8 SD; 26 PD (1–20+ week)
Davis et al. [50]	cG250	Metastatic RCC (9)	Therapy	1 CR; 9 SD; 3 PD, 6–66 week
Davis et al. [50]	cG250 + IL-2	Metastatic RCC (9)	Therapy	2 SD; 7 PD (6 and 12 week)
Bleumer et al. [52]	cG250 + IL-2	Metastatic RCC (35)	Therapy	2 PR; 6 SD; 27 PD (24+ week)
Siebels et al. [79]	cG250 + IFN-2alpha	Metastatic RCC (31)	Therapy	1 CR 17+ mo, 9 SD (24+ week)
Divgi et al. [42]	131I-mG250	Metastatic RCC (33)	Therapy	17 SD, 16 PD (2–3 month)
Steffens et al. [80]	131I-cG250	Metastatic RCC (12)	Therapy	1 PR; 1 SD; 10 PD (3 and 9+ month)
Divgi et al. [81]	131I-cG250	Metastatic RCC (15)	Therapy	7 SD; 8 PD (2–11 month)
Brouwers et al. [55]	131I-cG250, 2 doses	Metastatic RCC (27)	Therapy	5 SD; 22 PD (3–12 month)
ARISER	cG250	High risk nephrectomized patients (864)	Adjuvant	No benefit for whole population, high CAIX expression correlated with risk of recurrence reduction
Muselaers et al. [46]	111In-cG250	Primary and metastasized RCC (29)	Imaging	Positive image corresponds to active disease
Stillebroer et al. [59]	177Lu-cG250, multiple doses	Metastatic RCC (23)	Therapy	1 PR (9 + month), 17SD (3 + month)
Lamers et al. [71]	CAIX-specific CTL	Metastatic RCC (6)	Dose escalation	No benefit

undergo renal mass sampling before nephrectomy [35]. As a consequence surgical interventions are performed that could have been prevented. Thus, there is a clear need for improved RCC detection.

In view of the high number of CAIX expressing ccRCC tumors many animal and clinical studies have been performed to investigate the possibility to use CAIX imaging as new diagnostic tool. Biodistribution studies in animals—almost invariably in mice xenografted with human RCC—demonstrated that CAIX can accurately be targeted with mAbG250 (e.g., [36–40]). Different radionuclides have been studied, and the main difference observed between the various conjugated mAbG250 products has been related to antibody internalization and subsequent cellular retention of the radionuclide used.

In the first clinical trial with mAbG250 clear visualization of primary and metastatic RCC was observed [41]. G250 targeting to non-tumor lesions was very low, and occult metastatic lesions, not recognized by other imaging modalities were visualized. Importantly, mAbG250 accumulation in CAIX-positive normal tissues such as stomach was almost absent. This difference is most likely due to the different tissue vasculature: IgG extravasation in normal tissues is low, certainly in comparison to IgG extravasation in the aberrant tumor vasculature that permits easier IgG extravasation. Because histological confirmed CAIX-negative tumors did not image, it was concluded that mAbG250 accumulation was CAIX specific [41].

However, the administration of murine mAbG250 led to the formation of human anti-mouse antibodies (HAMA) in all patients, preventing multiple administrations [42]. To prevent HAMA, a chimeric variant of mAbG250 (cG250) was constructed and tested in a protein dose escalation trial in patients suspect for RCC [43]. The pharmacokinetics and biodistribution of the chimerised mAb was almost identical to the mouse mAbG250, and the optimal protein dose was found to be 5–10 mg. Clear visualization of all 13 CAIX-positive tumors as well as metastases was noted. The focal uptake of cG250 in primary RCC lesions was much higher than uptake observed with other mAbs in other solid tumors. Once more, this trial suggested that CAIX imaging might be a useful tool in diagnostic imaging of RCC. Figure 10.3 shows a typical example of CAIX imaging in a mRCC patient.

In the previous studies 131-iodine-labeled cG250 was used, but the imaging characteristics of 131I are suboptimal, and immunoPET—that is, PET scanning that combines the favorable characteristics of PET (higher spatial resolution, three-dimensional imaging, and superior quantitative estimations of uptake) with mAb cG250—may be superior to image (m)RCC lesions. Unfortunately combining the most commonly used positron emitters with mAb is not possible, because they have a too-short half-live that cannot be combined with the relatively slow pharmacokinetics of intravenously injected radiolabeled mAb. Optimal tumor uptake and therefore optimal imaging is reached after several days. In view of the favorable characteristics of 124I which has a half-life of 100 h, Divgi et al. studied 25 patients with suspect renal lesions scheduled for nephrectomy [44]. They hypothesized that 124I-cG250 PET imaging might allow visualization of ccRCC in patients with suspect renal masses versus non-ccRCC. In this proof-of-principle trial 15/16 patients

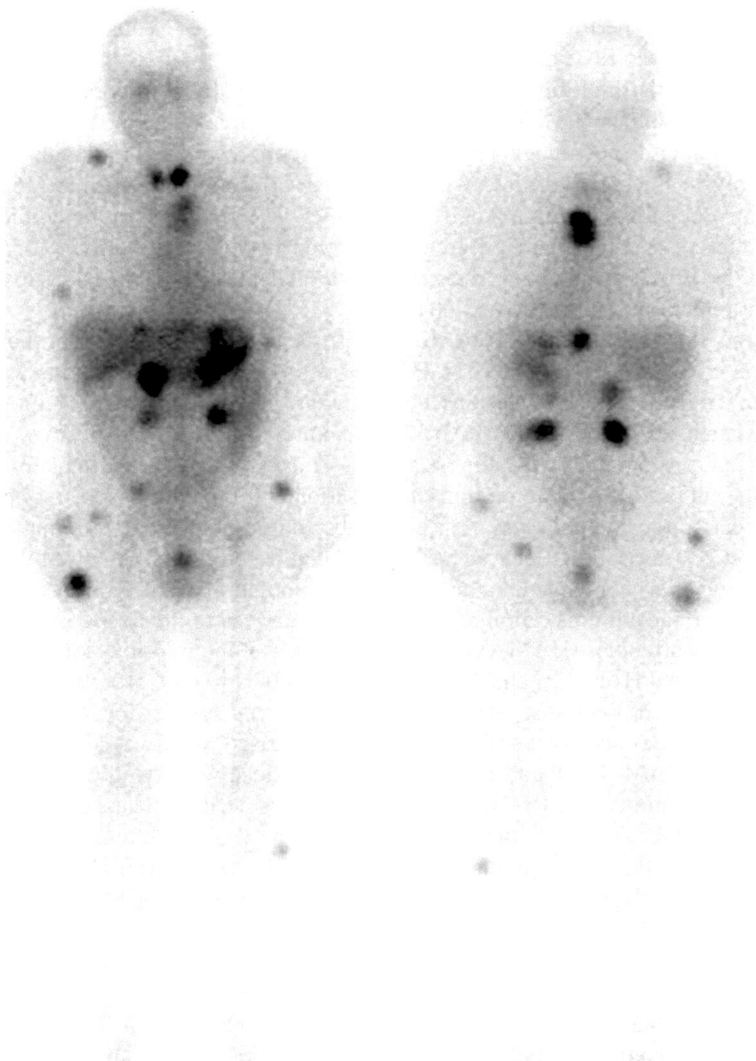

Fig. 10.3 Typical image of patient with mRCC injected with mAbG250, 5 days post-injection. Please note the clear imaging of metastatic lesions and absence of positive images of CAIX-positive organs

with histologically confirmed ccRCC after surgery had a positive scan. None of the nine patients with non-ccRCC showed any 124I-cG250 uptake. This study showed that immunoPET might be helpful in clinical decision-making and might aid in the surgical management of small renal masses scheduled for partial nephrectomy. Subsequently, a trial was designed to compare the sensitivity and specificity of

124I-G250 PET/CT to that of multiphasic contrast-enhanced CT (CECT). In total 226 patients were studied, of which 195 were evaluable. The average sensitivity was higher for G250 PET/CT (86.2 % versus 75.5 %) and the average specificity was 85.9 % for PET/CT and 46.8 % for CECT ($P = 0.005$). The authors concluded that 124I-labeled G250 PET/CT accurately and noninvasively identified ccRCC, with potential utility for designing best management approaches for patients with renal masses [45].

The centralized production of 124I-labeled G250 (124I is produced in very few sites world-wide) harbors a potential problem. As an alternative SPECT with 111In-labelled G250 was investigated [46]. Similar to 124I-G250 PET it is non-invasive and does not require the use of intravenous contrast agents, which makes it suitable for patients with an impaired renal function. In addition, it is easier to produce as an off-the-shelf agent. Of the 22 patients who presented with a renal mass of unknown origin, clear preferential uptake of 111In-G250 was observed in the lesions of 16 patients. Importantly, histopathologic examination confirmed active ccRCC in 15 of these patients resulting in a positive predictive value of 111In-cG250 imaging for ccRCC was 94 %.

In six patients presenting with a renal mass no targeting of 111In-G250 was observed. Four of these patients underwent surgery as planned and histopathologic evaluation revealed two cases of benign oncocytoma, a chromophobe and a mucinous tubular spindle cell carcinoma subtype tumor. The remaining two patients were closely monitored with repeated CT scans. No growth of the renal mass occurred in the follow-up period (> 24 months). The high and specific targeting of the radiolabelled antibody in patients suspected of (recurrent) ccRCC indicated that targeting of 111In-cG250 calls for surgical intervention [46].

6 CAIX Targeting in RCC for Therapy

The therapeutic potential of CAIX targeting with mAbG250 has been studied in numerous clinical trials (see Table 10.1). Roughly, these can be divided into trials with "naked" antibody alone or in combination with cytokines and radioimmunotherapy trials. In vitro studies showed that cG250 could initiate cell lysis through antibody-dependent cellular cytotoxicity (ADCC) of CAIX-positive cells [47, 48] and in vivo studies had shown that antibody treatment of CAIX-positive tumors resulted in a significant tumor growth reduction [49]. Based on these results a phase 1 study with escalating doses of 5–50 mg/m^2 of cG250, with weekly infusions for 6 weeks was initiated. One patient showed a complete response (CR), and nine patients had stable disease (SD) after one treatment cycle [50].

This protein dose escalation study put to rest one of the main concerns: does therapy directed at CAIX induce fulminant toxicities because of the expression of CAIX in normal tissue—predominantly in the epithelial structures. In the subsequent phase 2 study, patients received 50 mg of cG250 weekly for 12 week and 36 mRCC patients were included. Before treatment, 80 % of patients were progressive. After one treatment cycle, 28 % of patients had SD; during follow-up, one CR and one partial response (PR) were noted. The median survival of 15 month suggested that cG250 might be able to modulate the natural course of mRCC [51]. Based on these results, a large adjuvant phase 3 trial was initiated (ARISER), that aimed to reduce the recurrence of disease in nephrectomised RCC patients who have a high risk of relapse. Unfortunately, the trial did not meet its primary endpoint: the analysis showed no improvement in median Disease Free Survival (DFS) (approximately 72 months) following G250 treatment compared with placebo (http://www.wilex.de/portfolio-english/rencarex/phase-iii-ariser/). A biomarker analysis showed that with increasing density of CAIX expression in tumor tissue, as quantified by a CAIX score, the more significant the treatment effect became. DFS showed a clinically and statistically significant improvement in the patient population with a high CAIX level treated with cG250 compared to both placebo and patients with a low CAIX score (press release Wilex Februari 26th 2013). Therefore, an immunotherapy for ccRCC in the adjuvant setting would appear to be still an option.

After in vitro studies showed that IL-2–activated effector cells were more effective in eliciting cG250 ADCC than non-activated cells [47, 48] cG250 was combined with low-dose IL-2 treatment in a phase 2 trial. Clinical benefit was noted in 8 of 35 patients, with 2 PR, 6 long-lasting (>24 week) stabilizations of disease and a mean survival of 24 months. The increased survival (as compared to historic controls) was deemed to be cG250-related and not related to the IL-2, as a 6-fold decrease of the normal IL-2 dose was used [52]. The favorable outcome might have been due to a synergistic effect of cG250 and IL-2 as was observed in the in vitro studies. Because patients were not randomized it is difficult to judge the value of this observation. A randomized trial is needed to determine the true effect of the IL-2/cG250 treatment.

From the very first clinical studies onwards it has been apparent that the accumulation of cG250 in RCC lesions is very high: up to 10 times higher than uptake seen with mAbs in other solid tumors [41, 43]. Based on the assumption that tumor-sterilising radiation doses to ccRCC could be delivered by mAbG250 several radioimmunotherapy (RIT) trials have been performed. Almost invariably patients with clear cell histology of their primary tumor have been included, in view of the almost ubiquitous expression in ccRCC.

The first RIT studies were all performed with iodine-131 labeled mAbG250 [42]. In the first RIT trial, patients experienced hepatic toxicity, most likely the result of mAbG250 accumulation in the liver. i.e., this toxicity was truly target-related. This was a remarkable finding in view of the absence of any mAb-related toxicity in the non-RIT trials. This finding demonstrated that CAIX was available to mAbG250 in the liver, a highly perfused organ with a particular vascular tree which allows much easier extravasation of molecules than in other organs. In subsequent trials, patients first received a 131I-cG250 scout dose to prevent administration of high doses of 131I-cG250 to patients with CAIX-negative tumors, despite a ccRCC histology. Unexpectedly, as a corollary, liver toxicity was avoided, most likely because CAIX sites in the liver compartment were saturated by mAbG250 administered at the scout dose. The maximum tolerated dose (MTD) that could be administered was 2,220 MBq/m^2, with haematologic toxicity as the dose-limiting factor [42]. Of the eight patients receiving treatment, one patient showed SD, and one patient had a PR. In the phase II arm of the trial, 15 patients were treated at MTD, to determine efficacy, but no major responses were noted. However, overall survival of patients treated with 131I-mG250 seemed to be increased in comparison with that of historic control patients: 17/33 SD and 2 minor responses.

Following this RIT with the murine version of mAbG250, RIT with cG250 were designed because HAMA prevented multiple administrations. The dose-limiting toxicity of cG250 was significantly lower than DLT observed for the murine version [53]. Almost certainly the extended serum half-life is responsible for the enhanced hematopoietic toxicity, as this leads to extended radiation of the bone marrow compartment. Targeting of 131I-cG250 to disease was outstanding in patients with positive diagnostic scans. Nevertheless, quite disappointingly, no clinical response was observed. Therefore a study was performed to evaluate the effect of two sequential high-dose 131I-cG250 treatments [54]. Similar to the earlier trial, progressive mRCC patients received a scout dose of 185 MBq of 131I-cG250 to demonstrate tumor targeting. This dose was followed by 2,220 MBq/m^2 of 131I-cG250. After 3 month, if disease progression had halted and tumor targeting with another scout dose was established, patients received the second high-dose injection of 131I-cG250. The MTD of the second RIT was set at 1,665 MBq/m^2, again because of hematologic toxicity. None of the 18 evaluable patients showed an objective response, but five patients had stabilization of their disease lasting 3–12 month. An inverse correlation between the size of metastases and radiation-absorbed dose was observed and dosimetric analyses showed that therapeutic radiation doses ($>$50 Gy) were only guided to lesions smaller than 5 g. It was concluded that RIT in mRCC patients could best be given in the setting of small-volume disease, as adjuvant therapy, or other, more potent radionuclides should be used [55].

Animal studies performed with cG250 labeled with more potent radionuclides (177Lu, 90Y or 186Re) showed that tumor growth was most effectively inhibited by 177Lu-cG250, followed by 90Y, 186Re, and the least by 131IcG250 [56]. These preclinical RIT studies clearly showed the superiority of 177Lu- and 90Y-based RIT, in line with other studies. They also substantiated that trapped radionuclides are superior to the non-trapped iodine. In view of this observation subsequent studies have focused on the possibility to use 90Y or 177Lu in RIT [57].

First a dual-labeling study with cG250 labeled with the residualising radionuclide 111In (as substitute for 177Lu and 90Y) and 131I was performed in metastatic RCC patients to study whether 111In-cG250 was superior to 131I-cG250 [58]. This permitted the direct comparison of quantitatively calculated activities in RCC metastases of a iodinated form of cG250 with the calculated activities of cG250 labeled with 111In within the same patient. This comparison clearly showed that accumulation of 111In-cG250 was higher than 131I-cG250 and more metastatic lesions were visualized in four of the five patients studied [58].

The results of the phase I trials of 177Lu-cG250 have been published recently [59]. Identical to earlier studies, progressive mRCC patients with proven ccRCC received a scout dose of cG250 to establish adequate tumor accumulation followed by a dose of 177Lu-cG250 1 week after 111In-cG250. The predictive value of the 111In-cG250 scout dose for 177Lu-cG250 accumulation was demonstrated by super-imposable 111In-cG250/177Lu-cG250 images [60]. Some patients received second (13 of 23) and third (4 of 23) treatment cycles. This trial demonstrated that, as anticipated, guidance of 177Lu-cG250 to tumor tissue was superior to 131I-cG250 accumulation in earlier studies. Most patients demonstrated stable disease 3 months after the first treatment, and one patient showed a partial response that lasted for 9 months. The authors concluded that RIT with 177Lu-cG250, targeting CAIX may stabilize previously progressive metastatic ccRCC [59].

7 Carbonic Anhydrase IX as Biomarker in Renal Cell Carcinoma Nomograms

For the clinical management of RCC patients it is important to predict RCC disease-free survival rates after nephrectomy or to predict survival in metastatic patients. Most nomograms are based on clinical symptoms, age, performance status, TNM stage, nuclear grade, and tumor necrosis [61]. In an effort to improve the nomogram for RCC patients, a number of molecular markers have been studied, including CAIX [62]. In the first nomogram in which CAIX was included, the prognostic accuracy of these biomarkers was examined in patients with primary RCC, regardless of the presence of synchronous metastases, albeit that this is one of the strongest prognostic parameters. Inclusion of the biomarkers improved the prognostic accuracy over the two most common nomograms [63]. The second nomogram was designed to predict the survival of mRCC patients and also incorporated CAIX status. This nomogram, based on the combination of biomarkers and clinical parameters, proved to be a more accurate prognostic model for mRCC patients than other nomograms [64]. Unfortunately, although these studies clearly demonstrated that incorporation of CAIX expression in the RCC nomogram improved prognostic accuracy, CAIX expression is generally not included, because reliable commercial antibody with CAIX specificity that can be used on fromalin-fixed paraffin embedded tissues is not available. M75 can detect CAIX reliably on formalin-fixed paraffin embedded tissues [65, 66], but is not commercially available. mAbG250 can reliable detect CAIX, but only on snap-frozen tissues [32]. Commercially available mAb NB100-417 is reported to be a CAIX-specific mAb with similar staining characteristics as M75, recognizing the membranous epitope of CAIX [67], but cross-reactivity at high dilutions with beta-tubulin makes specific CAIX detection by mAb NB100-417 impossible [68]. It would be unfortunate if CAIX expression could not be included as biomarker because CAIX-specific agents are not available.

8 Carbonic Anhydrase IX Targeting: Non-antibody Mediated Approaches

Fluorescent sulfonamides with high affinity for CA IX have been developed and tested in vitro [69]. Remarkably, sulfonamide binding to CA IX was only observed during conditions of hypoxia in several cell lines, irrespective of the levels of CA IX or HIF-1α activity, indicating that the active site of CA IX is only available for sulfonamide binding during hypoxia [70]. Because CAIX expression in ccRCC is not driven by hypoxia, it is therefore not likely that these interesting agents will be applicable in the RCC setting.

Finally, CAIX has been used as target in gene therapy studies with modified cytotoxic T cells (CTL) [71]. CTL harvested from mRCC patients were transfected with a retroviral vector encoding for a modified T cell receptor with CAIX specificity, based on mAbG250-specificity. Patients received escalating doses of gene modified autologous CTL. Similar to the very first mAb trials patients experienced liver toxicity that resolved in time. Although the toxicity was not life threatening and the toxicity indicated that the infused CTL were highly active, the trial was put on hold [71]. Analysis of liver biopsies established that CAIX-specific CTL were located at CAIX-positive cells in the biliary trees [71]. In an effort to alleviate the liver toxicity patients were pre-dosed with cG250 before receiving the CAIX-specific CTL, similar to the RIT antibody trials. This pre-dosing prevented the on-target liver toxicity, again showing that liver toxicity can be prevented by pre-dosing [72].

9 Conclusions and Future Prospects

Carbonic anhydrases tightly control the acid–base balance in the kidney. Due to the unique molecular basis of ccRCC, CAIX is regarded as an excellent target for diagnosis and possibly for therapy. Clinical trials have unambiguously demonstrated that CAIX can be targeted to RCC tissues without damage to normal tissues expressing CAIX. The introduction of several anti-angiogenic small molecule drugs (e.g. Sutent®, Pazopanib, Axitinib, Temsirolimus, and Evirolimus) has rapidly changed the treatment of metastatic RCC [73]. Although the impact on disease progression is encouraging, a substantial proportion of patients do not respond adequately, and therapy resistance almost inevitably occurs [73]. Possibly combination treatment aimed at different, non-related pathways may be advantageous [74]. Finally, the new therapeutic options have lead to investigations that examine whether CAIX can be used in serum assays or as imaging target to study whether CAIX monitoring can be useful to predict responses [75–77]. The future will show whether combination of CAIX targeting with e.g. small molecule drugs can be useful in the treatment and clinical management of (metastatic) RCC.

References

1. Purkerson JM, Kittelberger AM, Schwartz GJ (2007) Basolateral carbonic anhydrase IV in the proximal tubule is a glycosylphosphatidylinositol-anchored protein. Kidney Int 71:407–416
2. Schwartz GJ (2002) Physiology and molecular biology of renal carbonic anhydrase. J Nephrol 15(Suppl 5):S61–S74
3. Brown D, Zhu XL, Sly WS (1990) Localization of membrane-associated carbonic anhydrase type IV in kidney epithelial cells. Proc Natl Acad Sci U S A 87:7457–7461
4. Tureci O, Sahin U, Vollmar E, Siemer S, Gottert E, Seitz G, Parkkila AK, Shah GN, Grubb JH, Pfreundschuh M, Sly WS (1998) Human carbonic anhydrase XII: cDNA cloning, expression, and chromosomal localization of a carbonic anhydrase gene that is overexpressed in some renal cell cancers. Proc Natl Acad Sci U S A 95:7608–7613
5. Jemal A, Bray F, Center MM, Ferlay J, Ward E, Forman D (2011) Global cancer statistics. CA Cancer J Clin 61:69–90
6. Siegel R, Naishadham D, Jemal A (2012) Cancer statistics, 2012. CA Cancer J Clin 62:10–29
7. Oosterwijk E, Rathmell WK, Junker K, Brannon AR, Pouliot F, Finley DS, Mulders PF, Kirkali Z, Uemura H, Belldegrun A (2011) Basic research in kidney cancer. Eur Urol 60:622–633
8. Jonasch E, Futreal PA, Davis IJ, Bailey ST, Kim WY, Brugarolas J, Giaccia AJ, Kurban G, Pause A, Frydman J, Zurita AJ, Rini BI, Sharma P, Atkins MB, Walker CL, Rathmell WK (2012) State of the science: an update on renal cell carcinoma. Mol Cancer Res MCR 10:859–880
9. Linehan WM, Bratslavsky G, Pinto PA, Schmidt LS, Neckers L, Bottaro DP, Srinivasan R (2010) Molecular diagnosis and therapy of kidney cancer. Annu Rev Med 61:329–343
10. Pastorek J, Pastorekova S, Callebaut I, Mornon JP, Zelnik V, Opavsky R, Zat'ovicova M, Liao S, Portetelle D, Stanbridge EJ et al (1994) Cloning and characterization of MN, a human tumor-associated protein with a domain homologous to carbonic anhydrase and a putative helix-loop-helix DNA binding segment. Oncogene 9:2877–2888
11. Grabmaier K, Vissers JL, De Weijert MC, Oosterwijk-Wakka JC, Van Bokhoven A, Brakenhoff RH, Noessner E, Mulders PA, Merkx G, Figdor CG, Adema GJ, Oosterwijk E (2000) Molecular cloning and immunogenicity of renal cell carcinoma-associated antigen G250. Int J Cancer 85:865–870
12. Zatovicova M, Sedlakova O, Svastova E, Ohradanova A, Ciampor F, Arribas J, Pastorek J, Pastorekova S (2005) Ectodomain shedding of the hypoxia-induced carbonic anhydrase IX is a metalloprotease-dependent process regulated by TACE/ADAM17. Br J Cancer 93:1267–1276
13. Wingo T, Tu C, Laipis PJ, Silverman DN (2001) The catalytic properties of human carbonic anhydrase IX. Biochem Biophys Res Commun 288:666–669
14. Liao SY, Lerman MI, Stanbridge EJ (2009) Expression of transmembrane carbonic anhydrases, CAIX and CAXII, in human development. BMC Dev Biol 9:22
15. Kaluz S, Kaluzova M, Liao SY, Lerman M, Stanbridge EJ (2009) Transcriptional control of the tumor- and hypoxia-marker carbonic anhydrase 9: a one transcription factor (HIF-1) show? Biochim Biophys Acta 1795:162–172
16. Grabmaier K, A de Weijert MC, Verhaegh GW, Schalken JA, Oosterwijk E (2004) Strict regulation of CAIX(G250/MN) by HIF-1alpha in clear cell renal cell carcinoma. Oncogene 23:5624–5631
17. Jaakkola P, Mole DR, Tian YM, Wilson MI, Gielbert J, Gaskell SJ, von Kriegsheim A, Hebestreit HF, Mukherji M, Schofield CJ, Maxwell PH, Pugh CW, Ratcliffe PJ (2001) Targeting of HIF-alpha to the von Hippel-Lindau ubiquitylation complex by O_2-regulated prolyl hydroxylation. Science 292:468–472
18. Schofield CJ, Ratcliffe PJ (2005) Signalling hypoxia by HIF hydroxylases. Biochem Biophys Res Commun 338:617–626
19. Wang GL, Jiang BH, Rue EA, Semenza GL (1995) Hypoxia-inducible factor 1 is a basic-helix-loop-helix-PAS heterodimer regulated by cellular O_2 tension. Proc Natl Acad Sci U S A 92:5510–5514

20. Makino Y, Kanopka A, Wilson WJ, Tanaka H, Poellinger L (2002) Inhibitory PAS domain protein (IPAS) is a hypoxia-inducible splicing variant of the hypoxia-inducible factor-3alpha locus. J Biol Chem 277:32405–32408
21. O'Rourke JF, Tian YM, Ratcliffe PJ, Pugh CW (1999) Oxygen-regulated and transactivating domains in endothelial PAS protein 1: comparison with hypoxia-inducible factor-1alpha. J Biol Chem 274:2060–2071
22. Ivanov S, Liao SY, Ivanova A, Danilkovitch-Miagkova A, Tarasova N, Weirich G, Merrill MJ, Proescholdt MA, Oldfield EH, Lee J, Zavada J, Waheed A, Sly W, Lerman MI, Stanbridge EJ (2001) Expression of hypoxia-inducible cell-surface transmembrane carbonic anhydrases in human cancer. Am J Pathol 158:905–919
23. Latif F, Tory K, Gnarra J, Yao M, Duh FM, Orcutt ML, Stackhouse T, Kuzmin I, Modi W, Geil L, Schmidt L, Zhou F, Li H, Wei MH, Chen F, Glenn G, Choyke P, Walther MM, Weng Y, Duan D-SR, Dean M, Glava D, Richards FM, Crossey PA, Ferguson-Smith MA, Le Paslier D, Chumakov I, Cohen D, Chinault AC, Maher ER, Linehan WM, Zbar B, Lerman MI (1993) Identification of the von Hippel-Lindau disease tumor suppressor gene. Science 260:1317–1320
24. Gnarra JR, Tory K, Weng Y, Schmidt L, Wei MH, Li H, Latif F, Liu S, Chen F, Duh FM, Lubensky I, Duan DR, Florence C, Pozzatti R, Walther MM, Bander NH, Grossman HB, Brauch H, Pomer S, Brooks JD, Isaacs WB, Lerman MI, Zbar B, Linehan WM (1994) Mutations of the VHL tumour suppressor gene in renal carcinoma. Nat Genet 7:85–90
25. Herman JG, Latif F, Weng Y, Lerman MI, Zbar B, Liu S, Samid D, Duan DS, Gnarra JR, Linehan WM (1994) Silencing of the VHL tumor-suppressor gene by DNA methylation in renal carcinoma. Proc Natl Acad Sci U S A 91:9700–9704
26. Ihnatko R, Kubes M, Takacova M, Sedlakova O, Sedlak J, Pastorek J, Kopacek J, Pastorekova S (2006) Extracellular acidosis elevates carbonic anhydrase IX in human glioblastoma cells via transcriptional modulation that does not depend on hypoxia. Int J Oncol 29:1025–1033
27. Kaluz S, Kaluzova M, Chrastina A, Olive PL, Pastorekova S, Pastorek J, Lerman MI, Stanbridge EJ (2002) Lowered oxygen tension induces expression of the hypoxia marker MN/carbonic anhydrase IX in the absence of hypoxia-inducible factor 1 alpha stabilization: a role for phosphatidylinositol 3′-kinase. Cancer Res 62:4469–4477
28. Barathova M, Takacova M, Holotnakova T, Gibadulinova A, Ohradanova A, Zatovicova M, Hulikova A, Kopacek J, Parkkila S, Supuran CT, Pastorekova S, Pastorek J (2008) Alternative splicing variant of the hypoxia marker carbonic anhydrase IX expressed independently of hypoxia and tumour phenotype. Br J Cancer 98:129–136
29. Ivanov SV, Kuzmin I, Wei MH, Pack S, Geil L, Johnson BE, Stanbridge EJ, Lerman MI (1998) Down-regulation of transmembrane carbonic anhydrases in renal cell carcinoma cell lines by wild-type von Hippel-Lindau transgenes. Proc Natl Acad Sci U S A 95:12596–12601
30. Parkkila S, Parkkila AK, Saarnio J, Kivela J, Karttunen TJ, Kaunisto K, Waheed A, Sly WS, Tureci O, Virtanen I, Rajaniemi H (2000) Expression of the membrane-associated carbonic anhydrase isozyme XII in the human kidney and renal tumors. J Histochem Cytochem 48:1601–1608
31. Uemura H, Nakagawa Y, Yoshida K, Saga S, Yoshikawa K, Hirao Y, Oosterwijk E (1999) MN/CA IX/G250 as a potential target for immunotherapy of renal cell carcinomas. Brit J Cancer 81:741–746
32. Oosterwijk E, Ruiter DJ, Hoedemaeker PJ, Pauwels EK, Jonas U, Zwartendijk J, Warnaar SO (1986) Monoclonal antibody G 250 recognizes a determinant present in renal-cell carcinoma and absent from normal kidney. Int J Cancer 38:489–494
33. Chiche J, Ilc K, Laferriere J, Trottier E, Dayan F, Mazure NM, Brahimi-Horn MC, Pouyssegur J (2009) Hypoxia-inducible carbonic anhydrase IX and XII promote tumor cell growth by counteracting acidosis through the regulation of the intracellular pH. Cancer Res 69:358–368
34. Patard JJ (2009) Incidental renal tumours. Curr Opin Urol 19:454–458
35. Volpe A, Finelli A, Gill IS, Jewett MA, Martignoni G, Polascik TJ, Remzi M, Uzzo RG (2012) Rationale for percutaneous biopsy and histologic characterisation of renal tumours. Eur Urol 62:491–504

36. Kranenborg MH, Boerman OC, de Weijert MC, Oosterwijk-Wakka JC, Corstens FH, Oosterwijk E (1997) The effect of antibody protein dose of anti-renal cell carcinoma monoclonal antibodies in nude mice with renal cell carcinoma xenografts. Cancer 80:2390–2397
37. Steffens MG, Kranenborg MH, Boerman OC, Zegwaart-Hagemeier NE, Debruyne FM, Corstens FH, Oosterwijk E (1998) Tumor retention of 186Re-MAG3, 111In-DTPA and 125I labeled monoclonal antibody G250 in nude mice with renal cell carcinoma xenografts. Cancer Biother Radiopharm 13:133–139
38. Steffens MG, Oosterwijk E, Kranenborg MH, Manders JM, Debruyne FM, Corstens FH, Boerman OC (1999) In vivo and in vitro characterizations of three 99mTc-labeled monoclonal antibody G250 preparations. J Nucl Med 40:829–836
39. Stillebroer AB, Oosterwijk E, Franssen GM, Oyen WJG, Boerman OC, Mulders PFA (2011) Optical imaging of renal cell carcinoma using the Anti-Caix monoclonal antibody Cg250. Eur Urol Suppl 10:105–105
40. van Schaijk FG, Oosterwijk E, Molkenboer-Kuenen JD, Soede AC, McBride BJ, Goldenberg DM, Oyen WJ, Corstens FH, Boerman OC (2005) Pretargeting with bispecific anti-renal cell carcinoma x anti-DTPA(In) antibody in 3 RCC models. J Nucl Med 46:495–501
41. Oosterwijk E, Bander NH, Divgi CR, Welt S, Wakka JC, Finn RD, Carswell EA, Larson SM, Warnaar SO, Fleuren GJ et al (1993) Antibody localization in human renal cell carcinoma: a phase I study of monoclonal antibody G250. J Clin Oncol 11:738–750
42. Divgi CR, Bander NH, Scott AM, O'Donoghue JA, Sgouros G, Welt S, Finn RD, Morrissey F, Capitelli P, Williams JM, Deland D, Nakhre A, Oosterwijk E, Gulec S, Graham MC, Larson SM, Old LJ (1998) Phase I/II radioimmunotherapy trial with iodine-131-labeled monoclonal antibody G250 in metastatic renal cell carcinoma. Clin Cancer Res 4:2729–2739
43. Steffens MG, Boerman OC, Oosterwijk-Wakka JC, Oosterhof GO, Witjes JA, Koenders EB, Oyen WJ, Buijs WC, Debruyne FM, Corstens FH, Oosterwijk E (1997) Targeting of renal cell carcinoma with iodine-131-labeled chimeric monoclonal antibody G250. J Clin Oncol 15:1529–1537
44. Divgi CR, Pandit-Taskar N, Jungbluth AA, Reuter VE, Gonen M, Ruan S, Pierre C, Nagel A, Pryma DA, Humm J, Larson SM, Old LJ, Russo P (2007) Preoperative characterisation of clear-cell renal carcinoma using iodine-124-labelled antibody chimeric G250 (124I-cG250) and PET in patients with renal masses: a phase I trial. Lancet Oncol 8:304–310
45. Divgi CR, Uzzo RG, Gatsonis C, Bartz R, Treutner S, Yu JQ, Chen D, Carrasquillo JA, Larson S, Bevan P, Russo P (2013) Positron emission tomography/computed tomography identification of clear cell renal cell carcinoma: results from the REDECT trial. J Clin Oncol 31:187–194
46. Muselaers CH, Boerman OC, Oosterwijk E, Langenhuijsen JF, Oyen WJ, Mulders PF (2013) Indium-111-labeled Girentuximab ImmunoSPECT as a diagnostic tool in clear cell renal cell carcinoma. Eur Urol 63:1101–1106
47. Surfus JE, Hank JA, Oosterwijk E, Welt S, Lindstrom MJ, Albertini MR, Schiller JH, Sondel PM (1996) Anti-renal-cell carcinoma chimeric antibody G250 facilitates antibody-dependent cellular cytotoxicity with in vitro and in vivo interleukin-2-activated effectors. J Immunother Emphasis Tumor Immunol 19:184–191
48. Liu Z, Smyth FE, Renner C, Lee FT, Oosterwijk E, Scott AM (2002) Anti-renal cell carcinoma chimeric antibody G250: cytokine enhancement of in vitro antibody-dependent cellular cytotoxicity. Cancer Immunol Immunother 51:171–177
49. van Dijk J, Uemura H, Beniers AJ, Peelen WP, Zegveld ST, Fleuren GJ, Warnaar SO, Oosterwijk E (1994) Therapeutic effects of monoclonal antibody G250, interferons and tumor necrosis factor, in mice with renal-cell carcinoma xenografts. Int J Cancer 56:262–268
50. Davis ID, Wiseman GA, Lee FT, Gansen DN, Hopkins W, Papenfuss AT, Liu Z, Moynihan TJ, Croghan GA, Adjei AA, Hoffman EW, Ingle JN, Old LJ, Scott AM (2007) A phase I multiple dose, dose escalation study of cG250 monoclonal antibody in patients with advanced renal cell carcinoma. Cancer Immun 7:13
51. Bleumer I, Knuth A, Oosterwijk E, Hofmann R, Varga Z, Lamers C, Kruit W, Melchior S, Mala C, Ullrich S, De Mulder P, Mulders PF, Beck J (2004) A phase II trial of chimeric monoclonal antibody G250 for advanced renal cell carcinoma patients. Br J Cancer 90:985–990

52. Bleumer I, Oosterwijk E, Oosterwijk-Wakka JC, Voller MC, Melchior S, Warnaar SO, Mala C, Beck J, Mulders PF (2006) A clinical trial with chimeric monoclonal antibody WX-G250 and low dose interleukin-2 pulsing scheme for advanced renal cell carcinoma. J Urol 175:57–62
53. Steffens MG, Boerman OC, Oosterwijk E, Oyen WJG, De Mulder PHM, Witjes JA, Oosterhof GON, Debruyne FMJ, Corstens FHM (1998) Phase I/II radioimmunotherapy with I-131 labeled chimeric monoclonal antibody cG250 in patients with metastasized renal cell carcinoma. J Nucl Med 39:247p–247p
54. Brouwers AH, Oyen WJ, Van den Broek WJ, Buijs WC, De Mulder PH, Mala C, Boerman OC, Booltink E, Oosterwijk E, Mulders PA, Corstens FH (2003) Efficacy of two sequential I-131-cG250 treatments at the maximum tolerated dose (MTD) in patients with advanced renal cell carcinoma (RCC). J Nucl Med 44:33p–33p
55. Brouwers AH, Mulders PF, de Mulder PH, van den Broek WJ, Buijs WC, Mala C, Joosten FB, Oosterwijk E, Boerman OC, Corstens FH, Oyen WJ (2005) Lack of efficacy of two consecutive treatments of radioimmunotherapy with 131I-cG250 in patients with metastasized clear cell renal cell carcinoma. J Clin Oncol 23:6540–6548
56. Brouwers AH, van Eerd JE, Frielink C, Oosterwijk E, Oyen WJ, Corstens FH, Boerman OC (2004) Optimization of radioimmunotherapy of renal cell carcinoma: labeling of monoclonal antibody cG250 with 131I, 90Y, 177Lu, or 186Re. J Nucl Med 45:327–337
57. Sharkey RM, Behr TM, Mattes MJ, Stein R, Griffiths GL, Shih LB, Hansen HJ, Blumenthal RD, Dunn RM, Juweid ME, Goldenberg DM (1997) Advantage of residualizing radiolabels for an internalizing antibody against the B-cell lymphoma antigen, CD22. Cancer Immunol Immunother 44:179–188
58. Brouwers AH, Buijs WC, Oosterwijk E, Boerman OC, Mala C, De Mulder PH, Corstens FH, Mulders PF, Oyen WJ (2003) Targeting of metastatic renal cell carcinoma with the chimeric monoclonal antibody G250 labeled with (131)I or (111)In: an intrapatient comparison. Clin Cancer Res 9:3953S–3960S
59. Stillebroer AB, Boerman OC, Desar IM, Boers-Sonderen MJ, van Herpen CM, Langenhuijsen JF, Smith-Jones PM, Oosterwijk E, Oyen WJ, Mulders PF (2012) Phase 1 radioimmunotherapy study with lutetium 177-labeled anti-carbonic anhydrase IX monoclonal antibody girentuximab in patients with advanced renal cell carcinoma. Eur Urol doi:pii: S0302-2838(12)00959-1. 10.1016/j.eururo.2012.08.024
60. Stillebroer AB, Zegers CM, Boerman OC, Oosterwijk E, Mulders PF, O'Donoghue JA, Visser EP, Oyen WJ (2012) Dosimetric analysis of 177Lu-cG250 radioimmunotherapy in renal cell carcinoma patients: correlation with myelotoxicity and pretherapeutic absorbed dose predictions based on 111In-cG250 imaging. J Nucl Med 53:82–89
61. Galfano A, Novara G, Iafrate M, Cavalleri S, Martignoni G, Gardiman M, D'Elia C, Patard JJ, Artibani W, Ficarra V (2008) Mathematical models for prognostic prediction in patients with renal cell carcinoma. Urol Int 80:113–123
62. Bui MH, Visapaa H, Seligson D, Kim H, Han KR, Huang Y, Horvath S, Stanbridge EJ, Palotie A, Figlin RA, Belldegrun AS (2004) Prognostic value of carbonic anhydrase IX and KI67 as predictors of survival for renal clear cell carcinoma. J Urol 171:2461–2466
63. Kim HL, Seligson D, Liu X, Janzen N, Bui MH, Yu H, Shi T, Figlin RA, Horvath S, Belldegrun AS (2004) Using protein expressions to predict survival in clear cell renal carcinoma. Clin Cancer Res 10:5464–5471
64. Kim HL, Seligson D, Liu X, Janzen N, Bui MH, Yu H, Shi T, Belldegrun AS, Horvath S, Figlin RA (2005) Using tumor markers to predict the survival of patients with metastatic renal cell carcinoma. J Urol 173:1496–1501
65. Bartosova M, Parkkila S, Pohlodek K, Karttunen TJ, Galbavy S, Mucha V, Harris AL, Pastorek J, Pastorekova S (2002) Expression of carbonic anhydrase IX in breast is associated with malignant tissues and is related to overexpression of c-erbB2. J Pathol 197:314–321
66. Zavada J, Zavadova Z, Pastorek J, Biesova Z, Jezek J, Velek J (2000) Human tumour-associated cell adhesion protein MN/CA IX: identification of M75 epitope and of the region mediating cell adhesion. Br J Cancer 82:1808–1813

67. Al-Ahmadie HA, Alden D, Qin LX, Olgac S, Fine SW, Gopalan A, Russo P, Motzer RJ, Reuter VE, Tickoo SK (2008) Carbonic anhydrase IX expression in clear cell renal cell carcinoma: an immunohistochemical study comparing 2 antibodies. Am J Surg Pathol 32:377–382
68. Li Y, Wang H, Oosterwijk E, Selman Y, Mira JC, Medrano T, Shiverick KT, Frost SC (2009) Antibody-specific detection of CAIX in breast and prostate cancers. Biochem Biophys Res Commun 386:488–492
69. Supuran CT (2008) Development of small molecule carbonic anhydrase IX inhibitors. BJU Int 101(Suppl 4):39–40
70. Dubois L, Peeters S, Lieuwes NG, Geusens N, Thiry A, Wigfield S, Carta F, McIntyre A, Scozzafava A, Dogne JM, Supuran CT, Harris AL, Masereel B, Lambin P (2011) Specific inhibition of carbonic anhydrase IX activity enhances the in vivo therapeutic effect of tumor irradiation. Radiother Oncol 99:424–431
71. Lamers CH, Sleijfer S, Vulto AG, Kruit WH, Kliffen M, Debets R, Gratama JW, Stoter G, Oosterwijk E (2006) Treatment of metastatic renal cell carcinoma with autologous T-lymphocytes genetically retargeted against carbonic anhydrase IX: first clinical experience. J Clin Oncol 24:e20–e22
72. Lamers CH, Sleijfer S, van Steenbergen S, van Elzakker P, van Krimpen B, Groot C, Vulto A, den Bakker M, Oosterwijk E, Debets R, Gratama JW (2013) Treatment of Metastatic Renal Cell Carcinoma With CAIX CAR-engineered T cells: clinical evaluation and management of on-target toxicity. Mol Ther J Am Soc Gene Ther 21:904–912
73. Escudier B, Szczylik C, Porta C, Gore M (2012) Treatment selection in metastatic renal cell carcinoma: expert consensus. Nat Rev Clin Oncol 9:327–337
74. Oosterwijk-Wakka JC, Kats-Ugurlu G, Leenders WP, Kiemeney LA, Old LJ, Mulders PF, Oosterwijk E (2011) Effect of tyrosine kinase inhibitor treatment of renal cell carcinoma on the accumulation of carbonic anhydrase IX-specific chimeric monoclonal antibody cG250. BJU Int 107:118–125
75. Pena C, Lathia C, Shan M, Escudier B, Bukowski RM (2010) Biomarkers predicting outcome in patients with advanced renal cell carcinoma: results from sorafenib phase III treatment approaches in renal cancer global evaluation trial. Clin Cancer Res 16:4853–4863
76. Choueiri TK, Regan MM, Rosenberg JE, Oh WK, Clement J, Amato AM, McDermott D, Cho DC, Atkins MB, Signoretti S (2010) Carbonic anhydrase IX and pathological features as predictors of outcome in patients with metastatic clear-cell renal cell carcinoma receiving vascular endothelial growth factor-targeted therapy. BJU Int 106:772–778
77. Choueiri TK, Cheng S, Qu AQ, Pastorek J, Atkins MB, Signoretti S (2012) Carbonic anhydrase IX as a potential biomarker of efficacy in metastatic clear-cell renal cell carcinoma patients receiving sorafenib or placebo: analysis from the treatment approaches in renal cancer global evaluation trial (TARGET). Urol Oncol doi:pii: S1078-1439(12)00246-3. 10.1016/j.urolonc.2012.07.004
78. Steffens MG, Boerman OC, Oyen WJ, Kniest PH, Witjes JA, Oosterhof GO, van Leenders GJ, Debruyne FM, Corstens FH, Oosterwijk E (1999) Intratumoral distribution of two consecutive injections of chimeric antibody G250 in primary renal cell carcinoma: implications for fractionated dose radioimmunotherapy. Cancer Res 59:1615–1619
79. Siebels M, Rohrmann K, Oberneder R, Stahler M, Haseke N, Beck J, Hofmann R, Kindler M, Kloepfer P, Stief C (2011) A clinical phase I/II trial with the monoclonal antibody cG250 (RENCAREX(R)) and interferon-alpha-2a in metastatic renal cell carcinoma patients. World J Urol 29:121–126
80. Steffens MG, Boerman OC, de Mulder PH, Oyen WJ, Buijs WC, Witjes JA, van den Broek WJ, Oosterwijk-Wakka JC, Debruyne FM, Corstens FH, Oosterwijk E (1999) Phase I radioimmunotherapy of metastatic renal cell carcinoma with 131I-labeled chimeric monoclonal antibody G250. Clin Cancer Res 5:3268s–3274s
81. Divgi CR, O'Donoghue JA, Welt S, O'Neel J, Finn R, Motzer RJ, Jungbluth A, Hoffman E, Ritter G, Larson SM, Old LJ (2004) Phase I clinical trial with fractionated radioimmunotherapy using 131I-labeled chimeric G250 in metastatic renal cancer. J Nucl Med 45:1412–1421

Chapter 11
Carbonic Anhydrase IX: Regulation and Role in Cancer

Martin Benej, Silvia Pastorekova, and Jaromir Pastorek

Abstract Tumor microenvironment substantially influences the process of tumorigenesis. In many solid tumors, imbalance between the demand of rapidly proliferating cancer cells and the capabilities of the vascular system generates areas with insufficient oxygen supply. In response to tumor hypoxia, cancer cells modulate their gene expression pattern to match the requirements of the altered microenvironment. One of the most significant adaptations to this milieu is the shift towards anaerobic glycolysis to keep up the energy demands. This oncogenic metabolism is often maintained also in aerobic cells. Lactic acid, its metabolic end-product, accumulates hand-in-hand with carbon dioxide, leading to acidification of the extracellular environment. Carbonic anhydrase IX (CA IX) is the most widely expressed gene in response to hypoxia. Its crucial role in intracellular pH maintenance represents the means by which cancer cells adapt to the toxic conditions of the extracellular milieu. Furthermore, the activity of CA IX stimulates the migratory pathways of cancer cells and is connected with the increase of the aggressive/invasive phenotype of tumors. CA IX expression in many types of tumors indicates its relevance as a general marker of tumor hypoxia. Moreover, its expression is closely related to prognosis of the clinical outcome in several tumor types. All above mentioned facts support the strong position of CA IX as a potential drug therapy target. Here, we summarize the state-of-the-art knowledge on its regulation and role in cancer development.

Susan C. Frost and Robert McKenna (eds.) Carbonic Anhydrase: Mechanism, Regulation, Links to Disease, and Industrial Applications

M. Benej • S. Pastorekova • J. Pastorek (✉)
Department of Molecular Medicine, Institute of Virology, Slovak Academy of Sciences, Bratislava, Slovakia
e-mail: virumabe@savba.sk; virusipa@savba.sk; virupast@savba.sk

S.C. Frost and R. McKenna (eds.), *Carbonic Anhydrase: Mechanism, Regulation, Links to Disease, and Industrial Applications*, Subcellular Biochemistry 75,
DOI 10.1007/978-94-007-7359-2_11,

Keywords Hypoxia • Cancer • Carbonic anhydrase IX • Tumor acidosis • pH regulation

1 Cancer and Tumor Microenvironment

Solid tumors are characterized by distinct attributes that directly lead to massive changes both in genotype and phenotype of cancer cells. First, mutations-driven uncontrolled cell growth and proliferation support spatial de-organization and a rapid increase of the tumor mass. As the tumor growth advances, areas with insufficient oxygen supply arise due to inadequate vasculature [1]. *Tumor hypoxia*, a condition defined by less than 5 % oxygen in the surrounding tissue, conducts dramatic alterations in cancer cell behavior including changes in gene expression, metabolism, genetic stability, proliferation and survival [1–3]. To address the limited availability of oxygen, the cancer cells shift their energetic metabolism from oxidative phosphorylation towards anaerobic glycolysis [4, 5]. Glycolytic metabolism generates large amounts of lactic acid and carbon dioxide that contribute to a decrease of the extracellular pH (pH_e), leading to the development of acidic milieu within the tumor microenvironment [6]. As the maintenance of vital biological functions of the cell is extremely sensitive to intracellular pH (pH_i) consistency, the survival of cancer cells is dependent on their ability to adapt to these limiting conditions [6]. One of the key players involved in the process of adaptation is carbonic anhydrase IX (CA IX) protein.

2 Carbonic Anhydrase IX: Introduction

CA IX is a transmembrane zinc metalloenzyme functioning as the catalyst of reversible hydration of carbon dioxide to bicarbonate ions and protons [7]. The protein belongs to the α class of the family of carbonic anhydrases (CAs) expressed in higher vertebrates [8]. CA IX (initially denominated "MN antigen") was discovered as a cell surface protein on HeLa cervical cancer cells using monoclonal M75 antibody [9, 10]. Following its identification, the cDNA and genomic sequences of the protein were analyzed, observing the presence of a 257 amino acid (aa) long catalytic domain specific for the family of CAs [11, 12]. Hence, the protein joined the family of carbonic anhydrases as the ninth member and was therefore renamed to CA IX.

3 Biochemical Structure of CA IX

The *CA9* gene is located on chromosome 9p 12–13 and encodes a 459 aa residues long transmembrane protein [13]. The N-terminal extracellular domain (ECD, 414 aa) comprises: (i) 37 aa signal peptide; (ii) 59 aa long proteoglycan-like (PG)

domain bearing features homological to that of keratan sulfate attachment domain of proteoglycan aggrecan; and finally, (iii) the 257 aa catalytic (CA) domain [12]. The presence of the PG domain is unique to CA IX among the CA family. The M75 antibody recognizes a linear epitope within the very PG domain, guaranteeing that the antibody binds CA IX exclusively [14].

The protein spans the plasma membrane by a 20 aa long hydrophobic transmembrane (TM) domain helix and the C-terminal domain of CA IX is protruding into the cytosol in the form of the 25 aa long intracellular (IC) tail [11, 12]. In addition, *CA9* mRNA undergoes alternative splicing (AS) generating the CA IX AS isoform lacking the IC domain, TM domain and a part of the CA domain, due to a frameshift-generated STOP codon at position 1,119 bp [15]. The truncated CA IX AS isoform was observed in cancer cells as well as in normal cells, regardless of the hypoxic conditions [15, 16].

Mature CA IX is represented by a 54/58 kDa double band on western blots in reducing conditions. In non-reducing conditions, a 153 kDa band was observed, possibly corresponding to CA IX homotrimer [9]. Subsequent analyses of the biochemical structure of the protein were carried out in vitro on recombinant CA IX forms lacking the IC tail and TM domain, expressed using baculovirus expression system. Intriguingly, mass spectrometry results suggested the presence of intermolecular S-S bridge occurring between Cys^{41} residues, indicating towards a homodimeric, rather than homotrimeric state of CA IX [7, 17]. The plausible CA IX homodimer might be furthermore stabilized by a number of non-covalent interactions involving two hydrogen bonds and additional van der Waals bonds [7, 18]. Additionally, intramolecular stabilization by disulfide bridge between Cys^{119} and Cys^{299} was identified [7]. Finally, N- and O-linked glycosylations were identified at Asn^{309} and Thr^{78} residues, respectively [7].

4 Catalytic Activity of CA IX

Members of the family of CAs catalyze the hydration of carbon dioxide to bicarbonate ions and protons. This is achieved by deprotonation of a water molecule that is bound to the active site present in the CA domain. To investigate the kinetic properties of CA IX, the aforementioned recombinant CA IX forms expressed in Sf9 insect cells, along with recombinant CA IX from *E.coli* expression system were analyzed using stopped-flow spectrophotometry [7]. Upon comparison with kinetic properties of other mammalian CAs, CA IX and CA II were found to be the most active [7, 19]. The optimal pH value for this process was observed around pH 6.49, compared to the optimal values close to pH 7, which are typical for other CA isoenzymes [19]. Experiments with ΔCA and ΔPG CA IX isoforms revealed that the CA domain alone is sufficient for the catalytic activity of CA IX although a possibility of cooperation with the PG domain in the function of CA IX is not excluded [20]. Interestingly, the unique PG domain might function as an intrinsic buffer, rendering CA IX a suitable pH_i regulator in the conditions of environmental acidosis [19].

The activity of CA IX addresses the rising requirement of the cell to neutralize its pH_i in an acidic environment. Although the expression of the protein is connected with several physiological processes including lipo- and ureagenesis, gluconeogenesis and bone resorption, its most prominent role is connected with pH_i regulation in cancer cells exposed to acidic environment caused by metabolic products of glycolysis [7, 21, 22].

Oncogenic metabolism in highly proliferative cancer cells generates excessive amounts of lactate, carbon dioxide and protons. To maintain optimal pH_i, these metabolites are extruded to the extracellular milieu where they accumulate due to poor vasculature [23]. Due to the continuously decreasing pH_e, cancer cells need to employ another adaptive mechanism. To neutralize their pH_i, bicarbonate anions generated by CA IX-catalyzed hydration of carbon dioxide are transported into the cells by anion transporters to interact with protons liberated by the glycolytic metabolism (described in more detail later in this chapter) [24, 25]. The process generates carbon dioxide, which diffuses through the plasma membrane [26].

CA IX catalytic activity thus contributes to (i) extracellular acidification by elevating the extracellular levels of carbon dioxide and protons; (ii) intracellular neutralization by generating bicarbonate ions, as well as by active participation in the process of their intake by a mechanism described later in this chapter.

Remarkably, a detailed analysis of CA IX interactome revealed an additional, hitherto unanticipated role of CA IX. Among the identified interacting partners, a substantial part was represented by proteins facilitating nucleo-cytoplasmic transport including importin TNPO1 and exportin XPO1 [27]. This finding suggested nuclear localization of CA IX, a notion which was subsequently confirmed by confocal immunofluorescence assay on various cell lines. It seems plausible that the nuclear moiety of CA IX functions by interaction with partner proteins rather than by direct binding to DNA [27]. Although the exact role of nuclear CA IX awaits its identification, these findings imply that the activity of CA IX goes far beyond its catalytic function.

5 CA IX Regulation

5.1 Regulation of CA IX Expression

Transcriptional activity of CA IX is subject to numerous regulatory elements employing various mechanisms of action. However, the principal regulator of CA IX expression is hypoxia via its executive officer, hypoxia-induced factor 1 (HIF-1) [28].

HIF-1 is a transcription factor originally identified as a *trans*-acting component mediating erythropoietin expression in hypoxia [29, 30]. The active transcription factor is a heterodimer composed of oxygen sensitive subunit α (HIF-1α) and constitutively expressed subunit β (HIF-1β) [31]. In normoxia, the fate of HIF-1α

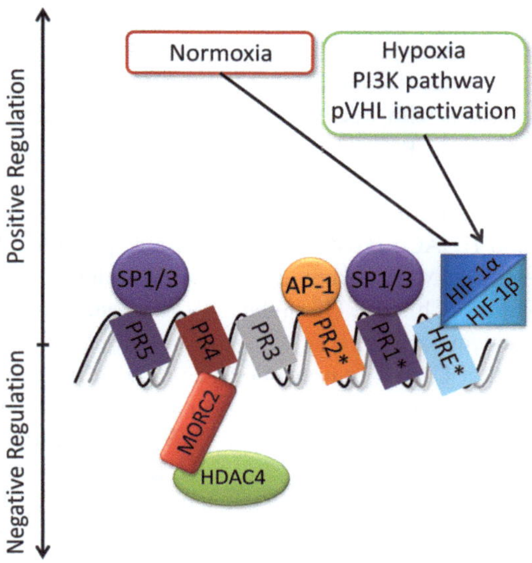

Fig. 11.1 *CA9* promoter regulation. Six distinct *cis* elements were identified within the *CA9* promoter sequence; the hypoxia responsive element (HRE) and five regions identified by DNaseI footprinting (Protected regions, PRs). The HRE element is recognized by HIF-1 heterodimer and its binding is crucial for the *CA9* promoter activity. PR1 and PR5 regions are recognized by the Specificity protein (SP) 1 and/or SP3; PR2 is bound by the Activator protein 1 (AP-1). Regions essential for the *CA9* promoter activity in hypoxic conditions are marked with an asterisk (*). The PR4 region acts as a repressor of the *CA9* promoter upon binding of the Microrchidia2 (MORC2) and Histone deacetylase 4 (HDAC4) complex

subunit is sentenced by von Hippel-Lindau tumor suppressor protein (pVHL)-mediated polyubiquitylation and subsequent proteasome degradation [32]. The process is dependent on pVHL recognition of hydroxylated proline residues Pro^{402} and Pro^{564} located in the oxygen-dependent degradation domain (ODDD) of HIF-1α. Both proline residues are hydroxylated by prolyl-hydroxylase domain proteins (PHDs) only in normoxia. In addition, HIF-1α activity is guarded by asparaginyl hydroxylase enzyme referred to as ‘factor inhibiting HIF-1’ (FIH-1) [33]. FIH-1 hydroxylates Asn^{803} present in the C-terminal activation domain of HIF-1α and thus prevents it from binding to its co-activators [34]. In low levels of oxygen, PHDs, as well as FIH-1 are inactivated and HIF-1α is transported into the nucleus where it forms a heterodimer with HIF-1β, binds with co-activators and activates expression of a vast array of genes having one common denominator: the presence of the hypoxia responsive element (HRE) 5’G/ACGTG3’ in their promoter sequence.

The HRE element, located upstream of transcription start site of the *CA9* promoter is the main *cis*-acting regulator of CA IX expression [28]. Furthermore, five additional regulatory DNA sequences, PR1-5 (‘Protected regions’, denominated after regions identified in DNaseI footprinting assays) were identified (Fig. 11.1) [35]. PR1-3 and PR5 were proven to regulate CA IX expression in a positive manner [35, 36]. These *cis*-acting elements are recognized by following *trans*-acting factors: PR1 and PR5 regions interact with SP1/SP3 (Specificity protein1 and 3) transcription factors; PR2 is bound by AP-1 (Activator protein-1) transcription factor and PR3 interacts with nuclear extract proteins [31, 35–37]. According to promoter deletion analysis, PR1 and PR2 are essential for CA IX transcription [35]. In contrast, PR4 is a negative regulatory element acting as a silencer of CA IX expression [35]. The region is bound by a repressor complex comprising MORC2 (Microrchidia 2) and HDAC4 (Histone deacetylase 4) proteins [38]. The proposed mechanism suggests that MORC2 recognizes the PR4 element and HDAC4 in turn catalyzes deacetlyation of histone H3 in the *CA9* promoter region and thus perturbs its transcriptional activation [38].

It is noteworthy that CA IX transcription can be also initiated in normoxia, by high density of in vitro cultivated cells. Higher cell density causes a decrease of pericellular oxygen levels defined by less than 5 % and more than 1 % of oxygen; a condition referred to as 'mild hypoxia' [36]. Such levels of oxygen are sufficient to trigger proteasome degradation of HIF-1α but nevertheless, CA IX is expressed [39]. This phenomenon stimulated the search for signaling pathways that may potentially influence CA IX expression in these conditions.

In mild hypoxia, CA IX expression is induced by PI3K pathway mechanism [39]. To trigger CA IX expression, only the HRE element and PR1 region are required; however, the process is still dependent on minimal HIF-1α activity [39]. This conclusion supports the ultimate dependence of CA IX expression on HIF-1 activation. In addition, high cell density-induced mild hypoxia has additive effect on true hypoxic conditions, considerably increasing the CA IX expression levels [39]. Furthermore, the expression of CA IX is regulated by components of MAPK pathway both in true hypoxic and mild hypoxic conditions [40]. Simultaneous inhibition of components of the PI3K and MAPK signaling pathways leads to a dramatic drop of CA IX transcription activity [40].

5.2 Regulation of CA IX Activity

The capacity of CA IX to acidify the extracellular environment is dependent on specific post-translational modifications within its IC tail.

The cytosol-protruding CA IX IC tail is involved in cell membrane targeting of assembled CA IX [41]. Deletion, as well as targeted mutagenesis experiments revealed the major role of CA IX IC tail in cell surface localization of the protein, along with the management of its pH regulating and ECD shedding capacity [41]. The CA IX IC tail contains several putative phosphorylation sites. Among these, Tyr^{449}, Thr^{443} and Ser^{448} bear the potential to regulate CA IX activity [42, 43]. Epidermal growth factor (EGF)-induced Tyr^{449} phosphorylation mediates cross-talk between CA IX and PI3K and in this manner contributes to Akt kinase activation [42]. However, the impact on CA IX activity is yet undetermined [43].

In vitro substitution of Thr^{443} by a non-phosphorylatable amino acid perturbs the extracellular acidification capacity of CA IX [43]. On the contrary, phosphorylation of Ser^{448} affects CA IX activity in a negative manner. Subsequent experiments revealed that Thr^{443} residue is recognized and phosphorylated by Protein kinase A (PKA) in vivo [43]. The relation of CA IX activity and PKA phosphorylation illustrates another aspect of the definitive dependency of CA IX on hypoxia, since: (i) elevated intracellular concentration of cAMP required for PKA activity; and (ii) elevated levels of active PKA (Catalytic subunit of PKA phosphorylated on Thr^{197} residue) were observed in hypoxia compared to the normoxic conditions [43].

As for the phosphorylation status of Ser^{448} as a negative regulator, the in vivo evidence is at the moment missing, and so is the putative kinase involved in the process, therefore further experimentation will be required in this matter [43].

5.3 Regulation of CA IX Abundance

Total cell surface population of CA IX is regulated by metalloproteinase-mediated cleavage of CA IX ECD in a process termed 'ectodomain shedding' [44, 45]. The CA IX ECD comprises CA and PG domains and is represented by a 50/54 kDa twin band on Western blots [44]. The basal rate of 'shed' ECD represents up to 1/5 of the membrane-bound CA IX population [45]. Experimental data revealed that CA IX ECD shedding is induced by TNFα-converting enzyme (TACE/ADAM17) [45]. However, the biological impact of ECD shedding still awaits its identification.

6 Role of CA IX in Cancer

6.1 CA IX Expression in Normal and Cancer Tissues

CA IX is limitedly distributed among normal tissues with almost a total exclusivity in the gastrointestinal tract epithelium, namely glandular gastric mucosa, epithelium of the gallbladder and cryptic enterocytes of duodenum, jejunum and ileum [3, 46, 47]. Also, the expression in ovarian coelomic epithelium, pancreatic ductal cells, cells of the hair follicles, and fetal rete testis has been observed [48, 49]. In contrast, the protein is ectopically expressed in a variety of cancer tissues. The expression of CA IX was observed in many tumors including malignancies of the brain, head/neck, breast, lung, bladder, cervix uteri, colon/rectum and kidney [48, 49]. cDNA comparison of CA IX sequence obtained from normal and tumor tissues showed no difference between these sequences assuming regulatory intervention into CA IX expression [47]. Indeed, the overexpression of CA IX in these solid tumors is driven by hypoxia, with the exception of clear cell renal cell carcinoma (CCRCC), where the overexpression is triggered by a different mechanism.

As mentioned earlier in this chapter, in normoxia, the level of HIF-1α transcription factor is regulated by pVHL-mediated polyubiquitylation and subsequent degradation in the proteasome. However, in CCRCC, CA IX expression throughout the tumor tissue was detected even in normoxia [50]. The responsible mechanism was revealed after the two seemingly independent lines of research were brought together.

6.2 CA IX as a Diagnostic Marker

In 1986, using monoclonal antibody G250, a novel marker specific for renal cell carcinoma (RCC) was discovered [51]. The marker showed homogenous expression at a high level throughout the tumor, in contrast to almost no expression in normal tissues despite a few exceptions. Interestingly, the marker was discovered in other types of tumors as well, although the expression was rather heterogenous and at a lower rate [52]. The identity had not been elucidated until 2000, when G250 cDNA was analyzed, revealing 100 % identity with CA IX [13, 52]. By a fusion of the two lines of research, it became evident that hypoxia and namely HIF-1α are responsible for the heterogeneous expression in many types of solid tumors, whereas inactivating mutations of the VHL tumor-suppressor determine the normoxic expression in RCC [52].

Approximately 60–80 % of all RCC patients carry inactivating mutations of the VHL tumor-suppressor gene in their genome [53]. In this manner, HIF-1α subunit is stabilized even in normoxia and constitutive CA IX expression is administered. Among different RCC subtypes, CCRCC is associated with the highest CA IX expression. The protein is expressed in 97 % of CCRCC cases with apparently no expression in normal renal tissue [48]. Therefore, CA IX bears the potential of being a diagnostic marker for renal malignancies. However, its expression is not only connected with numerous RCC subtypes, yet with a whole spectrum of solid tumors. Large-scale studies of tumor-derived cell line and tissue specimens revealed CA IX expression in specific carcinomas of the cervix uteri and uterine corpus, ovary, gastrointestinal tract, liver, pancreas, lung, head/neck, salivary gland, body cavity and skin [48, 49]. CA IX is thus rather a general marker of tumor hypoxia in many solid tumors, than a specific marker for distinct types/subtypes of malignancies.

6.3 CA IX as a Prognostic Marker

The importance of detecting CA IX ectopic expression in human tumors lies not only in its diagnostic nature. Interestingly, the level of CA IX expression is also associated with staging and survival prognosis in several types of human tumors. Higher levels of CA IX expression are associated with poor clinical outcome in cervical, rectal, breast, lung, and brain tumors [54–58]. In contrary, *low* expression levels of CA IX seem to indicate poor prognosis in CCRCC [59]. However, this notion has been opposed by a larger-scale study suggesting that there is no relevant connection between low CA IX expression and poorer prognosis in CCRCC [48]. The discrepancy might be related to a very high cut-off value (of 85 % CA IX-positive cells in a tissue), which was proposed to discriminate between high and low expression of CA IX [59].

Generally, the very impact of CA IX pH regulating activity on tumor behavior might be connected with poorer outcome, rendering the protein a promising biomarker and drug therapy target.

6.4 CA IX and pH Regulation in Solid Tumors

The events leading to formation of two hallmarks of solid tumors, *hypoxia* and relative *tumor acidosis* have been described earlier in this chapter. Briefly, rapid advancement of tumor tissue can be characterized by several phenomena summing up to the final attributes of the tumor microenvironment. Uncontrolled growth and proliferation lead to rapid increase of the tumor tissue. As the rate of angiogenesis ceases to match its demand, areas with true hypoxia develop in parts of the tumor most distal from the functional blood vessels. To address the reduced oxygen supply, HIF-1 induces the expression of HRE element-carrying target genes that organize the reconstruction of cellular behavior under true hypoxic conditions. The cancer cells shift their metabolism to anaerobic glycolysis to maintain energy production required for continuous proliferation. One of the key characteristics of glycolytic metabolism in hypoxic conditions is increased production of acidic metabolites that are poorly removed from extracellular space due to insufficient vasculature.

To summarize, hypoxia, and namely its right hand, tumor acidosis, represent toxic environmental conditions to which normal cells do not possess the means of adapting. Hence, the ability to regulate pH_i is the key selective advantage enabling to overcome the life-limiting stress represented by acidic tumor microenvironment.

7 Introduction to Tumor Acidosis

It is noteworthy that low pH in the tumor microenvironment supports the growth of cancer cells by several independent lines of impact. Acidic milieu is unfavorable for normal cells, as well as for immune cells that could recognize tumor antigens [23]. Decrease of pH_i leads to several major consequences ranging from disruption of ATP synthesis and alternative splicing of various proteins to p53-mediated apoptosis by the activation of caspases [5, 60]. It has also been discussed that acidic pH_e increases invasive behavior of cancer cells thus enhancing their metastatic potential [61]. In addition, low pH_e induces angiogenesis by upregulation of vascular endothelial growth factor (VEGF) expression in cancer cells [62].

Principally, excessive concentrations of lactic acid emerging from the glycolytic cycle were believed to acidify pH_e [63, 64]. However, experiments on lactate dehydrogenase (LDH) deficient cells proved that carbon dioxide may represent equally potent cause of pH_e decrease [64]. Either way, acidic pH_e is the direct result of the activity of pH_i regulators in cancer cells.

8 CA IX and pH Regulation in Cancer Cells

Simplified, two main processes represent the pH_i regulating mechanisms of tumor cells: pH_i buffering and acid extrusion. To position the place of CA IX within the complex pH regulating machinery in cancer cells, its enzymatic reaction needs to be highlighted. As mentioned earlier in this chapter, CA IX reversibly catalyzes carbon dioxide hydration to bicarbonate ions and protons by the reaction that is taking place in the extracellular space. To maintain acid extrusion from cells, carbon dioxide needs to be hydrated in the extracellular space and not in the cytosol [65]. In order to achieve this, the hydration of carbon dioxide needs to be catalyzed by CA IX which effectively facilitates the reaction so that the carbon dioxide concentration gradient is directed extracellularly [65]. Bicarbonate ions are transported into the cell by Na^+/HCO_3^- cotransporter (NBC), Na^+-dependent Cl^-/HCO_3^- (NDCBE) exchanger and by members of the anion exchanger (AE) family [24]. Interestingly, several lines of evidence indicate that CA IX may directly interact with the AE2 and NBC according to a metabolon concept, thus representing even more interconnected and adaptive pH regulating system (Fig. 11.2) [6, 66, 67]. This,

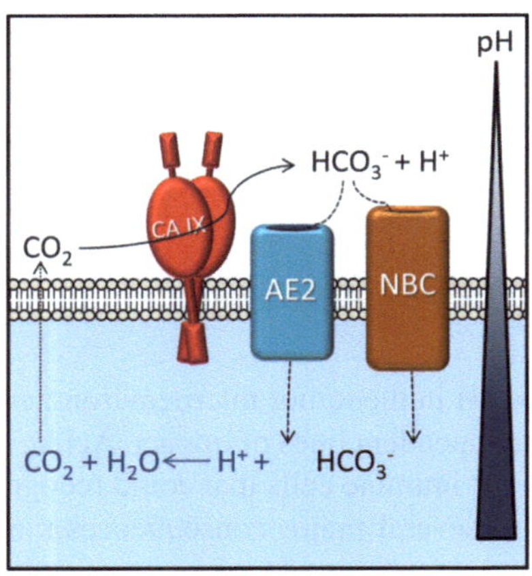

Fig. 11.2 The metabolon concept. CA IX catalyzes the hydration of carbon dioxide to bicarbonate ions and protons. By direct interaction with Na^+/HCO_3^- cotransporter (NBC) and Anion exchanger 2 (AE2) transporters, bicarbonate ions are immediately transported into the cell, where they interact with protons liberated by glycolysis. The reaction represents the basis of intracellular pH alkalinization. Carbon dioxide, the final product, then diffuses through the plasma membrane where it contributes to extracellular pH acidification

along with the fact that downregulation of individual proton transporters potentially perturbs tumor growth, demonstrates the complexity and extensive dynamics of pH balance maintenance in cancer cells [24].

The pH_i buffering activity of cancer cells is reinforced by acid extrusion. In line with CA IX-mediated carbon dioxide hydration, several additional acid extrusion mechanisms contribute to pH_i alkalinization and simultaneous extracellular pH acidification. One of the most important pH regulators functioning by H^+ extrusion is NHE isoform 1 (NHE-1) [65, 68]. Its crucial role in tumorigenesis was demonstrated in NHE-1 activity deficient cells that showed significantly decreased rates of tumor growth [24]. Proton extrusion is also maintained by vacuolar-type (V-type) H^+ ATPases [69]. The expression of V-type ATPases furthermore confers anti-apoptotic attributes to the cancer cells [65]. Finally, monocarboxylate transporters (MCTs) play an essential role in acid extrusion by exporting pyruvate and lactate along with protons, thereby being inevitable for pH_i alkalinization of cancer cells [70].

As mentioned previously, consequently to the above-mentioned processes the pH of the extracellular milieu is decreasing. In addition to poor removal of extruded acidic metabolites and protons, the activity of CA IX was proven to significantly reduce extracellular pH in hypoxia [20]. Experiments on immortalized CA IX-negative MDCK (Madin-Darby canine kidney) cell line compared to MDCK cells constitutively expressing CA IX after stable transfection revealed a major impact of CA IX CA domain on pH_e decrease. The concept was validated by experiments using selective CA IX inhibitors on HeLa (cervical cancer) and SiHa (cervical cancer) cancer cell lines and ΔCA CA IX isoform in HeLa cells [20].

Excessive acidification of pH_e due to the activity of cellular acid extrusion mechanisms and the activity of CA IX are not only connected with cancer cell survival in conditions that are unfavorable for normal cells. By several lines of action, tumor acidosis stimulates the metastatic potential of cancer cells.

8.1 The CA IX Involvement in the Metastatic Cascade

Accumulation of genetic disorders that led to the creation of cancer cell genotype may not alone define the aggressive phenotype of most cancers, as the possibilities of primary tumor growth are limited [5]. Invasion to adjacent tissues and distant organs, i.e. the metastatic process represents one of the conditions defining poor curability of patients [71]. Metastatic process is gradual and comprises several steps including reduction of cell-cell adhesion, degradation of extracellular matrix (ECM) components, invasion and migration into surrounding tissues or lymphatic and blood vessels, re-adhesion and finally proliferation and angiogenesis [71, 72].

Tumor acidosis, fuelled by the activity of pH_i regulating mechanisms of cancer cells stimulates their potential to metastasize. The involvement of low pH_e in the metastatic cascade is supposed to be both indirect and direct; by p53-mediated apoptosis of normal cells surrounding the tumor tissue and by promoting basement membrane degradation in acidic regions of the tumor, respectively [5, 61]. The degradation of the basement membrane represents the onset of cancer cell dissemination into distant organs and tissues, impersonating the barrier between metastatic and non-metastatic cancers. By overcoming this barrier, cancer cells are free to proliferate into the surrounding tissues. Acidic tumor microenvironment stimulates secretion of cathepsin B, a protein which plays a significant role in degradation of the components of the ECM including the basement membrane [73]. In addition, members of the matrix-metalloproteinase (MMP) family and the urokinase-like plasminogen activator (uPa) are believed to be secreted in acidic tumors [73, 74]. Finally, acidic pH_e alters the size, abundance and intracellular distribution of lysosomes thereby enhancing the secretion of ECM degrading proteinases [61, 74, 75].

It is noteworthy that CA IX also actively participates in the metastatic process. The protein is connected with the reduction of cell-cell adhesion and the migratory phenotype of cancer cells.

8.1.1 CA IX and Cell Migration

Cell adhesion is mediated by adhesion molecules, including cadherins, selectins, integrins and immunoglobulin protein family members. Of these, E-cadherin is in the spotlight because its level of expression often correlates with prognosis of the malignancy [76]. E-cadherin is a 120 kDa transmembrane protein expressed on the basolateral surface of epithelial cells [76, 77]. Its extracellular domain plays a key role in cell-cell adhesion by homotypic interaction with another E-cadherin molecule protruding from adjacent cell. The cytoplasmic domain of E-cadherin interacts with proteins of the catenin family, namely with β-catenin [78]. β-catenin is recognized and bound by α-catenin which in turn mediates the connection with actin cytoskeleton [79]. Furthermore, the E-cadherin-catenin complex is modified by WNT-1 protein, which recognizes β-catenin [76]. The catenin-mediated connection between E-cadherin and actin cytoskeleton is the essence of cell-cell adhesion.

Disruption of cell-cell adhesion is the initiating step of the metastatic cascade. In order to detach from the primary tumor mass, cancer cells undergo a series of morphological and behavioral changes, most prominent of which is the epithelial-mesenchymal transition (EMT). By EMT the morphology of polarized epithelial cells is rebuilt to that of migration prone mesenchymal cells by the activity of various transcription factors including Slug, Twist, Snail or by environmental conditions such as hypoxia [72, 80]. The morphological changes include cytoskeletal reorganization; polarization of the cancer cell along the axis of the movement and the formation of lamellipodia and invadopodia [81]. Formation of the leading edge of the migrating cell is followed by its detachment from the matrix and the formation of a trailing edge [82].

In the process of cell migration, intracellular as well as extracellular pH plays a major role. Thus, tumor acidosis caused by the activity of pH_i regulating mechanisms of cancer cells directly influences their migration capabilities. According to several studies, NHE-1 transporter has the most prominent role in affecting cancer cell migration [82–84]. Before migration, the re-localization of pH regulators including NHE-1, NBC, AE2, MCT and finally CA IX from their original position into the leading edge takes place [81]. Their activity generates pericellular as well as intracellular pH gradients. The pericellular pH gradient is directed from the leading edge to the trailing edge, while being the most acidic at the leading edge region. In contrary, due to the extensive activity of pHi regulating mechanisms at the leading edge, the intracellular pH gradient is rising from the trailing edge to the leading edge. The formation of pH_i gradient is one of the prerequisites of cell migration [82]. Interestingly, increased levels of cAMP were detected in lamellipodia and PKA was observed in the leading edge of migrating cells [85]. This, hand-in-hand with the metabolon concept, indicates that a complex network of closely related events takes place before migration is initiated; a network in which CA IX is one of the key players.

Indeed, the effect of CA IX on the promotion of cancer cell migration goes beyond its contribution to pH_i regulation. As observed in polarized MDCK cells, CA IX responds to calcium depletion by internalization along with E-cadherin, β-catenin and other adhesion molecules [86]. This behavior, typical for adhesion molecules was also observed in hypoxia. Upon further investigation of this phenomenon, it was revealed that CA IX interacts with β-catenin and thus interferes with the E-cadherin attachment to the actin cytoskeleton, hence, it perturbs E-cadherin-mediated adherent junctions between epithelial cells [86]. In other words, this competitive interaction destabilizes cell-cell adhesions and thus significantly supports cell migration. Experimentally, it was observed that ectopically expressed CA IX significantly increases migration rate of MDCK and HeLa cells in vitro, whereas ΔCA CA IX isoform lacks this ability [67]. Obviously, the catalytic activity of CA IX is required for this process. Furthermore, CA IX is presumably involved in activation of PI3K/Akt signaling pathway and the c-Src/FAK axis, and cross-talks with the Rho/ROCK signaling pathway, all of which are believed to increase the migration rate [42, 87, 88]. However, this function of CA IX might be regulated by DKK-1 protein interaction with the PG domain of CA IX, positioning the DKK-1 protein into the role of anti-tumorigenic agent [87].

Overall, the catalytic activity of CA IX, along with its role as cross-talk-mediator with essential migration-regulating pathways confirm not only its major contribution to cancer cell survival in conditions of environmental acidosis; the above mentioned functions of CA IX underline its direct involvement with aggressive and invasive phenotype of cancer cells.

8.2 CA IX as a Therapeutic Target

The above-mentioned characteristics of CA IX with respect to its significant role in tumorigenesis highlight the immense potential of CA IX-targeted cancer therapy. The discovery of its tumorigenesis-related activities has led to extensive search for selective CA IX inhibitors. On the level of all CAs, acetazolamide is one of the best studied inhibitors of CA activity, with proven contribution to the reduction of invasiveness of renal cell carcinoma-derived cell lines [89]. For more specific targeting, the drug-design strategy was focused on compounds bearing positive or negative charge, thus being unable to enter the cell [90]. Hence, the compounds can interact only with extracellular CA isoforms. Furthermore, mass spectrometric and X-ray crystallographic analyses of CA IX structure opened the door for site-specific drug design [7, 17, 18]. Determining that the active site (region 125–137, in particular) of the CAs is the most variable, these studies enabled to direct the activity of CA inhibitors against specific CA isoforms [17].

Throughout the years, many CA IX-selective compounds with various mechanisms of action were identified, most prominent of which are sulfonamides, their isoesthers (sulfamates, sulfamides) and coumarins [90]. Sulfonamides, both aromatic and heterocyclic, e.g. homosulfonilamide and 1,3,4-thiadizole-2-sulfonamide, respectively, were proven to inhibit specifically CA IX activity [22]. The mechanism of sulfonamide inhibiting action is based on binding of the sulfonamide anion to the zinc ion inside the enzyme active site [91, 92]. In contrast, the mechanism of coumarin-mediated CA IX inhibition is based on spatial blocking of the CA IX active site, upon binding of the hydrolyzed form of coumarin to its entrance [90, 91].

A different approach to target exclusively CA IX is the utilization of CA IX-specific antibodies. Monoclonal anti-CA IX antibodies have been successfully utilized for CA IX detection and visualization of hypoxic areas in solid tumors. Among the numerous CA IX specific antibodies used for this purpose, it is important to mention M75, G250 and its chimeric derivative (cG250) conjugated with radionuclides, and high–affinity A3 and CC7 human monoclonal antibodies [51, 93, 94]. A step towards anti-CA IX immunotherapy was the introduction of mouse monoclonal antibody VII/20, which possesses internalizing capabilities [95]. The antibody may serve as a carrier for targeted anti-cancer therapy. Furthermore, using phage display technology, novel CA IX-specific antibodies with internalizing and inhibitory effect on CA IX activity were developed [96, 97]. Finally, RENCAREX® (Wilex AG), a cG250 antibody underwent phase I-III clinical trials for treatment of patients with RCC. Phase I-II showed promising effects on metastatic renal cell carcinoma patients' survival rate and tolerance of the antibody [98]. Although the recently finished phase III clinical trial (ARISER), targeted at patients with non-metastatic renal cell carcinoma showed no significant improvement of disease-free survival among patients treated with RENCAREX® compared to placebo, more careful evaluation of data revealed clinically and statistically significant improvement in disease-free survival in the patient population with a high CA IX level treated with RENCAREX® compared to both placebo and patients with a low CA IX score (see http://www.wilex.de/press-investors/announcements/press-releases/20130226-2/). This is in agreement with earlier investigations, which showed that the efficient targeting of tumor cells is affected by antigen heterogeneity and requires certain CA IX antigen density [99].

9 Conclusion and Future Perspectives

After almost two decades of extensive research, the role of carbonic anhydrase IX in cancer has been found to be far more complex than originally imagined. The wealth of experimental data demonstrates the ultimate dependency of CA IX expression, assembly and activity on the hypoxic microenvironment. It is remarkable that its direct connection with tumor hypoxia and microenvironmental acidosis was experimentally proven to contribute to development of stress conditions that are toxic for normal cells but not for adaptable cancer cells. The involvement of CA IX in cell-cell adhesion, cell migration as a part of the metastatic cascade, as well as the observed increase of the invasiveness of cancer cells designate its strong position in the process of tumorigenesis. In this manner, CA IX represents an ideal target for anti-tumor therapy. The potential of CA IX-targeted anti-tumor therapy is even broadened by its rare expression in normal tissues and cell-surface localization. The advancements in the design of CA IX inhibitors, which enable to target CA IX only in hypoxic conditions, along with the prospect of CA IX-mediated immunotherapy represent a challenge for the following research. Both approaches follow perspective principles that are promising for translation into the clinic. At the same time, the knowledge of CA IX activators, putative interacting partners, as well as its potential role in other tumor-related phenomena acquired *in silico* and in vitro might bring interesting insights into new horizons of CA IX-targeted therapy in the near future.

Acknowledgments The authors' research is supported by grants from the 7th Framework program of EU (Collaborative project METOXIA), from the Research and Development Support Agency (DO7RP-0017-09, APVV-0893-11 and APVV-0658-11) and from the Research and Development Operational Program funded by the ERDF (project ITMS 26240120027).

References

1. Mucaj V, Shay J, Simon M (2012) Effects of hypoxia and HIFs on cancer metabolism. Int J Hematol 95(5):464–470. doi:10.1007/s12185-012-1070-5
2. Wilson WR, Hay MP (2011) Targeting hypoxia in cancer therapy. Nat Rev Cancer 11(6):393–410. doi:10.1038/nrc3064
3. McDonald PC, Winum JY, Supuran CT, Dedhar S (2012) Recent developments in targeting carbonic anhydrase IX for cancer therapeutics. Oncotarget 3(1):84–97
4. Gatenby RA, Smallbone K, Maini PK, Rose F, Averill J, Nagle RB, Worrall L, Gillies RJ (2007) Cellular adaptations to hypoxia and acidosis during somatic evolution of breast cancer. Br J Cancer 97(5):646–653. doi:10.1038/sj.bjc.6603922
5. Gatenby RA, Gillies RJ (2008) A microenvironmental model of carcinogenesis. Nat Rev Cancer 8(1):56–61. doi:10.1038/nrc2255
6. Chiche J, Ilc K, Laferriere J, Trottier E, Dayan F, Mazure NM, Brahimi-Horn MC, Pouyssegur J (2009) Hypoxia-inducible carbonic anhydrase IX and XII promote tumor cell growth by counteracting acidosis through the regulation of the intracellular pH. Cancer Res 69(1):358–368. doi:10.1158/0008-5472.can-08-2470

7. Hilvo M, Baranauskiene L, Salzano AM, Scaloni A, Matulis D, Innocenti A, Scozzafava A, Monti SM, Di Fiore A, De Simone G, Lindfors M, Janis J, Valjakka J, Pastorekova S, Pastorek J, Kulomaa MS, Nordlund HR, Supuran CT, Parkkila S (2008) Biochemical characterization of CA IX, one of the most active carbonic anhydrase isozymes. J Biol Chem 283(41):27799–27809. doi:10.1074/jbc.M800938200
8. Tu C, Foster L, Alvarado A, McKenna R, Silverman DN, Frost SC (2012) Role of zinc in catalytic activity of carbonic anhydrase IX. Arch Biochem Biophys 521(1–2):90–94. doi:10.1016/j.abb.2012.03.017
9. Pastorekova S, Zavadova Z, Kostal M, Babusikova O, Zavada J (1992) A novel quasi-viral agent, MaTu, is a two-component system. Virology 187(2):620–626
10. Zavada J, Zavadova Z, Pastorekova S, Ciampor F, Pastorek J, Zelnik V (1993) Expression of MaTu-MN protein in human tumor cultures and in clinical specimens. Int J Cancer 54(2):268–274
11. Pastorek J, Pastorekova S, Callebaut I, Mornon JP, Zelnik V, Opavsky R, Zat'ovicova M, Liao S, Portetelle D, Stanbridge EJ, Zavada J, Burny A, Kettman R (1994) Cloning and characterization of MN, a human tumor-associated protein with a domain homologous to carbonic anhydrase and a putative helix-loop-helix DNA binding segment. Oncogene 9(10):2877–2888
12. Opavsky R, Pastorekova S, Zelnik V, Gibadulinova A, Stanbridge EJ, Zavada J, Kettmann R, Pastorek J (1996) Human MN/CA9 gene, a novel member of the carbonic anhydrase family: structure and exon to protein domain relationships. Genomics 33(3):480–487. doi:10.1006/geno.1996.0223
13. Grabmaier K, Vissers JL, De Weijert MC, Oosterwijk-Wakka JC, Van Bokhoven A, Brakenhoff RH, Noessner E, Mulders PA, Merkx G, Figdor CG, Adema GJ, Oosterwijk E (2000) Molecular cloning and immunogenicity of renal cell carcinoma-associated antigen G250. Int J Cancer 85(6):865–870
14. Zavada J, Zavadova Z, Pastorek J, Biesova Z, Jezek J, Velek J (2000) Human tumour-associated cell adhesion protein MN/CA IX: identification of M75 epitope and of the region mediating cell adhesion. Br J Cancer 82(11):1808–1813. doi:10.1054/bjoc.2000.1111
15. Barathova M, Takacova M, Holotnakova T, Gibadulinova A, Ohradanova A, Zatovicova M, Hulikova A, Kopacek J, Parkkila S, Supuran CT, Pastorekova S, Pastorek J (2008) Alternative splicing variant of the hypoxia marker carbonic anhydrase IX expressed independently of hypoxia and tumour phenotype. Br J Cancer 98(1):129–136. doi:10.1038/sj.bjc.6604111
16. Malentacchi F, Simi L, Nannelli C, Andreani M, Janni A, Pastorekova S, Orlando C (2009) Alternative splicing variants of carbonic anhydrase IX in human non-small cell lung cancer. Lung Cancer 64(3):271–276. doi:10.1016/j.lungcan.2008.10.001
17. Alterio V, Hilvo M, Di Fiore A, Supuran CT, Pan PW, Parkkila S, Scaloni A, Pastorek J, Pastorekova S, Pedone C, Scozzafava A, Monti SM, De Simone G (2009) Crystal structure of the catalytic domain of the tumor-associated human carbonic anhydrase IX. Proc Natl Acad Sci U S A 106(38):16233–16238. doi:10.1073/pnas.0908301106
18. De Simone G, Supuran CT (2010) Carbonic anhydrase IX: biochemical and crystallographic characterization of a novel antitumor target. Biochim Biophys Acta 1804(2):404–409. doi:10.1016/j.bbapap.2009.07.027
19. Innocenti A, Pastorekova S, Pastorek J, Scozzafava A, De Simone G, Supuran CT (2009) The proteoglycan region of the tumor-associated carbonic anhydrase isoform IX acts as an intrinsic buffer optimizing CO_2 hydration at acidic pH values characteristic of solid tumors. Bioorg Med Chem Lett 19(20):5825–5828. doi:10.1016/j.bmcl.2009.08.088
20. Svastova E, Hulikova A, Rafajova M, Zat'ovicova M, Gibadulinova A, Casini A, Cecchi A, Scozzafava A, Supuran CT, Pastorek J, Pastorekova S (2004) Hypoxia activates the capacity of tumor-associated carbonic anhydrase IX to acidify extracellular pH. FEBS Lett 577(3):439–445. doi:10.1016/j.febslet.2004.10.043
21. Lopez-Lazaro M (2008) The Warburg effect: why and how do cancer cells activate glycolysis in the presence of oxygen? Anti-Cancer Agents in Medicinal Chemistry 8(3):305–312

22. Vullo D, Franchi M, Gallori E, Pastorek J, Scozzafava A, Pastorekova S, Supuran CT (2003) Carbonic anhydrase inhibitors: inhibition of the tumor-associated isozyme IX with aromatic and heterocyclic sulfonamides. Bioorg Med Chem Lett 13(6):1005–1009
23. Huber V, De Milito A, Harguindey S, Reshkin SJ, Wahl ML, Rauch C, Chiesi A, Pouyssegur J, Gatenby RA, Rivoltini L, Fais S (2010) Proton dynamics in cancer. J Transl Med 8:57. doi:10.1186/1479-5876-8-57
24. Parks SK, Chiche J, Pouyssegur J (2011) pH control mechanisms of tumor survival and growth. J Cell Physiol 226(2):299–308. doi:10.1002/jcp.22400
25. Chiche J, Ilc K, Brahimi-Horn MC, Pouyssegur J (2010) Membrane-bound carbonic anhydrases are key pH regulators controlling tumor growth and cell migration. Adv Enzyme Regul 50(1):20–33. doi:10.1016/j.advenzreg.2009.10.005
26. Swietach P, Wigfield S, Supuran CT, Harris AL, Vaughan-Jones RD (2008) Cancer-associated, hypoxia-inducible carbonic anhydrase IX facilitates CO_2 diffusion. BJU Int 4:22–24
27. Buanne P, Renzone G, Monteleone F, Vitale M, Monti SM, Sandomenico A, Garbi C, Montanaro D, Accardo M, Troncone G, Zatovicova M, Csaderova L, Supuran CT, Pastorekova S, Scaloni A, De Simone G, Zambrano N (2013) Characterization of carbonic anhydrase IX interactome reveals proteins assisting its nuclear localization in hypoxic cells. J Proteome Res 12(1):282–292. doi:10.1021/pr300565w
28. Wykoff CC, Beasley NJ, Watson PH, Turner KJ, Pastorek J, Sibtain A, Wilson GD, Turley H, Talks KL, Maxwell PH, Pugh CW, Ratcliffe PJ, Harris AL (2000) Hypoxia-inducible expression of tumor-associated carbonic anhydrases. Cancer Res 60(24):7075–7083
29. Wang GL, Semenza GL (1993) General involvement of hypoxia-inducible factor 1 in transcriptional response to hypoxia. Proc Natl Acad Sci U S A 90(9):4304–4308
30. Wang GL, Semenza GL (1993) Characterization of hypoxia-inducible factor 1 and regulation of DNA binding activity by hypoxia. J Biol Chem 268(29):21513–21518
31. Kaluz S, Kaluzova M, Liao SY, Lerman M, Stanbridge EJ (2009) Transcriptional control of the tumor- and hypoxia-marker carbonic anhydrase 9: a one transcription factor (HIF-1) show? Biochim Biophys Acta 1795(2):162–172. doi:10.1016/j.bbcan.2009.01.001
32. Aprelikova O, Chandramouli GV, Wood M, Vasselli JR, Riss J, Maranchie JK, Linehan WM, Barrett JC (2004) Regulation of HIF prolyl hydroxylases by hypoxia-inducible factors. J Cell Biochem 92(3):491–501. doi:10.1002/jcb.20067
33. Zhang N, Fu Z, Linke S, Chicher J, Gorman JJ, Visk D, Haddad GG, Poellinger L, Peet DJ, Powell F, Johnson RS (2010) The asparaginyl hydroxylase factor inhibiting HIF-1alpha is an essential regulator of metabolism. Cell Metab 11(5):364–378. doi:10.1016/j.cmet.2010.03.001
34. Lando D, Peet DJ, Whelan DA, Gorman JJ, Whitelaw ML (2002) Asparagine hydroxylation of the HIF transactivation domain a hypoxic switch. Science 295(5556):858–861. doi:10.1126/science.1068592
35. Kaluz S, Kaluzova M, Opavsky R, Pastorekova S, Gibadulinova A, Dequiedt F, Kettmann R, Pastorek J (1999) Transcriptional regulation of the MN/CA 9 gene coding for the tumor-associated carbonic anhydrase IX. Identification and characterization of a proximal silencer element. J Biol Chem 274(46):32588–32595
36. Kaluz S, Kaluzova M, Stanbridge EJ (2003) Expression of the hypoxia marker carbonic anhydrase IX is critically dependent on SP1 activity. Identification of a novel type of hypoxia-responsive enhancer. Cancer Res 63(5):917–922
37. Kaluzova M, Pastorekova S, Svastova E, Pastorek J, Stanbridge EJ, Kaluz S (2001) Characterization of the MN/CA 9 promoter proximal region: a role for specificity protein (SP) and activator protein 1 (AP1) factors. Biochem J 359(Pt 3):669–677
38. Hao Y, Li Y, Zhang J, Liu D, Liu F, Zhao Y, Shen T, Li F (2010) Involvement of histone deacetylation in MORC2-mediated down-regulation of carbonic anhydrase IX. Nucleic Acids Res 38(9):2813–2824. doi:10.1093/nar/gkq006
39. Kaluz S, Kaluzova M, Chrastina A, Olive PL, Pastorekova S, Pastorek J, Lerman MI, Stanbridge EJ (2002) Lowered oxygen tension induces expression of the hypoxia marker MN/carbonic anhydrase IX in the absence of hypoxia-inducible factor 1 alpha stabilization: a role for phosphatidylinositol 3'-kinase. Cancer Res 62(15):4469–4477

40. Kopacek J, Barathova M, Dequiedt F, Sepelakova J, Kettmann R, Pastorek J, Pastorekova S (2005) MAPK pathway contributes to density- and hypoxia-induced expression of the tumor-associated carbonic anhydrase IX. Biochim Biophys Acta 1729(1):41–49. doi:10.1016/j.bbaexp.2005.03.003
41. Hulikova A, Zatovicova M, Svastova E, Ditte P, Brasseur R, Kettmann R, Supuran CT, Kopacek J, Pastorek J, Pastorekova S (2009) Intact intracellular tail is critical for proper functioning of the tumor-associated, hypoxia-regulated carbonic anhydrase IX. FEBS Lett 583(22):3563–3568. doi:10.1016/j.febslet.2009.10.060
42. Dorai T, Sawczuk IS, Pastorek J, Wiernik PH, Dutcher JP (2005) The role of carbonic anhydrase IX overexpression in kidney cancer. Eur J Cancer 41(18):2935–2947. doi:10.1016/j.ejca.2005.09.011
43. Ditte P, Dequiedt F, Svastova E, Hulikova A, Ohradanova-Repic A, Zatovicova M, Csaderova L, Kopacek J, Supuran CT, Pastorekova S, Pastorek J (2011) Phosphorylation of carbonic anhydrase IX controls its ability to mediate extracellular acidification in hypoxic tumors. Cancer Res 71(24):7558–7567. doi:10.1158/0008-5472.can-11-2520
44. Zavada J, Zavadova Z, Zat'ovicova M, Hyrsl L, Kawaciuk I (2003) Soluble form of carbonic anhydrase IX (CA IX) in the serum and urine of renal carcinoma patients. Br J Cancer 89(6):1067–1071. doi:10.1038/sj.bjc.6601264
45. Zatovicova M, Sedlakova O, Svastova E, Ohradanova A, Ciampor F, Arribas J, Pastorek J, Pastorekova S (2005) Ectodomain shedding of the hypoxia-induced carbonic anhydrase IX is a metalloprotease-dependent process regulated by TACE/ADAM17. Br J Cancer 93(11):1267–1276. doi:10.1038/sj.bjc.6602861
46. Liao SY, Brewer C, Zavada J, Pastorek J, Pastorekova S, Manetta A, Berman ML, DiSaia PJ, Stanbridge EJ (1994) Identification of the MN antigen as a diagnostic biomarker of cervical intraepithelial squamous and glandular neoplasia and cervical carcinomas. Am J Pathol 145(3):598–609
47. Pastorekova S, Parkkila S, Parkkila AK, Opavsky R, Zelnik V, Saarnio J, Pastorek J (1997) Carbonic anhydrase IX, MN/CA IX: analysis of stomach complementary DNA sequence and expression in human and rat alimentary tracts. Gastroenterology 112(2):398–408
48. Leibovich BC, Sheinin Y, Lohse CM, Thompson RH, Cheville JC, Zavada J, Kwon ED (2007) Carbonic anhydrase IX is not an independent predictor of outcome for patients with clear cell renal cell carcinoma. J Clin Oncol 25(30):4757–4764. doi:10.1200/jco.2007.12.1087
49. Ivanov S, Liao SY, Ivanova A, Danilkovitch-Miagkova A, Tarasova N, Weirich G, Merrill MJ, Proescholdt MA, Oldfield EH, Lee J, Zavada J, Waheed A, Sly W, Lerman MI, Stanbridge EJ (2001) Expression of hypoxia-inducible cell-surface transmembrane carbonic anhydrases in human cancer. Am J Pathol 158(3):905–919. doi:10.1016/s0002-9440(10)64038-2
50. Tostain J, Li G, Gentil-Perret A, Gigante M (2010) Carbonic anhydrase 9 in clear cell renal cell carcinoma: a marker for diagnosis, prognosis and treatment. Eur J Cancer 46(18):3141–3148. doi:10.1016/j.ejca.2010.07.020
51. Oosterwijk E, Ruiter DJ, Hoedemaeker PJ, Pauwels EK, Jonas U, Zwartendijk J, Warnaar SO (1986) Monoclonal antibody G250 recognizes a determinant present in renal-cell carcinoma and absent from normal kidney. Int J Cancer 38(4):489–494
52. Oosterwijk E (2008) Carbonic anhydrase IX: historical and future perspectives. BJU Int 101(Suppl 4):2–7. doi:10.1111/j.1464-410X.2008.07641.x
53. Genega EM, Ghebremichael M, Najarian R, Fu Y, Wang Y, Argani P, Grisanzio C, Signoretti S (2010) Carbonic anhydrase IX expression in renal neoplasms: correlation with tumor type and grade. Am J Clin Pathol 134(6):873–879. doi:10.1309/ajcpppr57hnjmslz
54. Liao SY, Darcy KM, Randall LM, Tian C, Monk BJ, Burger RA, Fruehauf JP, Peters WA, Stock RJ, Stanbridge EJ (2010) Prognostic relevance of carbonic anhydrase-IX in high-risk, early-stage cervical cancer: a gynecologic oncology group study. Gynecol Oncol 116(3):452–458. doi:10.1016/j.ygyno.2009.10.062
55. Korkeila E, Talvinen K, Jaakkola PM, Minn H, Syrjanen K, Sundstrom J, Pyrhonen S (2009) Expression of carbonic anhydrase IX suggests poor outcome in rectal cancer. Br J Cancer 100(6):874–880. doi:10.1038/sj.bjc.6604949

56. Hussain SA, Ganesan R, Reynolds G, Gross L, Stevens A, Pastorek J, Murray PG, Perunovic B, Anwar MS, Billingham L, James ND, Spooner D, Poole CJ, Rea DW, Palmer DH (2007) Hypoxia-regulated carbonic anhydrase IX expression is associated with poor survival in patients with invasive breast cancer. Br J Cancer 96(1):104–109. doi:10.1038/sj.bjc.6603530
57. Simi L, Venturini G, Malentacchi F, Gelmini S, Andreani M, Janni A, Pastorekova S, Supuran CT, Pazzagli M, Orlando C (2006) Quantitative analysis of carbonic anhydrase IX mRNA in human non-small cell lung cancer. Lung Cancer 52(1):59–66. doi:10.1016/j.lungcan.2005.11.017
58. Jarvela S, Parkkila S, Bragge H, Kahkonen M, Parkkila AK, Soini Y, Pastorekova S, Pastorek J, Haapasalo H (2008) Carbonic anhydrase IX in oligodendroglial brain tumors. BMC Cancer 8:1. doi:10.1186/1471-2407-8-1
59. Bui MH, Seligson D, Han KR, Pantuck AJ, Dorey FJ, Huang Y, Horvath S, Leibovich BC, Chopra S, Liao SY, Stanbridge E, Lerman MI, Palotie A, Figlin RA, Belldegrun AS (2003) Carbonic anhydrase IX is an independent predictor of survival in advanced renal clear cell carcinoma: implications for prognosis and therapy. Clin Cancer Res 9(2):802–811
60. Park HJ, Lyons JC, Ohtsubo T, Song CW (1999) Acidic environment causes apoptosis by increasing caspase activity. Br J Cancer 80(12):1892–1897. doi:10.1038/sj.bjc.6690617
61. Gatenby RA, Gawlinski ET, Gmitro AF, Kaylor B, Gillies RJ (2006) Acid-mediated tumor invasion: a multidisciplinary study. Cancer Res 66(10):5216–5223. doi:10.1158/0008-5472.can-05-4193
62. Fukumura D, Xu L, Chen Y, Gohongi T, Seed B, Jain RK (2001) Hypoxia and acidosis independently up-regulate vascular endothelial growth factor transcription in brain tumors in vivo. Cancer Res 61(16):6020–6024
63. Pastorekova S, Ratcliffe PJ, Pastorek J (2008) Molecular mechanisms of carbonic anhydrase IX-mediated pH regulation under hypoxia. BJU Int 101(Suppl 4):8–15. doi:10.1111/j.1464-410X.2008.07642.x
64. Yamagata M, Hasuda K, Stamato T, Tannock IF (1998) The contribution of lactic acid to acidification of tumours: studies of variant cells lacking lactate dehydrogenase. Br J Cancer 77(11):1726–1731
65. Swietach P, Vaughan-Jones RD, Harris AL (2007) Regulation of tumor pH and the role of carbonic anhydrase 9. Cancer Metastasis Rev 26(2):299–310. doi:10.1007/s10555-007-9064-0
66. Morgan PE, Pastorekova S, Stuart-Tilley AK, Alper SL, Casey JR (2007) Interactions of transmembrane carbonic anhydrase, CAIX, with bicarbonate transporters. Am J Physiol 293(2):C738–C748. doi:10.1152/ajpcell.00157.2007
67. Svastova E, Witarski W, Csaderova L, Kosik I, Skvarkova L, Hulikova A, Zatovicova M, Barathova M, Kopacek J, Pastorek J, Pastorekova S (2012) Carbonic anhydrase IX interacts with bicarbonate transporters in lamellipodia and increases cell migration via its catalytic domain. J Biol Chem 287(5):3392–3402. doi:10.1074/jbc.M111.286062
68. Harguindey S, Orive G, Luis Pedraz J, Paradiso A, Reshkin SJ (2005) The role of pH dynamics and the Na^+/H^+ antiporter in the etiopathogenesis and treatment of cancer. Two faces of the same coin–one single nature. Biochim Biophys Acta 1756(1):1–24. doi:10.1016/j.bbcan.2005.06.004
69. Sennoune SR, Luo D, Martinez-Zaguilan R (2004) Plasmalemmal vacuolar-type H^+-ATPase in cancer biology. Cell Biochem Biophys 40(2):185–206. doi:10.1385/cbb:40:2:185
70. Halestrap AP, Meredith D (2004) The SLC16 gene family-from monocarboxylate transporters (MCTs) to aromatic amino acid transporters and beyond. Pflugers Archiv European J Physiol 447(5):619–628. doi:10.1007/s00424-003-1067-2
71. Bohle AS, Kalthoff H (1999) Molecular mechanisms of tumor metastasis and angiogenesis. Langenbecks Arch Surg 384(2):133–140
72. Misra A, Pandey C, Sze SK, Thanabalu T (2012) Hypoxia activated EGFR signaling induces epithelial to mesenchymal transition (EMT). PLoS One 7(11):e49766. doi:10.1371/journal.pone.0049766
73. Rozhin J, Sameni M, Ziegler G, Sloane BF (1994) Pericellular pH affects distribution and secretion of cathepsin B in malignant cells. Cancer Res 54(24):6517–6525

74. Moellering RE, Black KC, Krishnamurty C, Baggett BK, Stafford P, Rain M, Gatenby RA, Gillies RJ (2008) Acid treatment of melanoma cells selects for invasive phenotypes. Clin Exp Metastasis 25(4):411–425. doi:10.1007/s10585-008-9145-7
75. Glunde K, Guggino SE, Solaiyappan M, Pathak AP, Ichikawa Y, Bhujwalla ZM (2003) Extracellular acidification alters lysosomal trafficking in human breast cancer cells. Neoplasia 5(6):533–545
76. Beavon IR (1999) Regulation of E-cadherin: does hypoxia initiate the metastatic cascade? Mol Pathol 52(4):179–188
77. Berx G, Staes K, van Hengel J, Molemans F, Bussemakers MJ, van Bokhoven A, van Roy F (1995) Cloning and characterization of the human invasion suppressor gene E-cadherin (CDH1). Genomics 26(2):281–289
78. Drivalos A, Papatsoris AG, Chrisofos M, Efstathiou E, Dimopoulos MA (2011) The role of the cell adhesion molecules (integrins/cadherins) in prostate cancer. Int Braz J Urol 37(3):302–306
79. Czyzewska J, Guzinska-Ustymowicz K, Ustymowicz M, Pryczynicz A, Kemona A (2010) The expression of E-cadherin-catenin complex in patients with advanced gastric cancer: role in formation of metastasis. Folia Histochem Cytobiol 48(1):37–45. doi:10.2478/v10042-010-0017-z
80. Gupta GP, Massague J (2006) Cancer metastasis: building a framework. Cell 127(4):679–695. doi:10.1016/j.cell.2006.11.001
81. Stock C, Schwab A (2009) Protons make tumor cells move like clockwork. Pflugers Arch 458(5):981–992. doi:10.1007/s00424-009-0677-8
82. Martin C, Pedersen SF, Schwab A, Stock C (2011) Intracellular pH gradients in migrating cells. Am J Physiol 300(3):C490–C495. doi:10.1152/ajpcell.00280.2010
83. Stuwe L, Muller M, Fabian A, Waning J, Mally S, Noel J, Schwab A, Stock C (2007) pH dependence of melanoma cell migration: protons extruded by NHE1 dominate protons of the bulk solution. J Physiol 585(Pt 2):351–360. doi:10.1113/jphysiol.2007.145185
84. Stock C, Gassner B, Hauck CR, Arnold H, Mally S, Eble JA, Dieterich P, Schwab A (2005) Migration of human melanoma cells depends on extracellular pH and Na^+/H^+ exchange. J Physiol 567(Pt 1):225–238. doi:10.1113/jphysiol.2005.088344
85. Paulucci-Holthauzen AA, Vergara LA, Bellot LJ, Canton D, Scott JD, O'Connor KL (2009) Spatial distribution of protein kinase A activity during cell migration is mediated by A-kinase anchoring protein AKAP Lbc. J Biol Chem 284(9):5956–5967. doi:10.1074/jbc.M805606200
86. Svastova E, Zilka N, Zat'ovicova M, Gibadulinova A, Ciampor F, Pastorek J, Pastorekova S (2003) Carbonic anhydrase IX reduces E-cadherin-mediated adhesion of MDCK cells via interaction with beta-catenin. Exp Cell Res 290(2):332–345
87. Kim BR, Shin HJ, Kim JY, Byun HJ, Lee JH, Sung YK, Rho SB (2012) Dickkopf-1 (DKK-1) interrupts FAK/PI3K/mTOR pathway by interaction of carbonic anhydrase IX (CA9) in tumorigenesis. Cell Signal 24(7):1406–1413. doi:10.1016/j.cellsig.2012.03.002
88. Shin HJ, Rho SB, Jung DC, Han IO, Oh ES, Kim JY (2011) Carbonic anhydrase IX (CA9) modulates tumor-associated cell migration and invasion. J Cell Sci 124(Pt 7):1077–1087. doi:10.1242/jcs.072207
89. Parkkila S, Rajaniemi H, Parkkila AK, Kivela J, Waheed A, Pastorekova S, Pastorek J, Sly WS (2000) Carbonic anhydrase inhibitor suppresses invasion of renal cancer cells in vitro. Proc Natl Acad Sci U S A 97(5):2220–2224. doi:10.1073/pnas.040554897
90. Supuran CT (2012) Inhibition of carbonic anhydrase IX as a novel anticancer mechanism. World Journal of Clinical Oncology 3(7):98–103. doi:10.5306/wjco.v3.i7.98
91. Carta F, Maresca A, Scozzafava A, Supuran CT (2012) Novel coumarins and 2-thioxo-coumarins as inhibitors of the tumor-associated carbonic anhydrases IX and XII. Bioorg Med Chem 20(7):2266–2273. doi:10.1016/j.bmc.2012.02.014
92. Supuran CT, Manole G, Andruh M (1993) Carbonic anhydrase inhibitors. Part 11. Coordination compounds of heterocyclic sulfonamides with lanthanides are potent inhibitors of isozymes I and II. J Inorg Biochem 49(2):97–103
93. Pryma DA, O'Donoghue JA, Humm JL, Jungbluth AA, Old LJ, Larson SM, Divgi CR (2011) Correlation of in vivo and in vitro measures of carbonic anhydrase IX antigen expression in renal masses using antibody ^{124}I-cG250. J Nucl Med 52(4):535–540. doi:10.2967/jnumed.110.083295

94. Ahlskog JK, Schliemann C, Marlind J, Qureshi U, Ammar A, Pedley RB, Neri D (2009) Human monoclonal antibodies targeting carbonic anhydrase IX for the molecular imaging of hypoxic regions in solid tumours. Br J Cancer 101(4):645–657. doi:10.1038/sj.bjc.6605200
95. Zatovicova M, Jelenska L, Hulikova A, Csaderova L, Ditte Z, Ditte P, Goliasova T, Pastorek J, Pastorekova S (2010) Carbonic anhydrase IX as an anticancer therapy target: preclinical evaluation of internalizing monoclonal antibody directed to catalytic domain. Curr Pharm Des 16(29):3255–3263
96. Xu C, Lo A, Yammanuru A, Tallarico AS, Brady K, Murakami A, Barteneva N, Zhu Q, Marasco WA (2010) Unique biological properties of catalytic domain directed human anti-CAIX antibodies discovered through phage-display technology. PLoS One 5(3):e9625. doi:10.1371/journal.pone.0009625
97. Murri-Plesko MT, Hulikova A, Oosterwijk E, Scott AM, Zortea A, Harris AL, Ritter G, Old L, Bauer S, Swietach P, Renner C (2011) Antibody inhibiting enzymatic activity of tumour-associated carbonic anhydrase isoform IX. Eur J Pharmacol 657(1–3):173–183. doi:10.1016/j.ejphar.2011.01.063
98. Siebels M, Rohrmann K, Oberneder R, Stahler M, Haseke N, Beck J, Hofmann R, Kindler M, Kloepfer P, Stief C (2011) A clinical phase I/II trial with the monoclonal antibody cG250 (RENCAREX®) and interferon-alpha-2a in metastatic renal cell carcinoma patients. World J Urol 29(1):121–126. doi:10.1007/s00345-010-0570-2
99. Steffens MG, Boerman OC, Oyen WJ, Kniest PH, Witjes JA, Oosterhof GO, van Leenders GJ, Debruyne FM, Corstens FH, Oosterwijk E (1999) Intratumoral distribution of two consecutive injections of chimeric antibody G250 in primary renal cell carcinoma: implications for fractionated dose radioimmunotherapy. Cancer Res 59:1615–1619

Chapter 12
Carbonic Anhydrase IX as an Imaging and Therapeutic Target for Tumors and Metastases

Narges K. Tafreshi, Mark C. Lloyd, Marilyn M. Bui, Robert J. Gillies, and David L. Morse

Abstract Carbonic anhydrase IX (CAIX) which is a zinc containing metalloprotein, efficiently catalyzes the reversible hydration of carbon dioxide. It is constitutively up-regulated in several cancer types and has an important role in tumor progression, acidification and metastasis. High expression of CAIX generally correlates with poor prognosis and is related to a decrease in the disease-free interval following successful therapy. Therefore, it is considered as a prognostic indicator in oncology.

In this review, we describe CAIX regulation and its role in tumor hypoxia, acidification and metastasis. In addition, the molecular imaging of CAIX and its potential for use in cancer detection, diagnosis, staging, and for use in following therapy response is discussed. Both antibodies and small molecular weight compounds have been used for targeted imaging of CAIX expression. The use of CAIX expression as an attractive and promising candidate marker for systemic anticancer therapy is also discussed.

Susan C. Frost and Robert McKenna (eds.). Carbonic Anhydrase: Mechanism, Regulation, Links to Disease, and Industrial Applications

N.K. Tafreshi • R.J. Gillies • D.L. Morse (✉)
Department of Cancer Imaging and Metabolism, H. Lee Moffitt Cancer Center and Research Institute, Tampa, FL, USA
e-mail: Narges.Tafreshi@Moffitt.org; robert.gillies@moffitt.org; david.morse@moffitt.org

M.C. Lloyd
Analytical Microscopy Core Facility, H. Lee Moffitt Cancer Center and Research Institute, Tampa, FL, USA
e-mail: mark.lloyd@moffitt.org

M.M. Bui
Analytical Microscopy Core Facility and Department of Anatomic Pathology, H. Lee Moffitt Cancer Center and Research Institute, Tampa, FL, USA
e-mail: marilyn.bui@moffitt.org

S.C. Frost and R. McKenna (eds.), *Carbonic Anhydrase: Mechanism, Regulation, Links to Disease, and Industrial Applications*, Subcellular Biochemistry 75,
DOI 10.1007/978-94-007-7359-2_12,

Keywords Carbonic anhydrase IX • Carbonic anhydrase XII • Tumor hypoxia • Imaging • HIF-1 • Acid-mediated invasion • Cancer • Antibody • Small molecular probes • Fluorescent sulfonamide

1 Introduction

Carbonic anhydrases (CAs) are metalloproteins, usually containing Zinc (EC 4.2.1.1) that catalyze the reversible hydration of carbon dioxide. These are widespread in nature, being found in bacteria, plants and mammals [1, 2]. Sixteen CA isoforms have been characterized to date in mammals, which differ in their cellular localization and catalytic activity, susceptibility to different inhibitors, and tissue-specific distribution. Some isoforms are strictly cytosolic: (CAI, CAII, CAIII, CAVII and CAXIII), some are membrane-associated (CAIV, CAIX, CAXII, CAXIV and CAXV), some are mitochondrial (CAVA and CAVB), and one (CAVI) is secreted into saliva and milk. There are also three non-catalytic enzymes (CAVIII, CAX and CAXI), which are named as carbonic anhydrase related proteins, CARPs. Although structurally homologous, the inactivity of these CAs is due to the absence of one or more Zn-binding histidine residues [1, 3–6]. The structure and function of CA isozymes have been extensively reviewed by Hassan et al., 2012 [7].

CAs catalyze the reversible dehydration of bicarbonate ($CO_2 + H_2O \leftrightarrow HCO_3 - + H+$) [1, 2, 8]; Since CO_2 is the main byproduct of all oxidative processes, the CAs play a fundamental role in maintaining acid–base equilibrium [1, 2, 9], and are involved in crucial physiological processes such as respiration, electrolyte balance, biosynthetic reactions that require bicarbonate as a substrate (gluconeogenesis, lipogenesis, and ureagenesis), bone resorption, and calcification [1, 2]. Moreover, CAs are involved in pathological processes such as cancer progression, obesity and epilepsy and thus have been proposed as therapeutic targets.

A cDNA coding for the hCAIX protein was first cloned and investigated by Pastorek et al. [10], and the *CAIX* gene was further characterized by the same group [11]. Isoform CAIX, originally named the "MN protein", has been identified using a monoclonal antibody (M75) as a plasma membrane antigen detected in a human cervical carcinoma cell line, HeLa [12]. The expression of MN antigen also was detected in various tumor cell lines and surgical tumor specimens, but not in the corresponding normal tissues, suggesting its potential usefulness as a tumor marker [13]. Because it was the ninth mammalian CA identified, the MN protein was renamed CAIX. A renal cell cancer-associated antigen (G250) detected by mAb was then reported to be identical with MN/CAIX by another research group [14, 15].

CAIX is among the most active CAs for the CO_2 hydration reaction, and contains four domains on the basis of sequence similarity [11, 16]: an N-terminal proteoglycan-like (PG) domain, a CA catalytic domain, a transmembrane segment (TM), and an intracytoplasmic (IC) portion. Other CAs, such as the cytosolic CAII and the membrane-associated CAXII lack the PG-like domain (Fig. 12.1). It is

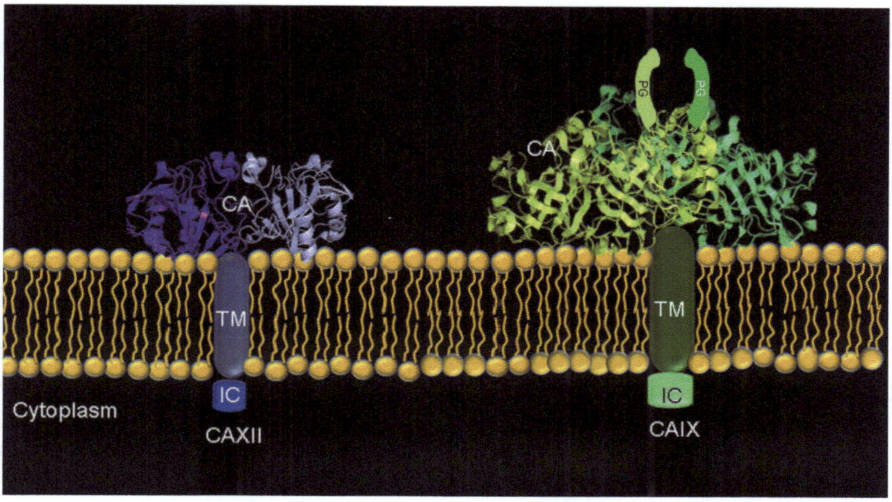

Fig. 12.1 Proposed model showing the structural arrangement of the CAXII and CAIX dimer on the cellular membrane [23, 24]. *PG* proteoglycan-like domain, *CA* catalytic domain, *TM* transmembrane segment, *IC* intracellular, cytosolic tail

worth mentioning that the M75 mAb binds to the repetitive epitope in the PG region, allowing selective detection of CAIX without cross-reactivity with other CAs [17]. The PG-like domain immediately adjacent to the catalytic domain allows the enzyme to act efficiently at acidic pH values [18, 19]. It has also been proposed that the PG domain is involved to cell adhesion and tumor invasion [17, 20, 21] via its interaction with β-catenin, which leads to a reduction in E-cadherin mediated cell-cell adhesion, thus promoting cell motility and invasion [20].

In contrast to the PG domain, the CAIX catalytic domain has significant sequence identity (from 30 % to 40 %) to catalytic domains of other hCA isozymes. Both the CA and PG domains are glycosylated [16]. The cytoplasmic tail of CAIX contains three potential phosphorylation sites (T443, S448 and Y449) that can participate in signal transduction. When tyrosine 449 of CAIX is phosphorylated, it can interact with the regulatory subunit of PI3K, which can transactivate the Ser/Thr protein kinase Akt, and have a direct impact on cellular glucose metabolism [22].

Acetazolamide is a clinically used sulfonamide-based CA inhibitor. An X-ray crystal structure of the catalytic domain of CAIX in complex with acetazolamide has been determined (Fig. 12.1). The crystal structure has identified significant differences between CAIX and the other α-CA isoforms that also have solved three-dimensional structures. These developments have led to increased interest in the rational drug design of CAIX isozyme-specific CA inhibitors with potential for use as anticancer drugs for systemic therapy targeting disseminated disease [23]. The CAIX protein assembles as a dimer which is stabilized by the formation of an intermolecular disulfide bond between the same Cys residue located on two CA catalytic domains [16, 23]. The active site clefts and the PG domains are thus located on one face of the dimer, while the C-termini are situated on the opposite face, facilitating protein anchoring to the cell membrane [23, 24]. It has been shown that neither lipid raft localization nor phosphorylation is likely required for CAIX dimerization or activity in MDA-mb-231 breast cancer cell [25].

CAIX expression is restricted to a few tissues including the epithelium of the stomach (gastric mucosa), biliary tree, crypt cells of the duodenum, and the epithelium of the small intestine [26, 27], but is overexpressed in response to tumor hypoxia in many tumor types and plays a critical role in hypoxia associated tumor acidosis and development of the metastatic phenotype. The comparison of CAIX cDNA sequences from cancer cell lines compared to normal cells does not reveal mutations or alternative splicing that would indicate a different function in cancer [27]. CAIX expression is associated with a poor prognosis [reviewed by [28]] and high CAIX expression is found in lymph node and distal metastases for a broad range of cancer types [see the "CAIX expression in cancer" section below]. Hence, CAIX is being pursued as a marker for development of targeted systemic therapies and diagnostic imaging probes.

2 CAIX Regulation

Expression of CAIX protein is primarily regulated at the level of transcription. The promoter of the *CA9* gene contains an HRE (hypoxia responsive element) that is located immediately upstream of the transcription start site at position −3/−10. HIF-1α binds to the HRE and induces transcription of the *CA9* gene in response to hypoxia and increased cell density [29]. The transcriptional regulation of CAIX has been reviewed in detail recently [30]. Alternative splicing of the *CA9* transcript has been observed [31]. Besides the predominant, full-length *CA9* mRNA (1.5 kb mRNA), an alternatively spliced variant, lacking exons 8/9 among 11 exons, has been detected by RT-PCR. This mRNA variant is constitutively expressed at very low levels and codes for a truncated, cytoplasmic/secreted form of CAIX without enzymatic activity. This is very important for studies related to hypoxia-related expression of *CAIX*, where the presence of the alternatively spliced variant is hypoxia-independent and can provide a false-positive signal [31]. Hence, primers for qRT-PCR should be carefully designed to detect the longer transcript containing exons 8/9. The physiological role of this variant is unknown. Finally, the CAIX protein can be post-translationally modified by metalloprotease mediated shedding of the ectodomain [32].

Under normoxic conditions (Fig. 12.2, right), HIF-1α is constitutively expressed but is rapidly degraded by the ubiquitin-28S proteasome pathway [33]. Specifically, two proline residues of HIF-1α (P402 and P564) and asparagine (N803) are hydroxylated by oxygen-dependent proline hydroxylases (PHD) and factor inhibiting HIF-1 (FIH-1), respectively. PHD and FIH-1 belong to the Fe(II) and 2-oxoglutarate dioxygenase super-family. Hydroxylated HIF-1α proteins bind to the E3 ubiquitin ligase, VHL, complex, leading to its rapid degradation by the

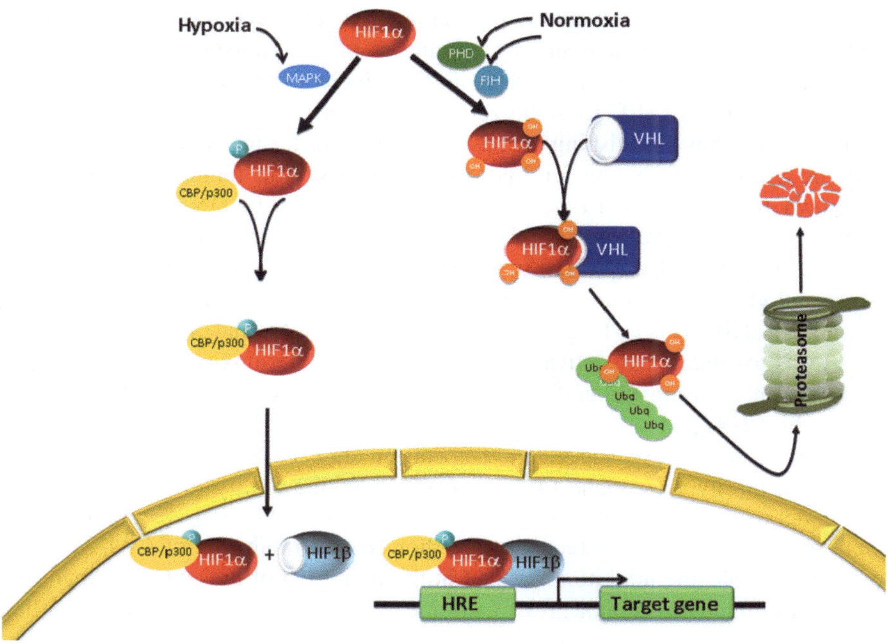

Fig. 12.2 Hypoxia-induced gene expression mediated by the HIF1 transcription factor heterodimer. HIF-1α is degraded in normoxia by the ubiquitin-proteasomal pathway. In hypoxia, it is phosphorylated by MAPK and bound by CBP/p300, inducing heterodimirezation with HIF-1β and transcription of variety of genes

proteasome [34]. Hydroxylation of N803 by FIH further inactivates HIF-1α by inhibiting its interaction with transcriptional co-activator, CBP/p300 [34].

Under hypoxic conditions (Fig. 12.2, left), lack of O_2 prevents the hydroxylation of HIF-1α by PHD and FIH (and thus degradation). Therefore, there is no VHL binding and HIF-1α is stabilized. Non-hydroxylated N803 of HIF-1α allows binding of CBP/p300 to HIF-1α and eventual translocation to the nucleus for heterodimerization with HIF-1β, producing the master transcription factor, HIF-1. Binding of HIF-1β to HIF-1α is enhanced by MAPK phosphorylation of HIF-1α [35, 36]. HIF-1 then binds to target genes that contain HRE sites, including, genes broadly related to iron metabolism (e.g. erythropoietin), glucose metabolism (e.g. GLUT-1 and GLUT-3), angiogenesis (e.g. vascular endothelial growth factor), and, last but not least, genes that are involved in pH regulation (e.g. CAIX and the sodium-hydrogen exchanger, NHE-1) and induces their transcription. Additional hypoxia-induced genes are involved in cell proliferation and viability, regulation of vascular remodeling and plasticity, cell adhesion, cell matrix metabolism, and other cellular processes [29, 35–37]. For a detailed list of target genes, see Table 2 of Ke and Costa [38].

HIF-1α stabilization is a common observation in advanced stage invasive cancer and metastases [39] potentially by a number of mechanisms, resulting in the constitutive expression of cancer related genes, including *CA9*. One cancer related mechanism involves HIF-1α stabilization by glycolytic end-products, lactate and pyruvate [40], thus establishing the basis for a positive feedback loop for glycolysis and the Warburg Effect under hypoxic conditions. The tumor suppressor p53 is known to signal HIF-1α degradation as a response to DNA damage, down regulating CAIX expression [41]. In most clear cell renal cell carcinomas (RCCs), the von Hippel-Lindau (pVHL) tumor suppressor is non-functional by mutation, hypermethylation or deletion. Non-functional VHL is unable to polyubiquitate HIF-1α and thus, HIF-1α is no longer recognized by the proteasomal degradation system. This leads to constitutive activation of HIF-1 even in normoxic conditions and, as a consequence, strong and uniform upregulation of CAIX and other HIF-1-induced genes [29, 42–44].

Recent studies have suggested that cancer stem cells (CSCs) preferentially survive in the hypoxic niche [45–47]. CSCs have the ability to generate tumors and undergo the epithelial–mesenchymal transition (EMT) and are implicated in metastasis. Interestingly, a recent study showed that inhibition of CAIX activity with small-molecule inhibitors resulted in the inhibition of breast CSC expansion, leading to diminished tumor growth and metastasis [48]. This result suggests that CAIX is an important therapeutic target for selectively depleting breast CSCs. The same study has identified the mTORC1 axis as a critical pathway downstream of CAIX in the regulation of CSC function and suggested that CAIX expression regulates CSC EMT by regulation of 'stemness' genes, such as Snail and Notch [48].

Our unpublished data and other studies have shown that expression levels of several hypoxia related proteins, including CAIX, are elevated at the invasive front of solid tumors at similar levels observed in central, perinecrotic areas [49]. This invasive front is also the observed location of CSC migration (Fig. 12.3) [50].

2.1 Cross Talk Between the CAs

Knockout (K/O) of *CA9* in transgenic animals resulted in homozygous animals that developed gastric hyperplasia, but no changes were observed in gastric pH, acid secretion or serum gastrin levels [51]. Also no tumorogenesis was observed in *CA9* K/O mice, although aged *CA9* null mice may suffer degenerative brain disease [52]. The knockout of *CA9* in mice interestingly led to the up-regulation of CAII, while expression of CAIX is upregulated about 2 fold in *CA2* null mice [53]. Chiche and his colleagues also observed that knockdown of *CA9* in LS174Tr colon carcinoma cells led to partial compensation by up-regulation of CAXII in vitro and in vivo [54], suggesting cross talk between CAs and that such cross talk is needed for cell pH homeostasis [51, 52]. The mechanism of this specific cross talk between the CAs remains unclear [55].

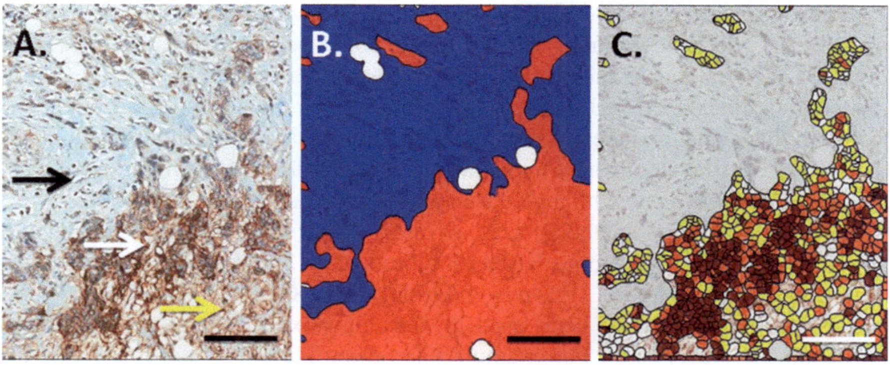

Fig. 12.3 Immunohistochemical staining of CAIX is an important tool to investigate 2-D localization of expression. (**a**) *Dark brown* CAIX overexpression at the edge of the tumor (*white arrow*) contrasts not only with the adjacent microenvironment in blue (*black arrow*), but also with the decreased expression of the tumor >200 μm from the edge (*yellow arrow*). (**b**) A mask depicts regions selected by computer learning which segment and classify the tumor (*red*) and non-tumor (*blue*) regions. (**c**) Single cell segmentation and classification of the tumor cells demonstrates the location of strong (*red*), moderate (*orange*), weak (*yellow*) and negative (*white*) expressing cells. These classifications allow investigators to quantify metrics of the IHC stain localization. Scale = 150 μm

3 CAIX Expression in Tumor Hypoxia

Hypoxia can be defined as the lack of oxygen supply in a tissue such that cellular mechanisms are compromised or an adaptive response occurs. Solid tumors usually have heterogenous regions of hypoxia due to insufficient supply of oxygen by irregular and functionally defective tumor vasculature. This situation creates selective pressure in favor of tumor cells that can adapt to this microenvironmental stress, with oncogenic alterations in metabolism as a result. These adaptive changes also result in expansion of cells with more aggressive phenotypes and increased metastatic potential [56].

The importance of hypoxia has been demonstrated clinically, where it is an independent predictor of poor survival in several types of cancer. In fact, hypoxia is correlated with malignant progression and resistance to conventional chemo- and radio-therapies in many cancer types [57]. This resistance is partially due to poor perfusion, which limits drug delivery and tumor penetration. Also, since chemotherapeutic agents target rapidly dividing cells, often they fail to target cells in hypoxic areas which are more slowly growing or non-proliferating. Additionally, upregulation of HIF-1 can lead to chemoresistance in tumors and metastases (e.g. HIF-1 is able to activate the multidrug resistance 1 (MDR1) gene in response to hypoxia) [58, 59]. Finally, radiotherapy-based treatments which rely on the generation of oxygen super-radicals also often fail to provide effective treatment because of inadequate oxygen within the tumor mass [60].

These properties of tumor hypoxia make relevant the ability to detect hypoxic regions in order to determine which patients could benefit from adjunctive anti-hypoxia therapy and to monitor hypoxia targeted therapy response [61]. Consequently, significant effort has been spent to qualify reliable hypoxia biomarkers for both diagnostic and therapeutic purposes. An exciting advance in this field has been the recent development of hypoxia-activated pro-drugs [62], which specifically target the hypoxic niche with high dose systemic chemotherapy.

Although different methods such as direct tumor pO_2 measurement by microelectrode [63], PET probes based on bioreductive 2-nitroimidazoles following intravenous administration (^{18}F-misonidazole, ^{18}F-HX4, ^{18}F-FAZA and ^{18}F-EF5 [reviewed by [64]]) have been used to detect hypoxia, each of these methods have some disadvantages. For example microelectrode use is constrained by the difficulty of accessing tumors, cost, invasiveness, inter-observer variability [65], failure to distinguish necrosis from hypoxia [66] and difficulty in calibration [61]. Bioreductive-based probes can be limited by poor tumor-to-background ratios and non-specific signal generated by oxygen dependent metabolism of these probes [67].

As mentioned above, at the molecular level, hypoxia inducible factor 1 (HIF-1) is up-regulated in response to oxygen deprivation, binds to the hypoxia responsive element (HRE), and controls the transcriptional activity of numerous downstream genes, such as CAIX, which has been emerged as one of the most promising endogenous markers of cellular hypoxia.

Notably, however, staining of CAIX is not in complete agreement with pimonidazole (a hypoxia marker) staining in patient tumors and human xenograft tumors in mice [68–72]. This difference may arise from short-lived fluctuations in tumor perfusion and slow rates of CAIX degradation [73], or that the pimonidazole threshold for labeling is <1.3 % oxygen, whereas up-regulation of molecular markers such as CAIX occur at pO_2 values over a larger range, 0.2–2 % oxygen [74].

Also, although CAIX expression correlates with the pattern of HIF-1α expression in most regions [54], exceptions have been observed where expression of CAIX is detected in the absence of HIF-1α staining [75], and these mismatches may be informative. These observations could be due to the different half-lives of HIF-1α and CAIX expression where CAIX protein is more stable [76, 77], i.e. HIF-1α is rapidly degraded in normoxic conditions and just as rapidly stabilized in hypoxic conditions. Therefore, cells that had been hypoxic and then were reoxygenated may stain for CAIX but not HIF-1α [30]. In addition, when cells or tissue have only recently become hypoxic, they may stain positive for HIF-1α but not CAIX, because those cells may have been analyzed before the full onset of CAIX expression [30]. Also, the duration and severity of hypoxia identified as acute, cycling (acute hypoxia followed by rapid reoxygenation), and chronic hypoxia can have differential effects on transcription and translation of HIF clients [78–80]. The use of CAIX as a biomarker for imaging of hypoxia is discussed in the following section.

4 The Role of CAIX in Metabolism and Metastasis: Acid Mediated Invasion

Tumor acidosis, low tumor pH, is a commonly observed consequence of dysangiogenesis and increased metabolism, which are 'hallmarks' of cancer. Acidosis results from enhanced rates of glucose fermentation (conversion to lactic acid) and the inefficient removal of acidic end products because of deficient vasculature in the tumor mass [81–83]. Tumor acidosis has been reported to favor metastasis [84]. Therefore it is considered an attractive therapeutic target.

Since intracellular (pHi) regulation controls multiple cellular functions such as energy production, proliferation and migration, tumor cells must develop strategies to regulate their pHi to survive and proliferate [85–87]. In fact, tumor cells have highly sophisticated mechanisms involving a variety of proteins and buffering systems that lead to maintenance of a moderately alkaline pHi in tumor cells while generating a markedly acidic extracellular environment, allowing them not only to survive and proliferate in hypoxia, but also favoring metastasis [reviewed by [88, 89]].

As described above, HIF-1 is activated under hypoxic conditions and in many cases it is constitutively activated in tumor cells. HIF-1 acts to regulate pH homeostasis by up-regulating expression of membrane located transporters, exchangers, pumps and ecto-enzymes such as Na^+/H^+ exchanger (NHE-1), $H+$ –ATPases and monocarboxylate transporters (MCTs), that actively export acids from tumor cells, Cl^-/HCO_3^- exchangers (AE) that make the cytoplasm alkaline by pumping the weak base HCO_3^- into the cells, and carbonic anhydrases (CAs) [83, 90–93].

CAs alone can contribute to pH regulation, i.e. raising intracellular pH and lowering extracellular pH. Intracellular CAs can mediate the conversion of intracellular HCO_3^- (a product of AE) into CO_2. This process consumes an intracellular proton. The resulting CO_2 then diffuses through the plasma membrane possibly through an aquaporin channel, wherein it is reconverted back into bicarbonate and protons by the extracellular carbonic anhydrases, e.g. CAIX, particularly in the context of a hypoxic microenvironment [94]. Thus, the net effect is the movement of a proton from the cytosol to the extracellular space. The produced protons contribute to extracellular acidosis, leading to aggressive tumor phenotype by promoting invasion and metastasis [95]. The newly generated HCO_3^- ions can then be transported back into the tumor cells by HCO_3^- transport proteins. The coupled transport process is probably essential for hypoxic cancer cells to buffer and maintain intracellular pH at near neutral conditions which is necessary for their biosynthetic reactions [26, 30, 54, 93, 94, 96]. It has been shown that knockdown of CAIX minimizes the gradient in extracellular pH that occurs across three dimensional tumor spheroids, which normally have an acidic core, with a corresponding decrease in the intracellular pH of hypoxic tumor cells [97].

As mentioned previously, the crystal structure of the CAIX catalytic domain has confirmed the dimeric nature of the enzyme and showed that the N-terminal regions of both monomers are located on the same face of the dimer, while both C-termini are located on the opposite face [23]. This structural organization allows for concomitant orientation of both PG domains toward the extracellular environment at the entrance to the active site clefts. This positioning of the PG domain at the border of the active site suggests a role in assisting the CA domain-mediated catalysis. In fact, the influence of the PG domain on the pH-dependent profile of the k_{cat}/K_M values for CO_2 hydration was demonstrated in a recent study [23]; while the CAIX catalytic domain alone presented an optimal pK_a at 7.01, this value changed to 6.49 in the presence of the PG domain. Therefore, the PG domain of CAIX enables the hydration of CO_2 to occur at the low pH levels observed in the tumor microenvironment [19, 98]. In this context, the PG domain may act as an intrinsic buffer due to the large number of COOH side groups in this region [19].

Since an acidic microenvironment is commonly observed in cancers, methods are needed to overcome low pHe drug resistance in order to improve therapeutic efficacy [99]. Targeted treatment with proton pump inhibitors (PPIs) disrupt transmembrane pH gradients and can inhibit growth in human B cell leukemia [100]. Also, inhibition of CAIX catalytic activity using monoclonal antibodies or specific small molecule inhibitors leads to intracellular acidification with the consequent impairment of the tumor growth and reduced invasive capacity of the tumors [101, 102].

Alternatively, acidosis can be inhibited directly with the ingestion of alkaline buffers, such as bicarbonate [103]. Alkalization of tumors by administration of bicarbonate in drinking water effectively increased the extracellular tumor pH without affecting the pHi or systemic pH and led to reduced metastases in human breast and prostate cancer mouse models and increased survival [104]. Interestingly, immunohistochemical staining for CAIX in hyperplastic regions of the C57BL/6 TRAMP (transgenic adenocarcinoma mouse prostate) model showed increased expression in controls versus the group that had 200 mM sodium bicarbonate in their drinking water [105]. However, these results do not generalize across all tumor cell lines, as fast-growing tumor cell lines, e.g. B-16 melanoma, showed resistance. A concern about buffer therapy is whether or not a balance can be reached between the acid load produced by the tumor and the quantity of bicarbonate that can be administrated. Thus the effectiveness of this therapy will be reduced in large tumors [106]. However, the strong advantage of this strategy would be to restore a more neutral tumor pHe that should slow down the metastatic process and diminish drug resistance [77]. Another approach that takes advantage of low tumor pHe involves the development by many groups of low pH activated micelle systems that penetrate into solid tumors and release their cytotoxic therapeutic payloads within the acidic microenvironment. However, additional in vivo studies are needed to determine their efficacy [107, 108].

5 CAIX Expression in Cancer

The use of CAIX as a prognostic indicator in oncology makes it a potentially important biomarker for the evaluation of tissues. CAIX is constitutively up-regulated in several cancer types such as the lung [109–111], colon [112, 113], colorectal [114], gastric [115], pancreatic [116], breast [117–121], cervix [122–125], bladder [68, 126], ovaries [127], brain [128, 129], head and neck [130–132], astrocytomas [133] and oral cavity [134–136]. The mechanism of upregulation has been determined in some cancers, e.g. clear cell carcinoma of the kidney, where the VHL gene is mutated leading to high levels of CAIX expression (up to 150-fold) as a consequence of constitutive HIF activation [37, 137–140].

High expression of CAIX has been shown to generally correlate with poor prognosis and is related to shorter disease-free and disease specific survival [reviewed [28]]. For example, CAIX is particularly highly expressed in the aggressive basal-like or triple negative breast cancers, a tumor subgroup with poor prognosis and known resistance to systemic therapy [119, 141], and high expression has been reported in patients with squamous cell carcinoma of the lung [111]. However, there is still much to be uncovered regarding the prognostic indications of CAIX in various cancers. For example, it has been reported that CAIX over-expression, as detected by IHC staining, may be an indicator of a good prognosis in clear cell renal cell carcinoma [142]. While, in other renal tumors such as chromophobe renal cell carcinoma and oncocytoma, CAIX is reported to be absent via IHC [143]. In other cases, such as squamous cell carcinoma, IHC of CAIX does not appear to add prognostic or predictive value [144]. Therefore, it is important to carefully characterize the prognostic value of CAIX in individual tumor types. Once established, the utility of CAIX expression in IHC has been shown to be a useful marker [145].

A correlation between CAIX expression and metastatic disease has been also observed. For example, a recent study that examined over 3,600 human breast cancers reported that CAIX expression in distant metastases is an independent biomarker of poor prognosis [141]. These results were similar to other studies [122–124, 146, 147]. A high level of CAIX expression in cervical cancer tumors is associated with the presence of lymph node metastases [124, 125]. In addition, high stromal CAIX expression was associated with nodal metastasis and decreased survival in patients with surgically-treated oral cavity squamous cell carcinoma [148].

CAIX is also reported to be a useful IHC biomarker for the quantification of response to therapy. For example, CAIX has been used to predict renal cell carcinoma response to IL-2 therapy [149] and epirubicin/tamoxifen therapy in breast cancer [150].

Another membrane associated, CA, CAXII, is co-expressed with CAIX in several tumor tissues and highly expressed in a number of cancer types, including breast cancer [26, 151]. Our group has recently reported co-expression of CAIX and CAXII in different types of breast cancer [152] (Fig. 12.4). CAIX and CAXII

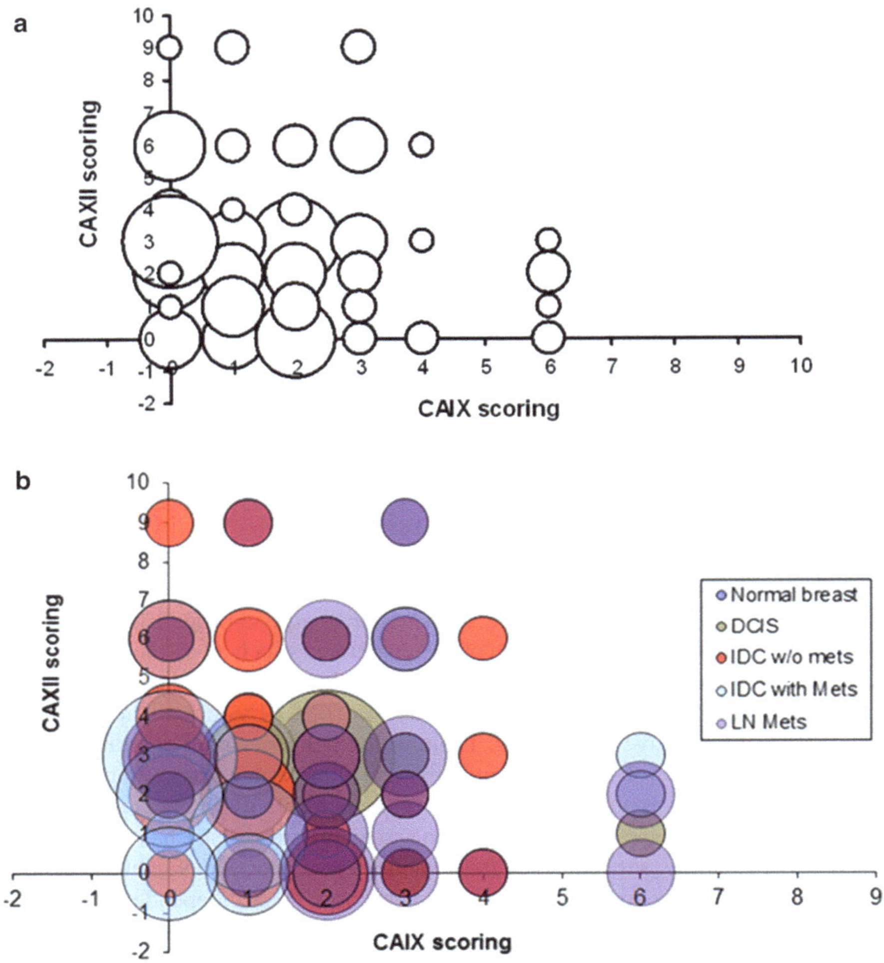

Fig. 12.4 Bubble plot of pathologist scores for CAIX and CAXII IHC staining (**a**) for samples from all breast cancer patients on the tissue microarray (TMA), n = 180; or (**b**) for 47 normal breast tissues, 42 ductal carcinoma in situ, 43 invasive ductal carcinomas without metastasis, 46 invasive ductal carcinomas with metastasis and 49 lymph node macrometastases of breast cancer on the TMA

staining was distributed in the cell membranes of tumor tissues. Interestingly, some samples were positive only for CAIX, some only for CAXII, and others had expression of both markers. This can be problematic when using agents that bind non-specifically to the CA active site if the goal is to target only CAIX (see discussions below). Lymph node metastases samples were positive for CAIX (71 %) and CAXII (76 %), both markers were expressed in 44 % of the samples, and all of the positive lymph nodes were found to express either CAIX or CAXII, i.e., the combination of CAIX and CAXII covers 100 % of patients with ALN metastases represented in our TMA. Heterogeneity of expression was found in 29 % and 35 % of the tumor samples for CAIX and CAXII, respectively, i.e. different areas of the same tumor on a histological section having different expression patterns for a given marker [152]. We have planned future studies to determine if there is a correlation of CAIX expression with adverse patient outcomes.

5.1 Detection of CAIX Expression in Tumors and Metastases

The extracellular domain of CAIX, including PG and CA domains, can be released into the culture medium of CCRCC (clear cell renal cell carcinoma) cells or in the body fluids (blood, urine) of patients with CCRCC. This release is most likely a result of proteolytic cleavage. The soluble 50/54 kDa form is termed s-CAIX and it is cleared from the blood within a few days after nephrectomy [153]. The concentration of s-CAIX is very low in healthy subjects [153, 154]. Therefore, it can be assayed by ELISA in serum, plasma and tissue for clinical detection and prognostic evaluation of patients with different types of cancers. ELISA could be a potential quantitative method for non-invasive diagnosis of renal tumors. Two different groups have reported similar values for serum s-CAIX in CCRCC (114.0–126.1 pg/mL), higher than in other types of renal cancer and healthy samples [143, 154]. Serum CAIX levels were significantly higher in CCRCC than in non-CCRCC and increased in conjunction with increased stage of tumor progression [154]. Serum s-CAIX levels also correlated with the tumor size in CCRCC patients [143]. In addition to serum s-CAIX, *CA9* mRNA may also be a promising molecular marker of CCRCC. For example, *CA9* mRNA was detected in 97 % of CCRCCs and highly correlated with IHC staining of corresponding samples [155].

Similar studies have also been performed for other cancer types. For example, urine was tested from 23 patients with transitional cell carcinoma (TCC) or squamous cell carcinoma (SCC) of the urinary bladder and renal pelvis, and the presence of urine s-CAIX correlated with tumor CAIX in most cases. Additionally, s-CAIX was detected in the urine of two patients with a suspected, but unconfirmed bladder tumor, who later developed tumors within 6 months of testing [156]. However, this correlation was not observed in TCC patients. Preoperative serum concentrations of CAIX were reported to be significantly higher in patients with high intratumoral expression of CAIX in vulvar cancer [157]. Serum levels of CAIX in metastatic breast cancer and vulvar cancer were correlated with poor prognosis [157, 158]. In non-small cell lung cancer, high plasma levels of s-CAIX were determined to be an independent prognostic biomarker, particularly for early-stage I or II carcinomas and were associated with significantly worse survival [111]. Also, detection of s-CAIX protein by ELISA and immunocytochemistry have been used to detect malignant pleural effusions [145]. The potential use of serum s-CAIX as a biomarker for therapy response was tested in patients with primary epithelial ovarian cancer, yet were not significantly altered during first-line therapy [159].

As mentioned above for CCRCC, immunohistochemical (IHC) staining may be also used to detect CAIX in tumor samples [145]. IHC of CAIX is an emerging field of interest and holds a great deal of promise for quantification of disease progression, use as a predictive or prognostic indicator, and as an indicator of therapy response. Like other IHC markers, it can be used to evaluate the expression or localization of CAIX in tissue. Using standard color producing enzymes (i.e. peroxidase coupled with DAB) and amplification techniques common to the practice of IHC staining, CAIX can be visualized in a range from negative to strong positive staining intensity. CAIX is generally regarded as a specific IHC biomarker in a myriad of tissue types [117]. As expected, CAIX is generally localized on the cell membrane but may also be found to a lesser degree in the cytoplasm.

6 Targeting of CAIX (Imaging and Therapy)

6.1 CAIX Imaging

The characterization and detection of hypoxic regions within solid tumor masses is an important issue in cancer care because hypoxia is an independent predictor of poor survival in several cancer types and hypoxic cancer cells are less sensitive to chemo- and radiation-therapies [160]. Therefore, a molecular imaging approach based on selective ligands to accessible proteins overexpressed at sites of hypoxia is desired. Such an agent could help physicians to decide which patients will benefit from adjunctive anti-hypoxia therapy [61, 161]. Also as discussed above, since CAIX has high and broad expression among a number of aggressive and late stage types of tumors and metastases, the non-invasive imaging of CAIX expression has potential for use in cancer diagnosis and staging, and to detect and follow the response of disseminated metastatic disease to systemic and targeted therapies. Both antibodies and small MW compounds have been used for targeted imaging of CAIX expression [28].

6.1.1 CAIX Specific Monoclonal Antibodies for Imaging

G250 was characterized as a monoclonal antibody in the 1980s by Oosterwijk et al. [14], and was shown to be a very strong biomarker for CCRCC due to its absence in normal kidney tissue. The specific epitope for the G250 antibody is unknown, but it has excellent specificity for CAIX in immunohistochemical analyses. To reduce human anti-mouse antibody responses (HAMA), a chimeric version of G250 (cG250) was developed in 1997, and was recently radiolabeled with iodine-124 or zirconium-89 for diagnosis, and with iodine-131 or lutetium-177 for therapy [162, 163]. Phase I of human clinical trials of positron emission tomography/computed tomography (PET/CT) imaging of CCRCC with Iodine-124 labeled cG250 have shown excellent sensitivity, specificity, positive predictive values, and negative predictive values (94 %,100 %, 100 %, and 90 %, respectively) [164]. Hence, CAIX targeted PET/CT-based molecular imaging has potential to alter the care and welfare of patients suspected of having CCRCC and could be used to monitor patients with a history of CCRCC. In addition, this approach allows for the non-invasive and non-surgical differentiation of clear cell and non-clear cell RCC. Currently, Wilex AG (Munich, Germany) is conducting phase III clinical trials using this PET imaging agent, under the name of Redectane®. Wilex is also developing a therapy application of G250 under the product name Rencarex® [165].

The M75 antibody is highly specific for CAIX [166]. It recognizes the extracellular proteoglycan-like domain and is useful for western blotting, immunoprecipitation, and immunohistochemistry. A radiolabeled derivative has been developed for pre-clinical imaging of CAIX in hypoxic tumors in mouse xenograft models [167, 168] and CAIX selective accumulation of the M75 antibody labeled with iodine-125 was demonstrated in HT-29 xenografts in nude mice [167].

Fully human CAIX single-chain variable fragment (scFv) mini-antibodies named A3 and CC7, were generated using phage-display technology by Ahlskog et al. [169]. These minibodies recognize the extracellular carbonic anhydrase (CA) domain of human CAIX and have low nanomolar affinity to CAIX. Because they are fully human, A3 and CC7 are expected to be non-immunogenic in patients with cancer. Also, because of their smaller size, approximately a factor of 10 smaller compared to full-length antibodies, these minibodies are expected to have faster tissue distribution and clearance relative to full-sized antibodies. The A3 and CC7 minibodies do not inhibit CAIX activity and do not bind to CAXII, which shares 39 % sequence identity with CAIX.

We have reported that at least one of the two cell-surface markers, either CAIX or CAXII, is expressed in 100 % of breast cancer axillary lymph node (ALN) metastases on our TMA. We have also developed CAIX and CAXII specific monoclonal antibody-based near-infrared fluorescent (NIRF) molecular imaging probes (CAIXAb-680 and CAXIIAb-68) using antibodies from R&D Systems for the non-invasive detection of breast cancer ALN metastases and we have confirmed the high selectivity of the probes in vitro and in vivo using preclinical breast cancer metastasis models [152] (Fig. 12.5).

6.1.2 Imaging CAIX Expression Using Small Molecule-Based Probes

Carbonic anhydrase inhibitors are divided to four groups: inorganic anions, sulfonamide based compounds (sulphonamides, sulphanilamides, sulphamates and their derivatives), phenols and coumarins [1, 170–172]. Inorganic anions and sulfonamides inhibit CAIX by coordinating to the zinc ion within the active site, with mM to μM Ki inhibition (the anions) or μM to nM Ki inhibition (sulfonamides

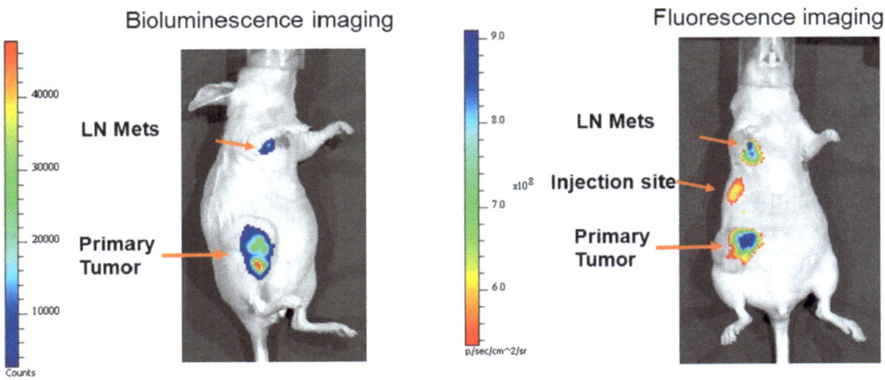

Fig. 12.5 In vivo bioluminescence and fluorescence images of a mouse following spontaneous metastasis to the axillary lymph node from a primary mammary fat pad (MFP) xenograft tumor. MDA-mb-231 cells were used that were engineered to overexpress luciferase and constitutively express CAIX. The fluorescence image was acquired 24 h after MFP injection of CA9Ab-680 probe

and their congeners) [1, 170, 173–175]. Phenols are generally low μM Ki CA inhibitors, and act by anchoring to the zinc-bound water molecule/hydroxide ion through a network of hydrogen bonds which stabilize the enzyme-inhibitor adduct [171]. Coumarins are a novel class of CA inhibitors and act differently compared to the other inhibitors, i.e. through a non-zinc mediated mechanism. Coumarins function as suicide inhibitors, undergoing hydrolysis to 2-hydroxycinnamic acids and anchoring irreversibly at the entrance to the active site cavity [172]. Although several low nanomolar and isoform-selective CAIX inhibitors based on the coumarin/thiocoumarin ring were recently reported [172], the best investigated and most robust class of inhibitors are the sulfonamides due to their high affinity, availability, and ease of chemical manipulation [1, 176]. Supuran's group has pioneered the development of specific CAIX inhibitors to target tumor growth and many sulfonamide CA inhibitors with low nM Ki inhibition have been identified in the last several years, demonstrating particular promise as potential anti-cancer agents [1, 94, 177, 178].

Interestingly, it has been demonstrated by testing several cell lines in hypoxic conditions, that sulfonamides bind to CAIX only when the enzyme is active, indicating that the active site of CAIX is only available for sulfonamide binding during hypoxia [94, 179]. Hence, CAIX sulfonamides can distinguish cells that are currently in hypoxic conditions from those that were previously hypoxic, while CAIX specific antibodies do not, which is a big advantage since CAIX expression is very stable (with a half-life of ~40 h) with CAIX remaining on the cell surface for a relatively long time after reoxygenation [76, 94]. Therefore, imaging with CAIX inhibitors might provide different imaging and prognostic information compared to monoclonal antibodies [98].

However, due to a high degree of homology among CA isotypes, sulfonamide-based inhibitors are generally not specific to the CAIX isoform, i.e. sulfonamides specifically bind to the catalytic sites of other CAs [1, 180]. This is a problem, because unmodified sulfonamides are able to pass through the plasma membrane and can thus interact with the physiologically dominant CAI and CAII isoforms which are soluble proteins located in the cytosol or mitochondria. Several strategies are being employed to develop selective CA-isozyme inhibitors without "off-target" inhibition of intracellular CAs. These strategies include the addition of charged species or bulky entities such as FITC, albumin or hydrophilic sugar moieties, all of which prevent transportation across the plasma membrane [166]. However, these strategies do not solve the problem of extracellular isoforms other than CAIX that are known to be overexpressed in cancers, e.g. CAXII. In fact, both CAIX and CAXII are transmembrane proteins with catalytic domains oriented extracellularly. Hence, the development of CAIX specific inhibitors is an area of significant interest and the design and synthesis of CAIX selective inhibitors has been recently reviewed [180, 181]. For example, sulfonamides conjugated to fluorescent dye (FITC) have been developed and are shown to have membrane-impermeant properties and high affinities for CAIX. These agents have been shown to bind to cells that express CAIX in hypoxic conditions, but not in normoxic conditions or to control cells that lack CAIX expression [94]. This finding suggests that hypoxia may alter the conformation of the CAIX active site, allowing access to the inhibitor. It has been proposed that interaction of the highly acidic PG domain with highly basic residues around the conserved zinc-binding histidines form a cover at the entrance to the catalytic site that open in hypoxic conditions. In support of this hypothesis, it was observed that a CAIX variant that lacks the PG domain binds to inhibitor in normoxic conditions at high levels that are comparable to binding in hypoxia [178].

In vivo studies with fluorescent sulfonamides have also shown promise for detecting CAIX expression in HT-29 and SK-RC-52 tumor xenografts [178, 182]. An inhibitor-fluorescent dye conjugate was also shown to inhibit the CAIX-mediated acidification of the tumor extracellular environment [141, 179, 183]. In addition, Dubois et al. reported significant accumulation of the inhibitor in the HT-29 colorectal tumors when animals were placed in hypoxic breathing conditions (7 % oxygen). Furthermore, the bound CA inhibitor fraction rapidly and significantly decreased upon return to normoxic conditions and tumor reoxygenation [177].

Also very recently, a series of acetazolamide derivatives conjugated to various near-infrared fluorescent (NIRF) dyes were synthesized with the aim of imaging hypoxia-induced hCAIX expression in tumor cells in vitro, ex vivo and in vivo. The compounds demonstrated up to 50-fold selectivity for hCAIX compared to the cytosolic hCAII isoform. Some of the compounds also showed inhibition selectivity for other transmembrane isoforms as well, e.g. hCAXII and hCAXIV. A pilot in vivo study in HT-29 tumor bearing mice quantified significant accumulation (10 % i.d.) in the tumor with little accumulation in other organs (except the kidneys), demonstrating the potential use of these new compounds for quantitative imaging of

CAIX via fluorescence molecular tomography (FMT) [184]. These results illustrate the potential of NIRF CAIX inhibitors and FMT imaging for pre-clinical studies to non-invasively quantify CAIX expression as an endogenous marker of tumor hypoxia. This capability is crucial to the study of the underlying biology of tumor hypoxia and the development and monitoring of novel anti-cancer therapies [184].

A 99mTc-labeled sulfonamide derivative has been synthesized and evaluated for visualization of CAIX expression by single photon emission computed tomography (SPECT) imaging. However despite the favorable Ki values of the tracer, a maximum tumor tracer uptake of only 0.1 % ID/g was observed at 30 min post injection, indicating that the tracer is not very promising for imaging of CAIX [185]. Although tracer uptake in tumors was minimal, tumor-to-blood ratios increased as a function of time and a different radionuclide with a slower decay rate may produce greater contrast at later time-points.

6.2 CAIX Targeted Therapy

CAIX is highly expressed in different tumor types and has relatively low expression in normal tissues; has an important role in tumor progression, acidification and metastasis; and is located on the extracellular surface of cell membranes, allowing for efficient targeting by antibodies or small molecule inhibitors. Therefore, CAIX constitutes an attractive and promising candidate marker for systemic anticancer therapy that targets disseminated metastatic disease. Two major therapeutic tools are being studied for targeting of CAIX: monoclonal antibodies and small molecule inhibitors [reviewed by [28]].

6.2.1 CAIX Specific Monoclonal Antibodies for Therapy

CAIX-specific monoclonal antibodies can be useful for anti-tumor therapy either by (a) antibody-dependent cell cytotoxicity (ADCC); (b) targeting the CAIX active site; or (c) receptor-mediated internalization for targeted delivery of various therapeutic agents, including cytotoxins and radionuclides [28].

(a) The first category of antibodies acts through ADCC. For example cG250 which targets the PG-like domain of CAIX and is marketed by WILEX AG under the trade name RENCAREX® has been extensively characterized as an anticancer immunotherapy [90]. This antibody acts via ADCC [186], and phase I and II trials have demonstrated safety and high efficacy of the antibody in either monotherapy or in combination with interferon (IFN)-α for treatment of renal cell carcinoma (RCC) [187, 188]. Phase III trials have just begun using cG250 as an adjuvant therapy for reducing recurrence in surgically-treated renal cell carcinoma (RCC) patients who have a high risk of relapse and in combination with interleukin 2 (IL-2) or IFN-α for metastatic RCC [28, 187].

(b) While cG250 is directed at the PG-like domain, recent studies have focused on developing antibodies that target the catalytic domain of CAIX, thus targeting its tumorigenic functions, including pH regulation [28]. For example a mAb directed against the CAIX catalytic domain has been generated using hybridoma technology and this antibody demonstrated efficient binding, internalization and persistence in cultured cells. In vivo evaluation of the antibody demonstrated an anti-cancer effect in a mouse xenograft model of colorectal carcinoma by limiting tumor growth. However, the effect was not observed when tumors were allowed to grow and become established prior to treatment [189]. Though this might be the expected result, since buffer therapy has no effect on the growth of established primary cancers, while exerting a significant inhibition on the development of spontaneous metastases [104].

    Antigen-binding antibody fragments (Fab) against human CAIX have been generated by phage-display and inhibition of CAIX demonstrated in spheroid cell cultures, suggesting that these Fabs could be used to inhibit CAIX therapeutically [190]. Two mAbs with high affinity to CAIX generated by phage display have also been described [169]. However more studies are needed for preclinical evaluation of these antibodies.

(c) The third category of CAIX antibodies are those that are internalized and can thus be used for delivery of therapeutic payloads, such as radionuclides and cytotoxic agents. For example, stability, biodistribution and therapeutic efficacy of various radio-immunoconjugates of cG250 have been tested in nude mice with subcutaneous renal cell cancer (RCC) tumors and high therapeutic efficacy has been demonstrated by attachment of [177] Lu, [90] Y or [186] Re compared to [131] I-cG250 [191]. A phase I/II trial in metastatic RCC patients is currently ongoing for [177] Lu-cG250 [192]. Also in another study a CAIX mAb which was identified by phage display was conjugated to monomethyl auristatin E (a synthetic antineoplastic agent) and high efficacy as a potent anti-tumor drug confirmed in several preclinical human xenograft tumor models [193].

Alternative combination therapies are being designed that include targeting of CAIX. An example is the tyrosine kinase inhibitor sunitinib (Sutent®). This approved first line therapy for metastatic renal cell carcinoma (mRCC) has had limited clinical success. Oosterwijk et al., are developing an agent that combines Sutent® with targets of CAIX for enhancement of therapeutic effect. A second example involves treatment of the same disease (mRCC) with autologous T-cells genetically retargeted against CAIX [194].

6.2.2 CAIX Targeted Small Molecule Inhibitors for Therapy

Sulfonamides are commonly used as diuretics, anti-glaucoma, anti-obesity, and anti-infective treatments. Additionally, they have potential for use as antitumor drugs and this is an active area of clinical research. In fact, with the help of X-ray crystallography, much recent progress has been made for the designing of selective CAIX inhibitors. For example, Indisulam, a sulfonamide which was recently investigated in phase II clinical trials, is considered one of the most potent anticancer sulfonamides yet and has demonstrated high anti-tumor activity in various preclinical cancer models [195]. The combination of CAIX inhibitors with conventional chemotherapy may yield even better efficacy [98]. Also, one of several potent bis-sulfonamide CAIX inhibitors identified by screening one million compounds in a DNA-encoded chemical library has demonstrated high and specific accumulation in tumor models [196]. In addition, due to high-expression in hypoxia, CAIX is an attractive target for the design of hypoxia-activatable prodrugs. This may lead to a completely new approach for developing less toxic anticancer agents [see Ref [181] for review]. Recently, carbonic anhydrase inhibitors coated with gold nanoparticles, which are membrane impermeant and therefore more effective in targeting the extracellular active site of CAIX, have been developed [197]. These nanoparticles exhibited more potent inhibition of CAIX compared to acetazolamide, and may be a candidate for RCC treatment or imaging and treatment of hypoxic tumors. Targeting of CAIX by small molecules in preclinical models of human cancer has been reviewed extensively in ref [7, 28].

7 CAIX and CAXII Selectivity Issues

CAXII is a transmembrane isoform with an extracellular active site that was originally identified because of its overexpression in renal cancer cells [43, 198]. CAIX and CAXII proteins share an overall amino acid sequence identity of 35 % and there is considerable homology at the catalytic site with the three key zinc-binding histidine residues conserved in both sequences [43]. As mentioned above, CAIX contains an additional proteoglycan-like domain that has been implicated in cell adhesion, which is not present on CAXII [17, 20, 22]. The lack of PG domains suggests that CAXII has a more neutral pH optimum for activity.

Similar to CAIX, CAXII is highly overexpressed in breast cancer [26] and other tumor types [for review see [77]], but it is also expressed in some normal tissues such as pancreas, kidney, gut and the gastrointestinal tract [199–202]. Several studies have shown that while CAIX expression is associated with a negative prognosis, including resistance to chemo- and radiation-therapy in breast cancer [117, 203, 204], CAXII has been associated with a favorable prognosis in invasive breast cancer [205].

Although expression of both CAIX and CAXII can be induced by hypoxia [29], CAXII is not as robustly regulated by hypoxia as CAIX and, if present, its HIF response element (HRE) has not been identified [26, 29, 54]. Thus, the mechanism for CAXII induction by hypoxia is still in question. Our group has studied CAIX and CAXII expression in breast cancer tumor xenografts using multiple tumor lines. CAIX was invariably induced in regions adjacent to necrosis, while CAXII

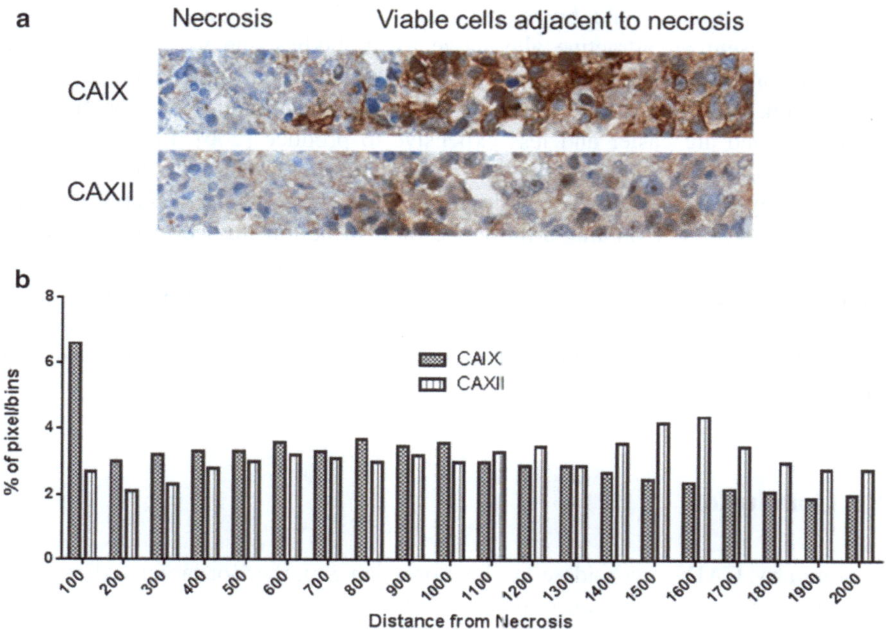

Fig. 12.6 (**a**) CAIX and CAXII IHC staining in a region of a representative MCF-7 tumor xenograft adjacent to necrosis. (**b**) Quantified staining intensity by distance from necrosis with pixels 0–100 being adjacent to a necrotic region

induction was generally lower compared to CAIX and that, although cell line dependent, both markers could have significant constitutive expression distal to the necrotic edge (Fig. 12.6).

Thus, imaging probes that specifically target CAIX may provide superior results compared to probes that non-specifically target the CA active site and can cross-react with expression of other cell-surface CA isoforms, e.g. CAXII. However in terms of therapeutic agents, the use of membrane impermeable sulfonamides, which display inhibitory constants in the low nanomolar range both towards CAIX and CAXII may be useful; because of high expression of both CAIX and CAXII is observed in several cancers [55, 152]. Simultaneous silencing of both CAIX and CAXII in transfected LS174T colorectal cancer xenograft models led to substantial growth retardation that was greater compared to HIF-1α silencing alone, and greater than knockouts of CAIX or CAXII individually, which yielded substantially lower tumor growth retardation [54]. Therefore drugs that can target both CAIX and CAXII may act as novel and potentially efficient drug approaches [55]. However, CAXII expression in normal tissues involved in agent clearance (e.g. kidney) and toxicity (e.g. pancreas and gut) is of some concern in the context of therapy.

Specific peptides could be a promising candidate for imaging of hypoxia. Peptides present an attractive alternative to antibodies due to their small size and therefore optimal pharmacokinetic properties and high effectiveness of tumor penetration. In addition, peptides do not possess the immunogenic potential of antibodies and are easier and less expensive to produce [206]. Recently, a new peptide (CaIX-P1) with specificity for the extracellular domain of CAIX was identified by phage display technology using a commercially available Ph.D.12 library and the CAIX binding affinity and biodistribution studied [207]. Although the stability and affinity of CaIX-P1 was recently optimized, organ distribution studies revealed low tumor-to-blood ratios and high background distribution, which is not favorable for imaging applications [208]. Therefore, further research is required for generation of peptide-based ligands with high target specificity for human carbonic anhydrase IX.

8 Conclusion

In summary, CAIX is considered to be a promising marker for targeted diagnostic imaging and systemic anticancer therapy due to its overexpression in many cancers and metastases with a limited presence in their normal tissue counterparts. CAIX expression is generally associated with a poor prognosis and is associated with tumor progression to metastasis due to its role in acidification of the extratumoral medium, which leads to both the acquisition of the metastasic phenotype and chemoresistance to weakly basic anticancer drugs.

Inhibition of CAIX activity by either specific antibodies or small molecules has been studied in the preclinical setting and more recently in clinical trials as a mode of therapy. Many classes of CAIX targeted therapeutic agents have been developed and pharmacological evaluation has demonstrated promising results. Some of these agents are now in phase III clinical trials. Concerns remain about off-target toxicities of agents that are not excluded to the extracellular environment due to interactions with intracellular CAs, and due to lack of selectivity among other extracellular CAs, e.g. CAXII which is expressed in normal tissues of concern.

There is significant interest in the development of diagnostic imaging tools for the non-invasive detection of tumor hypoxia via CAIX induced expression, for detection of CAIX expression in lymph node metastases for staging, for the diagnosis of disseminated disease and to follow response to systemic therapy during treatment of distal metastasis. In conjunction with the development of targeted therapies, there is now significant interest in the pharmaceutical industry in the development of companion targeted imaging agents for use during clinical trials. In terms of the use of CAIX as a marker for tumor hypoxia, it has been shown that CAIX is generally up-regulated in regions of hypoxia. However expression of CAIX in non-hypoxic regions of tumors has also been observed due to HIF-1 stabilization in cancer. Also, high constitutive CAXII expression in tumors and metastases may be a confounding factor for imaging agents that are not CAIX specific. However, for detection of elevated CA expression in disseminated disease, this requirement for specificity is not as important. Although, since CAXII expression is also observed in normal tissues, this could cause increased non-specific background signal in agents that are not CAIX specific.

References

1. Supuran CT (2008) Carbonic anhydrases: novel therapeutic applications for inhibitors and activators. Nat Rev Drug Discov 7:168–181
2. Supuran CT (2010) Carbonic anhydrase inhibitors. Bioorg Med Chem Lett 20:3467–3474
3. Okamoto N, Fujikawa-Adachi K, Nishimori I, Taniuchi K, Onishi S (2001) cDNA sequence of human carbonic anhydrase-related protein, CA-RP X: mRNA expressions of CA-RP X and XI in human brain. Biochim Biophys Acta 1518:311–316
4. Picaud SS, Muniz JR, Kramm A, Pilka ES, Kochan G, Oppermann U, Yue WW (2009) Crystal structure of human carbonic anhydrase-related protein VIII reveals the basis for catalytic silencing. Proteins 76:507–511
5. Saczewski F, Innocenti A, Brzozowski Z, Slawinski J, Pomarnacka E, Kornicka A, Scozzafava A, Supuran CT (2006) Carbonic anhydrase inhibitors. Selective inhibition of human tumor-associated isozymes IX and XII and cytosolic isozymes I and II with some substituted-2-mercapto-benzenesulfonamides. J Enzyme Inhib Med Chem 21:563–568
6. Saczewski F, Slawinski J, Kornicka A, Brzozowski Z, Pomarnacka E, Innocenti A, Scozzafava A, Supuran CT (2006) Carbonic anhydrase inhibitors. Inhibition of the cytosolic human isozymes I and II, and the transmembrane, tumor-associated isozymes IX and XII with substituted aromatic sulfonamides activatable in hypoxic tumors. Bioorg Med Chem Lett 16:4846–4851
7. Imtaiyaz Hassan M, Shajee B, Waheed A, Ahmad F, Sly WS (2012) Structure, function and applications of carbonic anhydrase isozymes. Bioorg Med Chem 21:1570–1582
8. Supuran CT, Scozzafava A, Casini A (2003) Carbonic anhydrase inhibitors. Med Res Rev 23:146–189
9. Swietach P, Wigfield S, Cobden P, Supuran CT, Harris AL, Vaughan-Jones RD (2008) Tumor-associated carbonic anhydrase 9 spatially coordinates intracellular pH in three-dimensional multicellular growths. J Biol Chem 283:20473–20483
10. Pastorek J, Pastorekova S, Callebaut I, Mornon JP, Zelnik V, Opavsky R, Zat'ovicova M, Liao S, Portetelle D, Stanbridge EJ (1994) Cloning and characterization of MN, a human tumor-associated protein with a domain homologous to carbonic anhydrase and a putative helix-loop-helix DNA binding segment. Oncogene 9:2877–2888
11. Opavsky R, Pastorekova S, Zelnik V, Gibadulinova A, Stanbridge EJ, Zavada J, Kettmann R, Pastorek J (1996) Human MN/CA9 gene, a novel member of the carbonic anhydrase family: structure and exon to protein domain relationships. Genomics 33:480–487
12. Pastorekova S, Zavadova Z, Kostal M, Babusikova O, Zavada J (1992) A novel quasi-viral agent, MaTu, is a two-component system. Virology 187:620–626
13. Zavada J, Zavadova Z, Pastorekova S, Ciampor F, Pastorek J, Zelnik V (1993) Expression of MaTu-MN protein in human tumor cultures and in clinical specimens. Int J Cancer 54:268–274
14. Oosterwijk E, Ruiter DJ, Hoedemaeker PJ, Pauwels EK, Jonas U, Zwartendijk J, Warnaar SO (1986) Monoclonal antibody G 250 recognizes a determinant present in renal-cell carcinoma and absent from normal kidney. Int J Cancer 38:489–494
15. Grabmaier K, Vissers JL, De Weijert MC, Oosterwijk-Wakka JC, Van Bokhoven A, Brakenhoff RH, Noessner E, Mulders PA, Merkx G, Figdor CG, Adema GJ, Oosterwijk E (2000) Molecular cloning and immunogenicity of renal cell carcinoma-associated antigen G250. Int J Cancer 85:865–870

16. Hilvo M, Baranauskiene L, Salzano AM, Scaloni A, Matulis D, Innocenti A, Scozzafava A, Monti SM, Di Fiore A, De Simone G, Lindfors M, Janis J, Valjakka J, Pastorekova S, Pastorek J, Kulomaa MS, Nordlund HR, Supuran CT, Parkkila S (2008) Biochemical characterization of CA IX, one of the most active carbonic anhydrase isozymes. J Biol Chem 283:27799–27809
17. Zavada J, Zavadova Z, Pastorek J, Biesova Z, Jezek J, Velek J (2000) Human tumour-associated cell adhesion protein MN/CA IX: identification of M75 epitope and of the region mediating cell adhesion. Br J Cancer 82:1808–1813
18. De Simone G, Supuran CT (2010) Carbonic anhydrase IX: biochemical and crystallographic characterization of a novel antitumor target. Biochim Biophys Acta 1804:404–409
19. Innocenti A, Pastorekova S, Pastorek J, Scozzafava A, De Simone G, Supuran CT (2009) The proteoglycan region of the tumor-associated carbonic anhydrase isoform IX acts as anintrinsic buffer optimizing CO2 hydration at acidic pH values characteristic of solid tumors. Bioorg Med Chem Lett 19:5825–5828
20. Svastova E, Zilka N, Zat'ovicova M, Gibadulinova A, Ciampor F, Pastorek J, Pastorekova S (2003) Carbonic anhydrase IX reduces E-cadherin-mediated adhesion of MDCK cells via interaction with beta-catenin. Exp Cell Res 290:332–345
21. Zavadova Z, Zavada J (2005) Carbonic anhydrase IX (CA IX) mediates tumor cell interactions with microenvironment. Oncol Rep 13:977–982
22. Dorai T, Sawczuk IS, Pastorek J, Wiernik PH, Dutcher JP (2005) The role of carbonic anhydrase IX overexpression in kidney cancer. Eur J Cancer 41:2935–2947
23. Alterio V, Hilvo M, Di Fiore A, Supuran CT, Pan P, Parkkila S, Scaloni A, Pastorek J, Pastorekova S, Pedone C, Scozzafava A, Monti SM, De Simone G (2009) Crystal structure of the catalytic domain of the tumor-associated human carbonic anhydrase IX. Proc Natl Acad Sci U S A 106:16233–16238
24. Whittington DA, Waheed A, Ulmasov B, Shah GN, Grubb JH, Sly WS, Christianson DW (2001) Crystal structure of the dimeric extracellular domain of human carbonic anhydrase XII, a bitopic membrane protein overexpressed in certain cancer tumor cells. Proc Natl Acad Sci U S A 98:9545–9550
25. Li Y, Wang H, Tu C, Shiverick KT, Silverman DN, Frost SC (2011) Role of hypoxia and EGF on expression, activity, localization and phosphorylation of carbonic anhydrase IX in MDA-MB-231 breast cancer cells. Biochim Biophys Acta 1813:159–167
26. Ivanov S, Liao SY, Ivanova A, Danilkovitch-Miagkova A, Tarasova N, Weirich G, Merrill MJ, Proescholdt MA, Oldfield EH, Lee J, Zavada J, Waheed A, Sly W, Lerman MI, Stanbridge EJ (2001) Expression of hypoxia-inducible cell-surface transmembrane carbonic anhydrases in human cancer. Am J Pathol 158:905–919
27. Pastorekova S, Parkkila S, Parkkila AK, Opavsky R, Zelnik V, Saarnio J, Pastorek J (1997) Carbonic anhydrase IX, MN/CA IX: analysis of stomach complementary DNA sequence and expression in human and rat alimentary tracts. Gastroenterology 112:398–408
28. McDonald PC, Winum JY, Supuran CT, Dedhar S (2012) Recent developments in targeting carbonic anhydrase IX for cancer therapeutics. Oncotarget 3:84–97
29. Wykoff CC, Beasley NJ, Watson PH, Turner KJ, Pastorek J, Sibtain A, Wilson GD, Turley H, Talks KL, Maxwell PH, Pugh CW, Ratcliffe PJ, Harris AL (2000) Hypoxia-inducible expression of tumor-associated carbonic anhydrases. Cancer Res 60:7075–7083
30. Kaluz S, Kaluzova M, Liao SY, Lerman M, Stanbridge EJ (2009) Transcriptional control of the tumor- and hypoxia-marker carbonic anhydrase 9: a one transcription factor (HIF-1) show? Biochim Biophys Acta 1795:162–172
31. Barathova M, Takacova M, Holotnakova T, Gibadulinova A, Ohradanova A, Zatovicova M, Hulikova A, Kopacek J, Parkkila S, Supuran CT, Pastorekova S, Pastorek J (2008) Alternative splicing variant of the hypoxia marker carbonic anhydrase IX expressed independently of hypoxia and tumour phenotype. Br J Cancer 98:129–136
32. Zatovicova M, Sedlakova O, Svastova E, Ohradanova A, Ciampor F, Arribas J, Pastorek J, Pastorekova S (2005) Ectodomain shedding of the hypoxia-induced carbonic anhydrase IX is a metalloprotease-dependent process regulated by TACE/ADAM17. Br J Cancer 93:1267–1276

33. Salceda S, Caro J (1997) Hypoxia-inducible factor 1alpha (HIF-1alpha) protein is rapidly degraded by the ubiquitin-proteasome system under normoxic conditions. Its stabilization by hypoxia depends on redox-induced changes. J Biol Chem 272:22642–22647
34. Metzen E, Berchner-Pfannschmidt U, Stengel P, Marxsen JH, Stolze I, Klinger M, Huang WQ, Wotzlaw C, Hellwig-Burgel T, Jelkmann W, Acker H, Fandrey J (2003) Intracellular localisation of human HIF-1 alpha hydroxylases: implications for oxygen sensing. J Cell Sci 116:1319–1326
35. Semenza GL (1999) Regulation of mammalian O2 homeostasis by hypoxia-inducible factor 1. Annu Rev Cell Dev Biol 15:551–578
36. Kallio PJ, Pongratz I, Gradin K, McGuire J, Poellinger L (1997) Activation of hypoxia-inducible factor 1alpha: posttranscriptional regulation and conformational change by recruitment of the Arnt transcription factor. Proc Natl Acad Sci U S A 94:5667–5672
37. Maxwell PH, Wiesener MS, Chang GW, Clifford SC, Vaux EC, Cockman ME, Wykoff CC, Pugh CW, Maher ER, Ratcliffe PJ (1999) The tumour suppressor protein VHL targets hypoxia-inducible factors for oxygen-dependent proteolysis. Nature 399:271–275
38. Ke Q, Costa M (2006) Hypoxia-inducible factor-1 (HIF-1). Mol Pharmacol 70:1469–1480
39. Semenza GL (2012) Hypoxia-inducible factors: mediators of cancer progression and targets for cancer therapy. Trends Pharmacol Sci 33:207–214
40. Lu H, Forbes RA, Verma A (2002) Hypoxia-inducible factor 1 activation by aerobic glycolysis implicates the Warburg effect in carcinogenesis. J Biol Chem 277:23111–23115
41. Kaluzova M, Kaluz S, Lerman MI, Stanbridge EJ (2004) DNA damage is a prerequisite for p53-mediated proteasomal degradation of HIF-1alpha in hypoxic cells and downregulation of the hypoxia marker carbonic anhydrase IX. Mol Cell Biol 24:5757–5766
42. Kondo K, Kaelin WG Jr (2001) The von Hippel-Lindau tumor suppressor gene. Exp Cell Res 264:117–125
43. Ivanov SV, Kuzmin I, Wei MH, Pack S, Geil L, Johnson BE, Stanbridge EJ, Lerman MI (1998) Down-regulation of transmembrane carbonic anhydrases in renal cell carcinoma cell lines by wild-type von Hippel-Lindau transgenes. Proc Natl Acad Sci U S A 95:12596–12601
44. Grabmaier K, A de Weijert MC, Verhaegh GW, Schalken JA, Oosterwijk E (2004) Strict regulation of CAIX(G250/MN) by HIF-1alpha in clear cell renal cell carcinoma. Oncogene 23:5624–5631
45. Das B, Tsuchida R, Malkin D, Koren G, Baruchel S, Yeger H (2008) Hypoxia enhances tumor stemness by increasing the invasive and tumorigenic side population fraction. Stem Cells 26:1818–1830
46. Mohyeldin A, Garzon-Muvdi T, Quinones-Hinojosa A (2010) Oxygen in stem cell biology: a critical component of the stem cell niche. Cell Stem Cell 7:150–161
47. Xing F, Okuda H, Watabe M, Kobayashi A, Pai SK, Liu W, Pandey PR, Fukuda K, Hirota S, Sugai T, Wakabayshi G, Koeda K, Kashiwaba M, Suzuki K, Chiba T, Endo M, Mo YY, Watabe K (2011) Hypoxia-induced Jagged2 promotes breast cancer metastasis and self-renewal of cancer stem-like cells. Oncogene 30:4075–4086
48. Lock FE, McDonald PC, Lou Y, Serrano I, Chafe SC, Ostlund C, Aparicio S, Winum JY, Supuran CT, Dedhar S (2012) Targeting carbonic anhydrase IX depletes breast cancer stem cells within the hypoxic niche. Oncogene 1–10
49. Horree N, van Diest PJ, Sie-Go DM, Heintz AP (2007) The invasive front in endometrial carcinoma: higher proliferation and associated derailment of cell cycle regulators. Hum Pathol 38:1232–1238
50. Brabletz T, Jung A, Spaderna S, Hlubek F, Kirchner T (2005) Opinion: migrating cancer stem cells – an integrated concept of malignant tumour progression. Nat Rev Cancer 5:744–749
51. Gut MO, Parkkila S, Vernerova Z, Rohde E, Zavada J, Hocker M, Pastorek J, Karttunen T, Gibadulinova A, Zavadova Z, Knobeloch KP, Wiedenmann B, Svoboda J, Horak I, Pastorekova S (2002) Gastric hyperplasia in mice with targeted disruption of the carbonic anhydrase gene Car9. Gastroenterology 123:1889–1903
52. Leppilampi M, Karttunen TJ, Kivela J, Gut MO, Pastorekova S, Pastorek J, Parkkila S (2005) Gastric pit cell hyperplasia and glandular atrophy in carbonic anhydrase IX knockout mice: studies on two strains C57/BL6 and BALB/C. Transgenic Res 14:655–663

53. Pan P, Leppilampi M, Pastorekova S, Pastorek J, Waheed A, Sly WS, Parkkila S (2006) Carbonic anhydrase gene expression in CA II-deficient (Car2−/−) and CA IX-deficient (Car9−/−) mice. J Physiol 571:319–327
54. Chiche J, Ilc K, Laferriere J, Trottier E, Dayan F, Mazure NM, Brahimi-Horn MC, Pouyssegur J (2009) Hypoxia-inducible carbonic anhydrase IX and XII promote tumor cell growth by counteracting acidosis through the regulation of the intracellular pH. Cancer Res 69:358–368
55. Chiche J, Ilc K, Brahimi-Horn MC, Pouyssegur J (2010) Membrane-bound carbonic anhydrases are key pH regulators controlling tumor growth and cell migration. Adv Enzyme Regul 50:20–33
56. Harris AL (2002) Hypoxia–a key regulatory factor in tumour growth. Nat Rev Cancer 2:38–47
57. Wouters A, Pauwels B, Lardon F, Vermorken JB (2007) Review: implications of in vitro research on the effect of radiotherapy and chemotherapy under hypoxic conditions. Oncologist 12:690–712
58. Comerford KM, Wallace TJ, Karhausen J, Louis NA, Montalto MC, Colgan SP (2002) Hypoxia-inducible factor-1-dependent regulation of the multidrug resistance (MDR1) gene. Cancer Res 62:3387–3394
59. Rohwer N, Cramer T (2011) Hypoxia-mediated drug resistance: novel insights on the functional interaction of HIFs and cell death pathways. Drug Resist Updat 14:191–201
60. Hockel M, Schlenger K, Mitze M, Schaffer U, Vaupel P (1996) Hypoxia and radiation response in human tumors. Semin Radiat Oncol 6:3–9
61. Evans SM, Koch CJ (2003) Prognostic significance of tumor oxygenation in humans. Cancer Lett 195:1–16
62. Sun JD, Liu Q, Wang J, Ahluwalia D, Ferraro D, Wang Y, Duan JX, Ammons WS, Curd JG, Matteucci MD, Hart CP (2012) Selective tumor hypoxia targeting by hypoxia-activated prodrug TH-302 inhibits tumor growth in preclinical models of cancer. Clin Cancer Res 18:758–770
63. Gatenby RA, Kessler HB, Rosenblum JS, Coia LR, Moldofsky PJ, Hartz WH, Broder GJ (1988) Oxygen distribution in squamous cell carcinoma metastases and its relationship to outcome of radiation therapy. Int J Radiat Oncol Biol Phys 14:831–838
64. Mees G, Dierckx R, Vangestel C, Van de Wiele C (2009) Molecular imaging of hypoxia with radiolabelled agents. Eur J Nucl Med Mol Imaging 36:1674–1686
65. Nozue M, Lee I, Yuan F, Teicher BA, Brizel DM, Dewhirst MW, Milross CG, Milas L, Song CW, Thomas CD, Guichard M, Evans SM, Koch CJ, Lord EM, Jain RK, Suit HD (1997) Interlaboratory variation in oxygen tension measurement by Eppendorf "Histograph" and comparison with hypoxic marker. J Surg Oncol 66:30–38
66. Jenkins WT, Evans SM, Koch CJ (2000) Hypoxia and necrosis in rat 9L glioma and Morris 7777 hepatoma tumors: comparative measurements using EF5 binding and the Eppendorf needle electrode. Int J Radiat Oncol Biol Phys 46:1005–1017
67. Cook GJ (2003) Oncological molecular imaging: nuclear medicine techniques. Br J Radiol 76:S152–S158
68. Hoskin PJ, Sibtain A, Daley FM, Wilson GD (2003) GLUT1 and CAIX as intrinsic markers of hypoxia in bladder cancer: relationship with vascularity and proliferation as predictors of outcome of ARCON. Br J Cancer 89:1290–1297
69. Jankovic B, Aquino-Parsons C, Raleigh JA, Stanbridge EJ, Durand RE, Banath JP, MacPhail SH, Olive PL (2006) Comparison between pimonidazole binding, oxygen electrode measurements, and expression of endogenous hypoxia markers in cancer of the uterine cervix. Cytometry B Clin Cytom 70:45–55
70. Li XF, Carlin S, Urano M, Russell J, Ling CC, O'Donoghue JA (2007) Visualization of hypoxia in microscopic tumors by immunofluorescent microscopy. Cancer Res 67:7646–7653
71. Olive PL, Aquino-Parsons C, MacPhail SH, Liao SY, Raleigh JA, Lerman MI, Stanbridge EJ (2001) Carbonic anhydrase 9 as an endogenous marker for hypoxic cells in cervical cancer. Cancer Res 61:8924–8929

72. Sobhanifar S, Aquino-Parsons C, Stanbridge EJ, Olive P (2005) Reduced expression of hypoxia-inducible factor-1alpha in perinecrotic regions of solid tumors. Cancer Res 65:7259–7266
73. Carlin S, Khan N, Ku T, Longo VA, Larson SM, Smith-Jones PM (2010) Molecular targeting of carbonic anhydrase IX in mice with hypoxic HT29 colorectal tumor xenografts. PLoS One 5:e10857
74. Beasley NJ, Wykoff CC, Watson PH, Leek R, Turley H, Gatter K, Pastorek J, Cox GJ, Ratcliffe P, Harris AL (2001) Carbonic anhydrase IX, an endogenous hypoxia marker, expression in head and neck squamous cell carcinoma and its relationship to hypoxia, necrosis, and microvessel density. Cancer Res 61:5262–5267
75. Mayer A, Hockel M, Vaupel P (2005) Carbonic anhydrase IX expression and tumor oxygenation status do not correlate at the microregional level in locally advanced cancers of the uterine cervix. Clin Cancer Res 11:7220–7225
76. Rafajova M, Zatovicova M, Kettmann R, Pastorek J, Pastorekova S (2004) Induction by hypoxia combined with low glucose or low bicarbonate and high posttranslational stability upon reoxygenation contribute to carbonic anhydrase IX expression in cancer cells. Int J Oncol 24:995–1004
77. Chiche J, Brahimi-Horn MC, Pouyssegur J (2010) Tumour hypoxia induces a metabolic shift causing acidosis: a common feature in cancer. J Cell Mol Med 14:771–794
78. Brown JM, Wilson WR (2004) Exploiting tumour hypoxia in cancer treatment. Nat Rev Cancer 4:437–447
79. Dewhirst MW, Cao Y, Moeller B (2008) Cycling hypoxia and free radicals regulate angiogenesis and radiotherapy response. Nat Rev Cancer 8:425–437
80. Koritzinsky M, Magagnin MG, van den Beucken T, Seigneuric R, Savelkouls K, Dostie J, Pyronnet S, Kaufman RJ, Weppler SA, Voncken JW, Lambin P, Koumenis C, Sonenberg N, Wouters BG (2006) Gene expression during acute and prolonged hypoxia is regulated by distinct mechanisms of translational control. EMBO J 25:1114–1125
81. Brahimi-Horn MC, Chiche J, Pouyssegur J (2007) Hypoxia and cancer. J Mol Med (Berl) 85:1301–1307
82. Brahimi-Horn MC, Chiche J, Pouyssegur J (2007) Hypoxia signalling controls metabolic demand. Curr Opin Cell Biol 19:223–229
83. Brahimi-Horn MC, Pouyssegur J (2007) Oxygen, a source of life and stress. FEBS Lett 581:3582–3591
84. Hanahan D, Weinberg RA (2000) The hallmarks of cancer. Cell 100:57–70
85. Chambard JC, Pouyssegur J (1986) Intracellular pH controls growth factor-induced ribosomal protein S6 phosphorylation and protein synthesis in the G0–G1 transition of fibroblasts. Exp Cell Res 164:282–294
86. Pouyssegur J, Sardet C, Franchi A, L'Allemain G, Paris S (1984) A specific mutation abolishing Na+/H + antiport activity in hamster fibroblasts precludes growth at neutral and acidic pH. Proc Natl Acad Sci U S A 81:4833–4837
87. Roos A, Boron WF (1981) Intracellular pH. Physiol Rev 61:296–434
88. Neri D, Supuran CT (2011) Interfering with pH regulation in tumours as a therapeutic strategy. Nat Rev Drug Discov 10:767–777
89. Parks SK, Chiche J, Pouyssegur J (2011) pH control mechanisms of tumor survival and growth. J Cell Physiol 226:299–308
90. Cardone RA, Casavola V, Reshkin SJ (2005) The role of disturbed pH dynamics and the Na+/H + exchanger in metastasis. Nat Rev Cancer 5:786–795
91. Counillon L, Pouyssegur J (2000) The expanding family of eucaryotic Na(+)/H(+) exchangers. J Biol Chem 275:1–4
92. Ullah MS, Davies AJ, Halestrap AP (2006) The plasma membrane lactate transporter MCT4, but not MCT1, is up-regulated by hypoxia through a HIF-1alpha-dependent mechanism. J Biol Chem 281:9030–9037

93. Karumanchi SA, Jiang L, Knebelmann B, Stuart-Tilley AK, Alper SL, Sukhatme VP (2001) VHL tumor suppressor regulates Cl−/HCO3− exchange and Na+/H + exchange activities in renal carcinoma cells. Physiol Genomics 5:119–128
94. Svastova E, Hulikova A, Rafajova M, Zat'ovicova M, Gibadulinova A, Casini A, Cecchi A, Scozzafava A, Supuran CT, Pastorek J, Pastorekova S (2004) Hypoxia activates the capacity of tumor-associated carbonic anhydrase IX to acidify extracellular pH. FEBS Lett 577:439–445
95. Raghunand N, Gatenby RA, Gillies RJ (2003) Microenvironmental and cellular consequences of altered blood flow in tumours. Br J Radiol 76:S11–S22
96. Li Y, Tu C, Wang H, Silverman DN, Frost SC (2011) Catalysis and pH control by membrane-associated carbonic anhydrase IX in MDA-MB-231 breast cancer cells. J Biol Chem 286:15789–15796
97. Swietach P, Vaughan-Jones RD, Harris AL (2007) Regulation of tumor pH and the role of carbonic anhydrase 9. Cancer Metastasis Rev 26:299–310
98. Pastorekova S, Ratcliffe PJ, Pastorek J (2008) Molecular mechanisms of carbonic anhydrase IX-mediated pH regulation under hypoxia. BJU Int 101(Suppl 4):8–15
99. Wojtkowiak JW, Verduzco D, Schramm KJ, Gillies RJ (2011) Drug resistance and cellular adaptation to tumor acidic pH microenvironment. Mol Pharm 8:2032–2038
100. De Milito A, Canese R, Marino ML, Borghi M, Iero M, Villa A, Venturi G, Lozupone F, Iessi E, Logozzi M, Della Mina P, Santinami M, Rodolfo M, Podo F, Rivoltini L, Fais S (2010) pH-dependent antitumor activity of proton pump inhibitors against human melanoma is mediated by inhibition of tumor acidity. Int J Cancer 127:207–219
101. Swietach P, Patiar S, Supuran CT, Harris AL, Vaughan-Jones RD (2009) The role of carbonic anhydrase 9 in regulating extracellular and intracellular ph in three-dimensional tumor cell growths. J Biol Chem 284:20299–20310
102. Swietach P, Hulikova A, Vaughan-Jones RD, Harris AL (2010) New insights into the physiological role of carbonic anhydrase IX in tumour pH regulation. Oncogene 29:6509–6521
103. Raghunand N, He X, van Sluis R, Mahoney B, Baggett B, Taylor CW, Paine-Murrieta G, Roe D, Bhujwalla ZM, Gillies RJ (1999) Enhancement of chemotherapy by manipulation of tumour pH. Br J Cancer 80:1005–1011
104. Robey IF, Baggett BK, Kirkpatrick ND, Roe DJ, Dosescu J, Sloane BF, Hashim AI, Morse DL, Raghunand N, Gatenby RA, Gillies RJ (2009) Bicarbonate increases tumor pH and inhibits spontaneous metastases. Cancer Res 69:2260–2268
105. Ibrahim-Hashim A, Cornnell HH, Abrahams D, Lloyd M, Bui M, Gillies RJ, Gatenby RA (2012) Systemic buffers inhibit carcinogenesis in TRAMP mice. J Urol 188:624–631
106. Silva AS, Yunes JA, Gillies RJ, Gatenby RA (2009) The potential role of systemic buffers in reducing intratumoral extracellular pH and acid-mediated invasion. Cancer Res 69:2677–2684
107. Gullotti E, Yeo Y (2009) Extracellularly activated nanocarriers: a new paradigm of tumor targeted drug delivery. Mol Pharm 6:1041–1051
108. Rios-Doria J, Carie A, Costich T, Burke B, Skaff H, Panicucci R, Sill K (2012) A versatile polymer micelle drug delivery system for encapsulation and in vivo stabilization of hydrophobic anticancer drugs. J Drug Deliv 2012:951741
109. Swinson DE, Jones JL, Richardson D, Wykoff C, Turley H, Pastorek J, Taub N, Harris AL, O'Byrne KJ (2003) Carbonic anhydrase IX expression, a novel surrogate marker of tumor hypoxia, is associated with a poor prognosis in non-small-cell lung cancer. J Clin Oncol 21:473–482
110. Kim SJ, Rabbani ZN, Dewhirst MW, Vujaskovic Z, Vollmer RT, Schreiber EG, Oosterwijk E, Kelley MJ (2005) Expression of HIF-1alpha, CA IX, VEGF, and MMP-9 in surgically resected non-small cell lung cancer. Lung Cancer 49:325–335
111. Ilie M, Mazure NM, Hofman V, Ammadi RE, Ortholan C, Bonnetaud C, Havet K, Venissac N, Mograbi B, Mouroux J, Pouyssegur J, Hofman P (2010) High levels of carbonic anhydrase IX in tumour tissue and plasma are biomarkers of poor prognostic in patients with non-small cell lung cancer. Br J Cancer 102:1627–1635

112. Saarnio J, Parkkila S, Parkkila AK, Haukipuro K, Pastorekova S, Pastorek J, Kairaluoma MI, Karttunen TJ (1998) Immunohistochemical study of colorectal tumors for expression of a novel transmembrane carbonic anhydrase, MN/CA IX, with potential value as a marker of cell proliferation. Am J Pathol 153:279–285
113. Korkeila E, Talvinen K, Jaakkola PM, Minn H, Syrjanen K, Sundstrom J, Pyrhonen S (2009) Expression of carbonic anhydrase IX suggests poor outcome in rectal cancer. Br J Cancer 100:874–880
114. Niemela AM, Hynninen P, Mecklin JP, Kuopio T, Kokko A, Aaltonen L, Parkkila AK, Pastorekova S, Pastorek J, Waheed A, Sly WS, Orntoft TF, Kruhoffer M, Haapasalo H, Parkkila S, Kivela AJ (2007) Carbonic anhydrase IX is highly expressed in hereditary nonpolyposis colorectal cancer. Cancer Epidemiol Biomarkers Prev 16:1760–1766
115. Chen J, Rocken C, Hoffmann J, Kruger S, Lendeckel U, Rocco A, Pastorekova S, Malfertheiner P, Ebert MP (2005) Expression of carbonic anhydrase 9 at the invasion front of gastric cancers. Gut 54:920–927
116. Juhasz M, Chen J, Lendeckel U, Kellner U, Kasper HU, Tulassay Z, Pastorekova S, Malfertheiner P, Ebert MP (2003) Expression of carbonic anhydrase IX in human pancreatic cancer. Aliment Pharmacol Ther 18:837–846
117. Chia SK, Wykoff CC, Watson PH, Han C, Leek RD, Pastorek J, Gatter KC, Ratcliffe P, Harris AL (2001) Prognostic significance of a novel hypoxia-regulated marker, carbonic anhydrase IX, in invasive breast carcinoma. J Clin Oncol 19:3660–3668
118. Trastour C, Benizri E, Ettore F, Ramaioli A, Chamorey E, Pouyssegur J, Berra E (2007) HIF-1alpha and CA IX staining in invasive breast carcinomas: prognosis and treatment outcome. Int J Cancer 120:1451–1458
119. Tan EY, Yan M, Campo L, Han C, Takano E, Turley H, Candiloro I, Pezzella F, Gatter KC, Millar EK, O'Toole SA, McNeil CM, Crea P, Segara D, Sutherland RL, Harris AL, Fox SB (2009) The key hypoxia regulated gene CAIX is upregulated in basal-like breast tumours and is associated with resistance to chemotherapy. Br J Cancer 100:405–411
120. Hussain SA, Ganesan R, Reynolds G, Gross L, Stevens A, Pastorek J, Murray PG, Perunovic B, Anwar MS, Billingham L, James ND, Spooner D, Poole CJ, Rea DW, Palmer DH (2007) Hypoxia-regulated carbonic anhydrase IX expression is associated with poor survival in patients with invasive breast cancer. Br J Cancer 96:104–109
121. Neumeister VM, Sullivan CA, Lindner R, Lezon-Geyda K, Li J, Zavada J, Martel M, Glazer PM, Tuck DP, Rimm DL, Harris L (2012) Hypoxia-induced protein CAIX is associated with somatic loss of BRCA1 protein and pathway activity in triple negative breast cancer. Breast Cancer Res Treat 136:67–75
122. Loncaster JA, Harris AL, Davidson SE, Logue JP, Hunter RD, Wycoff CC, Pastorek J, Ratcliffe PJ, Stratford IJ, West CM (2001) Carbonic anhydrase (CA IX) expression, a potential new intrinsic marker of hypoxia: correlations with tumor oxygen measurements and prognosis in locally advanced carcinoma of the cervix. Cancer Res 61:6394–6399
123. Kim JY, Shin HJ, Kim TH, Cho KH, Shin KH, Kim BK, Roh JW, Lee S, Park SY, Hwang YJ, Han IO (2006) Tumor-associated carbonic anhydrases are linked to metastases in primary cervical cancer. J Cancer Res Clin Oncol 132:302–308
124. Lee S, Shin HJ, Han IO, Hong EK, Park SY, Roh JW, Shin KH, Kim TH, Kim JY (2007) Tumor carbonic anhydrase 9 expression is associated with the presence of lymph node metastases in uterine cervical cancer. Cancer Sci 98:329–333
125. Woelber L, Kress K, Kersten JF, Choschzick M, Kilic E, Herwig U, Lindner C, Schwarz J, Jaenicke F, Mahner S, Milde-Langosch K, Mueller V, Ihnen M (2011) Carbonic anhydrase IX in tumor tissue and sera of patients with primary cervical cancer. BMC Cancer 11:12
126. Klatte T, Seligson DB, Rao JY, Yu H, de Martino M, Kawaoka K, Wong SG, Belldegrun AS, Pantuck AJ (2009) Carbonic anhydrase IX in bladder cancer: a diagnostic, prognostic, and therapeutic molecular marker. Cancer 115:1448–1458
127. Choschzick M, Oosterwijk E, Muller V, Woelber L, Simon R, Moch H, Tennstedt P (2011) Overexpression of carbonic anhydrase IX (CAIX) is an independent unfavorable prognostic marker in endometrioid ovarian cancer. Virchows Arch 459:193–200

128. Nordfors K, Haapasalo J, Korja M, Niemela A, Laine J, Parkkila AK, Pastorekova S, Pastorek J, Waheed A, Sly WS, Parkkila S, Haapasalo H (2010) The tumour-associated carbonic anhydrases CA II, CA IX and CA XII in a group of medulloblastomas and supratentorial primitive neuroectodermal tumours: an association of CA IX with poor prognosis. BMC Cancer 10:148
129. Proescholdt MA, Merrill MJ, Stoerr EM, Lohmeier A, Pohl F, Brawanski A (2012) Function of carbonic anhydrase IX in glioblastoma multiforme. Neuro Oncol 14:1357–1366
130. Hoogsteen IJ, Marres HA, Wijffels KI, Rijken PF, Peters JP, van den Hoogen FJ, Oosterwijk E, van der Kogel AJ, Kaanders JH (2005) Colocalization of carbonic anhydrase 9 expression and cell proliferation in human head and neck squamous cell carcinoma. Clin Cancer Res 11:97–106
131. De Schutter H, Landuyt W, Verbeken E, Goethals L, Hermans R, Nuyts S (2005) The prognostic value of the hypoxia markers CA IX and GLUT 1 and the cytokines VEGF and IL 6 in head and neck squamous cell carcinoma treated by radiotherapy +/− chemotherapy. BMC Cancer 5:42
132. Koukourakis MI, Bentzen SM, Giatromanolaki A, Wilson GD, Daley FM, Saunders MI, Dische S, Sivridis E, Harris AL (2006) Endogenous markers of two separate hypoxia response pathways (hypoxia inducible factor 2 alpha and carbonic anhydrase 9) are associated with radiotherapy failure in head and neck cancer patients recruited in the CHART randomized trial. J Clin Oncol 24:727–735
133. Haapasalo JA, Nordfors KM, Hilvo M, Rantala IJ, Soini Y, Parkkila AK, Pastorekova S, Pastorek J, Parkkila SM, Haapasalo HK (2006) Expression of carbonic anhydrase IX in astrocytic tumors predicts poor prognosis. Clin Cancer Res 12:473–477
134. Eckert AW, Kappler M, Schubert J, Taubert H (2012) Correlation of expression of hypoxia-related proteins with prognosis in oral squamous cell carcinoma patients. Oral Maxillofac Surg 16:189–196
135. Eckert AW, Lautner MH, Schutze A, Bolte K, Bache M, Kappler M, Schubert J, Taubert H, Bilkenroth U (2010) Co-expression of Hif1alpha and CAIX is associated with poor prognosis in oral squamous cell carcinoma patients. J Oral Pathol Med 39:313–317
136. Choi SW, Kim JY, Park JY, Cha IH, Kim J, Lee S (2008) Expression of carbonic anhydrase IX is associated with postoperative recurrence and poor prognosis in surgically treated oral squamous cell carcinoma. Hum Pathol 39:1317–1322
137. Pouyssegur J, Dayan F, Mazure NM (2006) Hypoxia signalling in cancer and approaches to enforce tumour regression. Nature 441:437–443
138. Liao SY, Aurelio ON, Jan K, Zavada J, Stanbridge EJ (1997) Identification of the MN/CA9 protein as a reliable diagnostic biomarker of clear cell carcinoma of the kidney. Cancer Res 57:2827–2831
139. Muriel Lopez C, Esteban E, Berros JP, Pardo P, Astudillo A, Izquierdo M, Crespo G, Sanmamed M, Fonseca PJ, Martinez-Camblor P (2012) Prognostic factors in patients with advanced renal cell carcinoma. Clin Genitourin Cancer 10:262–270
140. Jiang YD, Ren F, Zheng SB (2012) Value of MN/CAIX in the diagnosis of renal cell carcinoma. Nan Fang Yi Ke Da Xue Xue Bao 32:412–414
141. Lou Y, McDonald PC, Oloumi A, Chia S, Ostlund C, Ahmadi A, Kyle A, Auf dem Keller U, Leung S, Huntsman D, Clarke B, Sutherland BW, Waterhouse D, Bally M, Roskelley C, Overall CM, Minchinton A, Pacchiano F, Carta F, Scozzafava A, Touisni N, Winum JY, Supuran CT, Dedhar S (2011) Targeting tumor hypoxia: suppression of breast tumor growth and metastasis by novel carbonic anhydrase IX inhibitors. Cancer Res 71:3364–3376
142. Choueiri TK, Regan MM, Rosenberg JE, Oh WK, Clement J, Amato AM, McDermott D, Cho DC, Atkins MB, Signoretti S (2010) Carbonic anhydrase IX and pathological features as predictors of outcome in patients with metastatic clear-cell renal cell carcinoma receiving vascular endothelial growth factor-targeted therapy. BJU Int 106:772–778
143. Zhou GX, Ireland J, Rayman P, Finke J, Zhou M (2010) Quantification of carbonic anhydrase IX expression in serum and tissue of renal cell carcinoma patients using enzyme-linked immunosorbent assay: prognostic and diagnostic potentials. Urology 75:257–261

144. Eriksen JG, Overgaard J (2007) Lack of prognostic and predictive value of CA IX in radiotherapy of squamous cell carcinoma of the head and neck with known modifiable hypoxia: an evaluation of the DAHANCA 5 study. Radiother Oncol 83:383–388
145. Liao ND, Lee WY (2012) Detection of carbonic anhydrase IX protein in the diagnosis of malignant pleural effusion by enzyme-linked immunosorbent assay and immunocytochemistry. Cancer Cytopathol 120:269–275
146. Garcia S, Dales JP, Charafe-Jauffret E, Carpentier-Meunier S, Andrac-Meyer L, Jacquemier J, Andonian C, Lavaut MN, Allasia C, Bonnier P, Charpin C (2007) Poor prognosis in breast carcinomas correlates with increased expression of targetable CD146 and c-Met and with proteomic basal-like phenotype. Hum Pathol 38:830–841
147. Liao SY, Darcy KM, Randall LM, Tian C, Monk BJ, Burger RA, Fruehauf JP, Peters WA, Stock RJ, Stanbridge EJ (2010) Prognostic relevance of carbonic anhydrase-IX in high-risk, early-stage cervical cancer: a Gynecologic Oncology Group study. Gynecol Oncol 116:452–458
148. Brockton NT, Klimowicz AC, Bose P, Petrillo SK, Konno M, Rudmik L, Dean M, Nakoneshny SC, Matthews TW, Chandarana S, Lau HY, Magliocco AM, Dort JC (2012) High stromal carbonic anhydrase IX expression is associated with nodal metastasis and decreased survival in patients with surgically-treated oral cavity squamous cell carcinoma. Oral Oncol 48:615–622
149. Atkins M, Regan M, McDermott D, Mier J, Stanbridge E, Youmans A, Febbo P, Upton M, Lechpammer M, Signoretti S (2005) Carbonic anhydrase IX expression predicts outcome of interleukin 2 therapy for renal cancer. Clin Cancer Res 11:3714–3721
150. Generali D, Berruti A, Brizzi MP, Campo L, Bonardi S, Wigfield S, Bersiga A, Allevi G, Milani M, Aguggini S, Gandolfi V, Dogliotti L, Bottini A, Harris AL, Fox SB (2006) Hypoxia-inducible factor-1alpha expression predicts a poor response to primary chemoendocrine therapy and disease-free survival in primary human breast cancer. Clin Cancer Res 12:4562–4568
151. Goonewardene TI, Sowter HM, Harris AL (2002) Hypoxia-induced pathways in breast cancer. Microsc Res Tech 59:41–48
152. Tafreshi NK, Bui MM, Bishop K, Lloyd MC, Enkemann SA, Lopez AS, Abrahams D, Carter BW, Vagner J, Grobmyer SR, Gillies RJ, Morse DL (2012) Noninvasive detection of breast cancer lymph node metastasis using carbonic anhydrases IX and XII targeted imaging probes. Clin Cancer Res 18:207–219
153. Zavada J, Zavadova Z, Zat'ovicova M, Hyrsl L, Kawaciuk I (2003) Soluble form of carbonic anhydrase IX (CA IX) in the serum and urine of renal carcinoma patients. Br J Cancer 89:1067–1071
154. Li G, Feng G, Gentil-Perret A, Genin C, Tostain J (2008) Serum carbonic anhydrase 9 level is associated with postoperative recurrence of conventional renal cell cancer. J Urol 180:510–513
155. Murakami Y, Kanda K, Tsuji M, Kanayama H, Kagawa S (1999) MN/CA9 gene expression as a potential biomarker in renal cell carcinoma. BJU Int 83:743–747
156. Hyrsl L, Zavada J, Zavadova Z, Kawaciuk I, Vesely S, Skapa P (2009) Soluble form of carbonic anhydrase IX (CAIX) in transitional cell carcinoma of urinary tract. Neoplasma 56:298–302
157. Kock L, Mahner S, Choschzick M, Eulenburg C, Milde-Langosch K, Schwarz J, Jaenicke F, Muller V, Woelber L (2011) Serum carbonic anhydrase IX and its prognostic relevance in vulvar cancer. Int J Gynecol Cancer 21:141–148
158. Muller V, Riethdorf S, Rack B, Janni W, Fasching PA, Solomayer E, Aktas B, Kasimir-Bauer S, Zeitz J, Pantel K, Fehm T (2011) Prospective evaluation of serum tissue inhibitor of metalloproteinase 1 and carbonic anhydrase IX in correlation to circulating tumor cells in patients with metastatic breast cancer. Breast Cancer Res 13:R71
159. Woelber L, Mueller V, Eulenburg C, Schwarz J, Carney W, Jaenicke F, Milde-Langosch K, Mahner S (2010) Serum carbonic anhydrase IX during first-line therapy of ovarian cancer. Gynecol Oncol 117:183–188

160. Bertout JA, Patel SA, Simon MC (2008) The impact of O2 availability on human cancer. Nat Rev Cancer 8:967–975
161. Scheurer SB, Rybak JN, Rosli C, Neri D, Elia G (2004) Modulation of gene expression by hypoxia in human umbilical cord vein endothelial cells: a transcriptomic and proteomic study. Proteomics 4:1737–1760
162. Steffens MG, Boerman OC, Oosterwijk-Wakka JC, Oosterhof GO, Witjes JA, Koenders EB, Oyen WJ, Buijs WC, Debruyne FM, Corstens FH, Oosterwijk E (1997) Targeting of renal cell carcinoma with iodine-131-labeled chimeric monoclonal antibody G250. J Clin Oncol 15:1529–1537
163. Stillebroer AB, Oosterwijk E, Oyen WJ, Mulders PF, Boerman OC (2007) Radiolabeled antibodies in renal cell carcinoma. Cancer Imaging 7:179–188
164. Divgi CR, Pandit-Taskar N, Jungbluth AA, Reuter VE, Gonen M, Ruan S, Pierre C, Nagel A, Pryma DA, Humm J, Larson SM, Old LJ, Russo P (2007) Preoperative characterisation of clear-cell renal carcinoma using iodine-124-labelled antibody chimeric G250 (124I-cG250) and PET in patients with renal masses: a phase I trial. Lancet Oncol 8:304–310
165. Reichert JM (2011) Next generation and biosimilar monoclonal antibodies: essential considerations towards regulatory acceptance in Europe. February 3–4, 2011, Freiburg, Germany. MAbs 3:223–240
166. Li Y, Wang H, Oosterwijk E, Selman Y, Mira JC, Medrano T, Shiverick KT, Frost SC (2009) Antibody-specific detection of CAIX in breast and prostate cancers. Biochem Biophys Res Commun 386:488–492
167. Chrastina A, Zavada J, Parkkila S, Kaluz S, Kaluzova M, Rajcani J, Pastorek J, Pastorekova S (2003) Biodistribution and pharmacokinetics of 125I-labeled monoclonal antibody M75 specific for carbonic anhydrase IX, an intrinsic marker of hypoxia, in nude mice xenografted with human colorectal carcinoma. Int J Cancer 105:873–881
168. Chrastina A, Pastorekova S, Pastorek J (2003) Immunotargeting of human cervical carcinoma xenograft expressing CA IX tumor-associated antigen by 125I-labeled M75 monoclonal antibody. Neoplasma 50:13–21
169. Ahlskog JK, Schliemann C, Marlind J, Qureshi U, Ammar A, Pedley RB, Neri D (2009) Human monoclonal antibodies targeting carbonic anhydrase IX for the molecular imaging of hypoxic regions in solid tumours. Br J Cancer 101:645–657
170. Vullo D, Franchi M, Gallori E, Pastorek J, Scozzafava A, Pastorekova S, Supuran CT (2003) Carbonic anhydrase inhibitors. inhibition of cytosolic isozymes I and II and transmembrane, cancer-associated isozyme IX with anions. J Enzyme Inhib Med Chem 18:403–406
171. Innocenti A, Vullo D, Scozzafava A, Supuran CT (2008) Carbonic anhydrase inhibitors: interactions of phenols with the 12 catalytically active mammalian isoforms (CA I-XIV). Bioorg Med Chem Lett 18:1583–1587
172. Maresca A, Temperini C, Vu H, Pham NB, Poulsen SA, Scozzafava A, Quinn RJ, Supuran CT (2009) Non-zinc mediated inhibition of carbonic anhydrases: coumarins are a new class of suicide inhibitors. J Am Chem Soc 131:3057–3062
173. Pastorekova S, Casini A, Scozzafava A, Vullo D, Pastorek J, Supuran CT (2004) Carbonic anhydrase inhibitors: the first selective, membrane-impermeant inhibitors targeting the tumor-associated isozyme IX. Bioorg Med Chem Lett 14:869–873
174. Casey JR, Morgan PE, Vullo D, Scozzafava A, Mastrolorenzo A, Supuran CT (2004) Carbonic anhydrase inhibitors. Design of selective, membrane-impermeant inhibitors targeting the human tumor-associated isozyme IX. J Med Chem 47:2337–2347
175. Winum JY, Pastorekova S, Jakubickova L, Montero JL, Scozzafava A, Pastorek J, Vullo D, Innocenti A, Supuran CT (2005) Carbonic anhydrase inhibitors: synthesis and inhibition of cytosolic/tumor-associated carbonic anhydrase isozymes I, II, and IX with bis-sulfamates. Bioorg Med Chem Lett 15:579–584
176. Pacchiano F, Carta F, McDonald PC, Lou Y, Vullo D, Scozzafava A, Dedhar S, Supuran CT (2011) Ureido-substituted benzenesulfonamides potently inhibit carbonic anhydrase IX and show antimetastatic activity in a model of breast cancer metastasis. J Med Chem 54:1896–1902

177. Dubois L, Lieuwes NG, Maresca A, Thiry A, Supuran CT, Scozzafava A, Wouters BG, Lambin P (2009) Imaging of CA IX with fluorescent labelled sulfonamides distinguishes hypoxic and (re)-oxygenated cells in a xenograft tumour model. Radiother Oncol 92:423–428
178. Cecchi A, Hulikova A, Pastorek J, Pastorekova S, Scozzafava A, Winum JY, Montero JL, Supuran CT (2005) Carbonic anhydrase inhibitors. Design of fluorescent sulfonamides as probes of tumor-associated carbonic anhydrase IX that inhibit isozyme IX-mediated acidification of hypoxic tumors. J Med Chem 48:4834–4841
179. Dubois L, Douma K, Supuran CT, Chiu RK, van Zandvoort MA, Pastorekova S, Scozzafava A, Wouters BG, Lambin P (2007) Imaging the hypoxia surrogate marker CA IX requires expression and catalytic activity for binding fluorescent sulfonamide inhibitors. Radiother Oncol 83:367–373
180. Thiry A, Supuran CT, Masereel B, Dogne JM (2008) Recent developments of carbonic anhydrase inhibitors as potential anticancer drugs. J Med Chem 51:3051–3056
181. Winum JY, Rami M, Scozzafava A, Montero JL, Supuran C (2008) Carbonic anhydrase IX: a new druggable target for the design of antitumor agents. Med Res Rev 28:445–463
182. Ahlskog JK, Dumelin CE, Trussel S, Marlind J, Neri D (2009) In vivo targeting of tumor-associated carbonic anhydrases using acetazolamide derivatives. Bioorg Med Chem Lett 19:4851–4856
183. Helmlinger G, Sckell A, Dellian M, Forbes NS, Jain RK (2002) Acid production in glycolysis-impaired tumors provides new insights into tumor metabolism. Clin Cancer Res 8:1284–1291
184. Groves K, Bao B, Zhang J, Handy E, Kennedy P, Cuneo G, Supuran CT, Yared W, Peterson JD, Rajopadhye M (2012) Synthesis and evaluation of near-infrared fluorescent sulfonamide derivatives for imaging of hypoxia-induced carbonic anhydrase IX expression in tumors. Bioorg Med Chem Lett 22:653–657
185. Akurathi V, Dubois L, Lieuwes NG, Chitneni SK, Cleynhens BJ, Vullo D, Supuran CT, Verbruggen AM, Lambin P, Bormans GM (2010) Synthesis and biological evaluation of a 99mTc-labelled sulfonamide conjugate for in vivo visualization of carbonic anhydrase IX expression in tumor hypoxia. Nucl Med Biol 37:557–564
186. Surfus JE, Hank JA, Oosterwijk E, Welt S, Lindstrom MJ, Albertini MR, Schiller JH, Sondel PM (1996) Anti-renal-cell carcinoma chimeric antibody G250 facilitates antibody-dependent cellular cytotoxicity with in vitro and in vivo interleukin-2-activated effectors. J Immunother Emphasis Tumor Immunol 19:184–191
187. Siebels M, Rohrmann K, Oberneder R, Stahler M, Haseke N, Beck J, Hofmann R, Kindler M, Kloepfer P, Stief C (2011) A clinical phase I/II trial with the monoclonal antibody cG250 (RENCAREX(R)) and interferon-alpha-2a in metastatic renal cell carcinoma patients. World J Urol 29:121–126
188. Davis ID, Wiseman GA, Lee FT, Gansen DN, Hopkins W, Papenfuss AT, Liu Z, Moynihan TJ, Croghan GA, Adjei AA, Hoffman EW, Ingle JN, Old LJ, Scott AM (2007) A phase I multiple dose, dose escalation study of cG250 monoclonal antibody in patients with advanced renal cell carcinoma. Cancer Immun 7:13
189. Zatovicova M, Jelenska L, Hulikova A, Csaderova L, Ditte Z, Ditte P, Goliasova T, Pastorek J, Pastorekova S (2010) Carbonic anhydrase IX as an anticancer therapy target: preclinical evaluation of internalizing monoclonal antibody directed to catalytic domain. Curr Pharm Des 16:3255–3263
190. Murri-Plesko MT, Hulikova A, Oosterwijk E, Scott AM, Zortea A, Harris AL, Ritter G, Old L, Bauer S, Swietach P, Renner C (2011) Antibody inhibiting enzymatic activity of tumour-associated carbonic anhydrase isoform IX. Eur J Pharmacol 657:173–183
191. Brouwers AH, van Eerd JE, Frielink C, Oosterwijk E, Oyen WJ, Corstens FH, Boerman OC (2004) Optimization of radioimmunotherapy of renal cell carcinoma: labeling of monoclonal antibody cG250 with 131I, 90Y, 177Lu, or 186Re. J Nucl Med 45:327–337
192. Stillebroer AB, Zegers CM, Boerman OC, Oosterwijk E, Mulders PF, O'Donoghue JA, Visser EP, Oyen WJ (2012) Dosimetric analysis of 177Lu-cG250 radioimmunotherapy in renal cell carcinoma patients: correlation with myelotoxicity and pretherapeutic absorbed dose predictions based on 111In-cG250 imaging. J Nucl Med 53:82–89

193. Petrul HM, Schatz CA, Kopitz CC, Adnane L, McCabe TJ, Trail P, Ha S, Chang YS, Voznesensky A, Ranges G, Tamburini PP (2012) Therapeutic mechanism and efficacy of the antibody-drug conjugate BAY 79–4620 targeting human carbonic anhydrase 9. Mol Cancer Ther 11:340–349
194. Lamers CH, Sleijfer S, Vulto AG, Kruit WH, Kliffen M, Debets R, Gratama JW, Stoter G, Oosterwijk E (2006) Treatment of metastatic renal cell carcinoma with autologous T-lymphocytes genetically retargeted against carbonic anhydrase IX: first clinical experience. J Clin Oncol 24:e20–e22
195. Dittrich C, Zandvliet AS, Gneist M, Huitema AD, King AA, Wanders J (2007) A phase I and pharmacokinetic study of indisulam in combination with carboplatin. Br J Cancer 96:559–566
196. Buller F, Steiner M, Frey K, Mircsof D, Scheuermann J, Kalisch M, Buhlmann P, Supuran CT, Neri D (2011) Selection of carbonic anhydrase IX Inhibitors from one million DNA-encoded compounds. ACS Chem Biol 6:336–344
197. Stiti M, Cecchi A, Rami M, Abdaoui M, Barragan-Montero V, Scozzafava A, Guari Y, Winum JY, Supuran CT (2008) Carbonic anhydrase inhibitor coated gold nanoparticles selectively inhibit the tumor-associated isoform IX over the cytosolic isozymes I and II. J Am Chem Soc 130:16130–16131
198. Tureci O, Sahin U, Vollmar E, Siemer S, Gottert E, Seitz G, Parkkila AK, Shah GN, Grubb JH, Pfreundschuh M, Sly WS (1998) Human carbonic anhydrase XII: cDNA cloning, expression, and chromosomal localization of a carbonic anhydrase gene that is overexpressed in some renal cell cancers. Proc Natl Acad Sci U S A 95:7608–7613
199. Kivela A, Parkkila S, Saarnio J, Karttunen TJ, Kivela J, Parkkila AK, Waheed A, Sly WS, Grubb JH, Shah G, Tureci O, Rajaniemi H (2000) Expression of a novel transmembrane carbonic anhydrase isozyme XII in normal human gut and colorectal tumors. Am J Pathol 156:577–584
200. Parkkila S, Parkkila AK, Saarnio J, Kivela J, Karttunen TJ, Kaunisto K, Waheed A, Sly WS, Tureci O, Virtanen I, Rajaniemi H (2000) Expression of the membrane-associated carbonic anhydrase isozyme XII in the human kidney and renal tumors. J Histochem Cytochem 48:1601–1608
201. Kivela AJ, Parkkila S, Saarnio J, Karttunen TJ, Kivela J, Parkkila AK, Pastorekova S, Pastorek J, Waheed A, Sly WS, Rajaniemi H (2000) Expression of transmembrane carbonic anhydrase isoenzymes IX and XII in normal human pancreas and pancreatic tumours. Histochem Cell Biol 114:197–204
202. Kivela AJ, Parkkila S, Saarnio J, Karttunen TJ, Kivela J, Parkkila AK, Bartosova M, Mucha V, Novak M, Waheed A, Sly WS, Rajaniemi H, Pastorekova S, Pastorek J (2005) Expression of von Hippel-Lindau tumor suppressor and tumor-associated carbonic anhydrases IX and XII in normal and neoplastic colorectal mucosa. World J Gastroenterol 11:2616–2625
203. Brennan DJ, Jirstrom K, Kronblad A, Millikan RC, Landberg G, Duffy MJ, Ryden L, Gallagher WM, O'Brien SL (2006) CA IX is an independent prognostic marker in premenopausal breast cancer patients with one to three positive lymph nodes and a putative marker of radiation resistance. Clin Cancer Res 12:6421–6431
204. Span PN, Bussink J, Manders P, Beex LV, Sweep CG (2003) Carbonic anhydrase-9 expression levels and prognosis in human breast cancer: association with treatment outcome. Br J Cancer 89:271–276
205. Watson PH, Chia SK, Wykoff CC, Han C, Leek RD, Sly WS, Gatter KC, Ratcliffe P, Harris AL (2003) Carbonic anhydrase XII is a marker of good prognosis in invasive breast carcinoma. Br J Cancer 88:1065–1070
206. Marr A, Markert A, Altmann A, Askoxylakis V, Haberkorn U (2011) Biotechnology techniques for the development of new tumor specific peptides. Methods 55:215–222
207. Askoxylakis V, Garcia-Boy R, Rana S, Kramer S, Hebling U, Mier W, Altmann A, Markert A, Debus J, Haberkorn U (2010) A new peptide ligand for targeting human carbonic anhydrase IX, identified through the phage display technology. PLoS One 5:e15962
208. Rana S, Nissen F, Marr A, Markert A, Altmann A, Mier W, Debus J, Haberkorn U, Askoxylakis V (2012) Optimization of a novel peptide ligand targeting human carbonic anhydrase IX. PLoS One 7:e38279

Chapter 13
Carbonic Anhydrase IX (CAIX) as a Mediator of Hypoxia-Induced Stress Response in Cancer Cells

Paul C. McDonald and Shoukat Dedhar

Abstract The development of hypoxic microenvironments within many types of solid tumors imposes a significant stress on cancer cells to which they must respond appropriately in order to survive and grow. Tumor-specific, hypoxia-induced upregulation of Carbonic Anhydrase IX (CAIX) is a component of the complex response of cancer cells to the evolving low oxygen environment. Here, we discuss evidence from in vivo tumor models employing inhibition or enhancement of CAIX expression, using gene depletion or overexpression strategies, respectively, or inhibition of its catalytic activity, using CAIX-specific small molecules or antibodies, to demonstrate that CAIX is a functional mediator of tumor growth and metastasis. We also discuss the functional contribution of CAIX to several specific biological processes critical for cancer progression, including pH regulation and cell survival, adhesion, migration and invasion, the maintenance of cancer stem cell function, and the acquisition of chemo and radioresistant properties. The demonstration of CAIX as a functional mediator of cancer progression provides a biological rationale for its use as a cancer-specific, clinically relevant therapeutic target.

Keywords Hypoxia • Therapeutic target • Enzymatic activity • Metastasis • Tumor growth

---

Susan C. Frost and Robert McKenna (eds.). Carbonic Anhydrase: Mechanism, Regulation, Links to Disease, and Industrial Applications

P.C. McDonald
Department of Integrative Oncology, British Columbia Cancer Agency Centre, Vancouver, BC, Canada
e-mail: pmcdonal@bccrc.ca

S. Dedhar (✉)
Department of Biochemistry and Molecular Biology, University of British Columbia, Vancouver, BC, Canada
e-mail: sdedhar@bccrc.ca

S.C. Frost and R. McKenna (eds.), *Carbonic Anhydrase: Mechanism, Regulation, Links to Disease, and Industrial Applications*, Subcellular Biochemistry 75,
DOI 10.1007/978-94-007-7359-2_13, © Springer Science+Business Media Dordrecht 2014

1 Introduction

Hypoxia is a central environmental feature common to many types of solid cancers and stems from the inability of the tumor vasculature, which is composed of structurally abnormal and functionally unstable vessels, to adequately respond to the increased oxygen demands of the growing tumor [1, 2]. The evolution of an intratumoral hypoxic microenvironment imposes a significant stress on cancer cells and they must adapt if they are to survive and grow in this hostile environment. To cope effectively with reduced oxygen availability, cancer cells engage a highly conserved hypoxia-induced intracellular signaling cascade regulated by the transcription factors hypoxia-inducible factor 1 and 2 (HIF-1/2) [2]. The activation of the HIF pathway results in the modulation of many genes that regulate several critical processes involved in tumor progression, including cell survival and proliferation, metabolic reprogramming [3, 4], stem cell maintenance, growth factor signaling, epithelial-mesenchymal transition (EMT), angiogenesis, invasion, metastasis, and resistance to radiation therapy and chemotherapy.

Amongst the genes targeted for upregulation by the HIF pathway in cancer cells, Carbonic Anhydrase IX (CAIX) often shows the most dramatic transcriptional activation [5, 6]. The robust and systematic overexpression of CAIX by many types of cancer cells in response to hypoxia underscores its importance in mediating the response of these cells to aberrant oxygen status within tumors. Indeed, CAIX is now well-recognized as having a critical role in primary cancer development and progression, as well as in metastatic disease (Fig. 13.1) [6, 7] (also see Chaps. 10, 11, and 12). Furthermore, its membrane-bound, extracellular location and its tumor-specific overexpression, together with its highly restricted expression profile in normal tissue, have made CAIX a very attractive target for cancer therapy [6, 8–10]. However, while its striking upregulation in hypoxic regions and the well-established demonstration that high CAIX expression is poorly prognostic in many types of solid cancers [6] suggest that its expression is critically important and functionally relevant for tumor growth, the use of CAIX as a therapeutic target in cancer, particularly for therapies designed to interfere with its enzymatic activity, hinges on robust demonstration of CAIX as a functional mediator for the growth and spread of cancer.

In this Chapter, we will first critically review the data demonstrating that perturbation of CAIX expression using genetic strategies or inhibition of enzymatic function using small molecule inhibitors or antibodies results in the inhibition of tumor growth and metastasis in vivo, a key measure of the ability of CAIX to function as a mediator of cancer progression. We will then focus on data showing that CAIX contributes to several specific biological processes critical for cancer progression, including pH regulation and cell survival, adhesion, migration and invasion, the maintenance of cancer stem cell function, and the acquisition of chemo and radioresistant properties.

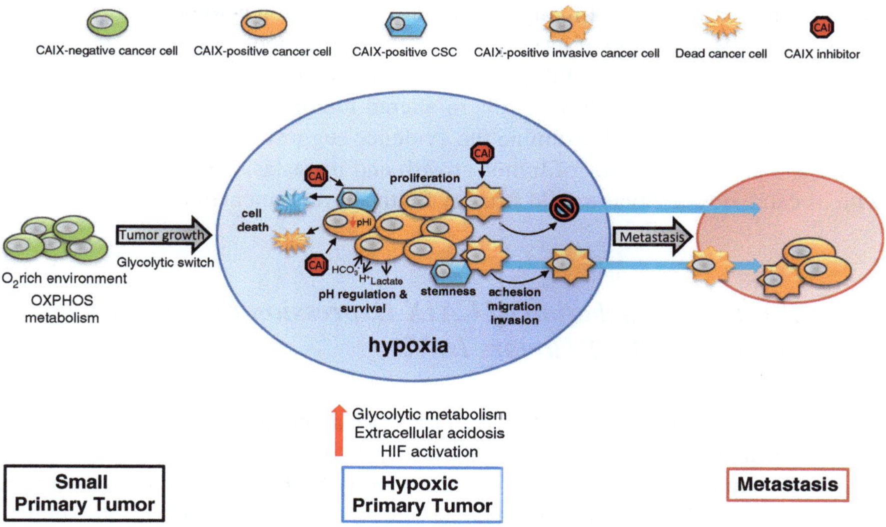

Fig. 13.1 Carbonic Anhydrase IX (*CAIX*) is a functional effector of several biological processes important for tumor growth and metastasis. The development of hypoxia by a growing tumor imposes a significant stress on cancer cells and results in HIF-mediated regulation of many genes, including the dramatic upregulation of CAIX. CAIX contributes to the maintenance of an intracellular pH favorable for tumor cell survival and growth, while also participating in the generation of an increasingly acidic extracellular environment, promoting survival, growth, migration and invasion of cancer cells. CAIX is also an important mediator of the maintenance and stemness properties of cancer stem cells. Inhibition of CAIX enzymatic activity results in reduced survival, migration and invasion in hypoxia. The stemness properties and survival of CSCs are also negatively impacted. Abbreviations. CAIX, Carbonic Anhydrase IX; CAI, Carbonic Anhydrase IX Inhibitor; CSC, cancer stem cell; H^+, proton; HCO_3^-, bicarbonate; HIF, hypoxia-inducible factor; O_2, oxygen; OXPHOS, oxidative phosphorylation; pHi, intracellular pH

2 Inhibition of CAIX Expression and/or Activity Impacts Tumor Growth and Metastasis in Preclinical Cancer Models

The recognition that CAIX is selectively and systematically upregulated by many types of cancer cells in response to tumor hypoxia, together with its cell surface localization and highly restricted normal tissue expression pattern, has solidified CAIX as a robust cancer-specific target [6] and has driven the development of antibodies and highly selective small molecules based on several chemical scaffolds designed to inhibit CAIX catalytic activity [8, 11]. However, the rationale that targeting CAIX enzymatic activity in the context of hypoxia-induced expression will lead to therapeutic benefit is valid only if CAIX is an effective functional mediator of tumor progression and metastasis. A critical measure of the capacity of CAIX to function as a mediator of malignant disease progression is to determine whether inhibition or enhancement of its expression, using gene depletion or overexpression strategies, respectively, or inhibition of its catalytic activity, using CAIX-specific small molecules or antibodies, results in altered tumor progression in vivo. In the following sections we will examine the evidence suggesting that hypoxia-induced CAIX is a critical mediator of tumor growth and metastasis, and that interference with its catalytic activity is a valid strategy for the development of targeted therapies.

2.1 *Effect of Perturbation of CAIX Expression by Cancer Cells on the Growth of Tumors In Vivo*

Data from several studies evaluating the effects of silencing hypoxia-induced CAIX expression in vivo have shown that stable depletion of CAIX in an appropriate microenvironmental context results in attenuation of tumor growth across multiple tumor types. Reduced tumor growth has been observed in mouse and human models of breast cancer [12], as well as in human models of colorectal cancer [13, 14] and glioblastoma [14]. Interestingly, data from these studies suggest that some tumor types, such as breast cancer, may be particularly dependent on CAIX expression for continued growth as regions of hypoxia become established [12], while tumors derived from other tissue types, notably colorectal cancer, may exhibit compensatory regulation related to the expression of extracellular CAs, particularly CAXII, in the event that CAIX function is compromised [13–15]. However, the degree to which tumors compensate for the loss of CAIX with the induction of CAXII expression is variable, as are the functional consequences. In studies of HT-29 colon cancer xenografts, induction of CAXII expression in the absence of CAIX was modest and the functional consequences were not specifically investigated [14]. In contrast, stable downregulation of CAIX together with CAXII in the LS174Tr colon carcinoma model resulted in an 85 % reduction in tumor growth, significantly more than the 40 % reduction in tumor growth attributed to CAIX silencing alone in this model system [13]. Therefore, while the extent to which the growth of hypoxic tumors is reduced in the absence of CAIX expression may be dependent on tumor type, the published data clearly demonstrate that the presence of CAIX in these tumors is of functional importance for their growth.

In addition to depletion of gene expression, the role of CAIX in mediating the growth of tumors has been interrogated using exogenous expression of human CAIX. In this context, it is important to recognize the inherent challenges associated with constitutive overexpression of proteins in situations where the production and functional relevance of the endogenous protein is dictated by intratumoral environmental conditions. Overexpression studies often employ tumors derived from cell lines that do not express endogenous CAIX in normoxia or hypoxia. However, the fact that these cells do not upregulate CAIX in response to appropriate environmental cues means that these cells may not require CAIX in vivo to gain a competitive advantage and often complicates the interpretation of data derived using these models. Despite these caveats, studies have demonstrated an impact of forced CAIX expression on tumor growth. For example, introduction of constitutive CAIX expression in HCT-116 colorectal cancer xenografts, a model that only weakly induces CAIX in vivo, resulted in an increased rate of tumor growth [14], suggesting that the introduction of CAIX expression may enhance cancer progression. Experiments have also been conducted recently wherein shRNA has been used to silence endogenous hypoxia-induced CAIX expression in 4T1 mouse breast cancer cells, a cell type that is dependent on CAIX upregulation for continued tumor growth in the presence of hypoxia in vivo [12]. Introduction of human CAIX into this CAIX-depleted system resulted in the rescue of growth of tumors established from these cells, compared to CAIX silenced controls [12], demonstrating that CAIX acts as a mediator of hypoxic tumor progression. Thus, based on data demonstrating that silencing CAIX gene expression reduces tumor growth, together with findings that CAIX overexpression results in enhancement of tumor growth or rescue of growth in a hypoxia-specific context, it is clear that the expression of CAIX plays a role in mediating tumor progression.

However, while manipulation of gene expression provides important insights into the functional consequences of protein depletion and demonstrates proof-of-principle for the importance of CAIX as a mediator of cancer growth, such strategies do not allow investigation of individual aspects (e.g. catalytic activity) of a multifunctional protein, nor are they currently viable as options for cancer therapy. Observations using specific small molecule inhibitors are equally important for addressing questions related to whether targeted inhibition of CAIX catalytic activity results in a reduction in tumor growth, thereby demonstrating the functional importance of CAIX enzymatic activity and providing an avenue for cancer treatment.

2.2 *The Effect of Treatment with CAIX-Specific Small Molecule Inhibitors on Tumor Growth In Vivo*

Similar to studies that demonstrate that altering CAIX gene expression has a measurable impact on tumor growth, the development of specific small molecule inhibitors allows the effect of selectively inhibiting CAIX catalytic activity on the growth of tumors to be ascertained. An important consideration for the use of targeted therapeutic agents against CAIX is the ability of the drug to reach regions of high CAIX expression, which are defined by poor vascular networks and diffusion characteristics [9]. Several studies have now demonstrated tumor-specific accumulation of CAIX-specific inhibitors in a number of tumor models [16–18], demonstrating that these agents can indeed access CAIX-expressing cancer cells in vivo.

CA inhibitors have been available for some time, but only recently have inhibitors that are selective for cancer associated extracellular CAs such as CAIX (and CAXII) over other, closely related "off-target" CA isoforms become available [8, 10] (also see Chap. 15). Initial strategies included efforts to reduce the membrane permeability of the pan CA inhibitor, acetazolamide, thereby making it more selective for CAIX [16]. Treatment of SK-RC-52 human renal cell carcinoma xenografts consitutively expressing CAIX with this inhibitor resulted in decreased tumor growth, but the effect was not reproduced in LS174tr colorectal xenograft tumors, a model of hypoxia-induced CAIX expression [13, 16]. Treatment of human breast tumors derived from MDA-MB-231 cells with a sulfamate inhibitor of CAIX has also been reported, but whereas this inhibitor reduced metastatic burden in the lung (see Sect. 3, below), no effect on primary tumor growth was observed [19]. Recently, however, several CAIX-selective sulfonamide inhibitors have been developed that directly and specifically inhibit the growth of hypoxic, CAIX-positive tumors in vivo. Administration of a fluorescently labelled sulfonamide inhibitor to mice harboring syngeneic breast tumors derived from highly metastatic 4T1 cells that express CAIX in hypoxia resulted in significant inhibition of tumor growth [12]. In contrast, similar treatment of syngeneic breast tumors derived from isogenic non-metastatic, CAIX-negative 67NR cells had no effect, demonstrating that CAIX inhibitors are effective specifically against tumors expressing the target in a biologically relevant context [12]. Furthermore, novel ureido sulfonamides [20] and indanesulfonamides [21] have been used successfully to treat preclinical models of hypoxic, CAIX-positive breast cancer [12] and colorectal cancer [21]. Novel glycosyl coumarin inhibitors of CAIX, which function by a mechanism distinct from that of the sulfonamides, have also been described recently and treatment of human and mouse breast cancer models with this class of inhibitors have resulted in significant inhibition of tumor growth [22, 23]. Taken together, these studies demonstrate, using specific inhibitors of CAIX enzymatic activity, that CAIX functions as an important mediator of the growth of hypoxic tumors and indicate that the anti-tumor effect observed with depletion of CAIX expression involves, at least in part, its enzymatic function.

3 CAIX as a Mediator of Cancer Metastasis

Given that the vast majority of cancer deaths are attributable to metastatic disease, understanding whether CAIX functions as a mediator in cancer metastasis is an important question. Indeed, several studies have evaluated CAIX as a target for treatment of metastatic disease [7]. A review of CAIX as a target of cancer metastasis can be found in Chap. 11 and 12. Here we will briefly discuss the evidence showing that gene depletion or inhibition with small molecules defines CAIX as a mediator in this process. Silencing of CAIX expression by 4T1 breast tumor cells resulted in inhibition of both spontaneous and experimental lung metastases [12]. Treatment of experimental models of breast cancer metastasis with novel ureido sulfonamide and glycosyl coumarin CAIX inhibitors resulted in a significant decrease in lung metastases [12, 20], suggesting that this new generation of selective CAIX inhibitors have the capacity to work as targeted therapeutics for the treatment of both primary tumor growth and metastasis. CAIX-specific inhibitors have also been used to assess spontaneous metastasis. It has been reported that a sulfamate CAIX inhibitor, while not affecting the growth of primary human breast tumors derived from MDA-MB-231 cells, did reduce the metastatic tumor burden in the lung, further demonstrating the capacity of CAIX inhibitors as anti-metastatic agents [19]. Recent data have also demonstrated that an ureido sulfonamide CAIX inhibitor, alone and especially when used in conjunction with conventional chemotherapy, effectively attenuated spontaneous lung metastases in a model of highly metastatic human breast cancer [22]. Moreover, tumor cells obtained from pleural effusions of breast cancer patients, a source of patient-derived metastatic cancer cells, demonstrated high CAIX expression and when these cells were cultured as tumorspheres in hypoxia in the presence of CAIX inhibitors, a decrease in the number of viable cells and impaired growth as aggregates was observed [22], suggesting that inhibition of CAIX is effective against human primary metastatic breast cancer cells.

Thus, from the discussion above, CAIX is not only overexpressed in aggressive, hypoxic tumors, but also is an effector of cancer growth and metastasis, and targeted inhibition of its expression using gene depletion strategies or its enzymatic function using CAIX specific inhibitors clearly and significantly reduces disease burden, demonstrating that its inhibition is a rational strategy for cancer therapy. However, while these in vivo data clearly show that CAIX is functionally important for tumor growth and metastasis, the mechanisms by which CAIX acts to effect these changes are equally important and evoke a number of questions. For example, do the data in cancer cells support the idea that regulation of pH to influence survival are at work in these models? Are alterations in proliferation, migration and invasion involved? Is CAIX expression and activity relevant to the maintenance of cancer stem cells? Does CAIX mediate functions related to chemo and radioresistance? Data supporting these tenants of CAIX function are reviewed and discussed in the following sections.

4 CAIX as a Mediator of Cell Survival and Growth Through Regulation of pH

It is now well-recognized that CAIX is an important component of a complex cellular system involving several proteins and buffer systems dedicated to ensuring the stability of intracellular pH in the presence of increasing extracellular acidosis resulting from the accumulation of byproducts produced by highly metabolic cancer cells engaged in glycolytic metabolism [3, 24]. The inability to control the intracellular pH has adverse consequences for critical cellular processes including ATP synthesis, proliferation, and survival (Fig. 13.1) [24, 25]. It is generally accepted that, through the conversion of CO_2 to bicarbonate and protons, CAIX contributes to the maintenance of a pHi favorable for tumor cell survival and growth, while also participating in the generation of an increasingly acidic extracellular environment, fueling survival and growth as well as breakdown of the extracellular matrix and consequent tumor cell invasion [24, 26]. The finding that depletion of CAIX expression or inhibition of its catalytic activity results in a reduction in tumor growth across multiple types of cancer suggests that CAIX is important for cancer cell survival and growth in hypoxic environments, and may be acting through this mechanism, as discussed below.

Using depletion of CAIX gene expression or inhibition of its catalytic activity in the context of hypoxia to dysregulate pH homeostasis, several studies have established a link between pH regulation by CAIX and the viability of several types of cancer cells in vitro. Silencing of hypoxia-induced CAIX expression in 4T1 breast cancer cells results in the inhibition of acidification of culture medium [12], together with reduced survival of CAIX-depleted cells in hypoxia [12]. Reductions in clonogenic survival and cell growth have also been reported for human breast cancer cells depleted of CAIX [27]. Studies using human tumor cell lines that constitutively express CAIX or induce the enzyme in hypoxia have demonstrated that CAIX-selective inhibitors reduce CAIX-mediated acidification of the extracellular environment [12, 21, 28, 29] or induce acidification of the intracellular pH [30], while negatively impacting cell survival [21, 30]. Recent studies have shown that U251 glioblastoma cells depleted of CAIX expression and grown in hypoxic and glycolytic conditions exhibited decreased ATP levels and reduced cell viability [31]. It should be noted that in some cancer cell types, for example human hepatocellular carcinoma (HCC) cells, neither transient depletion of CAIX expression using siRNA nor treatment with a CAIX inhibitor alone affected cell growth or induced apoptosis in hypoxia [32], indicating that some types of cancer cells may have evolved compensatory mechanisms designed to counter the effects of interference with CAIX function.

Interestingly, recent data derived from cells cultured as monolayers and as spheroids in three dimensions suggest that the impact of CAIX expression on pH homeostasis may be somewhat more complex and intimately links cell survival and cell proliferation in the hypoxic environment. There is also now accumulating data suggesting that CAIX, through its ability to maintain cancer cell survival in hypoxia, may allow cancer cells to sustain a robust proliferative capacity, providing continued tumor growth, but depleting available resources due to increased metabolic demand, ultimately leading to an increase in cell death and necrosis [14]. Silencing of CAIX in HT-29 cells grown in three dimensions reduced cell proliferation, but also reduced the amounts of apoptosis and necrosis [14]. Silencing hypoxia-induced CAIX or CAIX together with CAXII in LS174Tr colorectal cancer cells resulted in an increase in the proportion of cells in the radiosensitive G1/G2/M phases of the cell cycle, together with an increase in markers associated with reduced proliferation [15]. Furthermore, growth of CAIX-depleted and CAIX/CAXII-depleted LS174Tr cells as spheroids also resulted in decreased proliferation and, in the case of dual knockdown, an increase in cell death [15].

Examination of HT-29 xenografts grown in the presence or absence of CAIX has revealed that, while CAIX-depleted tumors grow more slowly, they also exhibit lower levels of necrosis, apoptosis and proliferation. Paradoxically, CAIX expression promotes necrosis, apoptosis and proliferation in spheroid cultures and in tumors in vivo [14], suggesting that the maintenance of cellular pH by CAIX overexpression allows cells to attain a high level of proliferation, thereby increasing the metabolic demand and depleting available resources in the hypoxic environment. However, CAIX expression ultimately results in larger numbers of viable cells, suggesting that the pro-survival and proliferative capacity imparted by CAIX outweighs the increase in apoptosis and necrosis.

5 CAIX as a Mediator of Adhesion, Migration and Invasion

The observation that depletion of CAIX gene expression or inhibition of its catalytic activity results in inhibition of metastasis in vivo suggests that CAIX may function as a mediator of cancer progression through an effect on the adhesive, migratory or invasive properties of cancer cells (Fig. 13.1). Indeed, the role of CAIX in maintaining pH homeostasis in cancer cells serves to increase the acid load in the extracellular space. The development of acidosis within the extracellular environment is thought to then drive local invasion through disruption of the extracellular matrix, activation of metalloproteases and increased cell invasiveness [33]. However, in addition to modulation of the extracellular environment, there is evidence that CAIX mediates adhesion, migration and invasion by directly modulating cancer cell behaviour. Differential gene expression profiling of cervical carcinoma cells transfected with CAIX or fibrosarcoma cells stably depleted of CAIX expression in normoxia have revealed perturbation of pathways related to cell-matrix adhesion [34], as well as cell growth and cytoskeletal organization [34, 35]. Recently, CAIX has been described as a "hypoxia-controlled catalyst of cell migration" [36] and several studies using models of constitutive CAIX expression have demonstrated that CAIX actively contributes to migration and invasion.

Studies using MDCK cells constitutively overexpressing CAIX in normoxia have suggested that CAIX may modulate E-cadherin-mediated cell adhesion through interactions with β-catenin, resulting in reduced adhesion [37]. Furthermore, this cell line exhibited increased scattering and migration properties in the presence of full length CAIX, but not when a CAIX variant lacking the catalytic domain was introduced, linking CAIX enzymatic activity with migration [38]. These studies further showed that CAIX controls cell migration by facilitating ion transport and pH control in the lamellopodia of motile cells, linking migration to CAIX mediated intracellular pH regulation [38]. Cytoskeletal remodelling, weakened cell-cell and cell-matrix adhesion, and increased migration were observed in cervical carcinoma cells overexpressing CAIX in normoxia, and these processes were shown to involve Rho-GTPase signaling in this cell model [34]. Combined knockdown of COX-2 and CAIX in colorectal cancer cells in normoxia resulted in blunted expression of active metalloproteinase-2 (MMP-2) and reduced ECM invasion [39], and a mild hypoxic environment induced by high cell density increased the expression of these genes and induced invasive potential [39]. However, these studies have employed constitutive expression of CAIX in normoxia, leaving open the question of whether CAIX modulates these processes in the context of hypoxia.

Findings from several recent studies now provide evidence for a role of CAIX in mediating cancer cell adhesion, migration and invasion in the context of hypoxia. Knockdown of CAIX expression in hypoxic glioblastoma cells results in reduced cell attachment and invasion in vitro [31]. Stable depletion of CAIX expression by highly metastatic MDA-231 LM2 human breast cancer cells in hypoxia reduced invasion of these cells through matrigel [22]. Hypoxia-mediated upregulation of CAIX expression by HeLa cells and subsequent inhibition of its activity by treatment with CA inhibitors or forced expression of a dominant negative isoform of CAIX resulted in reduced migration [38], suggesting the importance of catalytically active CAIX for migration of cancer cells. In agreement with these data, treatment of highly metastatic human breast cancer cells with ureido sulfonamide inhibitors of CAIX in hypoxia significantly reduced migration through Matrigel in 3 dimensions. Furthermore, a series of sulfamate CAIX inhibitors was evaluated and shown to inhibit the migration of human breast cancer cells cultured in conditions of very low oxygen [19] and re-expression of CAIX by gastric cancer cells in the context of 5-azadeoxycytidine treatment resulted in increased invasion. Thus, CAIX is important in the processes of migration and invasion, a functional phenotype typical of cancer cells having undergone EMT [22]. It should be noted that early studies in a few cancer cell types did not demonstrate a reduction in cell invasion with siRNA-mediated depletion of hypoxia-induced CAIX expression or overexpression of CAIX [27], possibly indicating certain cell type differences.

6 CAIX as a Mediator of CSC Stemness

The hypoxic niche is an environment that drives aggressive tumor cell behaviour and is a well-known source of cancer stem cells (CSCs), a subpopulation of tumor cells capable of self-renewal and important for continued growth and progression of solid malignancies. Furthermore, analysis of CSC markers in patients with invasive breast carcinoma showed that cells positive for stem cell markers, including CD133 and $CD24^{+}/CD24^{-/low}$, are associated with hypoxic, CAIX-positive regions of the tumor [40]. Recently published data now suggest that, in hypoxia, CAIX plays a critical role in the maintenance and stemness properties of CSCs (Fig. 13.1). Stable depletion of CAIX in the 4T1 mouse breast cancer cell model inhibited the hypoxia-driven expansion of the CSC population in tumorspheres, as well as attenuation of EMT and stemness markers, a hallmark of the EMT-mediated mesenchymal phenotype typical of CSCs in hypoxia [22]. Similarly, treatment with CAIX-specific inhibitors blocked the EMT/mesenchymal phenotype, suggesting a critical requirement of hypoxia-induced CAIX in the maintenance of stemness in the breast CSC population [22]. Interestingly, activation of mTORC1 signaling, a pathway known to be regulated by intracellular pH, in hypoxia was dependent on the expression of CAIX, both in cultured cells and in vivo [22], suggesting that CAIX may be an upstream regulator of this pathway. There is also evidence for a functional role of CAIX in maintenance of the CSC population in vivo. Treatment of orthotopic MDA-231 LM2-4 human breast cancer xenografts with either ureido sulfonamide or glycosyl coumarin inhibitors of CAIX significantly reduced the population of EpCAM + CSCs in these tumors, as determined by FACS [22]. Furthermore, immunostaining for ALDH1A3, a complementary marker of CSCs in this tumor type, was reduced in the inhibitor-treated animals [22]. These results suggest the functional maintenance of CSCs by CAIX.

7 Role of CAIX in Radioresistance and Chemoresistance

It is well-recognized that CAIX is most robustly expressed in regions of hypoxia, intratumoral areas often considered privileged microenvironments resistant to conventional chemo- and radiotherapy, and the site of residence of cancer stem cells (CSCs). Both conventional chemotherapeutic drugs and radiotherapy target rapidly dividing cells in well-oxygenated environments. However, these therapies are much less effective at killing tumor cells and CSCs in the hypoxic niche, allowing these cells to evade therapy and ultimately spur tumor growth, often resulting in more aggressive disease. Therefore, development of targeted therapeutics capable of killing cells in regions of hypoxia is critical if the ideal of treatment with curative intent is to be realized. Based on the premise that targeting CAIX will eliminate tumor cells resistant to conventional therapies, thereby producing a greater therapeutic benefit relative to either intervention alone, several studies have now evaluated the use of CAIX-targeted therapies in combination with radio- and chemotherapy. Importantly, these studies serve to demonstrate the ability of CAIX to mediate cancer progression in intratumoral regions protected from the effects of conventional cancer therapy.

The effects of silencing CAIX gene expression or inhibiting its catalytic activity using CAIX-specific small molecule inhibitors in conjunction with ionizing radiation in vivo have been reported in models of colorectal cancer. Silencing of CAIX gene expression, or tandem depletion of CAIX and CAXII expression in LS174Tr colorectal cancer xenografts compromised tumor growth when combined with ionizing radiation, an effect attributed to inhibition of the regulation of intracellular pH by CAIX [15]. Treatment of HT-29 colorectal cancer xenografts with radiotherapy in combination with an indanesulfonamide inhibitor of CAIX improved the therapeutic efficacy compared to single agents alone [21], indicating that CAIX enzymatic activity is important for the radioresistance of these tumors. In addition to these in vivo findings, depletion of hypoxia-induced CAIX expression by glioblastoma cells rendered them sensitive to radiation treatment and resulted in a higher rate of apoptotic cell death [31]. These studies suggest that CAIX is functionally important for imparting radioresistance on hypoxic tumor cells, thus helping to perpetuate tumor growth, and show that inhibition of CAIX can be used to augment the effects of radiotherapy.

While targeted agents are entering the clinic in ever increasing numbers, conventional chemotherapy remains the mainstay for systemic therapy. In fact, targeted therapies are often administered in combination with chemotherapy to increase therapeutic response. Studies have used this paradigm to evaluate CAIX inhibitors in combination with conventional chemotherapy or antiangiogenic therapy. For example, depletion of hypoxia-induced CAIX expression by glioblastoma cells enhanced the cytotoxic effect of temozoloide [31] and the inhibition of CA activity with acetazolamide in cancer cells highly expressing CAIX has been reported to modulate the toxicity of chemotherapeutic agents such as doxorubicin and melphalan [41], indicating that targeting CAIX function in combination with other agents may be effective in reducing tumor growth. Indeed, in vivo administration of CAIX inhibitors in combination with clinically available systemic therapeutic agents has demonstrated greater efficacy compared to single agent treatment. Specifically, combination of an ureido sulfonamide inhibitor of CAIX with paclitaxel provided benefit greater than that seen with either treatment alone in mouse and human orthotopic models of breast cancer [22], demonstrating the functional importance of CAIX catalytic activity in this hypoxic tumor setting. Moreover, it has been suggested recently that the use of antiangiogenic agents such as sunitinib and Bevacizumab result in increased hypoxia, and treatment of human breast cancer xenografts with these antiangiogenic agents was found to increase the population of CSCs through the generation a hypoxic environment [42]. These findings indicate the necessity of targeting additional tumor components concomitant with the use of angiogenic inhibitors. As part of the response to hypoxia triggered by these agents, CAIX is upregulated, offering an additional therapeutic target for inhibiting tumor growth in this context. While initial studies using the CAIX-selective albumin-binding derivative of acetazolamide described above in combination with sunitinib did not significantly outperform sunitinib alone [16], likely because of the high efficacy of sunitinib in the tumor model used, combination of CAIX inhibitors with anti-angiogenic agents effectively limited tumor growth in the HT-29 colorectal cancer model [14], demonstrating that CAIX-specific inhibitors can be used in conjunction with inhibitors of angiogenesis to exact a therapeutic advantage. Taken together, these findings suggest that CAIX is functionally important for the development of hypoxic tumor cell populations resistant to conventional therapies, thereby contributing to continued tumor growth, and that effective inhibition of its activity using CAIX-selective inhibitors can increase treatment efficacy.

8 Summary

As solid malignancies progress, they often become hypoxic, triggering a core, HIF-1-mediated signaling cascade that results in the activation of a plethora of genes vital for the adaptation of tumor cells to an increasingly stressful environment. One facet of this adaptive response is the marked upregulation of CAIX. Importantly, stable depletion of CAIX expression or inhibition of its enzymatic activity using selective small molecule inhibitors in the context of tumor hypoxia in vivo results in inhibition of primary tumor growth and metastasis, demonstrating a functionally active role of CAIX in mediating cancer progression. The regulation of extracellular and intracellular pH by CAIX is central to its ability to modulate a host of tumorigenic and metastatic processes, including cell survival, proliferation and invasion. Furthermore, hypoxia-induced CAIX plays an important role in the maintenance and stemness properties of CSCs, a population of tumor cells critical for continued growth and progression of solid malignancies as well as for resistance of cancers to chemotherapy and radiotherapy. The recognition that CAIX, especially in the biologically relevant context of hypoxia, is an important functional mediator of tumor growth and metastasis provides a solid rationale with which to specifically inhibit this cancer-specific target for therapy in patients, both as a monotherapy and in combination with conventional chemotherapy and radiotherapy.

References

1. Bailey KM, Wojtkowiak JW, Hashim AI, Gillies RJ (2012) Targeting the metabolic microenvironment of tumors. Adv Pharmacol 65:63–107
2. Lendahl U, Lee KL, Yang H, Poellinger L (2009) Generating specificity and diversity in the transcriptional response to hypoxia. Nat Rev Genet 10:821–832
3. Gillies RJ, Robey I, Gatenby RA (2008) Causes and consequences of increased glucose metabolism of cancers. J Nucl Med 49(Suppl 2):24S–42S
4. Brahimi-Horn MC, Bellot G, Pouyssegur J (2011) Hypoxia and energetic tumour metabolism. Curr Opin Genet Dev 21:67–72
5. Kaluz S, Kaluzova M, Liao SY, Lerman M, Stanbridge EJ (2009) Transcriptional control of the tumor- and hypoxia-marker carbonic anhydrase 9: a one transcription factor (HIF-1) show? Biochim Biophys Acta 1795:162–172
6. McDonald PC, Winum JY, Supuran CT, Dedhar S (2012) Recent developments in targeting carbonic anhydrase IX for cancer therapeutics. Oncotarget 3:84–97
7. Gieling RG, Williams KJ (2013) Carbonic anhydrase IX as a target for metastatic disease. Bioorg Med Chem 21:1470–1476
8. Neri D, Supuran CT (2011) Interfering with pH regulation in tumours as a therapeutic strategy. Nat Rev Drug Discov 10:767–777
9. Wilson WR, Hay MP (2011) Targeting hypoxia in cancer therapy. Nat Rev Cancer 11:393–410
10. Supuran CT (2012) Inhibition of carbonic anhydrase IX as a novel anticancer mechanism. World J Clin Oncol 3:98–103
11. Supuran CT (2008) Carbonic anhydrases: novel therapeutic applications for inhibitors and activators. Nat Rev Drug Discov 7:168–181
12. Lou Y, McDonald PC, Oloumi A, Chia S, Ostlund C, Ahmadi A, Kyle A, Auf dem Keller U, Leung S, Huntsman D, Clarke B, Sutherland BW, Waterhouse D, Bally M, Roskelley C, Overall CM, Minchinton A, Pacchiano F, Carta F, Scozzafava A, Touisni N, Winum JY, Supuran CT, Dedhar S (2011) Targeting tumor hypoxia: suppression of breast tumor growth and metastasis by novel carbonic anhydrase IX inhibitors. Cancer Res 71:3364–3376
13. Chiche J, Ilc K, Laferriere J, Trottier E, Dayan F, Mazure NM, Brahimi-Horn MC, Pouyssegur J (2009) Hypoxia-inducible carbonic anhydrase IX and XII promote tumor cell growth by counteracting acidosis through the regulation of the intracellular pH. Cancer Res 69:358–368

14. McIntyre A, Patiar S, Wigfield S, Li JL, Ledaki I, Turley H, Leek R, Snell C, Gatter K, Sly WS, Vaughan-Jones RD, Swietach P, Harris AL (2012) Carbonic anhydrase IX promotes tumor growth and necrosis in vivo and inhibition enhances anti-VEGF therapy. Clin Cancer Res 18:3100–3111
15. Doyen J, Parks SK, Marcie S, Pouyssegur J, Chiche J (2012) Knock-down of hypoxia-induced carbonic anhydrases IX and XII radiosensitizes tumor cells by increasing intracellular acidosis. Front Oncol 2:199
16. Ahlskog JK, Dumelin CE, Trussel S, Marlind J, Neri D (2009) In vivo targeting of tumor-associated carbonic anhydrases using acetazolamide derivatives. Bioorg Med Chem Lett 19:4851–4856
17. Dubois L, Lieuwes NG, Maresca A, Thiry A, Supuran CT, Scozzafava A, Wouters BG, Lambin P (2009) Imaging of CA IX with fluorescent labelled sulfonamides distinguishes hypoxic and (re)-oxygenated cells in a xenograft tumour model. Radiother Oncol 92:423–428
18. Buller F, Steiner M, Frey K, Mircsof D, Scheuermann J, Kalisch M, Buhlmann P, Supuran CT, Neri D (2011) Selection of carbonic anhydrase IX inhibitors from one million DNA-encoded compounds. ACS Chem Biol 6:336–344
19. Gieling RG, Babur M, Mamnani L, Burrows N, Telfer BA, Carta F, Winum JY, Scozzafava A, Supuran CT, Williams KJ (2012) Antimetastatic effect of sulfamate carbonic anhydrase IX inhibitors in breast carcinoma xenografts. J Med Chem 55:5591–5600
20. Pacchiano F, Carta F, McDonald PC, Lou Y, Vullo D, Scozzafava A, Dedhar S, Supuran CT (2011) Ureido-substituted benzenesulfonamides potently inhibit carbonic anhydrase IX and show antimetastatic activity in a model of breast cancer metastasis. J Med Chem 54:1896–1902
21. Dubois L, Peeters S, Lieuwes NG, Geusens N, Thiry A, Wigfield S, Carta F, McIntyre A, Scozzafava A, Dogne JM, Supuran CT, Harris AL, Masereel B, Lambin P (2011) Specific inhibition of carbonic anhydrase IX activity enhances the in vivo therapeutic effect of tumor irradiation. Radiother Oncol 99:424–431
22. Lock FE, McDonald PC, Lou Y, Serrano I, Chafe SC, Ostlund C, Aparicio S, Winum J-Y, Supuran CT, Dedhar S (2012) Targeting carbonic anhydrase IX depletes breast cancer stem cells within the hypoxic niche. Oncogene. doi:10.1038/onc.2012.1550
23. Touisni N, Maresca A, McDonald PC, Lou Y, Scozzafava A, Dedhar S, Winum JY, Supuran CT (2011) Glycosyl coumarin carbonic anhydrase IX and XII inhibitors strongly attenuate the growth of primary breast tumors. J Med Chem 54:8271–8277
24. Parks SK, Chiche J, Pouyssegur J (2011) pH control mechanisms of tumor survival and growth. J Cell Physiol 226:299–308
25. Gatenby RA, Gillies RJ (2008) A microenvironmental model of carcinogenesis. Nat Rev Cancer 8:56–61
26. Swietach P, Hulikova A, Vaughan-Jones RD, Harris AL (2010) New insights into the physiological role of carbonic anhydrase IX in tumour pH regulation. Oncogene 29:6509–6521
27. Robertson N, Potter C, Harris AL (2004) Role of carbonic anhydrase IX in human tumor cell growth, survival, and invasion. Cancer Res 64:6160–6165
28. Li Y, Wang H, Oosterwijk E, Tu C, Shiverick KT, Silverman DN, Frost SC (2009) Expression and activity of carbonic anhydrase IX is associated with metabolic dysfunction in MDA-MB-231 breast cancer cells. Cancer Invest 27:613–623
29. Dubois L, Douma K, Supuran CT, Chiu RK, van Zandvoort MA, Pastorekova S, Scozzafava A, Wouters BG, Lambin P (2007) Imaging the hypoxia surrogate marker CA IX requires expression and catalytic activity for binding fluorescent sulfonamide inhibitors. Radiother Oncol 83:367–373
30. Cianchi F, Vinci MC, Supuran CT, Peruzzi B, De Giuli P, Fasolis G, Perigli G, Pastorekova S, Papucci L, Pini A, Masini E, Puccetti L (2010) Selective inhibition of carbonic anhydrase IX decreases cell proliferation and induces ceramide-mediated apoptosis in human cancer cells. J Pharmacol Exp Ther 334:710–719
31. Proescholdt MA, Merrill MJ, Stoerr EM, Lohmeier A, Pohl F, Brawanski A (2012) Function of carbonic anhydrase IX in glioblastoma multiforme. Neuro Oncol 14:1357–1366

32. Yu SJ, Yoon JH, Lee JH, Myung SJ, Jang ES, Kwak MS, Cho EJ, Jang JJ, Kim YJ, Lee HS (2011) Inhibition of hypoxia-inducible carbonic anhydrase-IX enhances hexokinase II inhibitor-induced hepatocellular carcinoma cell apoptosis. Acta Pharmacol Sin 32:912–920
33. Estrella V, Chen T, Lloyd M, Wojtkowiak J, Cornnell HH, Ibrahim-Hashim A, Bailey K, Balagurunathan Y, Rothberg JM, Sloane BF, Johnson J, Gatenby RA, Gillies RJ (2013) Acidity generated by the tumor microenvironment drives local invasion. Cancer Res 73:1524–1535
34. Shin HJ, Rho SB, Jung DC, Han IO, Oh ES, Kim JY (2011) Carbonic anhydrase IX (CA9) modulates tumor-associated cell migration and invasion. J Cell Sci 124:1077–1087
35. Radvak P, Repic M, Svastova E, Takacova M, Csaderova L, Strnad H, Pastorek J, Pastorekova S, Kopacek J (2013) Suppression of carbonic anhydrase IX leads to aberrant focal adhesion and decreased invasion of tumor cells. Oncol Rep 29:1147–1153
36. Svastova E, Pastorekova S (2013) Carbonic anhydrase IX: a hypoxia-controlled "catalyst" of cell migration. Cell Adh Migr 7:226–231
37. Svastova E, Zilka N, Zat'ovicova M, Gibadulinova A, Ciampor F, Pastorek J, Pastorekova S (2003) Carbonic anhydrase IX reduces E-cadherin-mediated adhesion of MDCK cells via interaction with beta-catenin. Exp Cell Res 290:332–345
38. Svastova E, Witarski W, Csaderova L, Kosik I, Skvarkova L, Hulikova A, Zatovicova M, Barathova M, Kopacek J, Pastorek J, Pastorekova S (2012) Carbonic anhydrase IX interacts with bicarbonate transporters in lamellipodia and increases cell migration via its catalytic domain. J Biol Chem 287:3392–3402
39. Sansone P, Piazzi G, Paterini P, Strillacci A, Ceccarelli C, Minni F, Biasco G, Chieco P, Bonafe M (2009) Cyclooxygenase-2/carbonic anhydrase-IX up-regulation promotes invasive potential and hypoxia survival in colorectal cancer cells. J Cell Mol Med 13:3876–3887
40. Currie MJ, Beardsley BE, Harris GC, Gunningham SP, Dachs GU, Dijkstra B, Morrin HR, Wells JE, Robinson BA (2013) Immunohistochemical analysis of cancer stem cell markers in invasive breast carcinoma and associated ductal carcinoma in situ: relationships with markers of tumor hypoxia and microvascularity. Hum Pathol 44:402–411
41. Gieling RG, Parker CA, De Costa LA, Robertson N, Harris AL, Stratford IJ, Williams KJ (2013) Inhibition of carbonic anhydrase activity modifies the toxicity of doxorubicin and melphalan in tumour cells in vitro. J Enzyme Inhib Med Chem 28:360–369
42. Conley SJ, Gheordunescu E, Kakarala P, Newman B, Korkaya H, Heath AN, Clouthier SG, Wicha MS (2012) Antiangiogenic agents increase breast cancer stem cells via the generation of tumor hypoxia. Proc Natl Acad Sci U S A 109:2784–2789

Chapter 14
Carbonic Anhydrases and Brain pH in the Control of Neuronal Excitability

Eva Ruusuvuori and Kai Kaila

Abstract H^+ ions are remarkably efficient modulators of neuronal excitability. This renders brain functions highly sensitive to small changes in pH which are generated "extrinsically" via mechanisms that regulate the acid–base status of the whole organism; and "intrinsically", by activity-induced transmembrane fluxes and *de novo* generation of acid–base equivalents. The effects of pH changes on neuronal excitability are mediated by diverse, largely synergistically-acting mechanisms operating at the level of voltage- and ligand-gated ion channels and gap junctions. In general, alkaline shifts induce an increase in excitability which is often intense enough to trigger epileptiform activity, while acidosis has the opposite effect. Brain pH changes show a wide variability in their spatiotemporal properties, ranging from long-lasting global shifts to fast and highly localized transients that take place in subcellular microdomains. Thirteen catalytically-active mammalian carbonic anhydrase isoforms have been identified, whereof 11 are expressed in the brain. Distinct CA isoforms which have their catalytic sites within brain cells and the interstitial fluid exert a remarkably strong influence on the dynamics of pH shifts and, consequently, on neuronal functions. In this review, we will discuss the various roles of H^+ as an intra- and extracellular signaling factor in the brain, focusing on the effects mediated by CAs. Special attention is paid on the developmental expression patterns and actions of the neuronal isoform, CA VII. Studies on the

Susan C. Frost and Robert McKenna (eds.). Carbonic Anhydrase: Mechanism, Regulation, Links to Disease, and Industrial Applications

E. Ruusuvuori
Department of Biosciences, University of Helsinki, Helsinki, Finland
e-mail: Eva.Ruusuvuori@Helsinki.fi

K. Kaila (✉)
Department of Biosciences and Neuroscience Center, University of Helsinki, Helsinki, Finland
e-mail: Kai.Kaila@Helsinki.fi

S.C. Frost and R. McKenna (eds.), *Carbonic Anhydrase: Mechanism, Regulation, Links to Disease, and Industrial Applications*, Subcellular Biochemistry 75,
DOI 10.1007/978-94-007-7359-2_14,

various functions of CAs will shed light on fundamental mechanisms underlying neuronal development, signaling and plasticity; on pathophysiological mechanisms associated with epilepsy and related diseases; and on the modes of action of CA inhibitors used as CNS-targeting drugs.

Keywords $GABA_A$ receptor • Hippocampus • pH buffering • Neuronal development • Brain diseases • KCC2

1 Introduction

pH exerts a strong modulatory effect on the central nervous system (CNS) function and excitability. Changes in intracellular or extracellular pH (pH_i and pH_o, respectively) of 0.5 units or less are often sufficient to trigger or suppress paroxysmal activity and, accordingly, much smaller changes are needed for subtle modulation of neuronal excitability. The physiologically relevant pH range (pH 6.5–8.0) corresponds to a very low free H^+ concentration, from 10 to 300 nM. An interesting aspect is that this applies to both the intra- and extracellular compartments. Protons are thus eminently suited to affect H^+-sensitive targets both within and outside brain cells. However, studying the physiological and pathophysiological bases of H^+-modulation of neuronal functions is not a trivial task, because global and local pH transients are generated by multiple mechanisms operating at various levels of biological organization, from the whole organism to cellular and subcellular microdomains.

At the whole-organism level, the key elements in pH regulation are the lungs which control the partial pressure of CO_2 (P_{CO2}) in the blood, and the kidneys which are responsible for the net regulation of other important acid–base species, especially HCO_3^- and NH_4^+. With the major exception of chemosensitive neurons controlling breathing [1], the excitability of most central neurons and neuronal networks is enhanced by an alkalosis and suppressed by an acidosis. Exogenously-induced respiratory acidosis has a profound suppressing action on neuronal excitability and on seizures [2–5]. Respiratory alkalosis generated by hyperventilation is a standard technique used in the clinic for the precipitation of petit mal-type seizures [6]. Hyperventilation is also involved in the generation of febrile seizures in animal models [4] and most likely in children as well [7]. Metabolic alkalosis associated with renal dysfunction such as seen in the EAST syndrome is known to cause epileptiform activity [8].

The brain is protected by the blood–brain barrier (BBB) which is endowed by acid–base transporter molecules [9] and, as a diffusion barrier, prevents charged acid–base species from having direct access to brain interstitial fluid. Recent work has shown that in addition to this protective role, acid extrusion triggered by birth asphyxia across the BBB can lead to a brain-confined metabolic alkalosis and to consequent seizures [10, 11].

At the cellular level, pH_i regulation is based on plasmalemmal transporters of neurons and glia [12, 13]. A fascinating aspect of local H^+-signaling within the brain is that fast, robust and often highly localized pH shifts are evoked by electrical activity and by synaptic transmission [13–15]. These intrinsic shifts in pH_i and pH_o are largely generated by channel or transporter-mediated transmembrane fluxes of acid–base equivalents, and by accumulation of acid end products of energy metabolism such as CO_2 and lactate. In the former case, the transmembrane shifts of acid–base species generate pH changes of opposite direction within and outside neurons, while metabolic acidosis implies a fall of pH in both compartments. Distinct neuronal populations show a large heterogeneity in transporter and channel localization and expression [16], and intrinsic pH shifts are therefore likely to generate spatially restricted extra- and intracellular pH-microdomains.

By definition, the H^+-sensitive targets involved in pH-dependent modulation of neuronal activity consist of charged groups in proteins which are capable of binding and releasing H^+ ions in the physiologically and pathophysiologically relevant pH range. Such interactions affect the conformation and functional properties of a wide variety of membrane proteins involved in neuronal signaling, including voltage gated ion-channels [17, 18], $GABA_A$ receptors ($GABA_ARs$) [19, 20], N-methyl-D-aspartate receptors (NMDAR) [21, 22], gap junctions [23] and pH-sensing cation channels such as the acid-sensing ion channels [24] and TWIK-related acid-sensitive K^+ channels [25]. The pH sensitivity of ion channels and other key proteins that control neuronal excitability does not reflect a property common to *all* kinds of proteins. Rather, the specific, functionally synergistic patterns of "tuning" of the pK_a values of molecules underlying the pH-modulation of neuronal signalling suggests an evolutionary origin for the diverse but largely synergistic roles of H^+ as an *inter*cellular and *intra*cellular signalling agent in the brain.

Carbonic anhydrases (CAs) are a family of molecules with a key role in the control of pH at level of the whole organism (e.g. respiratory, energy-metabolic and renal functions), in the BBB, in neurons and glia, and in the interstitial fluid in the brain. For the physiologically ubiquitous CO_2/HCO_3^- buffering to act in a fast manner, the (de)hydration of CO_2 must be catalyzed by CA [26]. The 13 catalytically active CA isoforms identified so far in mammals differ in their tissue distribution, subcellular localization as well as in their enzymatic activity [27] providing a versatile molecular machinery for the modulation of pH. The acid–base equivalents that serve as substrates in the CO_2 dehydration-hydration reaction are also engaged in many carrier- and channel-mediated ion movements. In such processes, CA activity is in a key position to modulate transmembrane solute fluxes and their influence on local pH.

Recent findings further suggest that CAs, even if catalytically inactive, can act as 'proton collecting antennas' thereby increasing net transmembrane proton flux and suppressing the formation of H^+ microdomains [28]. CAs can also affect neuronal function in a manner not dependent on catalytic activity as shown in studies on mice devoid of isoform VIII [29]. Together with isoforms X and XI, isoform VIII belongs to the carbonic anhydrase related proteins (CARPs) that lack catalytic activity [27]. Mice with spontaneous mutation *Car8* show changes in e.g. the morphology and function of excitatory synapses in the cerebellum [30, 31].

Currently there are no pharmacological tools available that could be used for isoform-specific inhibition of CAs (see also Chap. 15 in this book). Hence, studies using genetic disruption of distinct CAs have provided much insight into the functional and spatial roles of specific isoforms. As will be discussed, mice devoid of the cytosolic CA II and VII and of the membrane attached isoforms IV and XIV as well as the double knock-outs of CA IV/XIV and CA II/VII have been used in studies focusing on the CA-dependent modulation of neuronal signaling [32–34].

The aim of the present chapter is to provide a general overview of the mechanisms and consequences of pH-mediated signalling in the brain, with an emphasis on the role of various CA isoforms. Despite the obvious, vast potential for elucidating novel physiological and pathophysiological mechanisms involved in of fundamental brain functions such as synaptic transmission and control of neuronal excitability, relatively little work has been done in this field of research. We hope that this review will act as a source of inspiration for further work on the diverse pH-sensitive and CA-dependent mechanisms that operate at the molecular, cellular and neuronal network level in the brain.

2 Generation and Maintenance of the Plasmalemmal pH Gradient

Before discussing the CA-dependent modulation of neuronal excitability in more detail, some basic aspects of pH homeostasis need to be addressed.

Passive equilibration of H^+ across the plasma membrane of a cell with a membrane potential at −60 mV and a pH_o of 7.3 would drive intracellular pH close to 6.3. However, neuronal pH_i (typically around 7.1) is only slightly more acidic than pH_o, which implies active regulation of pH_i by membrane-located acid–base transporters. Provided that the hydration-dehydration reaction of CO_2 and the transmembrane distribution of CO_2 are at equilibrium, the transmembrane HCO_3^- distribution is set by the pH gradient:

$$[HCO_3^-]_i = 10^{(pH_i - pH_o)} \times [HCO_3^-]_o$$

Under these conditions, the equilibrium potential of protons (E_H) and bicarbonate (E_{HCO3}) are equal, with a value of −12 mV set by the pH_o and pH_i given above [35]. The energy required for maintaining the electrochemical gradient (in our example of about −50 mV) is spent on combating intracellular acid loading that is generated by three fundamental mechanisms:

(i). Net transmembrane influx of acid equivalents by transporters working as "acid loaders", such as the Ca^{2+}/H^+ ATPase and the Cl^-/HCO_3^- exchanger.

(ii). All conductive pathways which are permeable for charged acid–base species. The concentration of H^+ ions is very low, and directly-measurable proton conductances have not been described in mammalian neurons [36]. However, a significant acid-loading HCO_3^- conductance is provided by $GABA_A$Rs and by glycine receptors.

(iii). Cellular metabolic processes that lead to *de novo* production of acid. Here, one should note, however, that weak organic acids such as lactate traverse the membrane in their neutral, H^+-bound form, and thus their generation by energy metabolism will not contribute to the long-term cellular acid–base budget.

Perturbations of pH_i are not always poised in the acid direction; alkaline loads are also known to take place following e.g. depolarization of the plasma membrane (especially in astrocytes [37]) or sudden removal of an acid load. When a cell is subject to an acid or alkaline load, the rate of pH_i change is proportional to the difference between the acid-extrusion and acid-loading rates, and inversely proportional to the total intracellular buffering capacity. A number of cells, including neurons, are equipped with several acid–base transporters which act as acid extruders or loaders. At first sight, such "push-pull" mechanisms look wasteful in terms of energy usage, but their concerted action brings about a much more stable set-point for pH_i under physiological conditions where the cell is subject to rapidly alternating acid and alkaline loads [12]. Moreover, a differential distribution of transporters in cells with complex geometry, such as neurons, is likely to bring about pH microdomains within the cell, thus enhancing the spatial precision of H^+ ions in intracellular signaling based on pH-sensitive proteins (see Introduction).

In the mammalian CNS, the predominant transporters involved in pH_i regulation are the secondary-active transporters that belong to the solute carrier gene families *Slc4* and *Slc9* [13, 16]. Some studies have reported acid extrusion in the nominal absence of Na^+ and CO_2/HCO_3^- suggesting that a putative H^+ pump contributes to neuronal and glial pH_i regulation [38–40]. The role of another primary active transporter, the Ca^{2+}/H^+-ATPase, has been described in much more detail. In neurons Ca^{2+}/H^+-ATPase, a major regulator in intracellular free calcium, works as an acid loader [41–43].

3 pH Buffering and CA Isoforms in Brain Tissue

3.1 pH Buffering Within and Outside Neurons

While transporters are needed for the active extrusion of acid–base equivalents, H^+ buffers determine the ability of the cytosol to suppress pH_i transients without any contribution by active transport. The total intracellular, cytoplasmic buffering capacity consists of a CO_2/HCO_3^- -dependent (β_{CO2}) and a non-bicarbonate buffering capacity (β_i). The latter mainly arises from phosphates and the imidazole groups of proteins. These buffers cannot cross the plasma membrane and therefore they form a *closed buffer system* within the cell [12, 44]. The extracellular fluid is practically devoid of non-bicarbonate buffers and thus relies on CO_2/HCO_3^- - dependent buffering.

In an ideal buffer which is open with respect to CO_2, β_{CO2} is given by $\beta_{CO2} = 2.3[HCO_3^-]$ [12]. However, attaining this value would require instantaneous equilibration of the system but, in reality, this is not achieved and β_{CO2} remains much lower than the theoretical maximum in response to fast acid/base perturbations both within and outside cells. For instance, in the hippocampal slice, stimulation-induced changes in pH_o indicated an extracellular buffering power that was less than 30 % of the theoretical maximum [45]. Interestingly, despite the presence of extracellular CA (CA_o), the amount of CA_o activity can also be rate-limiting for effective buffering of pH_o changes. Addition of CA II to the perfusion medium has been shown to curtail activity-induced extracellular alkalosis in brain slices [46, 47] and *in vivo* [48].

3.2 CA Isoforms with an Extracellular Catalytic Site

Since pH_o buffering is determined by the CO_2/HCO_3^- system, CA_o is in a key position to govern the kinetics of activity-generated pH_o transients. The membrane-bound CA isoforms IV and XIV which have their catalytic site located in the extracellular space are largely responsible for the CA_o activity detected in the rodent hippocampus [32, 49]. These isoforms differ in the way they are attached to the membrane and in their cell type-specific expression. CA IV is attached to plasma membrane by a glycosyl-phosphatidyl-inositol anchor [50] of both neurons and glia [51]. The more recently identified CA XIV has a membrane-spanning α-helix and a short intracellular C-terminus [52], and shows neuron-specific expression within the brain [53]. The possible contribution of the other membrane-attached isoforms, CA IX, XII and XV, in the CNS extracellular buffering is unclear as the regional localization of these isoforms has not been determined. The basal expression level of CA XII and IX in the CNS is low, but both isoforms are expressed at higher levels in malignant tumor cells [54, 55]. Their presence can be used as biomarkers for certain tumors with possible further diagnostic implications in the prognosis of malignization [56] (see also Chaps. 10, 11, 12, and 13 in this Book). Pathophysiological conditions such as seizures [57] and asphyxia [58] increase the expression of CA IV and CA XII also in brain cells with no apparent previous pathologies. This is expected to suppress the activity-dependent postsynaptic rise in pH_o and consequent NMDAR activation during synchronous neuronal activity, thus acting as a potential neuroprotective mechanism.

A wealth of data has shown that the developmental expression patterns of ion transporters, especially of cation-chloride cotransporters, have a major influence on the fundamental properties of neuronal signalling during brain ontogeny [59]. However, with respect to CAs, data of this kind is largely missing and there are no published reports on developmental changes of CA_o activity. Developmental expression patterns of CA_o isoforms might shape excitatory and inhibitory transmission in concert with the expression of glutamate as well as $GABA_AR$ subunits and the Cl^- transporters. For example, the expression of functional NMDARs precedes that of AMPA receptors (AMPARs) during postnatal maturation of rodent cortical structures [60]. Developmentally coinciding upregulation of functional AMPARs and CA_o activity would provide control over the NMDAR-modulating pH_o transients [21, 22] generated by excitatory transmission.

The contribution of the Cl^--HCO_3^- transporting anion exchangers (AEs) in neuronal Cl^- regulation has so far gained surprisingly little attention. In embryonic motoneurons the Cl^--HCO_3^- exchanger isoform 3 (AE3) acts as an important Cl^- uptake mechanism [61]. Since CA_o activity has been shown to enhance AE3 mediated Cl^--HCO_3^- exchange [49], developmental changes in CA_o activity might have a significant influence on GABAergic synaptic signalling.

3.3 CA Isoforms with a Cytoplasmic Catalytic Site

For long the intracellular CA (CA_i) activity in the CNS was thought to be mainly restricted to glial cells, endothelium of the capillaries and choroid plexus epithelial cells [62]. Now there is both functional and molecular biological evidence for the presence of intraneuronal CA in the mammalian CNS [63–70]. The first observations on the presence of cytosolic CA activity in CNS neurons, and that this activity promotes $GABA_AR$-mediated net HCO_3^- efflux in mammalian CNS neurons were made by us two decades ago [64].

Due to the lack of isoform-specificity of available cytosolic CA inhibitors, previous data on CA VII expression in rat pyramidal neurons [70] did not exclude the possible presence of other neuronal CA isoforms. Using a novel CA VII KO mouse together with a CA II KO and a CA II/VII double KO mouse we demonstrated that there is a sequential expression of *two* different isoforms in mouse pyramidal neurons (Fig. 14.1a, b) [34]. CA VII fully accounts for the up-regulation of neuronal CA activity detected at around P10 and is the only cytosolic isoform during the time window P10-18. After P18 pyramidal neurons start to express CA II in parallel with CA VII. A notable difference in the cellular expression patterns of the two isoforms was that CA VII is mainly found in the CNS where it localizes only to neurons. The ubiquitous CA II is present in a wide variety of tissues [71], and within the brain parenchyma it is expressed in both glia and neurons.

A possible explanation for the apparently redundant presence of two cytosolic CAs is that the two isoforms show differences in their biochemical functions other than (de)hydration of CO_2, such as esterase/phosphatase activity [72] or as oxygen radical scavengers [73]. The importance of the latter finding is underscored by the fact that a large developmental increase in cerebral oxidative energy

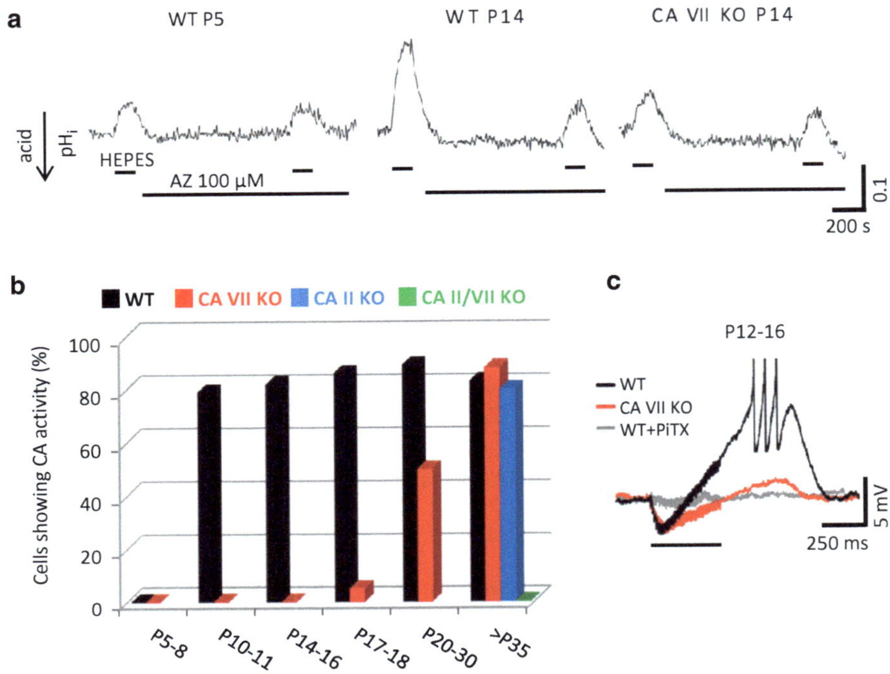

Fig. 14.1 Cytosolic CA activity in developing mouse CA1 pyramidal neurons is based on sequential expression of isoforms VII and II. (**a**) Original pH_i traces from WT (P5 and P14) and CA VII KO neurons on postnatal day 14 (P14). Replacing the CO_2/HCO_3^- buffer in the perfusion solution by HEPES (upper horizontal bars) evoked an acetazolamide-sensitive intracellular alkalinization only in the P14 WT neurons, thereby indicating the presence of CA_i activity (acetazolamide, AZ, 100 μM). (**b**) Developmental expression of the CA VII and CA II isoforms. Summary of the results obtained using the cytosolic CA activity detection method shown in a and quantified as the percentage of the cells showing cytosolic CA activity. (**c**) At P12–16, CA VII is solely responsible for promoting $GABA_A$R-mediated Cl^- accumulation and consequent depolarizing GABA responses. Intense stimulation of the interneuronal network evoked a biphasic GABAergic response in CA1 pyramidal neurons in WT slices that was abolished with picrotoxin (PiTX, 80 μM). The $GABA_A$-receptor mediated depolarization was large enough to trigger action potentials in WT but not in the CA VII KO neurons. Figure modified from ref. [34]

metabolism [74] coincides with the upregulation of neuronal CA VII expression. The existing data do not exclude the possibility that CA II and CA VII are located in distinct subcellular microdomains. Formation of isoform-specific metabolons with different acid–base transporters [75–77] would further promote the generation of developmentally and spatially distinct pH_i microdomains.

There are very few reports on intracellular CA expression changes after pathophysiological insults [57, 58] and, to our knowledge, none on CA activity changes. The co-operative functions of K-Cl cotransporter KCC2 and CA_i in the generation of HCO_3^- -dependent depolarizing GABA responses [78] (see also Sect. 5) raise the intriguing question whether changes in CA_i expression might take place in parallel to those of KCCs [59]. Parallel down-regulation of KCC2 and CA_i would lead to the suppression of excitatory GABAergic response (see also Fig. 14.1c) and, consequently, to suppression of seizures.

Here it is worth recalling that membrane-permeant CA blockers such as acetazolamide have a long history as antiepileptic compounds. The molecular targets and mechanisms of action of these broad-spectrum CA inhibitors at the neuronal network level are still poorly understood [27, 79].

4 Mechanisms and Consequences of Activity-induced H^+ Transients in Neurons and Neuronal Networks

Most experiments on activity-induced neuronal pH transients have been conducted using ion-selective electrodes. In mammalian brain tissue practically all data are on pH_o because of the obvious difficulties of impaling the neurons and maintaining them functionally intact. When evaluating these data, it is important to note that electrode recordings of pH_o are bound to reflect a spatiotemporal average because of the local damage caused by the tip and the relatively slow response time (at best in the range 0.1–1 s) of the electrodes [80].

Fluorescent pH indicators are optimally suited for intracellular recordings, and in this case, the signal-to-noise ratio will largely dictate the data sampling rate and set the temporal resolution. Interestingly, it seems to be possible to obtain useful recordings of pH_o transients occurring in the extremely narrow (around 20 nm) extracellular space in brain tissue using pH indicators. Such recordings have also shown that activity-induced pH_o changes are, indeed, fast enough to be able to modulate on-going synaptic transmission and neuronal activity [46].

Robust activity-induced (i.e., intrinsic) pH changes have been documented in a large number of studies on the CNS indicating that local pH shifts are a mandatory consequence of both electrical and synaptic signaling [13–15]. These pH transients are of different duration and magnitude, and they arise in the intracellular compartments of both neurons and glia as well as in the interstitial fluid surrounding them. It is evident that, depending on the cellular cytoarchitecture and synaptic connectivity of a given neuronal preparation/recording site and on the stimulation paradigm, the measured pH transients originate from distinct molecular and cellular sources. The activation patterns of multisynaptic circuits with both excitatory and inhibitory connections and the fractional volumes of neurons, glia and the extracellular space are among the key factors that shape local pH shifts. Thus, we have focused on studies made in the rodent hippocampus. Comprehensive overviews on activity-dependent pH-modulation in invertebrates, in other cell types and areas of the CNS are available [13, 81].

Intraneuronal pH measurements in rat hippocampal neurons have shown that both excitatory (glutamatergic) and inhibitory (GABAergic) signaling result in a fall of pH_i, but the underlying molecular generation mechanisms are completely

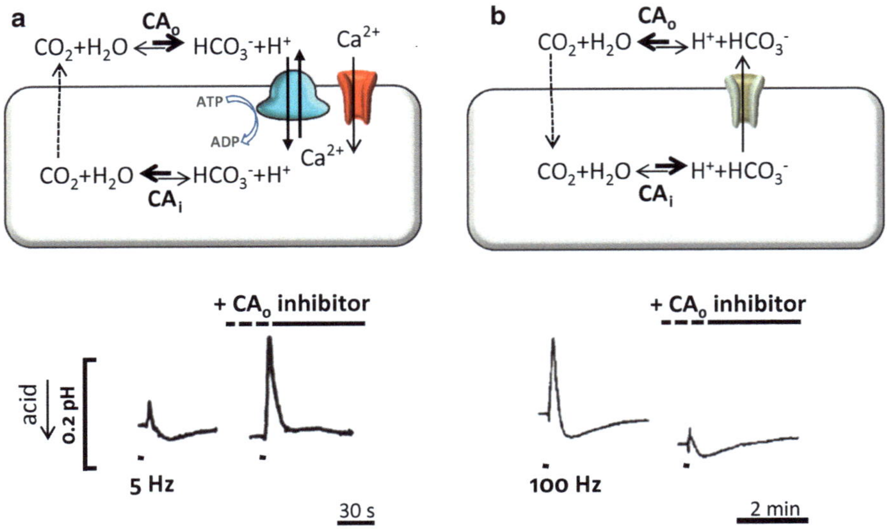

Fig. 14.2 Extracellular carbonic anhydrase activity (CA_o) modulates activity-dependent alkaline shifts. (**a**) The buffering provided by the CA_o-catalyzed CO_2/HCO_3 system attenuates the rapid alkaline pH_o transient generated by neuronal excitation-induced Ca^{2+} influx (*upper panel*). Inhibition of CA_o compromises CO_2/HCO_3 buffering by suppressing the rate of CO_2 hydration and, hence, boosts the alkalosis. (**b**) In contrast, CA_o activity is needed for the generation of $GABA_A$-receptor mediated alkalosis which is driven by the transmembrane CO_2/HCO_3 shuttle (*upper panel*). CA_o inhibitors prevent the fast replenishment of CO_2 in the extracellular space and largely block the alkalosis (for details, see text). The pH_o responses evoked by action potentials (lower panel in a) and stimulation of GABAergic interneurons (lower panel in b) were recorded using ion-sensitive microelectrodes. The small acid shift in the baseline pH_o after CA_o inhibition is likely due to a decrease in extracellular buffering capacity in face of continuous cellular acid extrusion. The illustrations in the lower panels are modified from ref. [90] (**a**) and ref. [43] (**b**)

different. As will be explained, activation of the anion-selective $GABA_A$Rs leads to a net efflux of HCO_3^- and influx of CO_2 which fully explains the GABA-induced intracellular acidosis as depicted in Fig. 14.2b [64, 82]. Because glutamate-gated postsynaptic channels are cation-selective, one might assume that conductive H^+ would lead to an intracellular acid load. However, this possibility has been excluded [83]. A key player in this context is the Ca^{2+}/H^+-ATPase which is activated by neuronal depolarization leading to an increase in intracellular Ca^{2+}, regardless of whether this is caused by synaptic excitation; application of glutamate agonists; or antidromic stimulation of neurons (Fig. 14.2a) in the continuous presence of blockers of synaptic transmission [83–85]. As might be expected, metabolic acid production is likely to contribute to excitation-linked pH shifts [86].

4.1 CA_o-Inhibitors as Tools in Mechanistic Analyses of pH_o Transients

In the presence of CA, the CO_2/HCO_3^- -system efficiently *attenuates* fast pH changes evoked by H^+ fluxes [26] (Fig. 14.2a). On the other hand, if the acid/base disturbance arises from a change in CO_2 and HCO_3^-, CA activity is instrumental for the *generation* of a rapid pH shift (Fig. 14.2b). A prime example of the latter case is the $GABA_AR$-mediated CO_2/HCO_3^- shuttle, which leads to a fall in the CO_2 concentration on the extracellular surface of the plasma membrane that is proportionally much higher than the increase in the local HCO_3^- concentration. Accordingly, inhibition of CA_o activity *abolished* the $GABA_AR$-mediated alkaline pH transient on the surface of crayfish muscle fibres [87].

Taking advantage of the opposing effects of CA_o-activity on the magnitude of pH transients of distinct origin, inhibitors of CA_os can be used as diagnostic tools in the analysis of mechanistically heterogeneous pH_o shifts triggered by simultaneous excitatory and inhibitory transmission, as will be discussed below. Selective CA_o inhibition can be achieved by the poorly-membrane permeable blocker benzolamide or by membrane-impermeant dextrane-bound sulfonamide derivatives [88].

In the extracellular space, activity-induced transmembrane fluxes of acid–base species produce pH_o changes which are qualitatively opposite to those seen in pH_i recordings. This is because the same molecular mechanisms that are involved in the intraneuronal acidosis contribute to the increase in pH_o [87–90]. Thus, a pronounced alkalinization is generally the immediate pH_o response to intense neuronal activity (induced by agonist application or electrical stimulation). Using CA_o-inhibitors, it was possible to dissect the relative contributions of the H^+ shifts caused by glutamatergic transmission and the HCO_3^- shifts caused by $GABA_AR$-mediated transmission to heterosynaptic (excitatory and inhibitory) stimulation-evoked pH_o responses in hippocampal slices [91]. The results showed that with Schaffer collateral stimulation delivered at low frequencies (5–20 Hz), the activity-evoked extracellular alkalinization has a predominantly glutamatergic origin. However, the dominance is gradually shifted to HCO_3^- -dependent, GABAergic alkalinisation by increasing the stimulation frequency to 50–100 Hz. In view of the pH sensitivity of NMDA and $GABA_ARs$, this frequency-dependence of the mechanisms underlying activity-induced pH_o changes is intriguing. For instance, it is possible that during high-frequency neuronal activity, typically used in paradigms for induction of long-term potentiation (LTP), the intense CO_2/HCO_3^- shuttle and consequent CA_o dependent rise in pH_o that will take place in the interstitial fluid close to $GABA_ARs$ would reduce the efficacy of inhibition, thereby facilitating LTP. In parallel with this, the alkalosis might directly enhance the LTP-inducing, pH-sensitive NMDA current [22, 47, 92–97]. Whether such H^+-mediated cross-talk between excitatory and inhibitory mechanisms takes place in the brain is an interesting question to be addressed in future studies. The idea that H^+ does act as a modulatory signal in microdomains of the brain extracellular space is supported by findings showing that transporter-mediated acidification of the synaptic microenvironment is sufficient to enhance GABAergic signaling [98].

The initial, fast activity-dependent alkalinization is typically followed by a slower and long-lasting acidification. There is strong evidence that depolarization-dependent activation of Na^+- HCO_3^- cotransport in glial cells [99], which is likely to be associated with an increase in P_{CO2}, accounts for the slow acid shift [100]. This kind of a mechanism will affect both neuronal pH_o and pH_i over a large area for a prolonged duration, and is tempting to speculate that it could provide a powerful control over gross network excitability. There is, indeed, evidence that such a feedback mechanism acts as an intrinsic antiepileptic mechanism by limiting the generation and propagation of seizure activity thereby contributing to seizure termination [101, 102].

5 Intraneuronal CA_i Activity Promotes Depolarizing and Excitatory GABAergic Transmission

The fact that $GABA_A$Rs show a substantial permeability to HCO_3^- leads to a unique and tight link between the functions of this transmitter system and the pH-regulatory machinery in brain cells and in the extracellular space [35, 103]. To fully understand the role of CA activity in GABAergic signalling and especially how it affects the 'ionic plasticity' of inhibitory transmission we need to go back and look at the basic properties of $GABA_A$Rs.

The transmembrane gradients of Cl^- and HCO_3^- determine the reversal potential of $GABA_A$Rs ($E_{GABA\text{-}A}$) [82]. pH_i -regulatory transporters maintain E_{HCO3} at a very positive level, around -10 to -15 mV (see Sect. 2) which means that the current component carried by HCO_3^- across $GABA_A$Rs is always depolarizing. In mature neurons, the K-Cl cotransporter KCC2 extrudes Cl^- which keeps E_{Cl} more negative than the resting membrane potential, thus providing the ionic basis for conventional hyperpolarizing IPSPs [104]. Thus, with the relative HCO_3^-/Cl^- permeability ratio of $GABA_A$Rs at around 0.2–0.4, $E_{HCO3} >> E_{GABA\text{-}A} > E_{Cl}$ [35]. However, especially during intense or prolonged activation of $GABA_A$Rs, a significant conductive uptake of Cl^- takes place [82] which produces a large, activity-dependent depolarizing shift in E_{Cl} and consequently, in $E_{GABA\text{-}A}$. In adult mammalian neurons, and especially in their dendrites which have a large surface-to-volume ratio [105], CA_i is generally thought to be necessary for maintaining the supply of intracellular HCO_3^- that drives GABAergic depolarizing and excitatory responses [34, 78, 106, 107].

In line with the above, intense activation of $GABA_A$ channels evokes biphasic GABAergic responses [106–111]. The early hyperpolarization, representing fused individual hyperpolarizing IPSPs, is followed by a prolonged depolarization that is often associated with pronounced spiking. The initial phase of the depolarization is generated by the fast shift in E_{Cl} driven by the HCO_3^- -dependent net uptake of Cl^-. Thereafter, extrusion of the accumulated Cl^- via KCC2 leads to a long-lasting increase in extracellular K^+ [78] and to a consequent *non-synaptically -induced* depolarization of both the neurons and the adjacent glial cells.

In the present context, the dual role of KCC2 described above is a very important topic because the depolarizing/excitatory action of GABA in mature pyramidal neurons shows a strict dependence on neuronal CA_i. This conclusion is based on the findings that (*i*) inhibition of intra- but not extracellular CA attenuates the post-tetanic GABAergic depolarization [107] and (*ii*) the HCO_3^- -dependent excitatory effects of GABA parallel the developmental upregulation of cytosolic CA activity [34, 70].

The extent of the activity-dependent rapid shifts in E_{GABA} to more depolarizing values is likely to differ between neuronal subpopulations because of differences in Cl^--HCO_3^- homeostasis [59, 61, 112, 113]. There is also experimental data suggesting that excitatory GABAergic transmission might contribute to seizure generation [78, 107, 114].

6 CA VII Contributes to the Generation of Febrile Seizures

We have recently conducted an extensive study on the CA VII KO mouse [34] where we addressed the role of CA VII in HCO_3^- -dependent depolarizing GABA responses and in the generation of experimental febrile seizures (eFS) induced by hyperthermia [4]. Given the distinct developmental expression profiles of CA VII and CA II, the novel CA VII KO mouse provided an excellent opportunity to examine how this neuron-specific isoform modulates excitability. Whole cell patch clamp studies on P12-16 WT and CA VII KO hippocampal pyramidal neurons showed that HCO_3^- -dependent net uptake of Cl^- via $GABA_A$Rs is strongly facilitated by CA VII activity. Consequently, the high-frequency stimulation – induced, long-lasting GABAergic depolarization was able to induce action potential firing in WT, but not in CA VII KO neurons (Fig. 14.1c). Boosting $GABA_A$R-mediated signalling with diazepam in P14 rat hippocampal slices prolonged the duration of excitatory GABAergic responses and increased the number of action potentials associated with the depolarization. After P35, when neuronal CA II expression has also taken place, both isoforms were equally efficient in promoting HCO_3^- -dependent GABAergic depolarization. As expected, in the CA II/CA VII double KO, i.e. in the absence of CA_i activity, GABAergic depolarization remained small also at adult stage.

The developmental stage of the rodent brain and especially its cortical structures at P13-P14 is generally thought to be relevant for comparisons to the human situation, where FS are first seen at an age of 6 months [7, 115]. A striking difference in seizure generation was found between WT and CA VII KO mice. Cortical EEG monitoring showed that electrographic seizures were present in WT but not in CA VII KO mice. Importantly, there were no genotype-dependent differences in hyperventilation and the consequent respiratory alkalosis which is a major trigger of eFS [4, 7].

Behavioural experiments on P14 rat pups showed that enhancing $GABA_A$R signalling by a low dose of diazepam facilitated the triggering of eFS without affecting breath rate. At higher concentration diazepam prevented the generation eFS, probably via suppressed breathing. In fact, the suppression of breathing and the consequent block of the FS-promoting respiratory alkalosis might be a major mechanism in the therapeutic actions of diazepam which is routinely given to children with FS.

In humans FS are the most common type of seizures during early childhood [116]. Consistent with a role in FS, exon array analysis showed a prenatal upregulation of CA VII in the human neocortex and hippocampus that precedes the postnatal time period during which FS are most commonly detected [34].

These data as a whole suggest that designing next-generation isoform-specific inhibitors of CA VII has much potential as a novel approach in the treatment of FS and possibly other epileptiform syndromes.

7 Conclusions

In comparison to Ca^{2+}, the multiple and evolutionarily ancient roles of H^+ ions in controlling neuronal signaling have received surprisingly little attention. For instance, the strikingly steep pH_o dependency of the gating of $GABA_A$ and NMDA channels has been recognized for decades, but the amount of work done on the functional impact of activity-evoked pH_o transients on synaptic transmission is sparse [46, 47, 96, 97, 117]. What *is* known about the actions of H^+ does indicate that it is one of the most important physiologically-active agents that exert a fundamental modulatory role in neuronal development, plasticity, as well as synaptic and electrical signalling.

Moreover, H^+ is an amazingly potent agent in the suppression of seizures [5, 101, 102]. Neuronal pH shifts exert also a strong influence on the outcome from disease states such as stroke and ischemia/anoxia [118]. Observations of this kind are consistent with the multiple physiological roles of H^+ signalling, and elucidating the underlying processes is likely to be useful in pre-clinical and clinical work on many other disease states, such as migraine and chronic pain [119]. In the context of pathophysiological mechanisms, strategies that target neuronal pH may turn out to be as, or even more relevant, than those designed for modulation of neuronal Cl^- homeostasis [102], an area which has recently attracted extensive attention within the neuroscience community [59]. Here, one should note that in addition to tight Ca^{2+}/H^+ interactions at the molecular and cellular level [120], pH and Cl^- regulation are closely linked, especially via HCO_3^- -dependent mechanisms [121].

The key role of CA isoforms in the suppression, generation and modulation of pH shifts in the brain and other parts of the CNS makes these molecules highly interesting in studies of the fundamental mechanisms underlying neuronal signalling. The developmental profiles of distinct CAs, as well as their strategic localization seen from the level of the whole organism to subcellular microdomains points to a high versatility of their regulatory functions thus providing an exciting subject for molecular, cellular, physiological, medical and pharmacological research [27, 119, 122]. Finally, it is obvious that CAs represent a promising family of targets for CNS drug research and design.

Acknowledgments The authors' original research work has been supported by the Academy of Finland, the Sigrid Jusélius Foundation, the Jane and Aatos Erkko Foundation, and the Letten Foundation. We thank Prof. Juha Voipio for discussions and constructive comments on the manuscript.

References

1. Nattie EE (2001) Central chemosensitivity, sleep, and wakefulness. Respir Physiol 129:257–268
2. Lee J, Taira T, Pihlaja P, Ransom BR, Kaila K (1996) Effects of CO2 on excitatory transmission apparently caused by changes in intracellular pH in the rat hippocampal slice. Brain Res 706:210–216
3. Dulla CG, Frenguelli BG, Staley KJ, Masino SA (2009) Intracellular acidification causes adenosine release during states of hyperexcitability in the hippocampus. J Neurophysiol 102:1984–1993
4. Schuchmann S, Schmitz D, Rivera C, Vanhatalo S, Salmen B, Mackie K, Sipilä ST, Voipio J, Kaila K (2006) Experimental febrile seizures are precipitated by a hyperthermia-induced respiratory alkalosis. Nat Med 12:817–823
5. Tolner EA, Hochman DW, Hassinen P, Otahal J, Gaily E, Haglund MM, Kubova H, Schuchmann S, Vanhatalo S, Kaila K (2011) Five percent CO_2 is a potent, fast-acting inhalation anticonvulsant. Epilepsia 52:104–114
6. Guaranha MS, Garzon E, Buchpiguel CA, Tazima S, Yacubian EM, Sakamoto AC (2005) Hyperventilation revisited: physiological effects and efficacy on focal seizure activation in the era of video-EEG monitoring. Epilepsia 46:69–75
7. Schuchmann S, Hauck S, Henning S, Gruters-Kieslich A, Vanhatalo S, Schmitz D, Kaila K (2011) Respiratory alkalosis in children with febrile seizures. Epilepsia 52:1949–1955
8. Bockenhauer D, Feather S, Stanescu HC, Bandulik S, Zdebik AA, Reichold M, Tobin J, Lieberer E, Sterner C, Landoure G, Arora R, Sirimanna T, Thompson D, Cross JH, Van't Hoff W, Al MO, Tullus K, Yeung S, Anikster Y, Klootwijk E, Hubank M, Dillon MJ, Heitzmann D, Arcos-Burgos M, Knepper MA, Dobbie A, Gahl WA, Warth R, Sheridan E, Kleta R (2009) Epilepsy, ataxia, sensorineural deafness, tubulopathy, and KCNJ10 mutations. N Engl J Med 360:1960–1970
9. Pedersen SF, O'Donnell ME, Anderson SE, Cala PM (2006) Physiology and pathophysiology of Na+/H+ exchange and Na+-K+-2Cl- cotransport in the heart, brain, and blood. Am J Physiol Regul Integr Comp Physiol 291:R1–R25
10. Helmy MM, Tolner EA, Vanhatalo S, Voipio J, Kaila K (2011) Brain alkalosis causes birth asphyxia seizures, suggesting therapeutic strategy. Ann Neurol 69:493–500
11. Helmy MM, Ruusuvuori E, Watkins PV, Voipio J, Kanold PO, Kaila K (2012) Acid extrusion via blood–brain barrier causes brain alkalosis and seizures after neonatal asphyxia. Brain 135:3311–3319
12. Roos A, Boron WF (1981) Intracellular pH. Physiol Rev 61:296–434
13. Chesler M (2003) Regulation and modulation of pH in the brain. Physiol Rev 83:1183–1221
14. Ballanyi K, Kaila K (1998) Activity-evoked changes in intracellular pH. In: Kaila K, Ransom BR (eds) pH and brain function. Wiley-Liss, New York, pp 291–308
15. Kaila K, Chesler M (1998) Activity-evoked changes in extracellular pH. In: Kaila K, Ransom BR (eds) pH and brain function. Wiley-Liss, New York, pp 309–337

16. Casey JR, Grinstein S, Orlowski J (2010) Sensors and regulators of intracellular pH. Nat Rev Mol Cell Biol 11:50–61
17. Tombaugh GC, Somjen GG (1996) Effects of extracellular pH on voltage-gated Na+, K+ and Ca2+ currents in isolated rat CA1 neurons. J Physiol 493(Pt 3):719–732
18. Tombaugh GC, Somjen GG (1997) Differential sensitivity to intracellular pH among high- and low- threshold Ca2+ currents in isolated rat CA1 neurons. J Neurophysiol 77:639–653
19. Pasternack M, Smirnov S, Kaila K (1996) Proton modulation of functionally distinct GABAA receptors in acutely isolated pyramidal neurons of rat hippocampus. Neuropharmacology 35:1279–1288
20. Wilkins ME, Hosie AM, Smart TG (2005) Proton modulation of recombinant GABA(A) receptors: influence of GABA concentration and the beta subunit TM2-TM3 domain. J Physiol 567:365–377
21. Traynelis SF, Cull-Candy SG (1990) Proton inhibition of N-methyl-D-aspartate receptors in cerebellar neurons. Nature 345:347–350
22. Makani S, Chen HY, Esquenazi S, Shah GN, Waheed A, Sly WS, Chesler M (2012) NMDA Receptor-dependent afterdepolarizations are curtailed by carbonic anhydrase 14: regulation of a short-term postsynaptic potentiation. J Neurosci 32:16754–16762
23. Spray DC, Harris AL, Bennet MVL (1981) Gap junctional conductance is a simple and sensitive function of intracellular pH. Science 211:712–715
24. Waldmann R, Champigny G, Bassilana F, Heurteaux C, Lazdunski M (1997) A proton-gated cation channel involved in acid-sensing. Nature 386:173–177
25. Duprat F, Lesage F, Fink M, Reyes R, Heurteaux C, Lazdunski M (1997) TASK, a human background K + channel to sense external pH variations near physiological pH. EMBO J 16:5464–5471
26. Maren TH (1967) Carbonic anhydrase: chemistry, physiology, and inhibition. Physiol Rev 47:595–781
27. Supuran CT (2008) Carbonic anhydrases: novel therapeutic applications for inhibitors and activators. Nat Rev Drug Discov 7:168–181
28. Stridh MH, Alt MD, Wittmann S, Heidtmann H, Aggarwal M, Riederer B, Seidler U, Wennemuth G, McKenna R, Deitmer JW, Becker HM (2012) Lactate flux in astrocytes is enhanced by a non-catalytic action of carbonic anhydrase II. J Physiol 590:2333–2351
29. Jiao Y, Yan J, Zhao Y, Donahue LR, Beamer WG, Li X, Roe BA, LeDoux MS, Gu W (2005) Carbonic anhydrase-related protein VIII deficiency is associated with a distinctive lifelong gait disorder in waddles mice. Genetics 171:1239–1246
30. Hirasawa M, Xu X, Trask RB, Maddatu TP, Johnson BA, Naggert JK, Nishina PM, Ikeda A (2007) Carbonic anhydrase related protein 8 mutation results in aberrant synaptic morphology and excitatory synaptic function in the cerebellum. Mol Cell Neurosci 35:161–170
31. Kaya N, Aldhalaan H, Al-Younes B, Colak D, Shuaib T, Al-Mohaileb F, Al-Sugair A, Nester M, Al-Yamani S, Al-Bakheet A, Al-Hashmi N, Al-Sayed M, Meyer B, Jungbluth H, Al-Owain M (2011) Phenotypical spectrum of cerebellar ataxia associated with a novel mutation in the CA8 gene, encoding carbonic anhydrase (CA) VIII. Am J Med Genet B Neuropsychiatr Genet 156B:826–834
32. Shah GN, Ulmasov B, Waheed A, Becker T, Makani S, Svichar N, Chesler M, Sly WS (2005) Carbonic anhydrase IV and XIV knockout mice: roles of the respective carbonic anhydrases in buffering the extracellular space in brain. Proc Natl Acad Sci U S A 102:16771–16776
33. Velisek L, Moshe SL, Xu SG, Cammer W (1993) Reduced susceptibility to seizures in carbonic anhydrase II deficient mutant mice. Epilepsy Res 14:115–121
34. Ruusuvuori E, Huebner AK, Kirilkin I, Yukin A, Blaesse P, Helmy MM, Kang HJ, Muayed M, Hennings JC, Sestan N, Hubner CA, Kaila K (2013) Neuronal carbonic anhydrase VII provides GABAergic excitatory drive to exacerbate febrile seizures. EMBO J 32:2275–2286
35. Kaila K (1994) Ionic basis of GABA(A) receptor channel function in the nervous system. Prog Neurobiol 42:489–537
36. Thomas RC, Meech RW (1982) Hydrogen ion currents and intracellular pH in depolarized voltage-clamped snail neurones. Nature 299:826–828

37. Deitmer JW, Rose CR (1996) pH regulation and proton signalling by glial cells. Prog Neurobiol 48:73–103
38. Pappas CA, Ransom BR (1993) A depolarization-stimulated, bafilomycin-inhibitable H+ pump in hippocampal astrocytes. Glia 9:280–291
39. Bevensee MO, Cummins TR, Haddad GG, Boron WF, Boyarsky G (1996) pH regulation in single CA1 neurons acutely isolated from the hippocampi of immature and mature rats. J Physiol 494:315–328
40. Yao H, Ma E, Gu XQ, Haddad GG (1999) Intracellular pH regulation of CA1 neurons in Na(+)/H(+) isoform 1 mutant mice. J Clin Invest 104:637–645
41. Schwiening CJ, Kennedy HJ, Thomas RC (1993) Calcium-hydrogen exchange by the plasma membrane Ca-ATPase of voltage-clamped snail neurons. Proc Biol Sci 253:285–289
42. Grichtchenko II, Chesler M (1996) Calcium- and barium-dependent extracellular alkaline shifts evoked by electrical activity in rat hippocampal slices. Neuroscience 75:1117–1126
43. Paalasmaa P, Kaila K (1996) Role of voltage-gated calcium channels in the generation of activity-induced extracellular pH transients in the rat hippocampal slice. J Neurophysiol 75:2354–2360
44. Burton RF (1978) Intracellular buffering. Respir Physiol 33:51–58
45. Chesler M, Chen JC, Kraig RP (1994) Determination of extracellular bicarbonate and carbon dioxide concentrations in brain slices using carbonate and pH-selective microelectrodes. J Neurosci Meth 53:129–136
46. Tong CK, Chen K, Chesler M (2006) Kinetics of activity-evoked pH transients and extracellular pH buffering in rat hippocampal slices. J Neurophysiol 95:3686–3697
47. Makani S, Chesler M (2007) Endogenous alkaline transients boost postsynaptic NMDA receptor responses in hippocampal CA1 pyramidal neurons. J Neurosci 27:7438–7446
48. Huang W, Smith SE, Chesler M (1995) Addition of carbonic anhydrase augments extracellular pH buffering in rat cerebral cortex. J Neurophysiol 74:1806–1809
49. Svichar N, Waheed A, Sly WS, Hennings JC, Hubner CA, Chesler M (2009) Carbonic anhydrases CA4 and CA14 both enhance AE3-mediated Cl-. J Neurosci 29:3252–3258
50. Zhu XL, Sly WS (1990) Carbonic anhydrase-IV from human lung-purification, characterization, and camparision with membrane carbonic-anhydrase from human kidney. J Biol Chem 265:8795–8801
51. Svichar N, Chesler M (2003) Surface carbonic anhydrase activity on astrocytes and neurons facilitates lactate transport. Glia 41:415–419
52. Mori K, Ogawa Y, Ebihara K, Tamura N, Tashiro K, Kuwahara T, Mukoyama M, Sugawara A, Ozaki S, Tanaka I, Nakao K (1999) Isolation and characterization of CA XIV, a novel membrane-bound carbonic anhydrase from mouse kidney. J Biol Chem 274:15701–15705
53. Parkkila S, Parkkila AK, Rajaniemi H, Shah GN, Grubb JH, Waheed A, Sly WS (2001) Expression of membrane-associated carbonic anhydrase XIV on neurons and axons in mouse and human brain. Proc Natl Acad Sci U S A 98:1918–1923
54. Tureci O, Sahin U, Vollmar E, Siemer S, Gottert E, Seitz G, Parkkila AK, Shah GN, Grubb JH, Pfreundschuh M, Sly WS (1998) Human carbonic anhydrase XII: cDNA cloning, expression, and chromosomal localization of a carbonic anhydrase gene that is overexpressed in some renal cell cancers. Proc Natl Acad Sci U S A 95:7608–7613
55. Ivanov S, Liao SY, Ivanova A, Danilkovitch-Miagkova A, Tarasova N, Weirich G, Merrill MJ, Proescholdt MA, Oldfield EH, Lee J, Zavada J, Waheed A, Sly W, Lerman MI, Stanbridge EJ (2001) Expression of hypoxia-lnducible cell-surface transmembrane carbonic anhydrases in human cancer. Am J Pathol 158:905–919
56. Haapasalo JA, Nordfors KM, Hilvo M, Rantala IJ, Soini Y, Parkkila AK, Pastorekova S, Pastorek J, Parkkila SM, Haapasalo HK (2006) Expression of carbonic anhydrase IX in astrocytic tumors predicts poor prognosis. Clin Cancer Res 12:473–477
57. Halmi P, Parkkila S, Honkaniemi J (2006) Expression of carbonic anhydrases II, IV, VII, VIII and XII in rat brain after kainic acid induced status epilepticus. Neurochem Int 48:24–30
58. Nogradi A, Domoki F, Degi R, Borda S, Pakaski M, Szabo A, Bari F (2003) Up-regulation of cerebral carbonic anhydrase by anoxic stress in piglets. J Neurochem 85:843–850

59. Blaesse P, Airaksinen MS, Rivera C, Kaila K (2009) Cation-chloride cotransporters and neuronal function. Neuron 61:820–838
60. Durand GM, Konnerth A (1996) Long-term potentiation as a mechanism of functional synapse induction in the developing hippocampus. J Physiol Paris 90:313–315
61. Gonzalez-Islas C, Chub N, Wenner P (2009) NKCC1 and AE3 appear to accumulate chloride in embryonic motoneurons. J Neurophysiol 101:507–518
62. Ridderstråle Y, Wistrand PJ (1998) Carbonic anhydrase isoforms in the mammalian nervous system. In: Kaila K, Ransom B (eds) pH and brain function. Wiley-Liss, New York, pp 21–44
63. Nogradi A, Mihaly A (1990) Light microscopic histochemistry of the postnatal-development and localization of carbonic anhydrase activity in glial and neuronal cell-types of the rat central nervous system. Histochemistry 94:441–447
64. Pasternack M, Voipio J, Kaila K (1993) Intracellular carbonic anhydrase activity and its role in GABA- induced acidosis in isolated rat hippocampal pyramidal neurones. Acta Physiol Scand 148:229–231
65. Lakkis MM, Bergenhem NCH, OShea KS, Tashian RE (1997) Expression of the acatalytic carbonic anhydrase VIII gene, car8, during mouse embryonic development. Histochem J 29:135–141
66. Nogradi A, Jonsson N, Walker R, Caddy K, Carter N, Kelly C (1997) Carbonic anhydrase II and carbonic anhydrase-related protein in the cerebellar cortex of normal and lurcher mice. Dev Brain Res 98:91–101
67. Munsch T, Pape HC (1999) Upregulation of the hyperpolarization-activated cation current in rat thalamic relay neurones by acetazolamide. J Physiol 519(Pt 2):505–514
68. Wang WG, Bradley SR, Richerson GB (2002) Quantification of the response of rat medullary raphe neurones to independent changes in pH(O) and P-CO_2. J Physiol 540:951–970
69. Willoughby D, Schwiening CJ (2002) Electrically evoked dendritic pH transients in rat cerebellar Purkinje cells. J Physiol-Lond 544:487–499
70. Ruusuvuori E, Li H, Huttu K, Palva JM, Smirnov S, Rivera C, Kaila K, Voipio J (2004) Carbonic anhydrase isoform VII acts as a molecular switch in the development of synchronous gamma-frequency firing of hippocampal CA1 pyramidal cells. J Neurosci 24:2699–2707
71. Spicer SS, Stoward PJ, Tashian RE (1979) The immunohistolocalization of carbonic anhydrase in rodent tissues. J Histochem Cytochem 27:820–831
72. Innocenti A, Scozzafava A, Parkkila S, Puccetti L, De Simone G, Supuran CT (2008) Investigations of the esterase, phosphatase, and sulfatase activities of the cytosolic mammalian carbonic anhydrase isoforms I, II, and XIII with 4-nitrophenyl esters as substrates. Bioorg Med Chem Lett 18:2267–2271
73. Truppo E, Supuran CT, Sandomenico A, Vullo D, Innocenti A, Di FA, Alterio V, De SG, Monti SM (2012) Carbonic anhydrase VII is S-glutathionylated without loss of catalytic activity and affinity for sulfonamide inhibitors. Bioorg Med Chem Lett 22:1560–1564
74. Erecinska M, Cherian S, Silver IA (2004) Energy metabolism in mammalian brain during development. Prog Neurobiol 73:397–445
75. Sterling D, Reithmeier RAF, Casey JR (2001) A transport metabolon. Functional interaction of carbonic anhydrase II and chloride/bicarbonate exchangers. J Biol Chem 276:47886–47894
76. Becker HM, Deitmer JW (2007) Carbonic anhydrase II increases the activity of the human electrogenic Na+/HCO3- cotransporter. J Biol Chem 282:13508–13521
77. Boron WF (2010) Evaluating the role of carbonic anhydrases in the transport of HCO3– related species. Biochim Biophys Acta 1804:410–421
78. Viitanen T, Ruusuvuori E, Kaila K, Voipio J (2010) The KCl -cotransporter KCC2 promotes GABAergic excitation in the mature rat hippocampus. J Physiol 588:1527–1540
79. Thiry A, Dogne J, Supuran CT, Masereel B (2007) Carbonic anhydrase inhibitors as anticonvulsant agents. Curr Top Med Chem 7:855–864
80. Fedirko N, Svichar N, Chesler M (2006) Fabrication and use of high-speed, concentric H+- and Ca2+-selective microelectrodes suitable for in vitro extracellular recording. J Neurophysiol 96:919–924

81. Kaila K, Ransom BR (1998) pH and brain function. Wiley-Liss, New York
82. Kaila K, Voipio J (1987) Postsynaptic fall in intracellular pH induced by GABA-activated bicarbonate conductance. Nature 330:163–165
83. Paalasmaa P, Taira T, Voipio J, Kaila K (1994) Extracellular alkaline transients mediated by glutamate receptors in the rat hippocampal slice are not due to a proton conductance. J Neurophysiol 72:2031–2033
84. Trapp S, Luckermann M, Kaila K, Ballanyi K (1996) Acidosis of hippocampal neurones mediated by a plasmalemmal Ca2+/H+ pump. Neuroreport 7:2000–2004
85. Luckermann M, Trapp S, Ballanyi K (1997) GABA- and glycine-mediated fall of intracellular pH in rat medullary neurons in situ. J Neurophysiol 77:1844–1852
86. Zhan RZ, Fujiwara N, Tanaka E, Shimoji K (1998) Intracellular acidification induced by membrane depolarization in rat hippocampal slices: roles of intracellular Ca2+ and glycolysis. Brain Res 780:86–89
87. Kaila K, Saarikoski J, Voipio J (1990) Mechanism of action of GABA on intracellular pH and on surface pH in crayfish muscle fibres. J Physiol 427:241–260
88. Voipio J, Paalasmaa P, Taira T, Kaila K (1995) Pharmacological characterization of extracellular pH transients evoked by selective synaptic and exogenous activation of AMPA, NMDA, and GABAA receptors in the rat hippocampal slice. J Neurophysiol 74:633–642
89. Chen JC, Chesler M (1992) Modulation of extracellular pH by glutamate and GABA in rat hippocampal slices. J Neurophysiol 67:29–36
90. Kaila K, Paalasmaa P, Taira T, Voipio J (1992) pH transients due to monosynaptic activation of GABAA receptors in rat hippocampal slices. Neuroreport 3:105–108
91. Taira T, Paalasmaa P, Voipio J, Kaila K (1995) Relative contributions of excitatory and inhibitory neuronal activity to alkaline transients evoked by stimulation of Schaffer collaterals in the rat hippocampal slice. J Neurophysiol 74:643–649
92. Traynelis SF, Hartley M, Heinemann SF (1995) Control of proton sensitivity of the NMDA receptor by RNA splicing and polyamines. Science 268:873–876
93. Krishek BJ, Amato A, Connolly CN, Moss SJ, Smart TG (1996) Proton sensitivity of the GABA(A) receptor is associated with the receptor subunit composition. J Physiol 492:431–443
94. Tang CM, Dichter M, Morad M (1990) Modulation of the N-methyl-D-aspartate channel by extracellular H+. Proc Natl Acad Sci U S A 87:6445–6449
95. Vyklicky LJ, Vlachov V, Krusek J (1990) The effect of external pH changes on responses to excitatory amino acids in mouse hippocampal neurones. J Physiol 430:497–517
96. Taira T, Smirnov S, Voipio J, Kaila K (1993) Intrinsic proton modulation of excitatory transmission in rat hippocampal slices. Neuroreport 4:93–96
97. Gottfried JA, Chesler M (1994) Endogenous H + modulation of NMDA receptor-mediated EPSCs revealed by carbonic anhydrase inhibition in rat hippocampus. J Physiol 478(Pt 3):373–378
98. Dietrich CJ, Morad M (2010) Synaptic acidification enhances GABAA signaling. J Neurosci 30:16044–16052
99. Pappas CA, Ransom BR (1994) Depolarization-induced alkalinization (DIA) in rat hippocampal astrocytes. J Neurophysiol 72:2816–2826
100. Voipio J (1998) Diffusion and buffering aspects of H+, HCO3-, and CO_2 movements in brain tissue. In: Kaila K, Ransom BR (eds) pH and brain function. Wiley-Liss, New York, pp 45–66
101. de Curtis M, Manfridi A, Biella G (1998) Activity-dependent pH shifts and periodic recurrence of spontaneous interictal spikes in a model of focal epileptogenesis. J Neurosci 18:7543–7551
102. Pavlov I, Kaila K, Kullmann DM, Miles R (2013) Cortical inhibition, pH and cell excitability in epilepsy: what are optimal targets for antiepileptic interventions? J Physiol 591:765–774
103. Farrant M, Kaila K (2007) The cellular, molecular and ionic basis of GABAA receptor signalling. Prog Brain Res 160:59–87

104. Rivera C, Voipio J, Payne JA, Ruusuvuori E, Lahtinen H, Lamsa K, Pirvola U, Saarma M, Kaila K (1999) The K+/Cl- co-transporter KCC2 renders GABA hyperpolarizing during neuronal maturation. Nature 397:251–255
105. Qian N, Sejnowski TJ (1990) When is an inhibitory synapse effective? Proc Natl Acad Sci U S A 87:8145–8149
106. Staley KJ, Soldo BL, Proctor WR (1995) Ionic mechanisms of neuronal excitation by inhibitory GABA(A) receptors. Science 269:977–981
107. Kaila K, Lamsa K, Smirnov S, Taira T, Voipio J (1997) Long-lasting GABA-mediated depolarization evoked by high- frequency stimulation in pyramidal neurons of rat hippocampal slice is attributable to a network-driven, bicarbonate-dependent K+ transient. J Neurosci 17:7662–7672
108. Alger BE, Nicoll RA (1982) Feed-forward dendritic inhibition in rat hippocampal pyramidal cells studied in vitro. J Physiol 328:105–123
109. Alger BE, Nicoll RA (1982) Pharmacological evidence for two kinds of GABA receptor on rat hippocampal pyramidal cells studied in vitro. J Physiol 328:125–141
110. Grover LM, Lambert NA, Schwartzkroin PA, Teyler TJ (1993) Role of HCO3- ions in depolarizing GABAA receptor-mediated responses in pyramidal cells of rat hippocampus. J Neurophysiol 69:1541–1555
111. Smirnov S, Paalasmaa P, Uusisaari M, Voipio J, Kaila K (1999) Pharmacological isolation of the synaptic and nonsynaptic components of the GABA-mediated biphasic response in rat CA1 hippocampal pyramidal cells. J Neurosci 19:9252–9260
112. Szabadics J, Varga C, Molnar G, Olah S, Barzo P, Tamas G (2006) Excitatory effect of GABAergic axo-axonic cells in cortical microcircuits. Science 311:233–235
113. Khirug S, Yamada J, Afzalov R, Voipio J, Khiroug L, Kaila K (2008) GABAergic depolarization of the axon initial segment in cortical principal neurons is caused by the Na-K-2Cl cotransporter NKCC1. J Neurosci 28:4635–4639
114. Stasheff SF, Mott DD, Wilson WA (1993) Axon terminal hyperexcitability associated with epileptogenesis in vitro. II. Pharmacological regulation by NMDA and GABAA receptors. J Neurophysiol 70:976–984
115. Berg AT, Shinnar S (1996) Complex febrile seizures. Epilepsia 37:126–133
116. Stafstrom CE (2002) The incidence and prevalence of febrile seizures. In: Baram TZ, Shinnar S (eds) Febrile seizures. Academic, San Diego, pp 1–25
117. Fedirko N, Avshalumov M, Rice ME, Chesler M (2007) Regulation of postsynaptic Ca2+ influx in hippocampal CA1 pyramidal neurons via extracellular carbonic anhydrase. J Neurosci 27:1167–1175
118. Orlowski P, Chappell M, Park CS, Grau V, Payne S (2011) Modelling of pH dynamics in brain cells after stroke. Interface Focus 1:408–416
119. Asiedu M, Ossipov MH, Kaila K, Price TJ (2010) Acetazolamide and midazolam act synergistically to inhibit neuropathic pain. Pain 148:302–308
120. Caldwell L, Harries P, Sydlik S, Schwiening CJ (2012) Presynaptic pH and vesicle fusion in *Drosophila* larvae neurones. Synapse 67:729–740
121. Romero MF, Fulton CM, Boron WF (2004) The SLC4 family of HCO3- transporters. Pflugers Arch 447:495–509
122. Ferrini F, Trang T, Mattioli TA, Laffray S, Del'Guidice T, Lorenzo LE, Castonguay A, Doyon N, Zhang W, Godin AG, Mohr D, Beggs S, Vandal K, Beaulieu JM, Cahill CM, Salter MW, de KY (2013) Morphine hyperalgesia gated through microglia-mediated disruption of neuronal Cl- homeostasis. Nat Neurosci 16:183–192

Chapter 15
Carbonic Anhydrase Inhibitors Drug Design

Robert McKenna and Claudiu T. Supuran

Abstract Inhibition of the metalloenzyme carbonic anhydrase (CA, EC 4.2.1.1) has pharmacologic applications in the field of antiglaucoma, anticonvulsant, antiobesity, and anticancer agents but is also emerging for designing anti-infectives (antifungal and antibacterial agents) with a novel mechanism of action. As a consequence, the drug design of CA inhibitors (CAIs) is a very dynamic field. Sulfonamides and their isosteres (sulfamates/sulfamides) constitute the main class of CAIs which bind to the metal ion in the enzyme active site. Recently the dithiocarbamates, possessing a similar mechanism of action, were reported as a new class of inhibitors. Other families of CAIs possess a distinct mechanism of action: phenols, polyamines, some carboxylates, and sulfocoumarins anchor to the zinc-coordinated water molecule. Coumarins and five/six-membered lactones are prodrug inhibitors, binding in hydrolyzed form at the entrance of the active site cavity. Novel drug design strategies have been reported principally based on the tail approach for obtaining all these types of CAIs, which exploit more external binding regions within the enzyme active site (in addition to coordination to the metal ion), leading thus to isoform-selective compounds. Sugar-based tails as well as click chemistry were the most fruitful developments of the tail approach. Promising compounds that inhibit CAs from bacterial and fungal pathogens, of the dithiocarbamate, phenol and carboxylate types have also been reported.

Susan C. Frost and Robert McKenna (eds.). Carbonic Anhydrase: Mechanism, Regulation, Links to Disease, and Industrial Applications

R. McKenna
Department of Biochemistry and Molecular Biology, University of Florida, Gainesville, FL, USA
e-mail: rmckenna@ufl.edu

C.T. Supuran (✉)
NEUROFARBA Department, Pharmaceutical Sciences Section, Universita degli Studi di Firenze, Polo Scientifico, Florence, Italy
e-mail: claudiu.supuran@unifi.it

S.C. Frost and R. McKenna (eds.), *Carbonic Anhydrase: Mechanism, Regulation, Links to Disease, and Industrial Applications*, Subcellular Biochemistry 75,
DOI 10.1007/978-94-007-7359-2_15,

1 Introduction

CO_2, bicarbonate and protons are essential molecules/ions in many important physiologic processes in all life kingdoms (Bacteria, Archaea, and Eukarya); hence, relatively high amounts of carbonic anhydrases (CAs, EC 4.2.1.1) are present in different tissues/cell compartments of most investigated organisms [1–9]. The α-CAs are present in vertebrates, protozoa, algae and cytoplasm of green plants and in some bacteria [1–9], the β-CAs are predominantly found in bacteria, algae and chloroplasts of both mono- as well as dicotyledons, but also in many fungi and some archaea [1–9]. The γ-CAs were found in archaea and some bacteria [3], whereas the δ- and ζ-CAs seem to be present only in marine diatoms [2]. In many organisms these enzymes are involved in crucial physiological processes connected with respiration and transport of CO_2/bicarbonate, pH and CO_2 homeostasis, electrolyte secretion in a variety of tissues/organs, biosynthetic reactions (e.g., gluconeogenesis, lipogenesis and ureagenesis), bone resorption, calcification, tumorigenicity, and many other physiologic or pathologic processes (thoroughly studied in vertebrates) [1, 4–7, 9]; whereas in algae, plants and some bacteria they play an important role in photosynthesis and other biosynthetic reactions [2, 3]. In diatoms δ- and ζ-CAs play a crucial role in carbon dioxide fixation [2]. Many such enzymes from vertebrates, fungi and bacteria are well-known drug targets [1, 4].

The CAs are a superfamily of metalloenzymes which catalyze the interconversion between CO_2 and bicarbonate by using a metal hydroxide nucleophilic mechanism [1–9]. Five distinct genetic CA families are known to date: the α-, β-, γ-, δ- and ζ-CAs, which differ in their preference for metal ions used within the active site for performing the catalysis. Zn(II) ions may be used by all five classes mentioned above but the γ-CAs are probably Fe(II) enzymes (being active also with Zn(II) or Co(II) ions) [3], whereas the ζ-class uses Cd(II) or Zn(II) to perform the physiologic reaction catalysis [2]. In all cases, the apo-enzymes are devoid of catalytic activity, the presence of the metal ion being essential both for catalysis as well as binding of inhibitors [1–9]. The catalytic/inhibition mechanisms are shown in Scheme 15.1.

In all CA classes a metal hydroxide species $\left(L_3\text{-}M^{2+}\text{-}OH^-\right)$ of the enzyme is the catalytically active nucleophile (working at neutral pH) which attacks the CO_2 molecule bound in a hydrophobic pocket nearby [10]. This metal hydroxide species is generated from a water coordinated to the metal ion, which is placed at the bottom of the active site cavity (Scheme 15.1, A). The active center normally comprises M(II) ions in a tetrahedral geometry, with three protein ligands (L) in addition to the water molecule/hydroxide ion coordinating the metal, but Zn(II) or Co(II) were also observed in trigonal bipyramidal or octahedral coordination geometries, at least in γ-CAs [11]. In many CAs, generation of the metal hydroxide species from the metal-coordinated water one, is the rate determining step of the catalytic turnover, which for some α- and ζ-CAs achieve k_{cat}/K_M values $> 10^8\ M^{-1} \times s^{-1}$, making CAs among the most effective catalysts known in nature [1, 2]. The metal

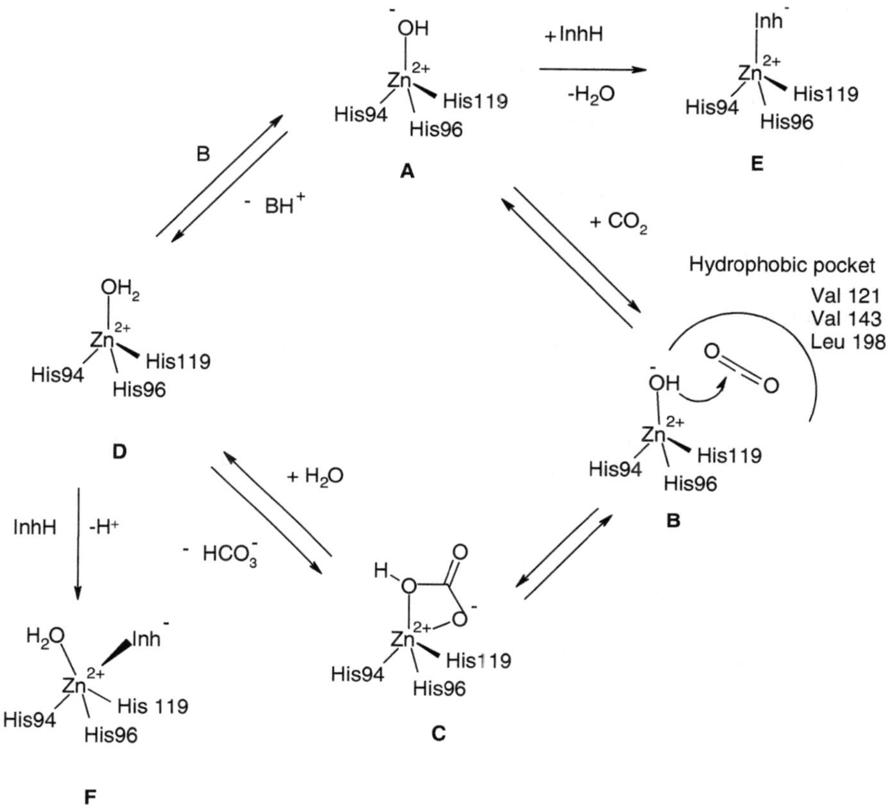

Scheme 15.1 Catalytic and inhibition mechanisms (with zinc ion binders) of α-CAs (hCA I amino acid numbering of the zinc ligands). A similar catalytic/inhibition mechanism is valid also for CAs from other classes (β-, γ- and ζ-CAs) but either the metal ion is coordinated by other amino acid residues or a Cd(II) ion is present instead of zinc at the active site

ion ligands are three His residues in α-, γ-, and δ-CAs or one His and two Cys residues in β- and ζ-CAs [1–9, 12]. Some β-class enzymes have four protein zinc ligands, i.e., one His, two Cys and one Asp coordinated to Zn(II) [13]. For these enzymes no water coordinated to the metal ion is present at $pH < 8$, as shown in an excellent crystallographic work from Jones' group on the mycobacterial enzymes Rv3558c and Rv1284 [13]. However, at $pH > 8$ a conserved Arg residue in all β-CAs investigated so far (belonging to a so-called catalytic dyad) makes a salt bridge with the Asp coordinated to Zn(II), liberating the fourth Zn(II) coordination position, which is then occupied by an incoming water molecule/hydroxide ion [13]. The substrate CO_2 was found bound (for the α-class enzymes) in a hydrophobic pocket near the Zn(II) ion, defined among others by residues Val121, Val143 and Leu198 [in the human (h) isoform hCA II, for which the X-ray crystal structure in complex with CO_2 was reported in a seminal paper by McKenna's group [10]], as represented in Scheme 15.1, B. Orientated in such a favorable position for the nucleophilic attack, CO_2 is transformed into bicarbonate which is coordinated bidentately to the Zn(II) ion, as shown in Scheme 15.1, C. There are X-ray crystal structures of adducts of hCA II with bicarbonate [10] (shown in Scheme 15.1, C) or thrithiocarbonate (CS_3^{2-}) [14] which is an interesting mimic of bicarbonate, having higher binding affinity for the metal ion compared to bicarbonate itself [14]. As the binding of bicarbonate to zinc is rather labile, intermediate C is readily transformed to D by reaction with water, which liberates the bicarbonate into solution. In this way, the acidic form of the enzyme is generated, with water coordinated to the metal ion (Scheme 15.1, D), which is catalytically ineffective. Generation of the nucleophilically active species of the enzyme A is achieved though a proton transfer reaction from the zinc-coordinated water (species D) to the buffer, which is also the rate-determining step of the entire process [15–17]. In many CA isoforms and probably in many enzyme classes (not only the α ones which are the most investigated CAs), this process is assisted by an active site residue able to participate in proton transfer processes, which for the specific case examined in Scheme 15.1 (i.e., hCA II) is a histidine residue placed in the middle of the active site [15–17]. For hCA II and several other human isoforms this residue is His64, which may be further assisted in the proton transfer process by a cluster of histidines prolonging from he middle of the cavity to its entrance and till the surface of the enzyme around the edge of the cavity [15]. In hCA II this cluster includes residues His4, His3, His10 and His15 [15].

The inhibition and activation of CAs are well understood processes, with most classes of inhibitors binding to the metal center [1–9], whereas activators bind at the entrance of the active site cavity and participate in proton shuttling processes between the metal ion – bound water molecule and the environment [15]. This leads to the enhanced formation of the metal hydroxide, catalytically active species of the enzyme [1, 15, 16]. Inhibitors generally bind to the metal ion from the enzyme active site in deprotonated state (as anions), as shown schematically in steps E and F of Scheme 15.1, for a tetrahedrally bound inhibitor (E) and for one in which the Zn(II) ion is in a trigonal bipyramidal geometry (F), case in which a water molecule in addition to the inhibitor is also coordinated to zinc [1–9]. Although the mechanisms of Scheme 15.1 were depicted for an α-CA, they are valid even if another metal ion is present within the active site cavity, i.e., Cd(II) or Fe(II), since the corresponding hydroxides have similar nucleophilicity with zinc hydroxide. The same is true for a different coordination pattern of the metal ion (i.e., two Cys and one His residues), as for the β- and ζ-class CAs [1–9].

It should be mentioned that recently other inhibition mechanisms than the binding to the metal center were reported for α-CAs, which do not directly involve the metal ion from the enzyme active site. For example polyamines bind to the enzyme by anchoring to the zinc-coordinated water/hydroxide ion [18], whereas coumarins act as prodrugs and bind at the entrance of the active site cavity, rather far away from the metal ion [19–21].

2 Drug Design Studies of Sulfonamides, Their Bioisosteres and Dithiocarbamates as CAIs

Sulfonamides are the most important class of CAIs [1, 4–9], with several compounds such as acetazolamide **1**, methazolamide **2**, ethoxzolamide **3**, sulthiame **4**, dichlorophenamide **5**, dorzolamide **6**, brinzolamide **7**, sulpiride **8** and zonisamide **9** in clinical use for many years, as diuretics, antiglaucoma agents, as well as antiepileptics [1] (Fig. 15.1). Sulfamates such as topiramate **10**, EMATE **12**, and irosustat **11** although developed independently of their potential CA inhibitory properties [1], are also potent CAIs and are clinically used as antiepileptics/antiobesity agents (topiramate [1, 4–9]) or are in advanced clinical trials as dual, steroid sulfatase inhibitors/CAIs with anticancer effects [22] (Fig. 15.1). Irosustat **12** is in fact the first-in-class steroid sulfatase inhibitor to be used clinically in patients with advanced hormone-dependent cancers [22] but is also a potent CAI [1, 22].

Sulfonamides and sulfamates strongly inhibit CAs belonging to most families, not only the α-class enzymes [23–28]. Other compounds possessing potent CA inhibitory activity and developed for other uses are celecoxib **13** and valdecoxib **14**, originally described as cyclooxygenase 2 (COX-2) inhibitors [29, 30]. Saccharin **15**, widely used sweetener, is also a CAI, as recently reported by Klebe's groups [31]. Compounds **1–15** may be considered as first/second generation CAIs. Their main problem is that they indiscriminately inhibit most of the human isoforms known to date [1]. Indeed, 16 such isozymes were described in non-primates, CAs I-XV with two V-type isoforms, CA VA and CA VB, and 15 isoforms are known in primates, as CA XV is not expressed in these mammals [32].

Both sulfonamides and sulfamates bind in deprotonated form, as anions, to the Zn(II) ion from the enzyme active site, which is in a tetrahedral geometry as shown in Scheme 15.1, step E. CA inhibition data of all mammalian CAs with compounds **1–15** (and other sulfonamides in clinica use such as among other the thiazide and high-ceiling diuretics) were published in ref. [1] and will be not discussed here in detail.

The side effects of many of the compounds belonging to the first/second generation CAIs of types **1–15** (among which metabolic acidosis, kidney stones, bone loss, etc. [1, 4]) are due to the potent inhibition of all mammalian CA isoforms, and not only of the target one [1, 4]. Indeed, among the many mammalian isoforms investigated to date it is generally considered that CAs II, IV and XII are the targets for antiglaucoma agents [1]; CAs VA and VB for antiobesity agents [1], CAs IX and XII for antitumor agents/diagnostic tools for imaging of hypoxic tumors [1, 4]. Whereas it is not at all clear which isoforms are targeted by the antiepileptic sulfonamides/sulfamates, but CAs VII and XIV seem to be among them [1]. There are several isoforms for which the physiologic role/targeting with inhibitors are not clearly established and among them are CAs I, III, VI and XIII [1]. It should be also noted that CAs VIII, X and XI are acatalytic isoforms, i.e., they are devoid of CO_2 hydrase activity [1].

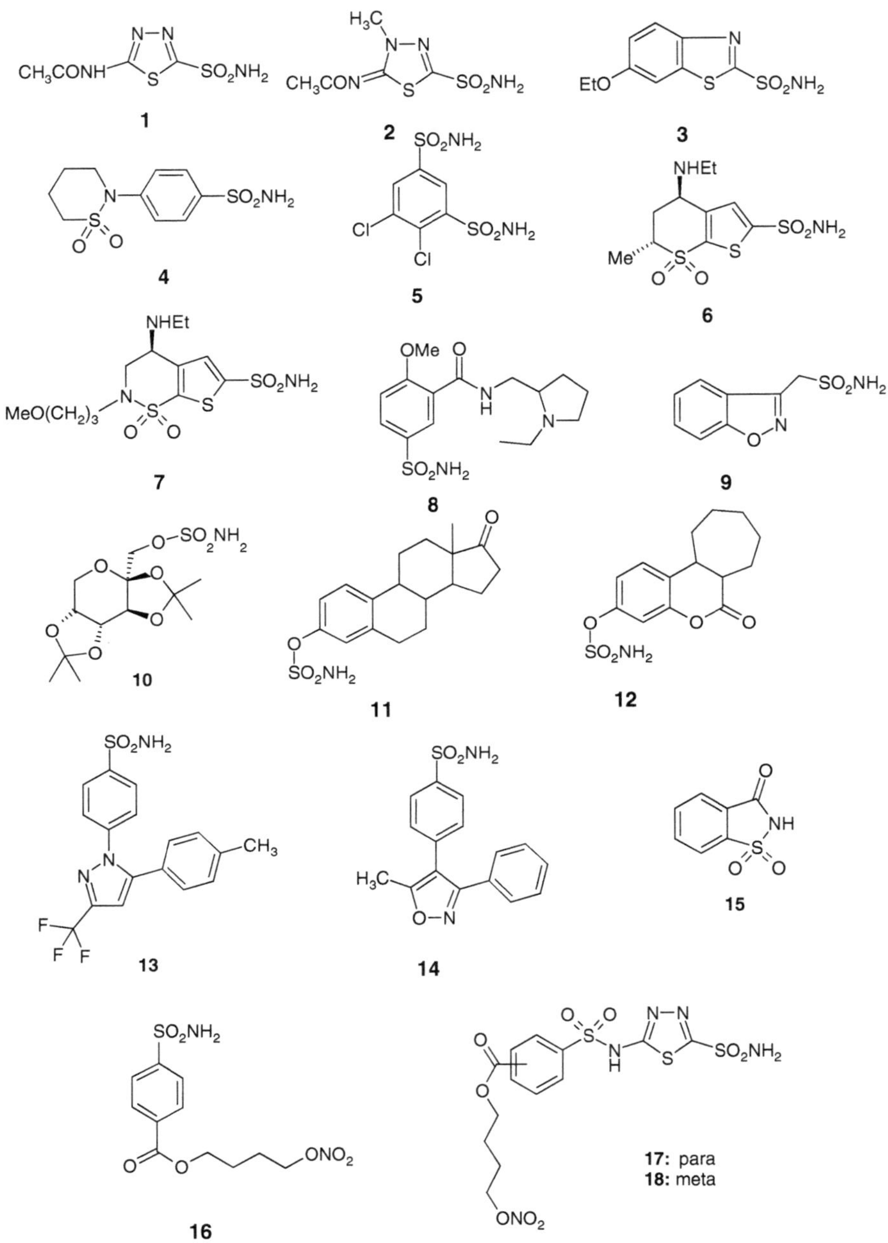

Fig. 15.1 Clinically used/preclinical sulfonamide and sulfamate CAIs **1–15** and compounds developed in the last period by the tail approach **(16–27)**

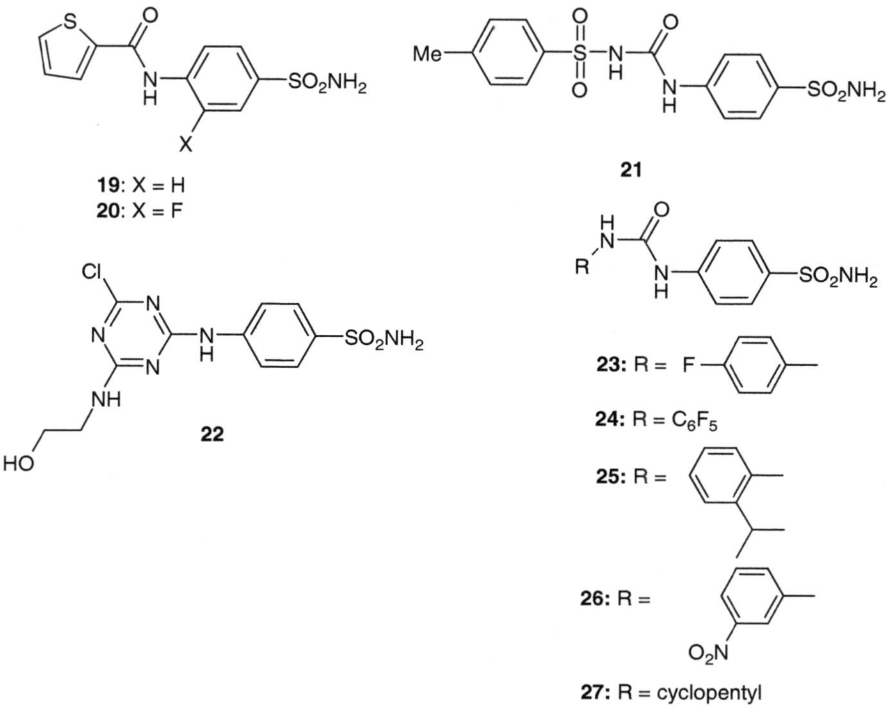

Fig. 15.1 (continued)

The main scope of the drug design campaigns in the last years has been to obtain isoform-selective CAIs for the various isoforms involved specifically in different pathologies, as detailed above. This is however not an easy task, considering that the 13 catalytically active isoforms have an active site architecture quite similar to each other [28, 33–36]. In fact, all human isoforms have the three conserved His residues (His94, 96 and 119) as zinc ligands, two other conserved residues acting as "gate keepers", i.e., Thr199 (hydrogen bonded through its -OH group with the water molecule/hydroxide ion coordinated to the zinc) and Glu106 (hydrogen bonded to Thr199, as well as half the active site mainly lined with hydrophobic residues and the opposite one with hydrophilic residues) [28, 33–36]. However, there are also important differences in amino acid residues mainly in the middle and towards the exit of the active site cavity [1, 4, 28, 33–36]. Most of the inhibitors mentioned above, of types **1–15**, do not make extensive contacts with them as their are rather compact molecules binding deep within the active site cavity which is rather similar in all mammalian isoforms [28, 33–36]. Thus, it appeared of considerable interest to devise alternative approaches for obtaining isoform-selective CAIs. One of the most successful one was termed "the tail approach" [37, 38]. It consists of appending "tails" to the scaffolds of aromatic/heterocyclic sulfonamides possessing derivable moieties of the amino, imino or hydroxy type, in such a way that an elongated molecule is obtained with its tail being able to interact with amino acid residues from the middle and the edge of the active site cavity [1, 37, 38]. By choosing tails with a diverse chemical nature, it was also possible to modulate the physico-chemical properties of such CAIs, which are crucial for their biological activity [1, 37, 38]. For example, antiglaucoma agents should possess an appropriate hydrophilicity and water solubility (in order to be formulated as eye drops) but also a balanced lipophilicity (for being able to penetrate through the plasma membranes and arrive at the ciliary processes where the enzymes responsible for aqueous humor secretion are found, such as CAs II, IV and XII) [37–39].

For targeting CAs IX and XII, which are transmembrane isoforms with an extracellular active site, and are predominantly found in tumors [40], charged, membrane-impermeable derivatives are preferred, which selectively inhibit only these and not the cytosolic/mitochondrial CAs [41–45]. These and other types of compounds could be obtained easily by this simple approach. Several examples will be provided in the following part of the chapter for the design of antiglaucoma hybrid drugs and antitumor sulfonamides targeting hypoxia. It should be noted that as this is a very dynamic field, in which many classes of compounds are being synthesized constantly and their structure in complex with various isoforms are reported every now and then. Due to space limitations, many interesting cases will be not discussed here. We concentrate the remaining part of this review on several relevant examples for which the structure-based drug design was also completed by in vivo data which may validate these derivatives as candidates for clinical trials.

Glaucoma is a disorder encompassing a group of ophthalmological diseases a common factor of which, is the occurrence of an optic neuropathy believed to be due to elevated intraocular pressure (IOP) [46, 47]. Treatment strategies to prevent glaucoma and the consequent irreversible vision loss are based on the reduction of IOP by using topically-acting or systemic hypotensive drugs [46, 47]. Sulfonamide CAIs, among which acetazolamide **1**, dichlorophenamide **5** (systemically-acting agents) as well as dorzolamide **6** and brinzolamide **7** (topically-acting drugs), are commonly used antiglaucoma agents [1, 47]. They reduce bicarbonate secretion in the aqueous humor by inhibiting CA isoforms present in the ciliary processes (CA II, IV and XII) with a consequennt reduction of IOP [46, 47]. Nitric oxide (NO), a radical gas produced by the enzyme nitric oxide synthase (NOS) is also involved in vasodilation, aqueous humour outflow within the eye, local modulation of ocular blood flow and retinal ganglion cells death by apoptosis [39, 48, 49]. It appeared thus of interest to combine these two pharmacophores, a sulfonamide CAI, and a moiety able to donate -NO, of the nitrate ester type, in the molecule of a hybrid drug by using the tail approach mentioned above [39, 48, 49]. In this way, a large number of sulfonamides with -NO donating properties were reported in the last several years, among which those of types **16–18** were the most interesting ones [39, 48, 49]. Several aromatic/heterocyclic sulfonamide scaffolds have been used to synthesize compounds incorporating NO-donating moieties of the nitrate ester type [48, 49]. Some of the new compounds showed effective in vitro inhibition of the

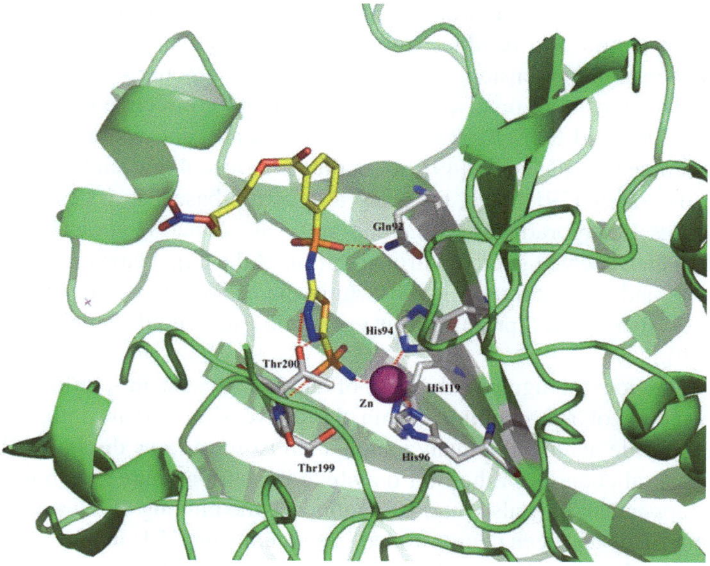

Fig. 15.2 hCA II in adduct with the sulfonamide incorporating an NO-donating moiety **18**, as obtained by X-ray crystallography [49]

target isoforms involved in glaucoma (in the low nanomolar range), and the X-ray crystal structure of one of them (compound **18**) revealed factors associated with this marked inhibitory activity (Fig. 15.2).

Compound **18** is a benzolamide derivative possessing a *meta*-COOH moiety on the phenyl ring which has been derivatized as a nitrate ester with an aliphatic, normal C_4 chain [49]. As for other hCA II-sulfonamide adducts investigated earlier [50–59], the deprotonated sulfonamide moiety of **18** was found coordinated to the Zn(II) ion at a distance of 1.96 Å [49]. The same NH group made a hydrogen bond with the OH of Thr199 (of 2.9 Å). Furthermore, the two endocyclic nitrogen of the 1,3,4-thidiazole ring participate in two hydrogen bonds (of 2.5–2.8 Å) with the -OH of Thr200, as reported earlier for a structurally related compounds [50–59]. One oxygen of the secondary SO_2 moiety of inhibitor **18** made a hydrogen bond (of 3.0 Å) with the NH_2 of Gln92. Due to the *meta*-substituent of the phenyl moiety present in **18** the conformation of the compound when bound to the hCA II active site is rather particular (Fig. 15.2). Indeed, it may be observed that the amino-1,3,4-thiadiazolyl-2-sulfamoyl moieties of this inhibitor is buried deep within the active site, as for other 1,3,4-thiadizoles for which the structure in adducts with various CAs have been reported [33, 34, 50]. However, the terminal fragment of the inhibitor (the 5-SO_2NH substituent of the thiadiazole ring and 3-substituted phenyl with the nitrate ester moiety incorporated in it) bind in an extended conformation which prolongs towards the external part of the active site, as expected for a molecule with such a long tail. This binding mode explains the potent hCA II inhibitory effects of the compound (K_I of 18 nM) which makes a large number of favorable interactions with various amino acid residues from the enzyme active site [49]. In an animal model of ocular hypertension, one of these new compounds incorporating NO-donating moieties, more precisely **16**, was twice more effective than dorzolamide in reducing elevated IOP characteristic of this disease, anticipating their potential for the treatment of glaucoma [49]. A detailed pharmacologic study of **16** was also reported thereafter [39]. Thus, by using a structure-based drug design approach, hybrid drugs incorporating sulfonamide and -NO donating moieties have been obtained, which showed good in vitro inhibition of the target enzymes, as they were observed bound in an interesting way within the active site of the enzyme. These compounds also showed promising in vivo action in animal models of glaucoma [39, 48, 49].

In the previous paragraphs the role of the tail present in a sulfonamide CAI for the binding of the inhibitor within the active site, and its influence on the inhibition profile against the many mammalian isoforms was discussed. However, small structural changes in the ring on which the sulfonamide zinc-binding group (ZBG) is appended, may also markedly influence the binding of the sulfonamide to the enzyme. This has been demonstrated in a recent work [54] in which the thienyl-carboxamido benzenesulfonamides **19** and **20** were crystallized bound to hCA II.

These two sulfonamides differ only by the presence of a fluorine atom in *meta* to the sulfonamide moiety in compound **20**. By means of X-ray crystallography it has been shown that the benzenesulfonamide fragments of the two compounds bind in a superposable manner to the enzyme [and in a similar fashion to most benzenesulfonamides crystallized up until now in complex with CAs [50–53]] but the orientation of the two thienylcarboxamide fragments was variable, as it also was the inhibition profile of these compounds against various CAs. For example, **20** was a weak hCA II inhibitor (K_I of 390 nM) whereas **19** was much more effective one (K_I of 50 nM) [60, 61]. Indeed, these compounds and some of their congeners showed isoforms selectivity for inhibiting CA VA/VB and CA VII [60, 61]. For example, **20** was a CA VII-selective inhibitor (over CA II), with a K_I for CA VII of 7 nM and a selectivity ratio for the inhibition of the two isoforms of 55.7 [60]. None of the clinically used sulfonamides shows such a selectivity ratio for inhibiting the brain enzyme CA VII [1].

Other interesting structure-based examples for obtaining isoform selective CAIs of the sulfonamide type are **21–27** which will be discussed here as tumor-associated, CA IX/XII-selective derivatives. Compound **21** incorporates a tosylureido tail [62]. This class of derivatives has been reported earlier [62] but only recently it was observed an interesting selectivity profile for inhibiting the transmembrane, tumor-associated isoforms (CAs IX and XII) over the cytosolic off-target CA II [56]. Indeed, **21** has K_Is of 12 nM against hCA II, 1.3 nM against hCA IX and 1.5 nM against hCA XII; whence, a selectivity ratio of 9.2 (hCA IX versus hCA II) and of 8.0 (hCA XII versus hCA II) [56]. However, the most CA IX/XII-selective compounds reported up until now are those incorporating triazinyl tails, of which **22** is an interesting example [56, 63, 64]. Compound **22** had inhibition constants of 1098 nM against hCA I, 37 nM against hCA II, 0.7 nM against hCA IX and 1.6 nM against hCA XII; whence, a selectivity ratios of 49.3 (CA IX versus CA II) and 23.1 (CA XII versus CA II) [56]. Some congeners of **22** showed even higher selectivity ratios (up to 700) for inhibiting the tumor-associated isoform hCA IX over hCA II [63, 64]. By comparing these inhibition data with the X-ray crystallographic structure of **21** and **22** in complex with hCA II [56], it was observed that the benzenesulfonamide fragment present in the two inhibitors binds in the expected manner (coordinating to the Zn(II) ion), being superposable for the two compounds. However, the tosylureido and substituted-triazinyl tails of the two compounds adopted extended conformation orientated towards opposing parts of the active site cavity [56]. This surely was reflected in the different inhibition/selectivity profiles of the two compounds, which were mentioned above.

Undoubtedly, the most interesting case of CA IX-selective compounds is furnished by derivatives **23–27**, which incorporate again the benzenesulfonamide head, but this time 4-aryl/alkylureido tails [45, 55]. A large series of such compounds has been reported [45] and for many of them good selectivity ratios (in the range of 16–53) for inhibiting CA IX over CA II were detected [45]. Another interesting feature for this series of compounds was that some of its members were also excellent hCA II inhibitors [in addition to strongly inhibiting the tumor-associated isoforms hCA IX and XII, in the low nanomolar range [45]]. For example, the inhibition constants of compounds **23–27** against hCA II were 96 nM, 50 nM, 3.3 nM, 15 nM and 226 nM, respectively [55]. The K_Is for the inhibition of hCA IX were 45 nM, 5.4 nM, 0.5 nM, 0.9 nM and 7.3 nM, respectively [45, 55]. By resolving the X-ray crystal structures of five such sulfonamides bound to hCA II, it was observed that the benzenesulfonamide fragments of the inhibitors were again quite superposable; whereas, the tails adopted a variety of conformations and orientations within the enzyme active site. This variability of binding was probably made possible by the flexible ureido linker between the benzenesulfonamide part and second moiety of the 1,3-disubstituted ureas **23–27**. Indeed, some of the groups from the terminal part of these molecules were observed in the hydrophobic pocket, others in the hydrophilic one and some of them between these two binding sites. Such different orientations may explain the range of inhibitory activity against hCA II (3.3–226 nM) and obviously hCA IX, as well as the selectivity ratios for inhibiting the two isoforms. However, a very interesting feature of some of these compounds (e.g., **26**) was their potent inhibition of growth of primary tumors and metastases in an animal model of breast cancer which strongly over-expresses CA IX [44, 45, 65]. In similar tumor cell lines without CA IX, no inhibition of the tumor growth has been observed, which demonstrated that the drug target of these compounds is indeed CA IX [44, 45, 65].

It may be observed from all data presented above that the sulfonamides dominated the drug design landscape of CAIs for many years. However, recently, new important chemotypes have emerged. Among them, the dithiocarbamates (DTCs) are undoubtedly very interesting ones [59, 66–68]. These compounds have been rationally discovered as CAIs after the report of trithiocarbonate (CS_3^{2-}) as a weak (milli – micromolar) CAI [14]. In the X-ray crystal structure of this inorganic anion bound to CA II, a monodentate coordination of the inhibitor is observed via one

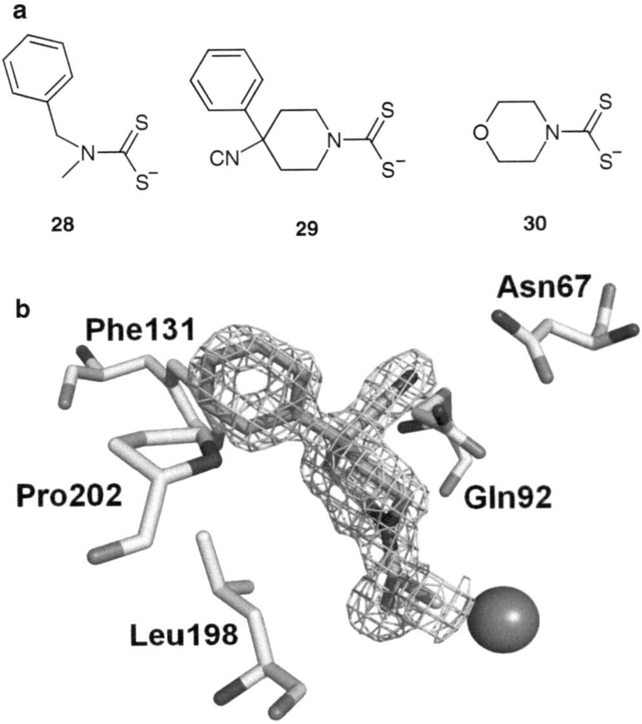

Fig. 15.3 **Panel A.** Structure of DTCs **28–30**. **Panel B.** Electronic density for the adduct of dithiocarbamate **29** bound within the active site of hCA II [59]. The zinc ion is shown as the central sphere and the amino acid residues involved in the binding are evidenced and numbered (hCA I numbering system) [59]

sulfur atom to the zinc ion from the enzyme active site, and a hydrogen bond in which another sulfur and the -OH of Thr199 were involved. Thus, the CS_2^- was discovered as a new ZBG. As DTCs incorporate this new ZBG, a rather large series of such compounds was then prepared and evaluated for their inhibitory activity against several mammalian, fungal and bacterial CAs [59, 66–68]. Several low nanomolar and subnanomolar CAIs were thus detected against all these isoforms. The X-ray crystal structures were also reported for three DTCs complexed to hCA II, compounds **28–30** (Fig. 15.3). DTCs **28–30** inhibited hCA II with K_Is of 25 nM, 41 nM and 0.95 nM, respectively, and hCA IX with K_Is of 53 nM, 757 nM and 6.2 nM, respectively [66]. As seen in Fig. 15.3, the binding mode of the ZBG present in **29** is identical to that of trithiocarbonate (with one sulfur coordinated to the metal ion), but the organic scaffold present in the investigated DTCs was observed to make extensive contacts with many amino acid residues from the active site, which may explain the wide range of inhibitory power of these derivatives [from the subnanomolar to the micromolar, for the entire series of around 30

DTCs reported so far [59, 66]]. Interestingly, the highly water soluble DTC **30** was also effective in vivo as an antiglaucoma agent when administered topically directly into the eye of hypertensive rabbits [66], a widely used animal model of glaucoma [39].

3 Drug Design Studies Targeting Fungal CAs with Sulfonamides, DTCs and Carboxylates

β-CAs were recently characterized in a high number of human pathogens, such as *Candida albicans, Candida glabrata, Cryptococcus neoformans, Saccharomyces cerevisiae* etc. [26, 69, 70]. Inhibition of such enzymes has been investigated with sulfonamides/sulfamates of types **1–14** among others [26, 27] and more recently with DTCs [67] and carboxylates [71]. Several ureidosulfonamides similar to those discussed above (compounds **23–27**) [27] or DTCs from the same series as compounds **28–30** [67] showed interesting inhibitory activity (low nanomolar inhibition) against many of the β-CAs from these fungal pathogens, but they will be not discussed here in detail. Instead, we will discuss some carboxylates as fungal β-CA inhibitors since the COO^- ZBG was less investigated in the drug design of CAIs until recently [11, 71]. Both aliphatic and aromatic carboxylates were investigated as inhibitors of α- and β-CAs in the recent period [11, 71]. The most interesting data have been obtained for the aromatic derivatives **31–46** shown in Table 15.1 [71].

It may be observed that these aromatic carboxylates (benzoic acid derivatives with various substitution patterns at the aromatic ring) were micromolar inhibitors of hCA I, hCA II (human off-target isoforms), Can2, Nce103, and β-CAs from the fungal pathogens. However, some of them showed low nanomolar activity against the fungal pathogenic enzymes. For example, 4-hydroxy- and 4-methoxy-benzoate inhibited Can2 with K_Is of 9.5–9.9 nM and Nce103 with inhibition constants of 0.1–0.2 μM. The methyl esters, hydroxamates, hydrazides and carboxamides of some of these derivatives were also effective inhibitors of these α- and β-CAs. Unfortunately, no X-ray crystal structures for the binding of these inhibitors to the fungal enzymes are available but Can2 has been co-crystallized in complex with acetate, a weak inhibitor [26]. Acetate was found directly bound to zinc in a monodentate manner, substituting the zinc-bound water molecule, with the metal ion in a slightly distorted tetrahedral geometry [26, 70]. Thus, the structure-based drug design of CAIs targeting the fungal enzymes is still in its infancy, due to the very limited number of X-ray structures of enzyme-inhibitor adducts available at this moment [26, 70]. It is anyhow rather probable that the benzoates/DTCs investigated later [67, 71] bind in a similar manner to the enzyme but the detailed interactions in which the organic scaffold participate are unknown for the moment.

Table 15.1 Inhibition constants of aromatic carboxylates **31–46** against isozymes hCA I and II (α-CA class), and β-CAs Can2 (from *Cryptococcus neoformans*) and Nce103 (from *Candida albicans*), for the CO_2 hydration reaction, at 20°C. The standard sulfonamide inhibitor acetazolamide (**1**) was also included as standard [71]

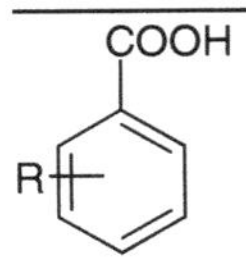

31-46

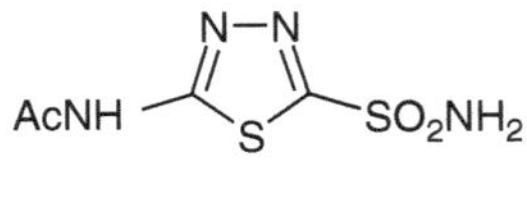

1

No	R[a]	hCA I[c]	K_I (μM)[b] hCA II[c]	Can2[d]	Nce103[d]
31	H	730	30	1,267[c]	24.4[c]
32	2,3,5,6-F_4-	74	6	49.1[c]	61.6[c]
33	4-F-	7.39	3.34	0.13	1.09
34	4-Cl-	7.61	3.81	3.40	2.65
35	4-Br-	8.00	4.70	4.24	4.18
36	4-I-	8.18	5.85	5.46	5.17
37	4-O_2N-	8.80	8.56	0.12	4.85
38	4-HO-	7.61	7.28	0.0095	0.11
39	4-MeO-	9.06	8.03	0.0099	0.15
40	4-H_2N-	8.40	8.13	0.12	2.95
41	3-H_2N-	8.79	8.63	3.06	3.40
42	4-NC-	6.92	6.79	0.14	4.47
43	4-AcHN-	4.04	7.28	0.32	0.34
44	4-$ClCH_2$-	3.88	4.36	3.65	4.71
45	4-Me_2N-	3.94	3.33	0.13	5.20
46	4-Cl_2NSO_2-	3.88	1.96	0.12	4.97
1		0.25	0.012	0.010	0.13

[a] As sodium salt
[b] Errors were in the range of 3–5 % of the reported values, from three different assays
[c] Human cloned isoforms, assay at pH 7.5
[d] Fungal enzymes, at pH 8.3

4 Targeting Bacterial CAs

Inhibition of bacterial CAs was recently reviewed [8]. Many pathogenic bacteria encode such enzymes belonging to the α-, β-, and/or γ-CA families [8]. In the last years, the α-CAs from *Neisseria spp.* and *Helicobacter pylori* as well as the β-class enzymes from *Escherichia coli, H. pylori, Mycobacterium tuberculosis, Brucella spp., Streptococcus pneumoniae, Salmonella enterica Haemophilus influenzae* and *Vibrio cholerae* have been cloned and characterized in detail [72–86]. For some of these enzymes the X-ray crystal structures were determined, and in vitro and in vivo inhibition studies with various classes of inhibitors, such as anions, sulfonamides and sulfamates reported, but no X-ray structures in complex with inhibitors are available so far. Bacterial CAs represent promising targets for obtaining antibacterials devoid of the resistance problems of the clinically used such agents but further studies are needed to validate these and other less investigated enzymes as novel drug targets [72–83].

Table 15.2 CAs from pathogenic bacteria cloned and characterized so far, and their inhibition studies [8]. The genome of many other bacteria contain CAs which have not yet been cloned and characterized

			Inhibition study	
Bacterium	Family	Name	In vitro	In vivo
Neisseria gonorrhoeae	α	–	Sulfonamides, anions	Sulfonamides
Neisseria sicca	α	–	Sulfonamides	Sulfonamides
Helicobacter pylori	α	hpαCA	Sulfonamides, anions	Sulfonamides
	β	hpβCA	Sulfonamides, anions	Sulfonamides
Escherichia coli	β	–	NI	NI
Haemophilus influenzae	β	HICA	Bicarbonate	NI
Mycobacterium tuberculosis	β	mtCA 1	Sulfonamides	Phenols
	β	mtCA 2	Sulfonamides	NA
	β	mtCA 3	Sulfonamides	NA
Brucella suis	β	bsCA 1	Sulfonamides	Sulfonamides
	β	bsCA2	Sulfonamides	Sulfonamides
Streptococcus pneumoniae	β	PCA	Sulfonamides, anions	NI
Salmonella enterica	β	stCA 1	Sulfonamides, anions	NI
	β	stCA 2	Sulfonamides, anions	NI
Vibrio cholerae	α	VchCA	Sulfonamides, anions	NI

- means not named, *NA* no activity in vivo (presumably due to penetration problems), *NI* not investigated

The enzymes which have been so far characterized in detail are shown in Table 15.2. It can be observed that there are detailed data regarding the CAs present in their genome only for nine bacterial pathogens and some of them cause serious infections worldwide, with important antibiotic resistance problems too [8]. Many of them contain more than one such enzyme (e.g., *M. tuberculosis* has 3 β-CAs; *B. suis* two such enzymes, *H. pylori* one α- and one β-CA, etc.), and in many cases there are members both from the α- and β-class in the same pathogen (e.g., *H. pylori*). However, most bacteria for which at least part of the genome was sequenced contain such enzymes and the near future will probably see important developments. As for fungi, it seems that the β-CAs are the most abundant such enzymes among bacteria.

As shown in Table 15.2, the inhibition of these enzymes has been studied in vitro with the main classes of inhibitors, i.e., inorganic anions and sulfonamides/sulfamates [the clinically used derivatives **1–14** as well as compounds which were designed on purpose to target the bacterial enzymes [72–81]]. The *M. tuberculosis* CAs are the most investigated ones due to the extensive drug resistance problems of this pathogen to the clinically used antibiotics. Thus, the three CAs from this bacterium were investigated also for their inhibition with DTCs [68] and carboxylates [83].

Many sulfonamide/sulfamate CAIs showed low nanomolar in vitro inhibition against many bacterial CAs [72–81], but in vivo, only for *Nessseria spp., H. pylori, B. suis*, and *S. pneumoniae* enzymes it has been possible to evidence inhibition of bacterial growth in vivo [8, 84, 85]. Probably the lack of correlation between the in vitro CA inhibition and in vivo inhibition of growth of the bacterial pathogen is due to penetration problems of the sulfonamides, which are highly polar molecules [8]. However, very recently, in vivo inhibition of growth of *M. tuberculosis* has been evidenced for several phenol type CAIs [87].

Indeed, in the last period, interesting chemotypes different of the sulfonamides were investigated as bacterial CAIs [68, 83]. Thus, a series of *N*-mono- and *N,N*-disubstituted DTCs of types **47–73**, recently discovered CAIs as shown earlier [59], were investigated as inhibitors of two β-CAs from the bacterial pathogen *M. tuberculosis*, mtCA 1 (Rv1284) and mtCA 3 (Rv3273) [68] – Table 15.3. Both enzymes were inhibited with efficacies between the subnanomolar to the micromolar one, depending on the substitution pattern at the nitrogen atom from the DTC ZBG. Aryl, arylalkyl-, heterocyclic as well as aliphatic and amino acyl such moieties led to potent mtCA 1 and 3 inhibitors in both the *N*-mono- and *N,N*-disubstituted dithiocarbamate series. For example, DTC **70** was a subnanomolar CAI of both mycobacterial enzymes [68]. This new class of β-CA inhibitors may have the potential for developing antimycobacterial agents with a diverse mechanism of action compared to the clinically used drugs for which many strains exhibit multi-drug/extensive multi-drug resistance [68]. No X-ray structures are available so far for adducts of these β-CAs with DTCs, but already the first drug design study identified extremely potent CAIs from this series. Work is in progress to evaluate the efficacy of such compounds in vivo.

The growth of *M. tuberculosis* is known to be strongly inhibited by weak acids although the mechanism by which these compounds act is not well understood [83]. A series of substituted benzoic acids, nipecotic acid, *ortho*- and *para*-coumaric acid, caffeic acid and ferulic acid, of types **74–83** were recently investigated as inhibitors of three β-CAs from this pathogen, mtCA 1 (Rv1284), mtCA 2 (Rv3588c) and mtCA 3 (Rv3273) [83] – Table 15.4. All three enzymes were inhibited with efficacies between the submicromolar to the micromolar one, depending on the scaffold present in the carboxylic acid. mtCA 3 was the isoform mostly inhibited by these compounds (K_Is in the range of 0.1–1.0. μM); followed by mtCA 2 (K_Is in the range of 0.6–8.1 μM) whereas against mtCA 1 these carboxylic acids showed inhibition constants in the range of 2.2–7.1 μM (Table 15.4). This class of relatively under explored β-CAIs warrants further, in vivo studies, as they may have the potential for developing antimycobacterial agents with a diverse mechanism of action compared to the clinically used drugs for which many strains exhibit multi-drug or extensive multi-drug resistance [83].

Table 15.3 Inhibition data of the human (h) α-CA isoforms hCA I and II, and mycobacterial β-CA isoforms mtCA 1 and 3 with dithiocarbamates **47–73** by a stopped-flow, CO_2 hydrase assay [68]

$R^1R^2N\text{-}CSS^-M^+$

47-73

			K_I (nM)[a]				
Cmpnd	R^1	R^2	hCA I	hCA II	mtCA 1	mtCA 3	M
47	H	Ph	4.8	4.5	5.6	2.5	Et_3NH
48	H	$O((CH_2CH_2))_2N$	4.8	3.6	6.1	2.4	K
49	H	$MeN((CH_2CH_2))_2N$	33.5	33.0	4.7	2.6	K
50	H	2-butyl	21.1	29.4	6.0	3.6	K
51	H	$O((CH_2CH_2))_2N(CH_2)_2$	31.8	36.3	7.1	2.8	K
52[b]	H	$N((CH_2CH_2)N)_3$	31.9	13.5	4.2	4.0	K
53	H	$PhCH_2$	4.1	0.7	7.1	87.3	Na
54	H	4-PyridylCH_2	3.5	16.6	5.4	5.7	Et_3NH
55	H	$((CH_2)_5N)CH_2CH_2$	4.5	20.3	9.1	8.8	K
56	H	2-thiazolyl	3.9	4.6	89.4	9.5	Et_3NH
57	H	$KOOCCH_2$	13.1	325	7.8	8.3	K
58	H	imidazol-1-yl-$(CH_2)_3$	8.6	24.7	5.3	8.7	A
59	Me	Me	699	6,910	893	659	Na
60	Et	Et	790	3,100	615	431	Na
61		$(CH_2)_5$	0.96	27.5	90.5	4.1	Na
62	n-Pr	n-Pr	1,838	55.5	74.8	80.0	Na
63	n-Bu	n-Bu	43.1	50.9	81.7	72.8	Na
64	iso-Bu	iso-Bu	0.97	0.95	86.2	43.0	Na
65	n-Hex	n-Hex	48.0	51.3	95.4	51.7	Na
66	Et	n-Bu	157	27.8	91.6	63.5	Na
67	$HOCH_2CH_2$	$HOCH_2CH_2$	9.2	4.0	7.5	6.0	Na
68	Me	Ph	39.6	21.5	25.2	46.8	Na
69	Me	$PhCH_2$	69.9	25.4	72.0	62.5	Na
70		$O((CH_2CH_2))_2$	0.88	0.95	0.94	0.91	Na
71		$NaS_2CN((CH_2CH_2))_2$	12.6	0.92	7.7	8.0	Na
72		$(NC)(Ph)C(CH_2CH_2)_2$	48.4	40.8	93.0	61.2	N
73[c]	(*S*)-$(CH_2CH_2CH_2$ $CH(COONa))$		2.5	17.3	7.1	6.4	Na
1	acetazolamide		250	12	481	104	–

A = imidazol-1-yl-$(CH_2)_3NH_3^+$

[a]Errors in the range of ± 10 % of the reported values, by a CO_2 hydrase assay method [68]

[b]Tris-dithiocarbamate

[c](*S*)-proline dithiocarbamate

Table 15.4 Inhibition data of mycobacterial β-CA isoforms mtCA 1–3 with carboxylic acids **74–83** and sulfanilamide **84** by a stopped-flow, CO_2 hydrase assay [83]

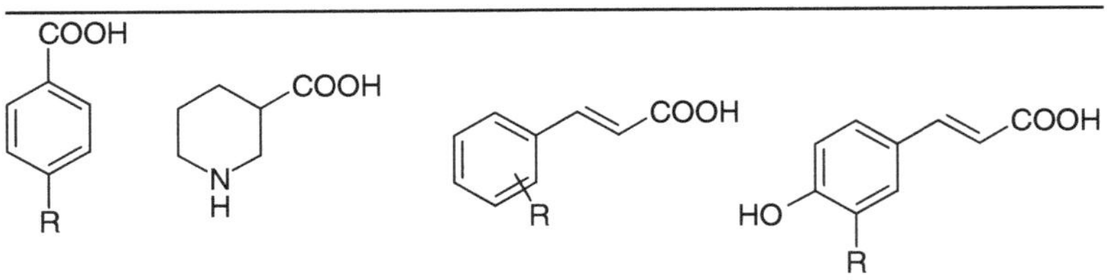

74-78 79 80,81 82,83

		K_I (μM)[a]		
Cmpnd	R	mtCA 1	mtCA 2	mtCA 3
74	Br	3.91	2.66	0.97
75	I	4.94	2.14	0.14
76	CN	6.20	2.12	0.19
77	AcNH	4.76	3.67	0.48
78	NO_2	5.87	8.10	0.52
79	–	4.89	5.27	0.11
80	2-OH	2.25	4.17	0.12
81	4-OH	3.03	1.56	0.60
82	OH	5.92	0.59	0.57
83	OMe	7.13	3.50	0.36
84	sulfanilamide	9.84	29.6	7.11

[a]Errors in the range of ± 10 % of the reported values, by a CO_2 hydrase assay method (from three different measurements) [30]

5 Alternative CA Inhibition Mechanisms: Phenols, Polyamines, Coumarins/Lactones, Sulfocoumarins and Proteinaceous Inhibitors

Phenol **85** was investigated as a CAI several decades ago, but its X-ray crystal structure in adducts with hCA II has been reported only in 1994 by Christianson's group [88]. As shown from the schematic representation of Fig. 15.4A, phenol has a totally new inhibition mechanism compared to the sulfonamides or inorganic anions (the only classes of CAIs known at that time), as this molecule is not coordinated to the Zn(II) ion but anchored to the zinc coordinated water molecule/hydroxide ion (it is most probable that the last species is the one present in the adduct, at the pH of 7.5 at which the experiments were performed [88]). Afterwards, a rather wide range of synthetic and natural product phenols were investigated for the inhibition of most mammalian CAs [89–93]. Many of them were medium potency CAIs, with K_Is typically in the millimolar range, although some more effective compounds (among the natural products with a more complex scaffold [92] were also detected). Recently, Martin and Cohen [94] reported the X-ray structure of several other such compounds (some of which also incorporate carboxylate moieties) which bind in a

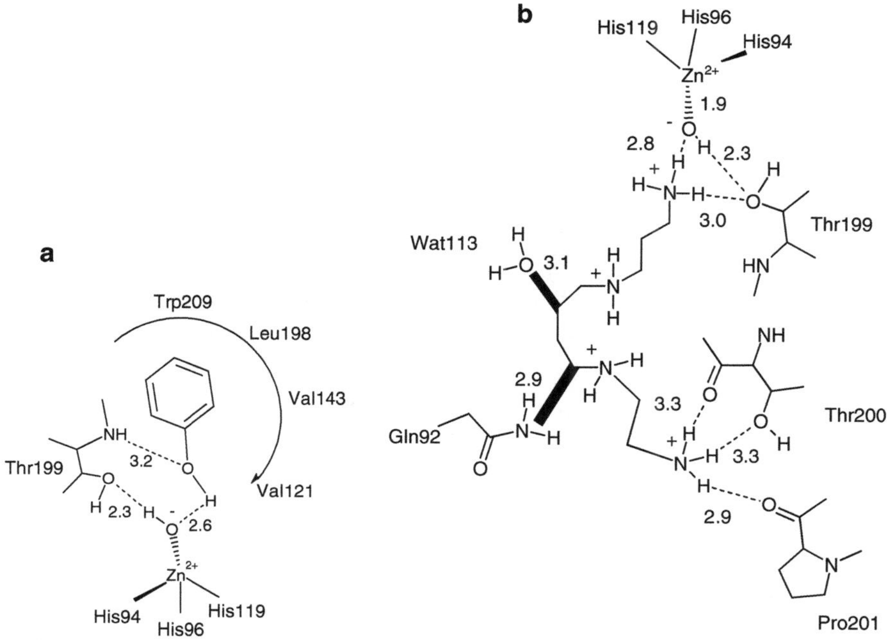

Fig. 15.4 Panel A. Schematic representation of the binding of phenol **85** to hCA II as determined by X-ray crystallography [88]. **Panel B.** Schematic representation of interactions in which spermine **86** (as tetracation) participates when bound to the hCA II active site. Figures represent distances (in Å). Hydrogen bonds are represented as dashed lines. In bold are shown two clashes involving some carbon atoms (C5 and C7) of the spermine scaffold with a water molecule (Wat113) and Gln92. The non-protein zinc ligand is represented as a hydroxide ion, which should be the preponderant species at the pH at which the experiments were done (7.5) [18]

fashion similar to the simple phenol **85**, i.e., anchoring to the non-protein zinc ligand by means of a hydrogen bond. Afterwards, the polyamine spermine **86** was also shown to act as a weak CAI and its crystal structure resolved in adduct with hCA II [18]. Unexpectedly, as seen from Fig. 15.4A, spermine has a similar mechanism of CA inhibition with phenol, as one of its ammonium terminal moieties is anchored to the zinc-coordinated water molecule/hydroxide ion by means of a hydrogen bond.

Figure 15.5 shows a superposition of three CAIs bound to hCA II, as determined by X-ray crystallography, the phenol **85**, spermine **86** and the hydrolysed coumarin (*trans*-2-hydroxycinnamic acid) **87** [18, 20, 88]. The first two compounds share the same inhibition mechanism but due to their chemical nature they occupy different parts of the active site. As will be discussed shortly, coumarins constitute a completely different class of CAIs.

Coumarins such as the simple derivative **87a** or the natural product **88a**, isolated from the Australian plant *Leionema ellipticum*, P.G. Wilson (Rutaceae) [19], were recently discovered to act as effective CAIs [19, 20]. Coumarins **87a** and **88a**

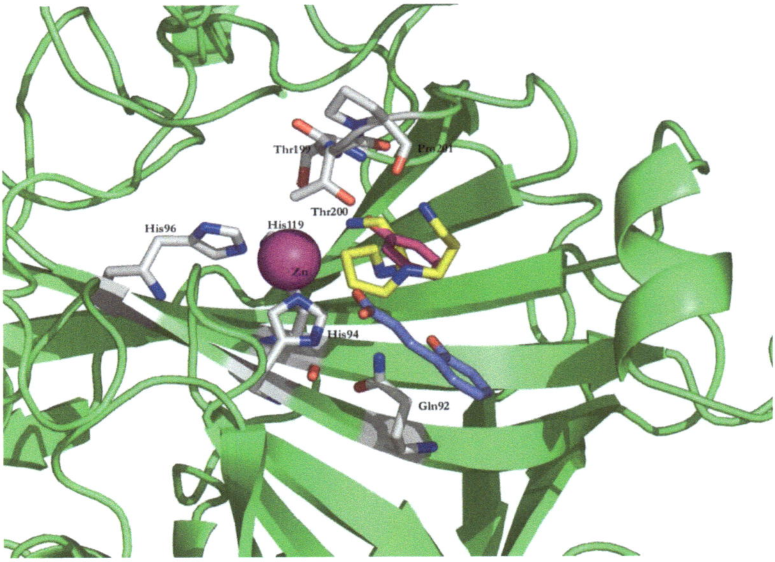

Fig. 15.5 Superimposition of the spermine **86** (*yellow*) [18], phenol **85** (*magenta*) [88] and *trans*-2-hydroxycinnamic acid **87b** [20] (*violet*) adducts with hCA II. The Zn(II) ion is the violet sphere. The Zn(II) histidine ligands as well as residues 92 and 199–201 involved in the binding of inhibitors are also evidenced. One terminal amino moiety of spermine (which hydrogen bonds the non-protein zinc ligand) is fully superimposed on the phenol OH moiety which is interacting with the same group of the enzyme. The phenyl moiety of phenol and the organic scaffold of spermine are not superimposable. 2-Hydroxycinnamic acid **87b** (in *violet*), the hydrolysis product of coumarin **87a** binds towards the exit of the active site and is not superposable with phenol **85** and spermine **86**

were shown to act as "prodrug" inhibitors, being hydrolyzed by the esterase CA activity to the corresponding 2-hydroxy-cinammic acids **87b** and **88b**, which are the de facto CAIs, as shown by kinetic, X-ray crystallographic and MS methods (Scheme 15.2) [19, 20]. Interestingly, these compounds were found bound at the entrance of the CA active site cavity in a region where only CA activators were observed earlier in the many CA – modulators of activity adducts reported in the literature [15, 16]. In this region of the enzyme active site, the classical CAIs such as the sulfonamides, sulfamates, sulfamides, DTCs or other (in)organic anions were never observed [1, 2, 9, 11, 12]. A large number of diversely substituted-coumarins was subsequently screened for their inhibitory activity against all the 13 catalytically active mammalian CA isoforms [95–99]. Thus, coumarins are a relevant class of CAIs because: (i) they are mechanism-based, "prodrug" inhibitors; (ii) they bind in a totally different active site region compared to the classical inhibitors (sulfonamides and congeners); and (iii) they led to highly isoform-selective compounds for many mammalian CA isoforms, such as CAs IX, XII, XIII and XIV among others. In addition, coumarins or they derivatives are easy to synthesize, being possible to incorporate in their molecule a large variety of substitution patterns, which

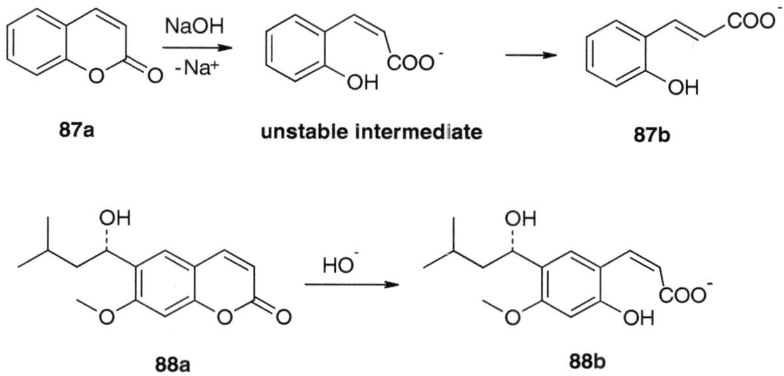

Scheme 15.2 Active site, CA-mediated hydrolysis of coumarins **87a** and **88a** affording the hydroxycinnamic acids **87b** or **88b,** as evidenced by X-ray crystallography or enzyme-inhibitor adducts [19, 20]

lead to the possibility of exploring a vast chemical space, difficultly accessible for other classes of CAIs. For example, thiocoumarins, 2-thioxocoumarins, dithiocoumarins, coumarin-N-oximes, 5-/6-membered lactones and thiolactones, or lactone oximes were recently investigated for their CA inhibitory properties [95–100] (Scheme 15.3). "Sulfocoumarins", i.e., 1,2-benzoxathiin 2,2-dioxides were also very recently explored as CAIs [101].

As for the coumarins, the sulfocoumarins are mechanism based inhibitors [101]. By means of X-ray crystallography it has been observed (in a mutant hCA II complexed with compound **89a**) that the hydrolysis product of **89a**, i.e., 2-dihydroxy-5-bromophenyl-vinyl sulfonic acid **89b** (Scheme 15.4) perfectly fitted within this electron density (Fig. 15.6).

As seen in Fig. 15.6, the electron density of **89b** is perfectly defined for all its atoms. It is probable that the sulfocoumarin **89a** is initially hydrolyzed (by the zinc hydroxide, nucleophilic species of the enzyme) to the *cis* vinyl sulfonic acid unstable intermediate which subsequently isomerizes spontaneously to the *trans* derivative **89b** (Scheme 15.4) which is observed in the electron density of the CA II/IX – **89a** adduct (Fig. 15.6). Interestingly, the sulfonic acid moiety (presumably as a sulfonate group, at the pH of 7.8 at which the crystallization was done) was not coordinated to the Zn(II) ion but was anchored to the zinc-bound water molecule/hydroxide ion (Fig. 15.6), a binding mode initially observed for phenol and spermine (and discussed above). With the X-ray structure of the sulfocoumarin adduct, it is obvious that anchoring to the zinc-coordinated water molecule may be considered as a quite general CA inhibition mechanism [we prefer to use the term "anchoring to the zinc-coordinated water" for the inhibition mechanism of these compounds, as originally proposed by Christianson [88] over the more recent term "nucleophile recognition" proposed by Martin and Cohen [94]]. The distance between the zinc-coordinated water/hydroxide ion and an oxygen atom of the sulfonate moiety of **89b** was of

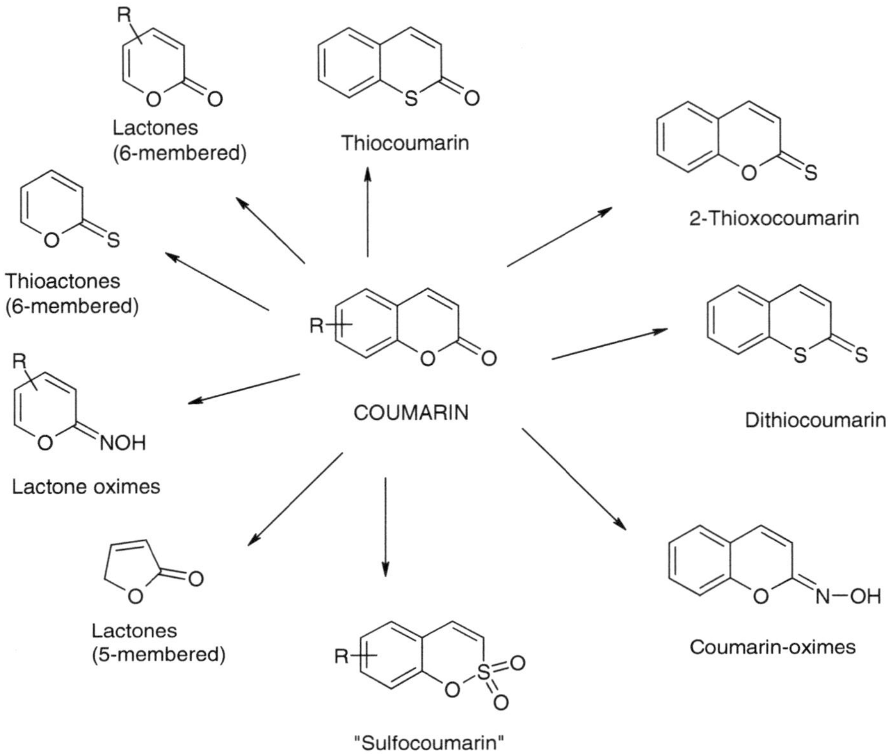

Scheme 15.3 CAIs obtained from coumarins as lead molecules [95–101]

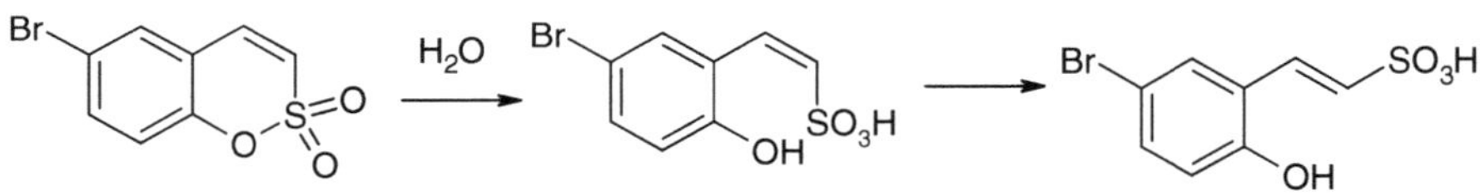

89a unstable intermediate **89b**

Scheme 15.4 Active site CA-mediated hydrolysis of **89a**

2.5 Å. Another oxygen of this moiety was at 2.8 Å from the metal ion, being "half-coordinated". The organic scaffold of **89b** did not participate in other interactions with residues from the enzyme active site, except for the -OH moiety in *ortho* to the ethenylsufonate group. This moiety participated in a bifurcated hydrogen bond with the hydroxyl of Thr200 (of 3.3 Å) and through a bridging water molecule, with the carbonyl oxygen of Pro201 (of 2.2 Å). The relatively few interactions between the inhibitor scaffold and the active site may in fact explain the not so good inhibitory power of this compound against the CA II/IX mimic [K_I of 0.9 μM [101]].

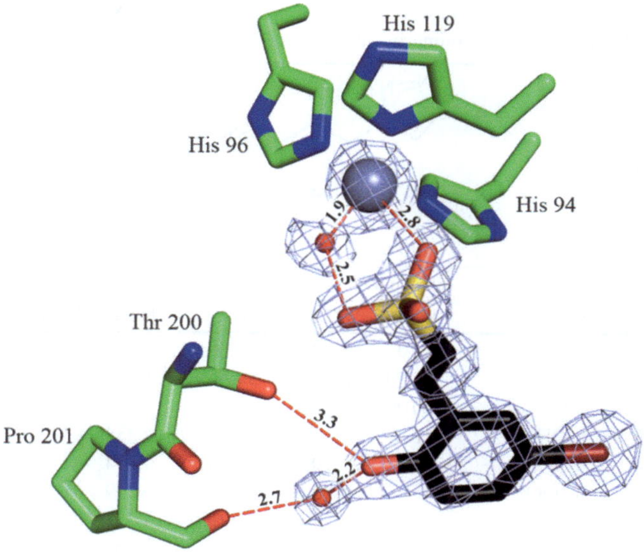

Fig. 15.6 Binding of compound **89b** within the active site of the CA II/IX mimic [101]. The $2F_o$-F_c map was calculated in the absence of ligand and contoured at 1σ. The hydroxyphenylvinylsulfonic acid (in *black*), the zinc ion (*grey sphere*), its ligands (His94, 96 and 119) and water molecules (*red spheres*) are shown. Polar interactions (with Thr200 and Pro201) are shown with dashed lines and the distances are indicated in Å

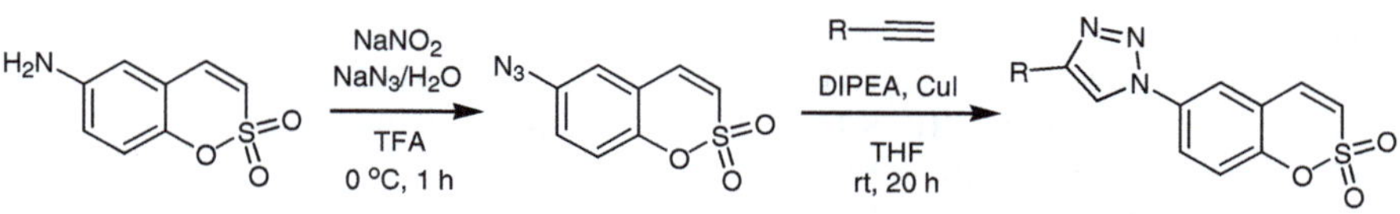

89a **89h** **90-99**

Scheme 15.5 Preparation of derivatives **90–99** by reaction of azide **89j** with alkynes (click chemistry) [101]

A series of mono- or disubstituted sulfocoumarins possessing various functionalities on the benzene ring, of types **89a-i** were reported recently and investigated as CAIs [101]. They were synthesized by an intramolecular aldol cyclization reaction of mesylsalicyl aldehydes in the presence of strong bases, such as 1,8-diazabicyclo(5.4.0)undec-7-ene (DBU). The amine **89 h** was diazotized and subsequently converted to the azide **89j**, which by reaction with alkynes was transformed to the 1,2,3-triazoles **90–99** by the classical click chemistry cycloaddition reaction (Scheme 15.5). Compounds **90–99** incorporated several groups at the triazole ring (such as substituted aryl, carboxyalkyl, substituted silyl, alkylaminomethyl, etc.), being structurally diverse when compared to compounds **89a-i**. These structural variations are essential in order to investigate the structure-activity relationship of this new class of CAIs. Indeed, as seen from Table 15.5, these compounds possessed

Table 15.5 Inhibition of isozymes hCA I, II, IX, XII and a CA II/IX active site mimic with coumarins **87a, 88a**, and sulfocoumarins **89a-89i** and **90–99** [101]

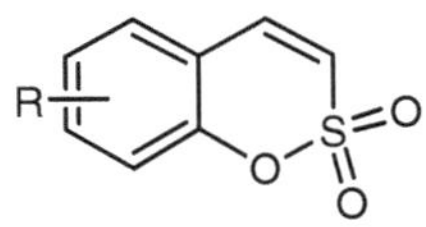

89a-i

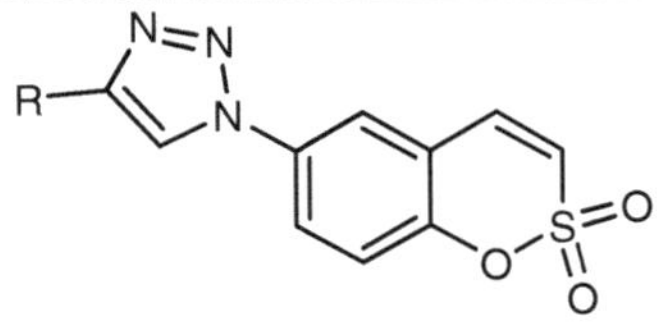

90-99

Compound	R	K_I (μM)^{a,b}				
		hCA I^c	hCA II^c	hCA IX^d	hCA XII^d	CA II/IX mimic
87a	–	3.10	9.20	>100	>100	–
88a	–	0.080	0.062	54.5	48.6	–
89a	6-Br	>100	>100	6.83	4.51	0.93
89c	6-OH	91	>100	0.300	0.234	0.80
89d	6-$MeSO_3$	99	>100	0.324	0.254	2.03
89e	6-BnO	93	>100	0.275	0.219	2.37
89f	5,6-Benzo	>100	>100	0.375	0.717	–
89 g	6-O_2N	92	>100	3.77	3.16	–
89 h	6-H_2N	6.78	8.89	0.046	0.023	–
89i	6,8-Cl_2	>100	>100	3.26	2.93	–
90	Ph	6.86	7.76	0.029	0.032	–
91	COOMe	8.05	6.33	0.095	0.012	–
92	COOEt	8.88	9.21	0.086	0.013	–
93	Me_3Si	6.00	7.20	0.060	0.009	–
94	$HOCH_2$	7.20	9.29	0.058	0.016	–
95	Et_2NCH_2	8.11	9.37	0.025	0.007	–
96	4-F_3CO-C_6H_4	8.43	9.64	0.074	0.014	–
97	4-MeO-C_6H_4	8.93	9.35	0.018	0.039	–
98	3-F_3C-C_6H_4	6.71	7.72	0.048	0.013	–
99	3-MeO-C_6H_4	7.47	8.61	0.049	0.021	–
Acetazolamide	–	0.25	0.012	0.025	0.005	–

[a]Errors in the range of ± 10 % of the reported data, from three different assays
[b]Pre-incubation of 6 h between enzyme and inhibitor
[c]Cytosolic full length, recombinant enzyme
[d]Catalytic domain, transmembrane, recombinant isoform

interesting CA inhibitory activity against four isoforms, hCAs I, II, IX and XII. As the coumarins [95–99], many sulfocoumarins were also relatively weak inhibitors of hCAs I and II, whereas they act as much more powerful hCA IX and XII inhibitors (Table 15.5): It may be observed that sulfocoumarins **90–99** were low nanomolar inhibitors of these tumor-associated isoforms, whereas their affinity for hCAs I and II was in the micromolar range. The simpler compounds **89a-g** and **89i** were submicromolar hCA IX/XII inhibitors and did not significantly inhibit hCAs I and II (Table 15.5). Thus, the sulfocoumarins, like the coumarins, lead to isoform-selective CAIs targeting the tumor-associated, transmembrane isoforms hCAs IX and XII [101].

6 Conclusion

By catalyzing the simple but highly important hydration of CO_2 to bicarbonate and protons, CAs are involved in critical steps of the life cycle of many organisms, including eukaryotes, bacteria and archaea. A wealth of X-ray structural data have been accumulated in the last 15 years for CA-inhibitor complexes, including the main classes of inhibitors: the pharmacologically relevant sulfonamides and their isosteres (sulfamates, sulfamides, ureates and hydroxamates) and the simple inorganic anions, but also the less investigated ones such as the carboxylates, phenols, polyamines, coumarins/sulfocoumarins and the newly identified DTCs. Although X-ray crystal structures are already available for the majority of the 12 catalytically active hCAs (i.e., isozymes I-VA, IX, XII, XIII and XIV), most of the reported complexes are for isozyme II (and to a less extent isozyme I). These data are relevant for the drug design of isozyme-selective CAIs, and important such advances have been made in the last years, mainly by rationalizing the various subpockets for the binding of sulfonamide/sulfamide inhibitors done by McKenna's group [102, 103]. In fact, the main problem with the classical, clinically used sulfonamides (including also the second generation agents dorzolamide and brinzolamide) is related to the fact that they are promiscuous inhibitors of all (or most of the) CA isozymes found in mammals. Some low levels of isozyme selectivity are shown by dorzolamide and brinzolamide, which have been designed in such a way as to act as much weaker CA I than CA II inhibitors, but similar to acetazolamide, methazolamide and ethoxzolamide, these two second generation inhibitors strongly inhibit the remaining ten CA isozymes [1, 3]. Thus, considering only the ZBGs and the organic scaffold, it is practically impossible to design isoform-selective CAIs, as the interactions around the metal ion and the organic scaffold (normally positioned at the bottom and in the middle of the active site cavity, respectively) are basically the same between the inhibitors and most CA isozymes with medicinal chemistry applications [1, 50]. This also explains why the first and second generation sulfonamide/sulfamate CAIs were devoid of any isozyme-selectivity. They are indeed rather small, compact molecules which bind deeply within the enzyme active site. However, 13 years ago the "tail-approach" was reported, which afforded a rather facile synthesis of a large number of CAIs starting from aromatic/heterocyclic scaffold also containing derivatizable amino, imino or hydroxy groups, to which various moieties (tails) were introduced by normal chemical modification reactions (acylation, alkylation, arylsulfonylation, condensation, cycloaddition, glycosylation, etc.) [1, 50]. In this way, it was possible both to modulate the physico-chemical properties of the synthesized inhibitors (for example by introducing tails which induce high water solubility, or enhanced lipophilicity, or positive/negative charges which lead to membrane impermeability, or fluorescence or spin-labeled groups, etc.), and also their affinity to the various isozymes, as the tail(s) usually interact with amino acid residues towards the exit of the active site or on its edge [50]. In fact, those are the amino acids which are less conserved among the various mammalian CAs, and this explains why most of these novel generation inhibitors showed much more interesting inhibition profiles as compared to the classical ones. X-Ray crystal structures and homology modeling are available for some of these compounds, which proved that both favorable interactions as well as clashes with particular amino acids present only in some isozymes, are critical for the inhibition profile and isozyme selectivity issues [50, 102, 103]. In fact, several interesting examples of inhibitors designed by the tail approach, which show efficient antiglaucoma and anticancer/antimetastatic activity in vivo have been discussed in this chapter. The tail approach has been in fact successfully applied also to generate other classes of CAIs, such as DTCs, coumarins/sulfocoumarins and even phenols. By using the diverse inhibition mechanisms of these compounds, these diverse scaffolds may offer advantages over the sulfonamides and their bioisosteres as CAIs targeting a variety of enzymes.

The carboxylate as an alternative ZBG to the sulfonamide one, was intensely studied in the last period, with several interesting developments, especially for the inhibition of fungal β-CAs from pathogenic fungi/yeasts [71]. Such enzymes were also highly inhibited by DTCs, a new class of CAIs targeting both α- and β-class CAs, recently discovered by our group [59]. X-ray crystal structures of DTCs complexed to hCA II allowed a deep understanding of the interactions between enzyme and inhibitor at atomic level. In fact the CS_2^- ZBG found in DTC is an excellent alternative to the sulfonamide ZBG present in the classical CAIs. Furthermore, such compounds are water soluble, easy to prepare and afford the exploration of a wide chemical space, which for the sulfonamides is not always possible (due to synthetic limitations).

In bacteria, many CAs were shown to be important for survival, invasion and pathogenicity. Bacteria encode CAs belonging to the α-, β-, and/or γ-families, but up to now only the first two classes have been investigated in some detail in different species. Indeed, the α-CAs from *Neisseria spp.* and *H. pylori* as well as the β-class enzymes from *E. coli, H. pylori, M. tuberculosis, Brucella spp., S. pneumoniae, S. enterica, V. cholerae* and *H. influenzae* have been cloned and characterized. For some of these enzymes the X-ray crystal structures were determined at rather high resolution, allowing for a good understanding of the catalytic/inhibition mechanisms. However, no adducts with inhibitors of these enzymes have been characterized so far; although in vitro and in vivo inhibition studies with various classes of inhibitors, such as anions, sulfonamides/sulfamates, DTCs and carboxylates have been reported. Efficient in vitro inhibitors have been reported for many such enzymes, and for various chemotypes, but only for *Nessseria spp., H. pylori, B. suis, M. tuberculosis* and *S. pneumoniae* CAs it has been possible to evidence inhibition of bacterial growth in vivo. Thus, bacterial CAs represent at this moment very promising targets for obtaining antibacterials devoid of the resistance problems of the clinically used such agents but further studies are needed to validate these and other less investigated enzymes as novel drug targets, and more importantly, structure-based drug design approached are needed for these targets.

Acknowledgments Research from C. T. S. laboratory was financed by several grants of the 6th and 7th Framework Programs of the European Union (DeZnIT, Metoxia and Dynano projects) and from R. M. laboratory by a grant from the NIH (GM25154).

References

1. Alterio V, Di Fiore A, D'Ambrosio K et al (2012) Multiple binding modes of inhibitors to carbonic anhydrases: how to design specific drugs targeting 15 different isoforms ? Chem Rev 112:4421–4468
2. Xu Y, Feng L, Jeffrey PD, Shi Y, Morel FM (2008) Structure and metal exchange in the cadmium carbonic anhydrase of marine diatoms. Nature 452:56–61
3. Supuran CT (2008) Carbonic anhydrases: novel therapeutic applications for inhibitors and activators. Nat Rev Drug Discov 7:168–181
4. Neri D, Supuran CT (2011) Interfering with pH regulation in tumours as a therapeutic strategy. Nat Rev Drug Discov 10:767–777
5. Supuran CT (2010) Carbonic anhydrase inhibitors. Bioorg Med Chem Lett 20:3467–3474
6. Pastorekova S, Parkkila S, Pastorek J, Supuran CT (2004) Carbonic anhydrases: current state of the art, therapeutic applications and future prospects. J Enzyme Inhib Med Chem 19:199–229
7. Supuran CT (2011) Carbonic anhydrase inhibitors and activators for novel therapeutic applications. Future Med Chem 3:1165–1180
8. Supuran CT (2011) Bacterial carbonic anhydrases as drug targets: towards novel antibiotics ? Front Pharmacol 2:34
9. Supuran CT, Scozzafava A, Casini A (2003) Carbonic anhydrase inhibitors. Med Res Rev 23:146–189
10. Domsic JF, Avvaru BS, Kim CU et al (2008) Entrapment of carbon dioxide in the active site of carbonic anhydrase II. J Biol Chem 283:30766–30771
11. De Simone G, Supuran CT (2012) (In)organic anions as carbonic anhydrase inhibitors. J Inorg Biochem 11:117–129
12. Supuran CT (2012) Structure-based drug discovery of carbonic anhydrase inhibitors. J Enzyme Inhib Med Chem 27:759–772
13. Suarez Covarrubias A, Bergfors T, Jones TA, Hogbom M (2006) Structural mechanics of the pH-dependent activity of the β-carbonic anhydrase from *Mycobacterium tuberculosis*. J Biol Chem 281:4993–4999
14. Temperini C, Scozzafava A, Supuran CT (2010) Carbonic anhydrase inhibitors. X-Ray crystal studies of the carbonic anhydrase II – trithiocarbonate adduct – An inhibitor mimicking the sulfonamide and urea binding to the enzyme. Bioorg Med Chem Lett 20:474–478
15. Briganti F, Mangani S, Orioli P, Scozzafava A, Vernaglione G, Supuran CT (1997) Carbonic anhydrase activators: X-ray crystallographic and spectroscopic investigations for the interaction of isozymes I and II with histamine. Biochemistry 36:10384–11039
16. Temperini C, Scozzafava A, Vullo D, Supuran CT (2006) Carbonic anhydrase activators. Activation of isozymes I, II, IV, VA, VII and XIV with L- and D-histidine and crystallographic analysis of their adducts with isoform II: engineering proton transfer processes within the active site of an enzyme. Chemistry 12:7057–7066
17. Maupin CM, Castillo N, Taraphder S, Tu C, McKenna R, Silverman DN, Voth GA (2011) Chemical rescue of enzymes: proton transfer in mutants of human carbonic anhydrase II. J Am Chem Soc 133:6223–6234

18. Carta F, Temperini C, Innocenti A, Scozzafava A, Kaila K, Supuran CT (2010) Polyamines inhibit carbonic anhydrases by anchoring to the zinc-coordinated water molecule. J Med Chem 53:5511–5522
19. Maresca A, Temperini C, Vu H, Pham NB, Poulsen SA, Scozzafava A, Quinn RJ, Supuran CT (2009) Non-zinc mediated inhibition of carbonic anhydrases: coumarins are a new class of suicide inhibitors. J Am Chem Soc 131:3057–3062
20. Maresca A, Temperini C, Pochet L, Masereel B, Scozzafava A, Supuran CT (2010) Deciphering the mechanism of carbonic anhydrase inhibition with coumarins and thiocoumarins. J Med Chem 53:335–344
21. Touisni N, Maresca A, McDonald PC, Lou Y, Scozzafava A, Dedhar S, Winum JY, Supuran CT (2011) Glycosyl coumarin carbonic anhydrase IX and XII inhibitors strongly attenuate the growth of primary breast tumors. J Med Chem 54:8271–8277
22. Woo LW, Ganeshapillai D, Thomas MP, Sutcliffe OB, Malini B, Mahon MF, Purohit A, Potter BV (2011) Structure-activity relationship for the first-in-class clinical steroid sulfatase inhibitor Irosustat (STX64, BN83495). ChemMedChem 6:2019–2034
23. Supuran CT, Scozzafava A (2007) Carbonic anhydrases as targets for medicinal chemistry. Bioorg Med Chem 15:4336–4350
24. Baranauskienė L, Hilvo M, Matulienė J, Golovenko D, Manakova E, Dudutienė V, Michailovienė V, Torresan J, Jachno J, Parkkila S, Maresca A, Supuran CT, Gražulis S, Matulis D (2010) Inhibition and binding studies of carbonic anhydrase isozymes I, II and IX with benzimidazo(1,2-c)(1,2,3)thiadiazole-7-sulphonamides. J Enzyme Inhib Med Chem 25:863–870
25. Zimmerman S, Innocenti A, Casini A, Ferry JG, Scozzafava A, Supuran CT (2004) Carbonic anhydrase inhibitors. Inhibition of the prokariotic beta and gamma-class enzymes from *Archaea* with sulfonamides. Bioorg Med Chem Lett 14:6001–6006
26. Schlicker C, Hall RA, Vullo D, Middelhaufe S, Gertz M, Supuran CT, Muhlschlegel FA, Steegborn C (2009) Structure and inhibition of the CO_2-sensing carbonic anhydrase Can2 from the pathogenic fungus *Cryptococcus neoformans*. J Mol Biol 385:1207–2020
27. Pacchiano F, Carta F, Vullo D, Scozzafava A, Supuran CT (2010) Inhibition of β-carbonic anhydrases with ureido-substituted benzenesulfonamides. Bioorg Med Chem Lett 20:102–105
28. Alterio V, Hilvo M, Di Fiore A et al (2009) Crystal structure of the extracellular catalytic domain of the tumor-associated human carbonic anhydrase IX. Proc Natl Acad Sci U S A 106:16233–16238
29. Weber A, Casini A, Heine A, Kuhn D, Supuran CT, Scozzafava A, Klebe G (2004) Unexpected nanomolar inhibition of carbonic anhydrase by COX-2 selective Celecoxib: new pharmacological opportunities due to related binding site recognition. J Med Chem 47:550–557
30. Di Fiore A, Pedone C, D'Ambrosio K, Scozzafava A, De Simone G, Supuran CT (2006) Carbonic anhydrase inhibitors: valdecoxib binds to a different active site region of the human isoform II as compared to the structurally related, cyclooxygenase II "selective" inhibitor celecoxib. Bioorg Med Chem Lett 16:437–442
31. Köhler K, Hillebrecht A, Schulze Wischeler J, Innocenti A, Heine A, Supuran CT, Klebe G (2007) Saccharin inhibits carbonic anhydrases: possible explanation for its unpleasant metallic aftertaste. Angew Chem Int Ed Engl 46:7697–7699
32. Hilvo M, Innocenti A, Monti SM, De Simone G, Supuran CT, Parkkila S (2008) Recent advances in research on the most novel carbonic anhydrases, CA XIII and XV. Curr Pharm Des 14:672–678
33. Di Fiore A, Monti SM, Hilvo M, Di Fiore A, Monti SM, Hilvo M, Parkkila S, Romano V, Scaloni A, Pedone C, Scozzafava A, Supuran CT, De Simone G (2009) Crystal structure of human carbonic anhydrase XIII and its complex with the inhibitor acetazolamide. Proteins 74:164–175
34. Di Fiore A, Truppo E, Supuran CT, Alterio V, Dathan N, Bootorabi F, Parkkila S, Monti SM, De Simone G (2010) Crystal structure of the C183S/C217S mutant of human CA VII in complex with acetazolamide. Bioorg Med Chem Lett 20:5023–5026

35. Whittington DA, Waheed A, Ulmasov B et al (2001) Crystal structure of the dimeric extracellular domain of human carbonic anhydrase XII, a bitopic membrane protein overexpressed in certain cancer tumor cells. Proc Natl Acad Sci U S A 98:9545–9550
36. Whittington DA, Grubb JH, Waheed A, Shah GN, Sly WS, Christianson DW (2004) Expression, assay, and structure of the extracellular domain of murine carbonic anhydrase XIV: implications for selective inhibition of membrane-associated isozymes. J Biol Chem 279:7223–7228
37. Casini A, Scozzafava A, Mincione F, Menabuoni L, Ilies MA, Supuran CT (2000) Carbonic anhydrase inhibitors: water-soluble 4-sulfamoylphenylthioureas as topical intraocular pressure-lowering agents with long-lasting effects. J Med Chem 43:4884–4892
38. Scozzafava A, Menabuoni L, Mincione F, Supuran CT (2002) Carbonic anhydrase inhibitors. A general approach for the preparation of water soluble sulfonamides incorporating polyamino-polycarboxylate tails and of their metal complexes possessing long lasting, topical intraocular pressure lowering properties. J Med Chem 45:1466–1476
39. Fabrizi F, Mincione F, Somma T, Scozzafava G, Galassi F, Masini E, Impagnatiello F, Supuran CT (2012) A new approach to antiglaucoma drugs: carbonic anhydrase inhibitors with or without NO donating moieties. Mechanism of action and preliminary pharmacology. J Enzyme Inhib Med Chem 27:138–147
40. Ebbesen P, Pettersen EO, Gorr TA, Jobst G, Williams K, Kieninger J, Wenger RH, Pastorekova S, Dubois L, Lambin P, Wouters BG, Van Den Beucken T, Supuran CT, Poellinger L, Ratcliffe P, Kanopka A, Görlach A, Gasmann M, Harris AL, Maxwell P, Scozzafava A (2009) Taking advantage of tumor cell adaptations to hypoxia for developing new tumor markers and treatment strategies. J Enzyme Inhib Med Chem 24(S1):1–39
41. Švastová E, Hulíková A, Rafajová M, Zat'ovicová M, Gibadulinová A, Casini A, Cecchi A, Scozzafava A, Supuran CT, Pastorek J, Pastoreková S (2004) Hypoxia activates the capacity of tumor-associated carbonic anhydrase IX to acidify extracellular pH. FEBS Lett 577:439–445
42. Dubois L, Lieuwes NG, Maresca A et al (2009) Imaging of CA IX with fluorescent labelled sulfonamides distinguishes hypoxic and (re)-oxygenated cells in a xenograft tumor model. Radiother Oncol 92:423–428
43. Ahlskog JKJ, Dumelin CE, Trüssel S, Marlind J, Neri D (2009) In vivo targeting of tumor-associated carbonic anhydrases using acetazolamide derivatives. Bioorg Med Chem Lett 19:4851–4856
44. Lou Y, McDonald PC, Oloumi A, Chia S, Ostlund C, Ahmadi A, Kyle A, Auf dem Keller U, Leung S, Huntsman D, Clarke B, Sutherland BW, Waterhouse D, Bally M, Roskelley C, Overall CM, Minchinton A, Pacchiano F, Carta F, Scozzafava A, Touisni N, Winum JY, Supuran CT, Dedhar S (2011) Targeting tumor hypoxia: suppression of breast tumor growth and metastasis by novel carbonic anhydrase IX inhibitors. Cancer Res 71:3364–3376
45. Pacchiano F, Carta F, McDonald PC, Lou Y, Vullo D, Scozzafava A, Dedhar S, Supuran CT (2011) Ureido-substituted benzenesulfonamides potently inhibit carbonic anhydrase IX and show antimetastatic activity in a model of breast cancer metastasis. J Med Chem 54:1896–1902
46. Mincione F, Scozzafava A, Supuran CT (2007) The development of topically acting carbonic anhydrase inhibitors as antiglaucoma agents. Curr Top Med Chem 7:849–854
47. Carta F, Supuran CT, Scozzafava A (2012) Novel therapies for glaucoma: a patent review 2007–2011. Expert Opin Ther Pat 22:79–88
48. Steele RM, Batugo MR, Benedini F, Borghi V, Carzaniga L, Impagnatiello F, Miglietta D, Chong WK, Rajapakse R, Cecchi A, Temperini C, Supuran CT (2009) Nitric oxide-donating carbonic anhydrase inhibitors for the treatment of open-angle glaucoma. Bioorg Med Chem Lett 19:6565–6570
49. Mincione F, Benedini F, Biondi S, Mincione F, Benedini F, Biondi S, Cecchi A, Temperini C, Formicola G, Pacileo I, Scozzafava A, Masini E, Supuran CT (2011) Synthesis and crystallographic analysis of new sulfonamides incorporating NO-donating moieties with potent antiglaucoma action. Bioorg Med Chem Lett 21:3216–3221

50. Alterio V, Di Fiore A, D'Ambrosio K, Supuran CT, De Simone G (2009) X-Ray crystallography of CA inhibitors and its importance in drug design. In: Supuran CT, Winum JY (eds) Drug design of zinc-enzyme inhibitors: functional, structural, and disease applications. Wiley, Hoboken, pp 73–138
51. Alterio V, Vitale RM, Monti SM, Pedone C, Scozzafava A, Cecchi A, De Simone G, Supuran CT (2006) Carbonic anhydrase inhibitors: X-ray and molecular modeling study for the interaction of a fluorescent antitumor sulfonamide with isozyme II and IX. J Am Chem Soc 128:8329–8335
52. Alterio V, De Simone G, Monti SM, Scozzafava A, Supuran CT (2007) Carbonic anhydrase inhibitors: inhibition of human, bacterial, and archaeal isozymes with benzene-1,3-disulfonamides-solution and crystallographic studies. Bioorg Med Chem Lett 17:4201–4207
53. Wagner J, Avvaru BS, Robbins AH, Scozzafava A, Supuran CT, McKenna R (2010) Coumarinyl-substituted sulfonamides strongly inhibit several human carbonic anhydrase isoforms: solution and crystallographic investigations. Bioorg Med Chem 18:4873–4878
54. Biswas S, Aggarwal M, Guzel O, Scozzafava A, McKenna R, Supuran CT (2011) Conformational variability of different sulfonamide inhibitors with thienyl-acetamido moieties attributes to differential binding in the active site of cytosolic human carbonic anhydrase isoforms. Bioorg Med Chem 19:3732–3738
55. Pacchiano F, Aggarwal M, Avvaru BS, Robbins AH, Scozzafava A, McKenna R, Supuran CT (2010) Selective hydrophobic pocket binding observed within the carbonic anhydrase II active site accommodate different 4-substituted-ureido-benzenesulfonamides and correlate to inhibitor potency. Chem Commun 46:8371–8373
56. Carta F, Garaj V, Maresca A, Wagner J, Avvaru BS, Robbins AH, Scozzafava A, McKenna R, Supuran CT (2011) Sulfonamides incorporating 1,3,5-triazine moieties selectively and potently inhibit carbonic anhydrase transmembrane isoforms IX, XII and XIV over cytosolic isoforms I and II: solution and X-ray crystallographic studies. Bioorg Med Chem 19:3105–3119
57. Hen N, Bialer M, Yagen B, Aggarwal M, Robbins AH, McKenna R, Scozzafava A, Supuran CT (2011) Anticonvulsant 4-aminobenzenesulfonamide derivatives with branched-alkylamide moieties: X-ray crystallography and inhibition studies of human carbonic anhydrase isoforms I, II, VII and XIV. J Med Chem 54:3977–3981
58. Kolayli S, Karahalil F, Sahin H, Dincer B, Supuran CT (2011) Characterization and inhibition studies of an α-carbonic anhydrase from the endangered sturgeon species *Acipenser gueldenstaedti*. J Enzyme Inhib Med Chem 26:895–900
59. Carta F, Aggarwal M, Maresca A, Scozzafava A, McKenna R, Supuran CT (2012) Dithiocarbamates: a new class of carbonic anhydrase inhibitors. Crystallographic and kinetic investigations. Chem Commun 48:1868–1870
60. Güzel Ö, Innocenti A, Scozzafava A, Salman A, Supuran CT (2009) Carbonic anhydrase inhibitors. Phenacetyl-, pyridylacetyl- and thienylacetyl-substituted aromatic sulfonamides act as potent and selective isoform VII inhibitors. Bioorg Med Chem Lett 19:3170–3173
61. Güzel Ö, Innocenti A, Scozzafava A, Salman A, Supuran CT (2009) Carbonic anhydrase inhibitors. Aromatic/heterocyclic sulfonamides incorporating phenacetyl-, pyridylacetyl- and thienylacetyl- tails act as potent inhibitors of human mitochondrial isoforms VA and VB. Bioorg Med Chem 17:4894–4899
62. Scozzafava A, Supuran CT (1999) Carbonic anhydrase inhibitors. Arylsulfonylureido and arylureido-substituted aromatic and heterocyclic sulfonamides: towards selective inhibitors of carbonic anhydrase isozyme I. J Enzyme Inhib 14:343–363
63. Garaj V, Puccetti L, Fasolis G, Winum JY, Montero JL, Scozzafava A, Vullo D, Innocenti A, Supuran CT (2004) Carbonic anhydrase inhibitors: synthesis and inhibition of cytosolic/tumor-associated carbonic anhydrase isozymes I, II and IX with sulfonamides incorporating 1,2,4-triazine moieties. Bioorg Med Chem Lett 14:5427–5433
64. Garaj V, Puccetti L, Fasolis G, Winum JY, Montero JL, Scozzafava A, Vullo D, Innocenti A, Supuran CT (2005) Carbonic anhydrase inhibitors. Novel sulfonamides incorporating 1,3,5-triazine moieties as inhibitors of the cytosolic and tumor-associated carbonic anhydrase isozymes I, II and IX. Bioorg Med Chem Lett 15:3102–3108

65. McDonald PC, Winum JY, Supuran CT, Dedhar S (2012) Recent developments in targeting carbonic anhydrase IX for cancer therapeutics. Oncotarget 3:84–97
66. Carta F, Aggarwal M, Maresca A, Scozzafava A, McKenna R, Masini E, Supuran CT (2012) Dithiocarbamates strongly inhibit carbonic anhydrases and show antiglaucoma action in vivo. J Med Chem 55:1721–1730
67. Monti SM, Maresca A, Viparelli F, Carta F, De Simone G, Mühlschlegel FA, Scozzafava A, Supuran CT (2012) Dithiocarbamates strongly inhibit the beta-class fungal carbonic anhydrases from *Cryptococcus neoformans, Candida albicans* and *Candida glabrata*. Bioorg Med Chem Lett 22:859–862
68. Maresca A, Carta F, Vullo D, Supuran CT (2013) Dithiocarbamates strongly inhibit the beta-class carbonic anhydrases from *Mycobacterium tuberculosis*. J Enzyme Inhib Med Chem 28 (in press)
69. Hall RA, Mühlschlegel FA (2009) Fungal and nematode carbonic anhydrases: their inhibition in drug design. In: Supuran CT, Winum JY (eds) Drug design of zinc-enzyme inhibitors: functional, structural, and disease applications. Wiley, Hoboken, pp 301–322
70. Ohndorf UM, Schlicker C, Steegborn C (2009) Crystallographic studies on carbonic anhydrases from fungal pathogens for structure-assisted drug development. In: Supuran CT, Winum JY (eds) Drug design of zinc-enzyme inhibitors: functional, structural, and disease applications. Wiley, Hoboken, pp 323–334
71. Carta F, Innocenti A, Hall RA, Mühlschlegel FA, Scozzafava A, Supuran CT (2011) Carbonic anhydrase inhibitors. Inhibition of the β-class enzymes from the fungal pathogens *Candida albicans* and *Cryptococcus neoformans* with branched aliphatic-/aromatic carboxylates and their derivatives. Bioorg Med Chem Lett 21:2521–2526
72. Nishimori I, Onishi S, Takeuchi H, Supuran CT (2008) The α and β classes carbonic anhydrases from *Helicobacter pylori* as novel drug targets. Curr Pharm Des 14:622–630
73. Minakuchi T, Nishimori I, Vullo D, Scozzafava A, Supuran CT (2009) Molecular cloning, characterization and inhibition studies of the Rv1284 β-carbonic anhydrase from *Mycobacterium tuberculosis* with sulfonamides and a sulfamate. J Med Chem 52:2226–2232
74. Nishimori I, Minakuchi T, Vullo D, Scozzafava A, Innocenti A, Supuran CT (2009) Carbonic anhydrase inhibitors. cloning, characterization, and inhibition studies of a new β-carbonic anhydrase from *Mycobacterium tuberculosis*. J Med Chem 52:3116–3120
75. Güzel Ö, Maresca A, Scozzafava A, Salman A, Balaban AT, Supuran CT (2009) Discovery of low nanomolar and subnanomolar inhibitors of the mycobacterial β-carbonic anhydrases Rv1284 and Rv3273. J Med Chem 52:4063–4067
76. Carta F, Maresca A, Suarez Covarrubias A, Mowbray SL, Jones TA, Supuran CT (2009) Carbonic anhydrase inhibitors. Characterization and inhibition studies of the most active β-carbonic anhydrase from *Mycobacterium tuberculosis*, Rv3588c. Bioorg Med Chem Lett 19:6649–6654
77. Nishimori I, Minakuchi T, Maresca A, Carta F, Scozzafava A, Supuran CT (2010) The β-carbonic anhydrases from *Mycobacterium tuberculosis* as drug targets. Curr Pharm Des 16:3300–3309
78. Winum JY, Kohler S, Supuran CT (2010) *Brucella* carbonic anhydrases: new targets for designing anti-infective agents. Curr Pharm Des 16:3310–3316
79. Vullo D, Nishimori I, Minakuchi T, Scozzafava A, Supuran CT (2011) Inhibition studies with anions and small molecules of two novel β-carbonic anhydrases from the bacterial pathogen *Salmonella enterica* serovar Typhimurium. Bioorg Med Chem Lett 21:3591–3595
80. Cronk JD, Endrizzi JA, Cronk MR, O'Neill JW, Zhang K (2001) Crystal structure of *E. coli* b-carbonic anhydrase, an enzyme with an unusual pH-dependent activity. Protein Sci 10:911–922
81. Cronk JD, Rowlett RS, Zhang KY, Tu C, Endrizzi JA, Lee J, Gareiss PC, Preiss JR (2006) Identification of a novel noncatalytic bicarbonate binding site in eubacterial β-carbonic anhydrase. Biochemistry 45:4351–4361
82. Maresca A, Vullo D, Scozzafava A, Supuran CT (2013) Inhibition of the alpha- and beta-carbonic anhydrases from the gastric pathogen *Helycobacter pylori* with anions. J Enzyme Inhib Med Chem 28:388–891

83. Maresca A, Vullo D, Scozzafava A, Manole G, Supuran CT (2013) Inhibition of the beta-class carbonic anhydrases from *Mycobacterium tuberculosis* with carboxylic acids. J Enzyme Inhib Med Chem 28 (in press)
84. Vullo D, Nishimori I, Scozzafava A, Köhler S, Winum JY, Supuran CT (2010) Inhibition studies of a β-carbonic anhydrase from *Brucella suis* with a series of water soluble glycosyl sulfanilamides. Bioorg Med Chem Lett 20:2178–2182
85. Nishimori I, Minakuchi T, Kohsaki T, Onishi S, Takeuchi H, Vullo D, Scozzafava A, Supuran CT (2007) Carbonic anhydrase inhibitors. The β-carbonic anhydrase from *Helicobacter pylori* is a new target for sulfonamide and sulfamate inhibitors. Bioorg Med Chem Lett 17:3585–3594
86. Del Prete S, Isik S, Vullo D, De Luca V, Carginale V, Scozzafava A, Supuran CT, Capasso C (2012) DNA cloning, characterization and inhibition studies of an α-carbonic anhydrase from the pathogenic bacterium *Vibrio cholerae*. J Med Chem 55:10742–10748
87. Buchieri MV, Riafrecha LE, Rodríguez OM, Vullo D, Morbidoni HR, Supuran CT, Colinas PA (2013) Inhibition of the β-carbonic anhydrases from *Mycobacterium tuberculosis* with *C*-cinnamoyl glycosides: identification of the first inhibitor with antimycobacterial activity. Bioorg Med Chem Lett 23 (in press)
88. Nair SK, Ludwig PA, Christianson DW (1994) Two-site binding of phenol in the active site of human carbonic anhydrase II: structural implications for substrate association. J Am Chem Soc 116:3659–3660
89. Innocenti A, Vullo D, Scozzafava A, Supuran CT (2008) Carbonic anhydrase inhibitors. Interactions of phenols with the 12 catalytically active mammalian isoforms (CA I – XIV). Bioorg Med Chem Lett 18:1583–1587
90. Innocenti A, Hilvo M, Scozzafava A, Parkkila S, Supuran CT (2008) Carbonic anhydrase inhibitors. Inhibition of the new membrane-associated isoform XV with phenols. Bioorg Med Chem Lett 18:3593–3596
91. Innocenti A, Vullo D, Scozzafava A, Supuran CT (2008) Carbonic anhydrase inhibitors. Inhibition of mammalian isoforms I – XIV with a series of substituted phenols including paracetamol and salicylic acid. Bioorg Med Chem 16:7424–7428
92. Bayram E, Senturk M, Kufrevioglu OI, Supuran CT (2008) In vitro effects of salicylic acid derivatives on human cytosolic carbonic anhydrase isozymes I and II. Bioorg Med Chem 16:9101–9105
93. Carta F, Vullo D, Maresca A, Scozzafava A, Supuran CT (2013) Mono-/dihydroxybenzoic acid esters and phenol pyridinium derivatives as inhibitors of the mammalian carbonic anhydrase isoforms I, II, VII, IX, XII and XIV. Bioorg Med Chem 21 (in press)
94. Martin DP, Cohen SM (2012) Nucleophile recognition as an alternative inhibition mode for benzoic acid based carbonic anhydrase inhibitors. Chem Commun 48:5259–5261
95. Carta F, Maresca A, Scozzafava A, Supuran CT (2012) Novel coumarins and 2-thioxo-coumarins as inhibitors of the tumor-associated carbonic anhydrases IX and XII. Bioorg Med Chem 20:2266–2273
96. Carta F, Vullo D, Maresca A, Scozzafava A, Supuran CT (2012) New chemotypes acting as isozyme-selective carbonic anhydrase inhibitors with low affinity for the offtarget cytosolic isoform II. Bioorg Med Chem Lett 22:2182–2185
97. Maresca A, Supuran CT (2010) Coumarins incorporating hydroxy- and chloro- moieties selectively inhibit the transmembrane, tumor-associated carbonic anhydrase isoforms IX and XII over the cytosolic ones I and II. Bioorg Med Chem Lett 20:4511–4514
98. Maresca A, Scozzafava A, Supuran CT (2010) 7,8-Disubstituted- but not 6,7-disubstituted coumarins selectively inhibit the transmembrane, tumor-associated carbonic anhydrase isoforms IX and XII over the cytosolic ones I and II in the low nanomolar/subnanomolar range. Bioorg Med Chem Lett 20:7255–7258
99. Davis RA, Vullo D, Maresca A, Supuran CT, Poulsen SA (2013) Natural product coumarins that inhibit human carbonic anhydrases. Bioorg Med Chem 21:1539–1543
100. Carta F, Maresca A, Scozzafava A, Supuran CT (2012) 5- and 6-Membered (thio)lactones are prodrug type carbonic anhydrase inhibitors. Bioorg Med Chem Lett 22:267–270

101. Tars K, Vullo D, Kazaks A, Leitans J, Lends A, Grandane A, Zalubovskis R, Scozzafava A, Supuran CT (2013) Sulfocoumarins (1,2-benzoxathiine 2,2-dioxides): a class of potent and isoform-selective inhibitors of tumor-associated carbonic anhydrases. J Med Chem 56:293–300
102. Aggarwal M, Boone CD, Kondeti B, McKenna R (2013) Structural annotations of human carbonic anhydrases. J Enzyme Inhib Med Chem 28:267–277
103. Aggarwal M, Kondeti B, McKenna R (2013) Insights towards sulfonamide drug specificity in α-carbonic anhydrases. Bioorg Med Chem 21:1526–1533

Chapter 16
Natural Products That Inhibit Carbonic Anhydrase

Sally-Ann Poulsen and Rohan A. Davis

Abstract The chemical diversity, binding specificity and propensity to interact with biological targets has inspired many researchers to utilize natural products as molecular probes. Almost all reported carbonic anhydrase inhibitors comprise a zinc binding group in their structure of which the primary sulfonamide moiety (-SO_2NH_2) is the foremost example and to a lesser extent the primary sulfamate (-O-SO_2NH_2) and sulfamide (-NH-SO_2NH_2) groups. Natural products that comprise these zinc binding groups in their structure are however rare and relatively few natural products have been explored as a source for novel carbonic anhydrase inhibitors. This chapter will highlight the recent and growing interest in carbonic anhydrase inhibitors sourced from nature, demonstrating that natural product chemical space presents a rich source of potential alternate chemotypes for the discovery of novel drug-like carbonic anhydrase inhibitors.

Keywords Carbonic anhydrase • Natural products • Coumarin • Phenol • Polyamine • Sulfonamide • Sulfamate

1 Introduction

Carbonic anhydrases (CAs) are zinc metalloenzymes that catalyze the reversible hydration of carbon dioxide to bicarbonate and a proton [1]. The active site zinc cation is the implied target for small molecule inhibitors to block the endogenous

Susan C. Frost and Robert McKenna (eds.). Carbonic Anhydrase: Mechanism, Regulation, Links to Disease, and Industrial Applications

S.-A. Poulsen (✉) • R.A. Davis
Eskitis Institute for Drug Discovery, Griffith University, Brisbane, QLD, Australia
e-mail: s.poulsen@griffith.edu.au; r.davis@griffith.edu.au

S.C. Frost and R. McKenna (eds.), *Carbonic Anhydrase: Mechanism, Regulation, Links to Disease, and Industrial Applications*, Subcellular Biochemistry 75,
DOI 10.1007/978-94-007-7359-2_16, © Springer Science+Business Media Dordrecht 2014

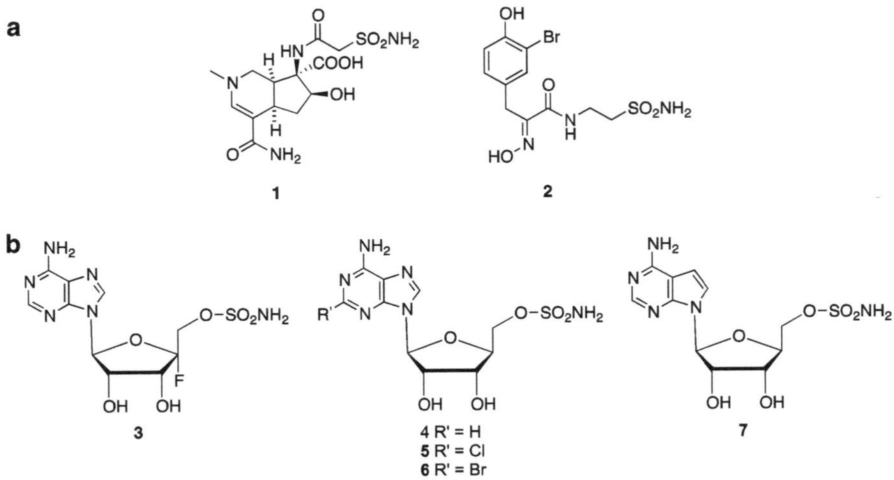

Fig. 16.1 Natural products (NPs) that comprise a primary sulfonamide or sulfamate moiety in their structure. (**a**) NP primary sulfonamides: (−)-altemicidin **1** and psammaplin C **2**. (**b**) NP primary sulfamates: nucleocidin **3**, 5′-*O*-sulfamoyl adenosine **4**, 5′-*O*-sulfamoyl 2-chloroadenosine **5**, 5′-*O*-sulfamoyl 2-bromoadenosine **6** and 5′-*O*-sulfamoyl tubercidin **7**

CA catalyzed reaction. Almost all reported CA inhibitors comprise a zinc binding group (ZBG) of which the primary sulfonamide moiety (-SO_2NH_2) is the foremost example, and to a lesser extent the primary sulfamate (-O-SO_2NH_2) and primary sulfamide (-NH-SO_2NH_2) groups. Compounds sourced from nature that comprise either a primary sulfonamide, sulfamate or sulfamide moiety in their structure are exceedingly rare. A literature search of the Dictionary of Natural Products (DNP) database [2] (a comprehensive and fully-edited database on natural products (NPs)) revealed just two NP primary sulfonamide compounds, (−)-altemicidin **1** and psammaplin C **2**, and five NP primary sulfamate compounds, nucleocidin **3** [3], 5′-*O*-sulfamoyl adenosine **4** [4], 5′-*O*-sulfamoyl 2-chloroadenosine **5** [5], 5′-*O*-sulfamoyl 2-bromoadenosine **6** [5, 6] and 5′-*O*-sulfamoyl tubercidin **7** (Fig. 16.1) [7, 8]. (−)-Altemicidin **1** is a marine alkaloid isolated from the actinomycete strain *Streptomyces sioyaensis* [9]. This compound exhibited potent acaricidal activity as well as strong inhibition of tumor cell growth [10]. A total synthesis of **1** as well as the isolation of two secondary sulfonamide analogues of **1** have subsequently been reported [9]. Psammaplin C **2** is a bromotyrosine amino acid derivative isolated from the marine sponge *Pseudoceratina purpurea* [11, 12]; no bioactivity for this alkaloid has been reported to date. The sulfamate nucleosides **3–7** were isolated from actinomycete species belonging to the genus *Streptomyces.* These structurally related sulfamates are reported to have a range of biological effects including cytotoxicity[4], herbicidal activity [6–8, 13], inhibition of blood platelet aggregation [5], antibacterial activity [5, 14] and antitrypanosomal activity [14]. Nucleocidin **3**, isolated from the fermentation broth of *Streptomyces calcus,* is of particular note since it was the first NP to contain either a fluorinated carbohydrate or a primary sulfamate group. This molecule became an attractive synthetic target owing to its novel structural features, with the first total synthesis reported in 1976 [15]. Whilst NP **3** has been shown to exhibit broad spectrum antibacterial effects as well as potent antitrypanosomal activity, its potential use in the clinic has been limited due to toxicity [16]. The NPs **1–7** have not been investigated for CA inhibition properties, however it is likely that these compounds would inhibit CA activity owing to the presence of an unhindered primary sulfonamide or primary sulfamate moiety within their structure. Finally, our search of the DNP failed to identify any NP primary sulfamides.

At the time of writing the Protein Data Bank (PDB) contained X-ray structures of ~160 sulfonamide ligands (R-SO_2NH_2) in complex with hCA II (h = human). The binding mode of the sulfonamide anion (R-SO_2NH^-) to the Zn^{2+} cation is invariant in these structures, with the sulfonamide anion coordinated to the active site Zn^{2+}. Primary sulfamates and sulfamides, ZBG isosteres of primary sulfonamides, contribute an additional ~35 X-ray structures of ligands in complex with hCA II in the PDB. Of the remaining PDB protein-ligand structures most comprise very simple ligands such as anions or small organic molecules, there are however several structures comprising more complex alternate CA ligands, some of which are NPs.

NPs comprise a vast collection of diverse chemical structures and have proven to be an invaluable source of new chemotherapies [17–22]. Plant NPs have been the basis of traditional medicine for thousands of years and continue to actively contribute to contemporary drug discovery [22]. In more recent times, marine macro- and micro-organisms along with terrestrial microbes have been the source of numerous lead molecules or drugs [23]. The significance of NPs in drug discovery is most evident in the anticancer and anti-infective therapeutic areas [20, 24–26]. For example, between 1940 and 2011 48.6 % of all new anticancer small molecule therapies approved by the FDA were either NPs or NP derivatives [26]. Furthermore between 1981 and 2011 75 % of all antibacterial new chemical entities were either NPs or their derivatives [26]. The success of NPs and their semi-synthetic derivatives as therapeutic agents is intrinsically linked to the fact that NPs have been biologically pre-validated and selected during evolution to bind to biosynthetic enzymes [27–32]. It has been hypothesized that this inherent capacity to bind in biological space allows NPs to also recognize human therapeutic targets [27, 29, 32]. Furthermore, computational studies have shown that NPs occupy complementary areas of chemical space compared with synthetic compounds, and thus should be implemented to increase the chemical complexity and drug-likeness of screening libraries [28–31]. The chemical diversity, binding specificity and efficiency, and propensity to interact with biological targets have inspired many researchers to utilize NPs as molecular probes. These studies go beyond the identification of potential new lead or drug molecules, and have increased our understanding of biological pathways and systems [33, 34]. While the primary sulfonamide and sulfamate moieties are poorly represented in NP chemical space, this space does provide a rich source of alternate chemotypes for the discovery of CA inhibitors with a different enzyme binding mode to the typical ZBGs. The remainder of this chapter will highlight the recent and growing interest in novel CA inhibitors sourced from nature.

2 Natural Products That Inhibit Carbonic Anhydrase

Several classes of novel CA inhibitors have been identified from screening collections of NPs, most notable are coumarin and phenol containing NPs [35]. NPs comprising these fragments display diverse profiles of CA inhibition when compared to classical ZBGs and these will be discussed further in the following sections. In addition a selection of NP-derived CA inhibitors have been synthesized where the NP scaffold has been synthetically modified to incorporate the classical ZBG of CA inhibitors i.e. sulfonamide, sulfamate or sulfamide. A summary of these hybrid molecules will also be presented.

2.1 *Coumarins*

Coumarin compounds are abundant secondary metabolites in plants and are found to a lesser extent in microorganisms and animal sources. Plant coumarins are phytoalexins, defense compounds produced when the plant is under threat from other organisms, and have attracted interest owing to a range of biological activities including antimicrobial, molluscicidal, acaricidal, antiviral, anticancer, antioxidant and anti-inflammatory properties [36]. The coumarin structure comprises a benzopyrone core, with NP coumarins categorized as (a) simple coumarins, (b) furanocoumarins, (c) pyranocoumarins, (d) bis- and triscoumarins, or (e) coumarinolignans [36]. Simple coumarins, including coumarin **8** (Fig. 16.2) are highly abundant in several plant species belonging to the taxonomic families Umbelliferae,

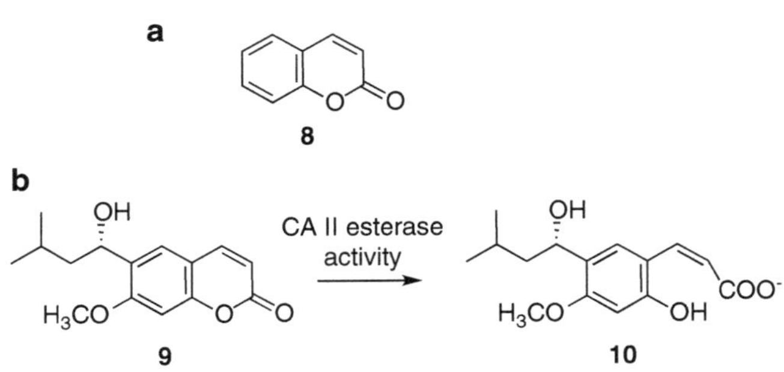

Fig. 16.2 NP coumarin CA inhibitors. (**a**) Coumarin **8**. (**b**) CA hydrolysis of coumarin **9** to form the cinnamic acid compound **10**

Rutaceae and Compositae [36]. A recent review of coumarin-based drugs highlights the growing interest in the coumarin compound class to deliver new therapeutics, [37] which is driving efforts towards the isolation and the structural characterization of further novel bioactive coumarin derivatives. In recent years it was discovered that NP coumarins inhibit CA enzymes via an alternate and unprecedented mechanism to classical sulfonamides [38], these findings are described next.

Nature Bank is a unique drug discovery resource that encompasses a diverse collection of >50,000 biota samples of plants, fungi and marine invertebrates collected from Australia, China and Papua New Guinea along with biota extracts, semi-purified fractions and pure compounds [39]. A selection of *Leionema ellipticum* (family Rutaceae) extracts was sourced from Nature Bank and screened using Fourier transform ion cyclotron resonance electrospray ionization mass spectrometry (FTICR ESI MS) for binding to bovine CA II (bCA II) [38]. From this study the NP coumarin, 6-(1*S*-hydroxy-3-methylbutyl)-7-methoxy-2H-chromen-2-one **9** was identified as a ligand for bCA II as it formed a noncovalent complex that could be detected by ESI MS. In follow on studies it was demonstrated that coumarin **9** inhibits a spectrum of human CAs in an unprecedented time dependent manner [40]. The usual enzyme assay conditions to investigate small molecule inhibition of CA activity is to incubate the test compound with the CA protein of interest for 15 min prior to monitoring the effect on CA-mediated CO_2 hydration. Under these conditions this coumarin had only weak CA inhibition prompting us to extend the pre-incubation time. Following 6 h of pre-incubation with hCA II the K_i of coumarin **9** dropped to 60 nM (with a similar reduction in K_i at other CA isozymes also observed). As the coumarin chemotype lacked a classic ZBG typical of known small molecule CA inhibitors and displayed unusual time-dependent inhibition it was important to understand how this chemotype binds to and inhibits CAs. Using protein X-ray crystallography the crystal structure of hCA II with this NP was obtained at a resolution of 2.0 Å (Fig. 16.3). The coumarin **9** was not observed, instead the hydrolysis product of **9**, the cinnamic acid derivative **10** was identified [40]. Esterase activity is known for CAs [41–43] and the observation of cinnamic acid **10** rather than NP coumarin **9**, although unexpected, could be rationalized as a consequence of hCA II esterase activity leading to hydrolysis of the lactone of **9**. The bulky hydrolysis product **10** then plugged the CA active site cavity entrance, exhibiting no interactions with the catalytic zinc ion. This unusual inhibition mode is previously unobserved for CAs and together with the coumarin pedigree in medicinal chemistry is suggestive of a potential new avenue for drug development compared to the ZBGs of classical CA inhibitors. Reactive Michael acceptors are a general structural alert in drug discovery, however it has been demonstrated that simple coumarins exhibit poor protein binding characteristics compared to other carbonyl containing Michael acceptors [44, 45]. The lower reactivity of the coumarin double bond compared with other Michael acceptors, has been attributed to it being part of a pseudoaromatic system [46].

Following the findings outlined above we performed a substructure search of the Nature Bank [39] pure compound repository against the bare coumarin scaffold **8**.

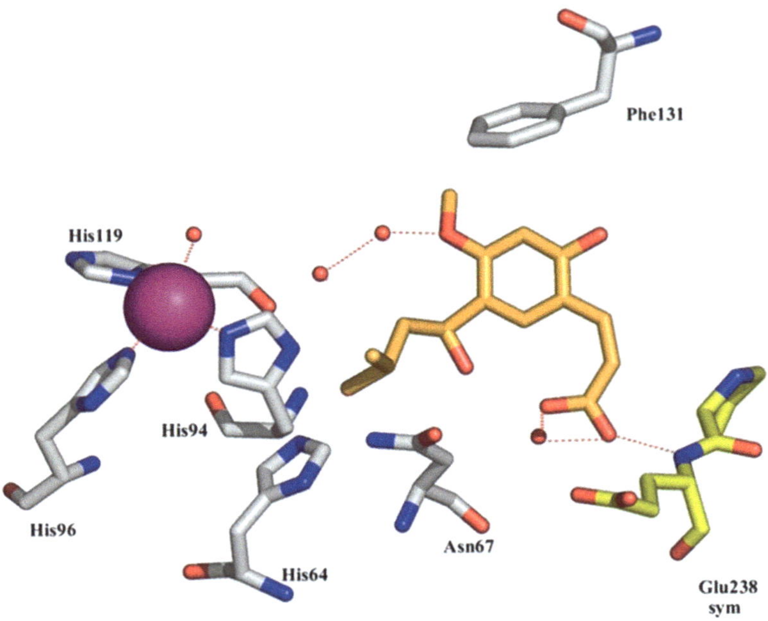

Fig. 16.3 Detailed interactions between hCA II and NP coumarin **9** hydrolysis product **10** from a protein X-ray crystal structure. The catalytic site showing the tetrahedral Zn^{2+} cation (*violet* sphere) with the three coordinated His ligands (His94, His96, and His119) and a water molecule (*red* sphere). The cinnamic acid **10** (*gold*) interacts with three active site ordered water molecules (*red* spheres), with Phe131 and Asn67 (CPK colors) from the active site as well as with Glu238sym (*yellow*) from a symmetry related enzyme molecule. The proton shuttle residue His64 is shown (CPK colors) (Reprinted with permission from Maresca, A.; Temperini, C.; Vu, H.; Pham, N. B.; Poulsen, S.-A.; Scozzafava, A.; Quinn, R. J.; Supuran, C. T. *J. Am. Chem. Soc.* 2009, *131*, 3057. Copyright 2009 American Chemical Society)

A set of 81 coumarins were identified and from this a subset of 27 coumarins were sourced in sufficient quantity and purity for follow up evaluation as CA inhibitors [47]. These NP coumarins, compounds **11–37** (Fig. 16.4) comprise 24 plant coumarins (compounds **11–34**) and three marine coumarins (compounds **35–37**). Specifically, the plant NPs comprise avicennin **11** [48, 49], trans-avicennol **12** [50, 51], calanolide B **13** [39, 52], dihydrogeiparvarin **14** [53], geiparvarin **15** [53, 54], dehydromarmin **16** [53], xanthyletin **17** [55], xanthoxyletin **18** [50, 55], ceylantin **19** [56], alloxanthoxyletin **20** [55], fraxidin **21** [57], fraxin **22** [58], scopoletin **23** [59], 6,7,8-trimethoxycoumarin **24** [60], 5,7,8-trimethoxycoumarin **25** [60], 7-hydroxy-8-methoxycoumarin **26** [59], isoscopoletin **27** [61], fraxoside **28** [62], scopolin **29** [63], murralongin **30** [64], (+)-isomurralonginol nicotinate **31** [65], isophellodenol C **32** [66], ellagic acid **33** [67] and nasutin B **34** [68]. The ascidian NP coumarins include lamellarins E **35** [69], B **36** [70], and G 8-sulfate **37** [71]. A variety of bioactivities have been reported for these coumarins, for example calanolide B **13** isolated from the tropical rainforest tree *Calophyllum*

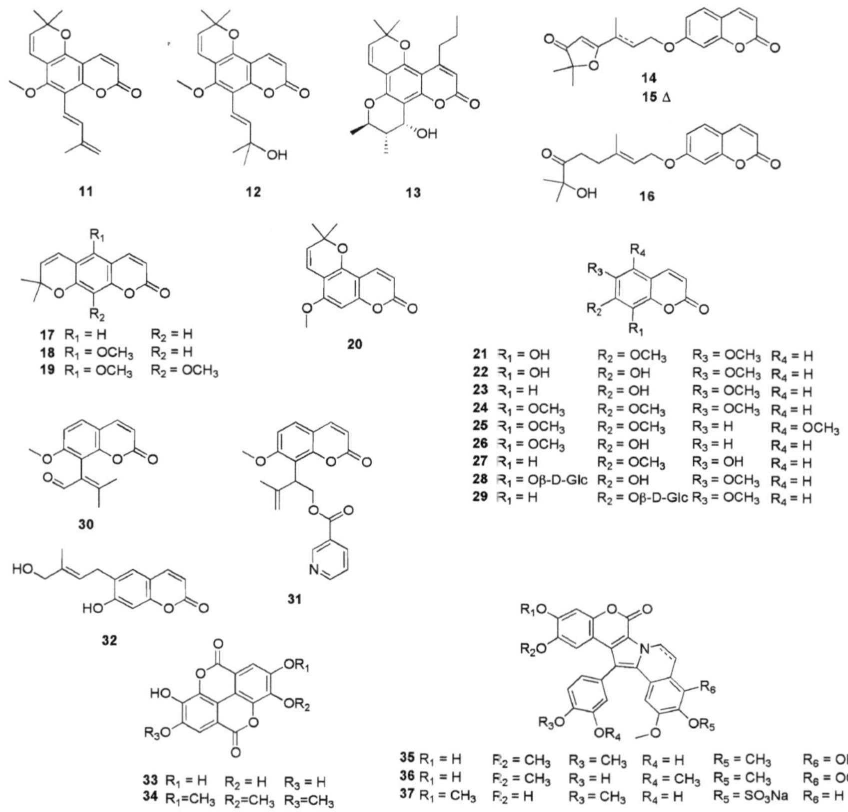

Fig. 16.4 NP coumarin library (**11–37**) (sourced from Nature Bank [39, 47])

lanigerum, displayed protection against HIV-1 replication and cytopathicity (EC_{50} = 0.4 μM) [52]. Dihydrogeiparvarin **14** and geiparvarin **15**, both isolated from *Geijera parviflora* [53, 54], possessed significant in vitro activity against human carcinoma of the nasopharynx [72, 73]. Xanthoxyletin **18** [50, 55], purified from a variety of *Citrus* species, acts as a DNA-damaging agent [74], while several synthetic derivatives have been shown to exhibit toxicity towards L-1210 leukemia cells with IC_{50} values ranging from 0.9 to 60.3 μM [74]. The inhibition activity data for the NP coumarins **9** and **11–37** against hCA I and II (off-target isozymes), as well hCA IX and XII (isozymes of interest in cancer drug development) is presented in Table 16.1. Data for the simplest coumarin **8** and standard CA inhibitors (Fig. 16.5) is included for development of structure-activity relationships (SAR).

Since the discovery of the NP coumarin **9** synthetic libraries of coumarins and thiocoumarins have been prepared and evaluated as CA inhibitors [79–81]. The complexity and diversity of NP coumarin structures far exceeds that described for synthetic coumarin CA inhibitors. Coumarin **8**, the simplest coumarin, is not an

Table 16.1 Inhibition data for coumarins **8, 9, 11–37** against hCA isozymes I, II, IX and XII [47]. Standard inhibitors (**AZA**, **ZNS** and **TPM**) are included for comparison [75–77]

	K_i (μM)[a–c]			
Compd	CA I	CA II	CA IX	CA XII
8	3.10	9.20	>1000	>1000
9	0.08	0.06	54.5	48.6
11	7.66	>100	0.62	0.79
12	8.46	>100	0.78	0.77
13	9.31	50.7	0.83	0.81
14	59.2	63.4	0.89	0.60
15	9.75	>100	0.60	0.83
16	7.81	>100	4.03	0.70
17	21.5	>100	7.51	25.7
18	7.71	>100	0.74	0.96
19	9.21	49.3	0.86	8.35
20	5.60	>100	3.50	9.10
21	9.89	>100	0.85	7.84
22	4.86	94.3	0.61	7.70
23	10.56	>100	0.96	4.05
24	0.0097	>100	6.58	18.2
25	4.31	9.65	0.76	0.83
26	36.4	>100	0.85	9.12
27	14.0	>100	7.37	4.14
28	5.04	>100	0.37	7.45
29	5.93	>100	8.72	0.78
30	9.11	>100	8.12	7.44
31	5.84	>100	0.67	7.39
32	7.52	78.9	9.75	0.77
33	68.2	>100	79.8	8.15
34	44.1	>100	17.4	7.42
35	6.45	>100	3.22	9.07
36	40.1	>100	6.33	8.51
37	6.55	>100	3.27	1.79
AZA	0.25	0.012	0.025	0.0057
ZNS	0.056	0.035	0.005	11
TPM	0.25	0.005	0.058	0.0038

[a] This inhibition data was acquired following a 6 h incubation time with enzyme using a stopped flow assay that monitors the CA catalyzed hydration of CO_2 [78]

[b] Errors in the range of ±5 % of the reported value, from three determinations

[c] All proteins were recombinant

appreciable inhibitor of CA IX or XII however it is a weak inhibitor of off-target CA I and CA II, with K_is of 3.1 and 9.2 μM, respectively. The NP coumarins are substituted at any of six available sites, with many fused to form tricyclic, tetracyclic or larger ring systems. This diversity does not readily allow simple SAR to be defined, however several trends surrounding CA inhibition are evident. Most

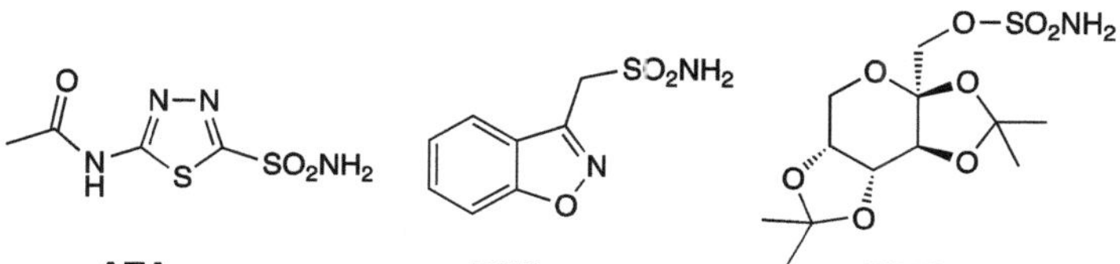

Fig. 16.5 Standard CA inhibitors: acetazolamide **AZA**, zonisamide **ZNS** and topiramate **TPM**

obvious is that the NP coumarin library members are very weak CA II inhibitors, most have K_is > 100 μM, the only exception being the trimethoxycoumarin **25** (K_i = 9.65 μM). When compared to the structurally related methoxy/hydroxy coumarins **21–24, 26** and **27**, compound **25** differs only in the pattern of substituents, this SAR indicates that it may be a combination of interacting substituents that directs the CA inhibition profile at CA II. At CA I, IX and XII many of the NP coumarins have K_is in the range of 1–10 μM, this tight grouping of K_is reflects minimal isozyme selectivity with these coumarins, however there are a few outliers to this general trend and these compounds represent interesting structures owing to their CA isozyme selectivity characteristics. At CA I there was one stand out compound being compound **24**, a nanomolar CA I inhibitor. This trimethoxy coumarin is the most potent of any of the NP coumarins at CA I and is a structural isomer of **25**, the only potent CA II coumarin of the study. Around half of the NP coumarins have submicromolar inhibition of the isozymes CA IX and XII, some of these coumarins (**11, 12, 13, 14, 15, 18** and **25**) are submicromolar at both CA IX and XII, while the remainder are submicromolar at either CA IX (**19, 21–23, 26, 28** and **31**) or CA XII (**16, 29** and **32**). This subset of NP coumarins has viable selectivity characteristics that warrant further studies in cell-based models of CA in cancer.

2.2 *Phenols*

The first single crystal X-ray structure of phenol **38** and a CA protein (hCA II) was reported in 1994 and identified that **38** binds in an unprecedented way within the enzyme active site [82]. It was shown that the phenolic OH interacts with the zinc-bound water molecule/hydroxide ion through a hydrogen bond while a second hydrogen bond formed between the phenolic OH and the NH amide of Thr199, an amino acid critical for the catalysis and inhibition of various CAs. NPs containing the phenol fragment **38** are highly abundant in nature. A substructure search of the DNP [2] against the phenol fragment identified >50,000 NPs from the 246,994 database entries that contain this fragment (~20 % of all entries). Early CA inhibitory studies focused on simple, commercially available mono-, di- or tri-substituted phenols that are also found in nature [2] such as pyrocatechol **39**, resorcinol **40**, hydroquinol **41**, salicylic acid **42**, *p*-hydroxybenzoic acid **43**, *p*-

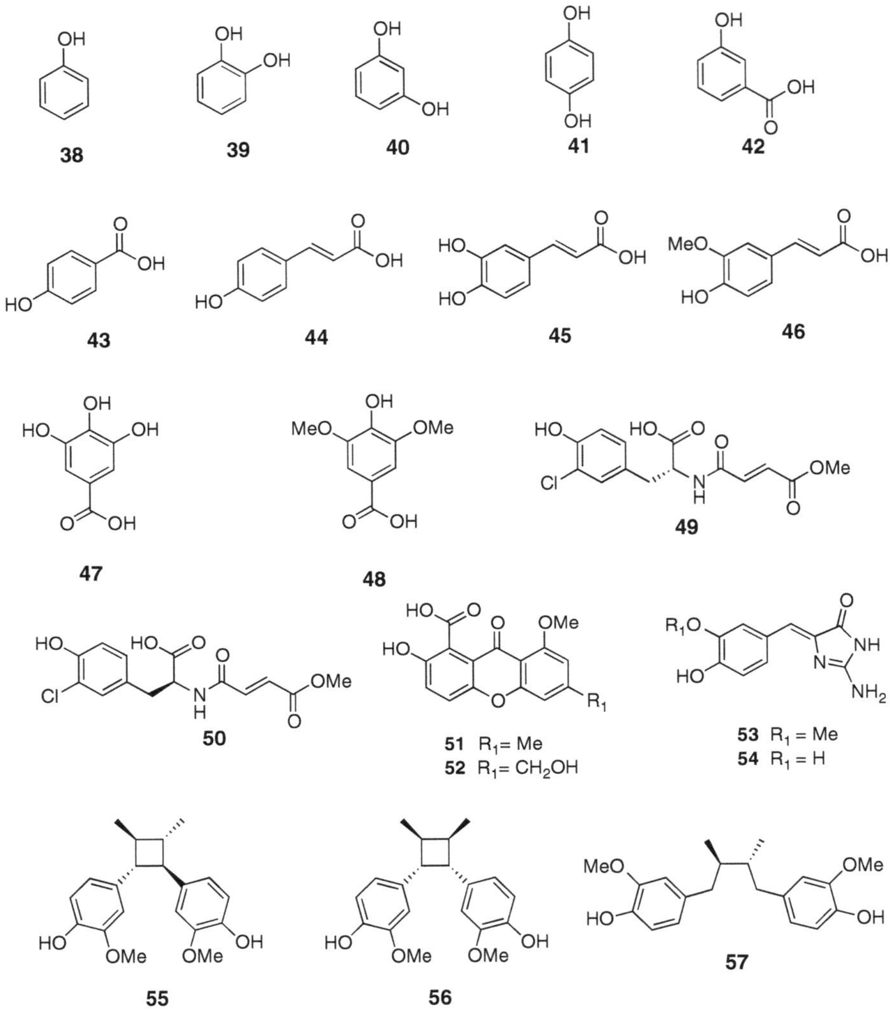

Fig. 16.6 NP phenols **38–57** tested as CA inhibitors [83, 84, 86, 87]

coumaric acid **44**, caffeic acid **45**, ferulic acid **46**, gallic acid **47** and syringic acid **48** (Fig. 16.6) [83, 84]. A number of phenolic-NPs containing more complex scaffolds **49–57** have since been sourced from the Davis open-access compound repository housed at the Queensland Compound Library (QCL) [85], and screened against selected CAs (Fig. 16.6) [86, 87]. These phenolic-derivatives include the endophytic fungal metabolites, (−)-xylariamide A **49** [88], and its synthetic enantiomer (+)-xylariamide A **50** [88], xanthones **51** and **52** [89], the marine ascidian-derived alkaloids, polyandrocarpamines A **53** and B **54** [90, 91] and the plant secondary metabolites, endiandrins A **55** [92] and B **56**, [93] and (−)-dihydroguaiaretic acid **57** [92, 93].

Table 16.2 Inhibition data for phenolic NPs **38–57** against hCA isozymes I, II, VA and VB [83, 84, 86, 87]. Standard inhibitors **AZA**, **ZNS** and **TPM** are included for comparison [75–77]

	K_i (μM)[a,b]			
Compd	CA I	CA II	CA VA	CA VB
38	10.2	5.5	218	543
39	4,003	9.9	55.1	4.2
40	795	7.7	8.7	7.1
41	10.7	0.090	14.1	12.5
42	9.9	7.1	678	355
43	9.8	10.6	9.2	10.5
44	1.07	0.98	5.9	7.7
45	2.38	1.61	6.5	9.1
46	2.89	2.40	7.0	10.5
47	3.20	2.25	4.1	9.9
48	4.15	3.19	6.3	35.4
49	239	8.3	0.095	0.114
50	231	8.0	0.108	0.102
51	201	8.4	0.093	0.103
52	374	9.2	0.094	0.102
53	10.5	9.6	0.099	0.070
54	355	13.1	0.101	0.076
55	363	11.7	0.093	0.069
56	354	12.1	0.098	0.079
57	307	230	0.085	0.071
AAZ	0.25	0.012	0.063	0.054
TPM	0.25	0.010	0.063	0.030
ZNS	0.056	0.035	0.020	6.3

[a]Errors in the range of ±5% of the reported value, from three determinations
[b]All proteins were recombinant

The phenols **38–57** have been evaluated for their inhibition of human cytosolic isoforms CA I and II (off-target) and mitochondrial isozymes CA VA and CA VB, Table 16.2. The latter have been recognized as potential targets for designing anti-obesity agents that act with a novel mechanism of action [94, 95]. The simple phenolic secondary metabolites **38–48** have also been tested against hCA III, IV, VI, VII, IX, XII, XIII and XIV [83, 84]. These data (not shown) indicate that the phenol class of NP CA inhibitor exhibits complex SAR, with small chemical changes leading to large effects on CA enzyme inhibition. The chemical diversity of phenolic NPs is vast; so far investigation of this chemotype for its interaction with CAs is in its infancy.

The β-CAs from *Helicobacter pylori, Candida albicans, Candida glabrata, Cryptococcus neoformans* and *Brucella suis* are essential for growth and have proven susceptible to inhibition with several compound classes including sulfonamides, carboxylates and boronic acids [96–103]. A positive correlation from enzyme assays to a cell-based anti-infective phenotype assay demonstrates that the β-CAs from these pathogens are potential druggable targets for anti-infective therapies. Mammals possess only α-CAs, whilst many pathogenic organisms, such

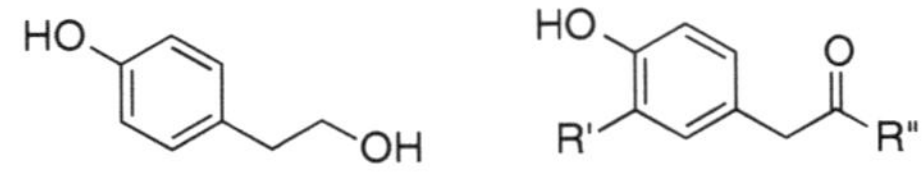

Fig. 16.7 Phenolic NPs **58–62** with activity against mycobacterial and fungal CAs [86]

Table 16.3 Enzyme inhibition of pathogenic *M. tuberculosis* β-CA isozymes Rv3273 and Rv1284, *C. albicans* isozyme Nce103, *C. neoformans* isozyme Can2 and human α-CA isozymes I and II with the NP phenols **49–62** [86] and standard CA Inhibitors **AZA**, **ZNS**, and **TPM**

	K_i (μM)[a,b]					
Compd	CA I	CA II	Rv3273	Rv1284	Nce103	Can2
38	10.1	5.5	79.0	64.0	17.3	25.9
49	239	8.3	11.3	0.84	1.03	1.15
50	231	8.0	10.9	0.71	1.06	1.11
51	201	8.4	11.4	10.5	1.06	1.12
52	374	9.2	10.9	0.99	1.01	1.08
53	10.5	9.6	0.91	11.8	0.92	0.89
54	355	13.1	0.92	0.91	0.90	0.95
55	368	11.7	8.92	0.82	0.73	0.77
56	354	12.1	0.89	0.80	0.70	0.95
57	307	230	9.10	0.85	0.62	0.81
58	430	8.7	12.1	0.85	1.10	1.08
59	309	10.3	11.4	10.8	1.02	0.90
60	309	11.2	9.12	0.85	0.91	0.84
61	265	8.6	10.8	10.3	1.08	1.12
62	237	131	11.2	10.5	1.00	0.85
AAZ	0.25	0.012	0.10	0.48	0.13	0.01
TPM	0.25	0.010	3.02	0.61	1.11	0.37
ZNS	0.056	0.035	0.21	286.8	0.94	0.97

[a]Errors in the range of ±5 % of the reported value, from three determinations
[b]All proteins were recombinant

as bacteria and fungi encode β-CAs. Similarly to α-CAs, a zinc cation defines the location of the active site of the β-CA enzymes. Phenols **49–57** [92, 93] along with the fungal NP phenols **58–62** [104–106] (Fig. 16.7) have been screened for enzyme inhibition against selected pathogen β-family CAs, Table 16.3. CAs from *Mycobacterium tuberculosis*, *Candida albicans* and *Cryptococcus neoformans* were studied and selectivity towards the pathogen isozymes over human CAs was assessed.

These studies showed that several phenolic NPs were selective inhibitors of mycobacterial and fungal β-CAs, with the two best performing NPs identified as (−)-dihydroguaiaretic acid **57** and 3-chloro-4-hydroxyphenylacetamide **62**. Specifically, **57** was a sub-micromolar β-CA inhibitor with up to 495-fold selectivity over hCA I and 371-fold selectivity over hCA II. Compound **62** was also a low micromolar inhibitor of the fungal CAs and displayed 130–280-fold selectivity over

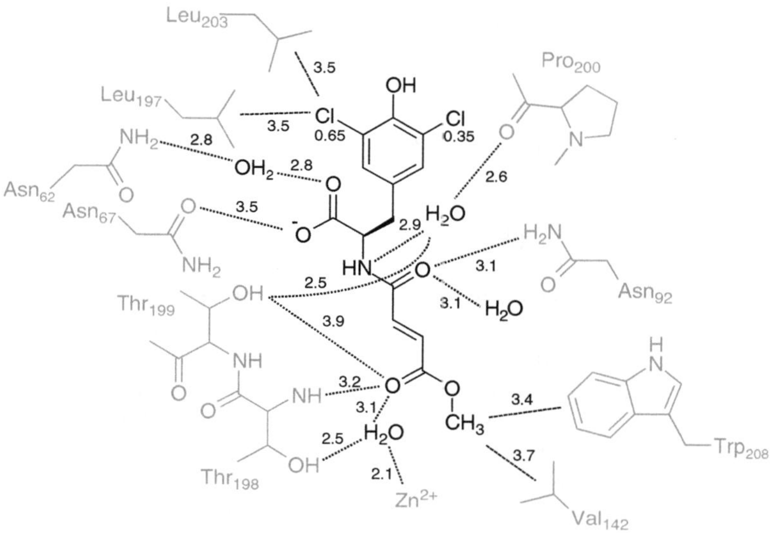

Fig. 16.8 The ligand-protein interactions are illustrated schematically for hCA II in complex with compound **50** [(+)-xylariamide A]. Protein residues are shown in *grey*; hydrogen bond interactions are shown as *dotted* lines, van der Waals interactions as *dashed* lines. Distances are given in Å. Two conformations of the phenol moiety were observed, giving rise to the appearance of the chloro substituent on both sides of the phenolic hydroxyl group; the refined occupancies for both positions are noted at each position (Reprinted with permission from Davis, R. A.; Hofmann, A.; Osman, A.; Hall, R. A.; Mühlschlegel, F. A.; Vullo, D.; Innocenti, A.; Supuran, C. T.; Poulsen, S. A. *J. Med. Chem.* 2011, *54*, 1682. Copyright 2011 American Chemical Society)

the two human CAs. These compounds were the first non-sulfonamide inhibitors that display β over α CA selectivity. In order to determine how the phenolic-based NPs **38–62** interacted with CAs, soaking and co-crystallization studies were undertaken with the readily available protein hCA II. While the most selective NPs **57** and **62** did not yield co-crystals with CAs suitable for X-ray diffraction studies, compound **50** [(+)-xylariamide A] did at a resolution of 2.0 Å. While it was predicted that the phenolic moiety present in **50** would play a major role in the hCA II binding it was discovered that instead the ester carbonyl of **50** interacts with a zinc-bound water molecule and is further engaged in a hydrogen bond donated by the backbone amide group of Thr198 (Fig. 16.8). In this crystal structure the electron density of the inhibitor is well defined, allowing unambiguous placement of the ligand. This was a totally new binding mode to CAs.

2.3 Polyamines

Polyamines belong to an alkaloid structure class and have been reported from various natural sources including terrestrial and marine animals, plants, fungi and

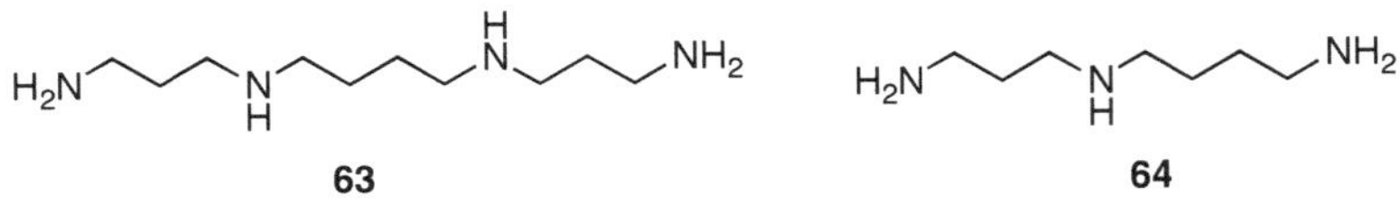

Fig. 16.9 NP polyamine CA inhibitors spermine **63** and spermidine **64** [114]

bacteria [2]. Two of the simplest polyamines isolated to date include spermine **63** and spermidine **64** (Fig. 16.9). A substructure search of the DNP against **63** and **64** identified >400 NPs from the 246,994 database entries that comprise these alkaloid fragments [2]. The polyamine chemotype has been shown to modulate multiple biological processes including gene expression, cell proliferation, translation, cell signaling, membrane stabilization and ion channel inhibition as well as antibacterial activity [107–113]. Despite the myriad bioactivities reported for polyamine NPs until recently no CA inhibition had been reported. Carta et al. showed that **63, 64** and several semi-synthetic polyamine analogues inhibited hCA I-XIV with K_i values ranging from low nanomolar to millimolar [114]. A single crystal X-ray structure of spermine **63** with hCA II (at a resolution of 2.0 Å) was also reported [114] showing compound **63** anchored to the zinc bound water ligand (as for phenol **38**) through a network of hydrogen bonds. The terminal amine moiety of **63** is hydrogen bonded with residues Thr200 and Pro201. Notably **63** binds differently to hCA II when compared to either sulfonamides, phenols or coumarins and thus polyamines have the potential for the identification and development of additional CA inhibitors with a unique mechanism of binding and CA selectivity profile. This alkaloid structure class warrants further investigation and we expect that NPs will provide future opportunities to study additional polyamine alkaloids for CA inhibition.

2.4 Semi-synthetic NPs Modified to Incorporate a ZBG and Inhibit CA

2.4.1 Carbohydrate-ZBG Hybrid Molecules

Carbohydrates represent an abundant group of NPs and a selection of naturally occurring mono- and disaccharides have been modified to incorporate CA recognizing ZBG's to give glycosyl primary sulfonamides (sugar-SO_2NH_2) [115], glycosyl primary sulfamides (sugar-NH-SO_2NH_2) [116], and glycoconjugate sulfamates (sugar-O-SO_2NH_2) [117]. Compounds **65–74** derived from the monosaccharides D-glucose, D-galactose, D-mannose and the disaccharide maltose are shown (Fig. 16.10). An aromatic group, which is typical for classical CA inhibitors, is absent from these compounds and instead they comprise the hydrophilic mono- or disaccharide fragment directly attached to the ZBG. These NP-ZBG hybrid molecules have been evaluated as CA inhibitors, Table 16.4. All carbohydrate-ZBG hybrid compounds behaved as weak inhibitors of hCA I, with the anomeric

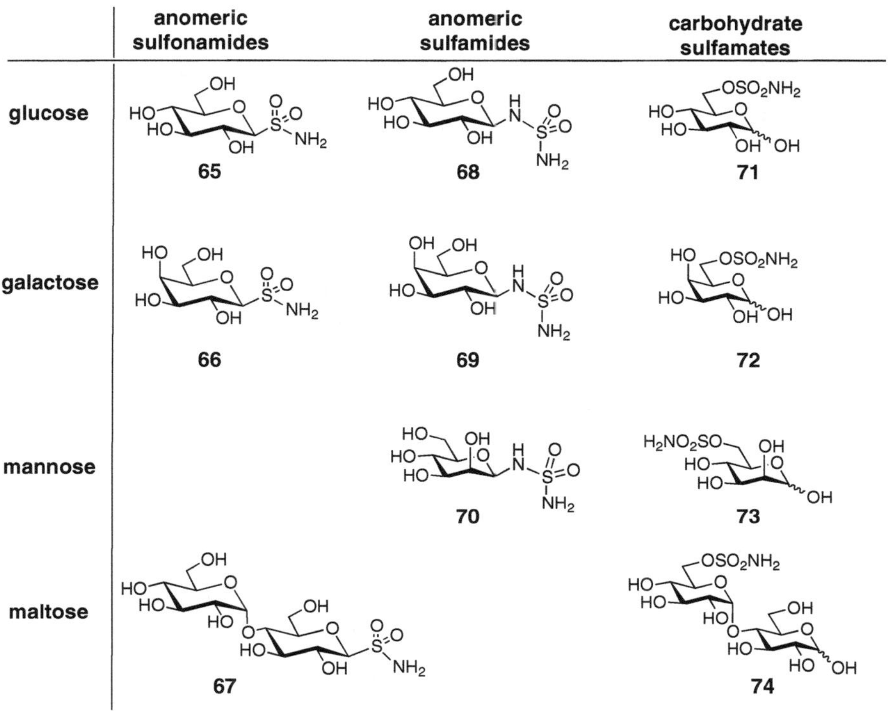

Fig. 16.10 Carbohydrate-ZBG hybrid molecules **65–74**: glycosyl primary sulfonamides (sugar-SO_2NH_2), glycosyl primary sulfamides (sugar-NH-SO_2NH_2) and glycoconjugate sulfamates (sugar-O-SO_2NH_2) [115–117]

sulfonamides **65–67** and sulfamides **68–70** also weak micromolar inhibitors of hCA II, IX and XII. In contrast the C-6 sulfamates **71–74** delivered good activity, particularly monosacchrides **71–73**, which showed K_i <10 nM against hCA XII. The glucose sulfamate **71** also has a K_i <10 nM at hCA IX and displayed selectivity for inhibiting the tumor-associated isoforms CA IX and XII over cytosolic CA I and II. The interested reader is directed to crystal structures for anomeric sulfonamides and glycoconjugate sulfamates in complex with hCA II in the PDB (accession codes: 3HKN, 3HKQ, 3HKT, 3HKU, 3T82, 3T83, 3T84, and 3T85).

The membrane permeability properties were measured for selected carbohydrate-ZBG hybrid CA inhibitors, the results confirm that the compounds are expected to have poor passive membrane permeability. cLog P is an indicator of passive diffusion through cell membranes and values <0 are indicative of molecules with poor membrane permeability. The cLog P values of the hybrid molecules **65–74** range from −2.7 for monosaccharides to −5.5 for dissacharides, Table 16.4. The compound design, employing a deliberate approach towards CA IX/XII isozyme selectivity by changing the physicochemical properties to impart poor membrane permeability, is consistent with these cLog P values. Membrane permeable ester

Table 16.4 Inhibition data of hCA Isozymes I, II, IX and XII and cLog P values for carbohydrate-ZBG hybrid molecules **65**–**74** [115–117]

Compd	K_i (nM)[a]				cLog P[d]
	CA I[b]	CA II[b]	CA IX[c]	CA XII[c]	
65	3,900	4,910	4,050	4,690	−2.8
66	3,930	4,550	4,190	4,800	−2.8
67	4,150	4,100	4,220	4,840	−5.0
68	75,400	4,680	6,470	1,970	−2.7
69	65,800	48,500	940	8,230	−2.7
70	91,900	21,200	1,790	5,440	−2.7
71	1,180	82	8.6	7.3	−3.3
72	4,500	93	62	7.6	−3.3
73	5,960	104	53	9.5	−3.3
74	8,750	513	497	138	−5.5

[a]Errors in the range of ±5–10 % of the reported value, from three determinations
[b]Human (cloned) isozymes
[c]Catalytic domain of human (cloned) isozymes
[d]cLog P data calculated using ChemBioDraw Ultra 11.0

'prodrugs' of the carbohydrate-ZBG hybrids were also synthesized, this allows for potential oral administration, with the polar carbohydrate-ZBG hybrid molecules 'unmasked' in vivo enabling targeting of extracellular CA IX and XII.

2.4.2 Coumarin-ZBG and Steroid-ZBG Hybrid Molecules

A selection of NP-derived sulfamates are potent inhibitors of the cancer drug target steroid sulfatase (STS) and are being developed as a therapy for hormone-dependent breast cancer [118]. This includes the steroidal sulfamate oestrone-3-*O*-sulphamate (EMATE) **75** and two coumarin based sulfamates, COUMATE-667 **76** and STX-118 **77** (Fig. 16.11), which at a simpler structural level may also be considered phenolic sulfamates. These NP hybrids, modified with the sulfamate ZBG, are also potent CA inhibitors, Table 16.5 [119–121]. It is hypothesized that dual steroid sulfatase/CA inhibitors may represent a novel method for treating hormone dependent breast cancer tumors, with the reversible binding of the sulfamates to erythrocyte CA II increasing the metabolic stability of the compounds by protecting the sulfamate moiety from rapid degradation [122]. This indirect improvement of biopharmaceutical properties may persist alongside the direct effect of modulating the activity of cancer-associated CA IX and XII. The X-ray crystal structure of hCA II with both **75** [123] and **76** [122] are reported. These structures conform to the classical ZBG interactions with the sulfamate moiety binding to the active site Zn^{2+} cation. The steroid fragment of **75** and the coumarin fragment of **76** interact with the residues in the hydrophobic half of the CA II active site.

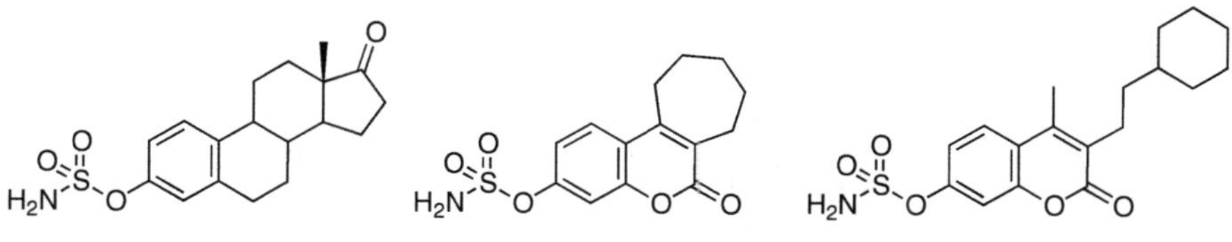

75 **76** **77**

Fig. 16.11 Steroidal sulfamate oestrone-3-*O*-sulphamate (EMATE) **75** and two coumarin based sulfamates, COUMATE-667 **76** and STX-118 **77**

Table 16.5 Inhibition data of hCA Isozymes I, II, IX and XII for steroid- and coumarin-ZBG hybrid molecules **75–77**

	K_i (nM)			
Compd	CA I	CA II	CA IX	CA XII
75 [119]	37	10	30	nd
76 [120]	3,450	21	34	12
77 [121]	nd	$IC_{50} = 59$	nd	nd

nd not determined

3 Conclusion

Contemporary drug discovery is under increased pressure to identify more suitable small molecules as chemical starting points for drug development and finding novel compounds as starting points for optimisation is one of the major challenges in drug discovery research. NPs already provide a significant portion of FDA approved drugs and have emerged as an effective way to sample chemical diversity. The chemical diversity within NPs is vast and while the investigation of NP chemotypes for interaction with CAs is in its infancy, an encouraging start has been made. The NP compounds presented here (phenols, coumarins and polyamines) are suggestive of a tremendous opportunity that NPs provide for the discovery of novel chemotypes for selectively targeting either human or pathogen CAs. It will be imperative for future efforts to further evaluate the NP or NP-hybrid compounds in cell-based models of CA associated disease alongside classical control compounds for validation. The identification of unique CA binding for any NPs might offer possibilities for future rational drug discovery design and development. Thus the use of NPs in the search for new CA inhibitors has a strategic advantage since nature's unique chemical diversity has only been superficially explored in this particular field of research. We predict additional NP structures classes will be identified as binding to and perturbing CA function as further research is undertaken.

References

1. Krishnamurthy VM, Kaufman GK, Urbach AR, Gitlin I, Gudiksen KL, Weibel DB, Whitesides GM (2008) Carbonic anhydrase as a model for biophysical and physical-organic studies of proteins and protein-ligand binding. Chem Rev 108:946–1051
2. (2012) Dictionary of Natural Products (DVD), Vol version 21.1. Taylor & Francis Group/CRC Press, London
3. Thomas SO, Singleton VL, Lowery JA, Sharpe RW, Pruess LM, Porter JN, Mowat JH, Bohonos N (1956) Nucleocidin, a new antibiotic with activity against *Trypanosomes*. Antibiot Annu 7:716–721
4. Rengaraju S, Narayanan S, Ganju PL, Amin MA, Iyengar MRS, Sasaki T, Miyadoh S, Shomura T, Sezaki M, Kojima M (1986) 5′-O-Sulfamoyladenosine (defluoronucleocidin) from a *Streptomyces*. Meiji Seika Kenkyu Nenpo 25:49–55
5. Takahashi E, Beppu T (1982) A new nucleosidic antibiotic AT-265. J Antibiot 35:939–947
6. Takahashi E, Osugi K (1988) Manufacture of 5′-sulfamoyl-2-bromoadenosine from *Streptomyces*. Kureha Chemical Industry, Japan, p 6
7. Iwata M, Sasaki T, Iwamatsu H, Miyadoh S, Tachibana K, Matsumoto K, Shomura T, Sezaki M, Watanabe T (1987) A new herbicidal antibiotic, SF 2494 (5′-O-sulfamoyltubercidin) produced by *Streptomyces mirabilis*. Meiji Seika Kenkyu Nenpo 26:17–22
8. Iwata, M., Sasaki, T., Miyaji, S., Tachibana, K., Matsumoto, K., Shomura, T., and Sezaki, M. (1988) Antibiotic SF 2494, its manufacture with *Streptomyces*, and its use as herbicide. Meiji Seika Kaisha, Japan, p 8
9. Takahashi A, Kurasawa S, Ikeda D, Okami Y, Takeuchi T (1989) Altemicidin, a new acaricidal and antitumour substance. I. Taxonomy, fermentation, isolation and physicochemical and biological properties. J Antibiot 42:1556–1561
10. Kende AS, Liu K, Brands KMJ (1995) Total synthesis of (-)-Altemicidin: a novel ecploitation of the Potier-Polonovski rearrangement. J Am Chem Soc 117:10597–10598
11. Jimenez C, Crews P (1991) Novel Marine Sponge derived Amino Acids 13. Additional Psammaplin derivatives from *Psammaplysilla purpurea*. Tetrahedron 47:2097–2102
12. Pina IC, Gautschi JT, Wang G-Y-S, Sanders ML, Schmitz FJ, Dennis France D, Cornell-Kennon S, Sambucetti LD, Remiszewski SW, Perez LB, Bair KW, Crews P (2003) Psammaplins from the sponge *Pseudoceratina purpurea*: inhibition of both Histone deacetylase and DNA methyltransferase. J Org Chem 68:3866–3873
13. Kristinsson K, Nebel K, O'Sullivan AC, Pachlatko JP, Yamaguchi Y (1995) Herbicidally active sulfamoyl nucleosides. Isolation and synthesis. ACS Symp Ser 584:206–219
14. Thomas SO, Singleton VL, Lowery JA, Sharpe RW, Pruess LM, Porter JN, Mowat JH, Bohonos N (1957) Nucleocidin, a new antibiotic with activity against *Trypanosomes*. Antibiot Annu 7:716–721
15. Jenkins ID, Verheyden JPH, Moffatt JG (1976) 4′-Substituted nucleosides. 2. Synthesis of the nucleoside antibiotic nucleocidin. J Am Chem Soc 98:3346–3357
16. Hewitt RI, Gumble AR, Taylor LH, Wallace WS (1957) Effectiveness of a new antibiotic, nucleocidin, in experimental infections with *Trypanosoma equiperdum*. Antibiot Annu 722–729
17. Camp D, Davis RA, Campitelli M, Ebdon J, Quinn RJ (2012) Drug-like properties: guiding principles for the design of natural product libraries. J Nat Prod 75:72–81
18. Camp D, Davis RA, Evans-Illidge EA, Quinn RJ (2012) Guiding principles for natural product drug discovery. Future Med Chem 4:1067–1084
19. Newman DJ, Cragg GM (2007) Natural products as sources of new drugs over the last 25 years. J Nat Prod 70:461–477
20. Newman DJ, Cragg GM (2009) Natural product scaffolds as leads to drugs. Future Med Chem 1:1415–1427
21. Lam KS (2007) New aspects of natural products in drug discovery. Trends Microbiol 15:279–289

22. Lachance H, Wetzel S, Kumar K, Waldmann H (2012) Charting, navigating, and populating natural product chemical space for drug discovery. J Med Chem 55:5989–6001
23. Williams DA, Lemcke TL (eds) (2002) Foye's principals of medicinal chemistry. Lippincott Williams and Wilkins, Boston
24. Butler MS, Cooper MA (2011) Antibiotics in the clinical pipeline in 2011. J Antibiot 64:413–425
25. Mann J (2002) Natural products in cancer chemotherapy: past, present and future. Nat Rev Cancer 2:143–148
26. Newman DJ, Cragg GM (2012) Natural products as sources of new drugs over the 30 years from 1981 to 2010. J Nat Prod 75:311–335
27. McArdle BM, Quinn RJ (2007) Identification of protein fold topology shared between different folds inhibited by natural products. Chembiochem 8:788–798
28. Ertl P, Roggo S, Schuffenhauer A (2008) Natural product-likeness score and its application for prioritization of compound libraries. J Chem Inf Model 48:68–74
29. Grabowski K, Baringhaus K-L, Schneider G (2008) Scaffold dversity of natural products: inspiration for combinatorial library design. Nat Prod Rep 25:892–904
30. Hert J, Irwin JJ, Laggner C, Keiser MJ, Shoichet BK (2009) Quantifying biogenic bias in screening libraries. Nat Chem Biol 5:479–483
31. Rosen J, Gottfries J, Muresan S, Backlund A, Oprea TI (2009) Novel chemical space exploration via natural products. J Med Chem 52:1953–1962
32. Kellenberger E, Hofmann A, Quinn RJ (2011) Similar interactions of natural products with biosynthetic enzymes and therapeutic targets could explain why nature produces such a large proportion of existing drugs. Nat Prod Rep 28:1483–1492
33. Piggott AM, Karuso P (2004) Quality, not quantity: the role of natural products and chemical proteomics in modern drug discovery. Comb Chem High Throughput Screen 7:607–630
34. Carlson EE (2010) Natural products as chemical probes. ACS Chem Biol 5:639–653
35. Supuran CT (2011) Carbonic anhydrase inhibition with natural products: novel chemotypes and inhibition mechanisms. Mol Divers 15:305–316
36. Borges F, Roleira F, Milhazes N, Santana L, Uriarte E (2005) Simple Coumarins and Analogues in medicinal chemistry: occurrence, synthesis and biological activity. Curr Med Chem 12:887–916
37. Kontogiorgis C, Detsi A, Hadjipavlou-Litina D (2012) Coumarin-based drugs: a patent review (2008 – present). Expert Opin Ther Pat 22:437–454
38. Vu H, Pham NB, Quinn RJ (2008) Direct screening of natural product extracts using mass spectrometry. J Biomol Screen 13:265–275
39. http://www.nature-bank.com.au
40. Maresca A, Temperini C, Vu H, Pham NB, Poulsen S-A, Scozzafava A, Quinn RJ, Supuran CT (2009) Non-zinc mediated inhibition of carbonic anhydrases: coumarins are a new class of suicide inhibitors. J Am Chem Soc 131:3057–3062
41. Lopez M, Vu H, Wang CK, Wolf MG, Groenhof G, Innocenti A, Supuran CT, Poulsen S-A (2011) Promiscuity of Carbonic Anhydrase II. Unexpected Ester hydrolysis of carbohydrate-based sulfamate Inhibitors. J Am Chem Soc 133:18452–18462
42. Krebs JF, Ippolito JA, Christianson DW, Fierke CA (1993) Structural and functional importance of a conserved hydrogen bond network in human carbonic anhydrase II. J Biol Chem 268:27458–27466
43. Innocenti A, Scozzafava A, Parkkila S, Puccetti L, De Simone G, Supuran CT (2008) Investigations of the esterase, phosphatase, and sulfatase activities of the cytosolic mammalian carbonic anhydrase isoforms I, II, and XIII with 4-nitrophenyl esters as substrates. Bioorg Med Chem Lett 18:2267–2271
44. Schultz TW, Rogers K, Aptula AO (2009) Read-across to rank skin sensitization potential: subcategories for the Michael acceptor domain. Contact Dermatitis 60:21–31
45. Wolan DW, Zorn JA, Gray DC, Wells JA (2009) Small-molecule activators of a proenzyme. Science 326:853–858

46. Aptula AO, Patlewicz G, Roberts DW (2005) Skin sensitization: reaction mechanistic applicability domains for structure – activity relationships. Chem Res Toxicol 18:1420–1426
47. Davis RA, Vullo D, Maresca A, Supuran CT, Poulsen S-A (2013) Natural product coumarins that inhibit human carbonic anhydrases. Bioorg Med Chem 21:1539–1543
48. Sarker SD, Armstrong JA, Gray AI, Waterman PG (1994) Pyranocoumarins from *Eriostemon apiculatus*. Biochem Syst Ecol 22:641–644
49. Sarker SD, Armstrong JA, Waterman PG (1994) Angular pyranocoumarins from *Eriostemon thryptomenoides*. Biochem Syst Ecol 22:863–864
50. Gray AI, Waigh RD, Waterman PG (1977) cis-Avicennol, a new pyranocoumarin from the root bark of *Zanthoxylum elephantiasis*. Phytochemistry 16:1017–1018
51. Sultana N, Sarker SD, Armstrong JA, Wilson PG, Waterman PG (2003) The coumarins of *Philotheca sensu lato*: distribution and systematic significance. Biochem Syst Ecol 31:681–691
52. Kashman Y, Gustafson KR, Fuller RW, Cardellina Ii JH, McMahon JB, Currens MJ, Buckheit RW Jr, Hughes SH, Cragg GM, Boyd MR (1992) The calanolides, a novel HIV-inhibitory class of coumarin derivatives from the tropical rainforest tree, *Calophyllum lanigerum*. J Med Chem 35:2735–2743
53. Dreyer DL, Lee A (1972) Chemotaxonomy of the Rutaceae. VIII. Extractives of *Geijera parviflora*. Phytochemistry 11:763–767
54. Lahey FN, MacLeod JK (1967) Coumarins of *Geijera pariflora*. Aust J Chem 20:1943–1955
55. Mujumdar RB, Rathi SS, Rao, AV R (1977) Heartwood constituents of *Chloroxylon swietenia DC*. Indian J Chem Sect B 15B: 200
56. Ahmad J, Shamsuddin KM, Zaman A (1984) A pyranocoumarin from *Atalantia ceylanica*. Phytochemistry 23:2098–2099
57. Spath E, Dobrovolny E (1938) Natural coumarins. XLII. Synthesis of fraxetin, fraxidin and isofraxidin. Ber Dtsch Chem Ges B 71B:1831–1836
58. Morikawa T, Tao J, Toguchida I, Matsuda H, Yoshikawa M (2003) Structures of new cyclic diarylheptanoids and iof nitric oxide production from Japanese folk medicine *Acer nikoense*. J Nat Prod 66:86–91
59. Barbera O, Alberto Marco J, Sanz JF, Sanchez-Parareda J (1986) 3-Methoxyflavones and coumarins from *Artemisia incanescans*. Phytochemistry 25:2357–2360
60. Gonzalez AG, Diaz Chico E, Lopez Dorta H, Medina JM, Rodriguez Luis F (1977) New sources of natural coumarins. XXXII. Chemical components of *Ruta* sp. Tene. 29662. An Quim 73:1015–1018
61. Zhang W-D, Kong D-Y, Li H-T, Gu Z-B, Qin L-P (1998) Chemical constituents of *Erigeron breviscapus*. Zhongguo Yiyao Gongye Zazhi 29:498–500
62. Lin S, Wang S, Liu M, Gan M, Li S, Yang Y, Wang Y, He W, Shi J (2007) Glycosides from the stem bark of *Fraxinus sieboldiana*. J Nat Prod 70:817–823
63. Wu TS, Tsang ZJ, Wu PL, Lin FW, Li CY, Teng CM, Lee KH (2001) New constituents and antiplatelet aggregation and anti-HIV principles of *Artemisia capillaris*. Bioorg Med Chem 9:77–83
64. Raj K, Misra SC, Kapil RS, Popli SP (1976) Coumarins from *Murraya paniculata*. Phytochemistry 15:1787–1787
65. Ito C, Furukawa H (1987) Three new coumarins from the leaves of *Murraya paniculata*. Heterocycles 26:2959–2962
66. Nakamori T, Taniguchi M, Shibano M, Wang N-H, Baba K (2008) Chemical studies on the root of *Heracleum candicans* WALL. J Nat Med 62:403–412
67. Bhatia IS, Bhatia MS, Sharma RS, Bajaj KL (1972) Polyphenolic constituents of the seeds and bark of *Callistemon lanceolatus*. Indian J Chem 10:959–959
68. Rao KV (1974) Toxic principles of *Hippomane mancinella*. Planta Med 25:166–171
69. Lindquist N, Fenical W, Van Duyne GD, Clardy J (1988) New alkaloids of the lamellarin class from the marine ascidian *Didemnum chartaceum* (Sluiter, 1909). J Org Chem 53:4570–4574
70. Andersen RJ, Faulkner DJ, He CH, Van Duyne GD, Clardy J (1985) Metabolites of the marine prosobranch mollusk *Lamellaria* sp. J Am Chem Soc 107:5492–5495

71. Davis RA, Carroll AR, Pierens GK, Quinn RJ (1999) New lamellarin alkaloids from the Australian ascidian, *Didemnum chartaceum*. J Nat Prod 62:419–424
72. Padmawinata K (1973) Isolation and identification of cancer delaying compounds from the leaves of *Geijera salicifolia*. Acta Pharm 4:1–9
73. Jerris PJ, Smith AB III (1981) Synthesis and configurational assignment of geiparvarin: a novel antitumor agent. J Org Chem 46:577–585
74. Gunatilaka L, Kingston D, Wijeratne K, Bandara R, Hofmann G, Johnson R (1994) Biological activity of some coumarins from Sri Lankan Rutaceae. J Nat Prod 57:518–520
75. Nishimori I, Vullo D, Innocenti A, Scozzafava A, Mastrolorenzo A, Supuran CT (2005) Carbonic anhydrase inhibitors. The mitochondrial isozyme VB as a new target for sulfonamide and sulfamate inhibitors. J Med Chem 48:7860–7866
76. Casini A, Antel J, Abbate F, Scozzafava A, David S, Waldeck H, Schäfer S, Supuran CT (2003) Carbonic anhydrase inhibitors: SAR and X-ray crystallographic study for the interaction of sugar sulfamates/sulfamides with isozymes I, II and IV. Bioorg Med Chem Lett 13:841–845
77. De Simone G, Di Fiore A, Menchise V, Pedone C, Antel J, Casini A, Scozzafava A, Wurl M, Supuran CT (2005) Carbonic anhydrase inhibitors. Zonisamide is an effective inhibitor of the cytosolic isozyme II and mitochondrial isozyme V: solution and X-ray crystallographic studies. Bioorg Med Chem Lett 15:2315–2320
78. Khalifah RG (1971) The carbon dioxide hydration activity of carbonic anhydrase. J Biol Chem 246:2561–2573
79. Carta F, Maresca A, Scozzafava A, Supuran CT (2012) Novel coumarins and 2-thioxocoumarins as inhibitors of the tumor-associated carbonic anhydrases IX and XII. Bioorg Med Chem 20:2266–2273
80. Maresca A, Supuran CT (2010) Coumarins incorporating hydroxy- and chloro-moieties selectively inhibit the transmembrane, tumor-associated carbonic anhydrase isoforms IX and XII over the cytosolic ones I and II. Bioorg Med Chem Lett 20:4511–4514
81. Maresca A, Temperini C, Pochet L, Masereel B, Scozzafava A, Supuran CT (2009) Deciphering the mechanism of carbonic anhydrase inhibition with coumarins and thiocoumarins. J Med Chem 53:335–344
82. Nair SK, Ludwig PA, Christianson DW (1994) Two-site binding of phenol in the active site of human carbonic anhydrase II: structural implications for substrate association. J Am Chem Soc 116:3659–3660
83. Innocenti A, Vullo D, Scozzafava A, Supuran CT (2008) Carbonic anhydrase inhibitors: inhibition of mammalian isoforms I-XIV with a series of substituted phenols including paracetamol and salicylic acid. Bioorg Med Chem 16:7424–7428
84. Innocenti A, Sarikaya SBO, Gulcin I, Supuran CT (2010) Carbonic anhydrase inhibitors. Inhibition of mammalian isoforms I-XIV with a series of natural product polyphenols and phenolic acids. Bioorg Med Chem 18:2159–2164
85. http://www.griffith.edu.au/science-aviation/queensland-compound-library
86. Davis RA, Hofmann A, Osman A, Hall RA, Mühlschlegel FA, Vullo D, Innocenti A, Supuran CT, Poulsen SA (2011) Natural product-based phenols as novel probes for mycobacterial and fungal carbonic anhydrases. J Med Chem 54:1682–1692
87. Davis RA, Innocenti A, Poulsen S-A, Supuran CT (2010) Carbonic anhydrase inhibitors. Identification of selective inhibitors of the human mitochondrial isozymes VA and VB over the cytosolic isozymes I and II from a natural product-based phenolic library. Bioorg Med Chem 18:14–18
88. Davis RA (2005) Isolation and structure elucidation of the new fungal metabolite (-)-Xylariamide A. J Nat Prod 68:769–772
89. Healy PC, Hocking A, Tran-Dinh N, Pitt JI, Shivas RG, Mitchell JK, Kotiw M, Davis RA (2004) Xanthones from a microfungus of the genus *Xylaria*. Phytochemistry 65:2373–2378
90. Davis RA, Baron PS, Neve JE, Cullinane C (2009) A microwave-assisted stereoselective synthesis of polyandrocarpamines A and B. Tetrahedron Lett 50:880–882

91. Davis RA, Aalbersberg W, Meo S, Moreira da Rocha R, Ireland CM (2002) The isolation and synthesis of polyandrocarpamines A and B. Two new 2-aminoimidazolone compounds from the *Fijian ascidian*, *Polyandrocarpa* sp. Tetrahedron 58:3263–3269
92. Davis RA, Carroll AR, Duffy S, Avery VM, Guymer GP, Forster PI, Quinn RJ (2007) Endiandrin A, a potent glucocorticoid receptor binder isolated from the Australian plant *Endiandra anthropophagorum*. J Nat Prod 70:1118–1121
93. Davis RA, Barnes EC, Longden J, Avery VM, Healy PC (2009) Isolation, structure elucidation and cytotoxic evaluation of endiandrin B from the Australian rainforest plant *Endiandra anthropophagorum*. Bioorg Med Chem 17:1387–1392
94. Supuran CT (2003) Carbonic anhydrase inhibitors in the treatment and prophylaxis of obesity. Expert Opin Ther Patents 13:1545–1550
95. De Simone G, Di Fiore A, Supuran CT (2008) Are carbonic anhydrase inhibitors suitable for obtaining antiobesity drugs? Curr Pharm Des 14:655–660
96. Nishimori I, Minakuchi T, Kohsaki T, Onishi S, Takeuchi H, Vullo D, Scozzafava A, Supuran CT (2007) Carbonic anhydrase inhibitors: the β-carbonic anhydrase from *Helicobacter pylori* is a new target for sulfonamide and sulfamate inhibitors. Bioorg Med Chem Lett 17:3585–3594
97. Nishimori I, Onishi S, Takeuchi H, Supuran CT (2008) The α and β classes carbonic anhydrases from *Helicobacter pylori* as novel drug targets. Curr Pharm Des 14:622–630
98. Isik S, Kockar F, Aydin M, Arslan O, Guler OO, Innocenti A, Scozzafava A, Supuran CT (2009) Carbonic anhydrase inhibitors: inhibition of the β-class enzyme from the yeast *Saccharomyces cerevisiae* with sulfonamides and sulfamates. Bioorg Med Chem 17:1158–1163
99. Innocenti A, Mühlschlegel FA, Hall RA, Steegborn C, Scozzafava A, Supuran CT (2008) Carbonic anhydrase inhibitors: inhibition of the β-class enzymes from the fungal pathogens *Candida albicans* and *Cryptococcus neoformans* with simple anions. Bioorg Med Chem Lett 18:5066–5070
100. Innocenti A, Hall RA, Schlicker C, Scozzafava A, Steegborn C, Mühlschlegel FA, Supuran CT (2009) Carbonic anhydrase inhibitors. Inhibition and homology modeling studies of the fungal β-carbonic anhydrase from *Candida albicans* with sulfonamides. Bioorg Med Chem 17:4503–4509
101. Innocenti A, Hall RA, Schlicker C, Mühlschlegel FA, Supuran CT (2009) Carbonic anhydrase inhibitors. Inhibition of the β-class enzymes from the fungal pathogens *Candida albicans* and *Cryptococcus neoformans* with aliphatic and aromatic carboxylates. Bioorg Med Chem 17:2654–2657
102. Joseph P, Turtaut F, Ouahrani-Bettache S, Montero J-L, Nishimori I, Minakuchi T, Vullo D, Scozzafava A, Kolhler S, Winum J-Y, Supuran CT (2010) Cloning, characterization, and inhibition studies of a β-carbonic anhydrase from *Brucella suis*. J Med Chem 53:2277–2285
103. Vullo D, Nishimori I, Scozzafava A, Köhler S, Winum J-Y, Supuran CT (2010) Inhibition studies of a β-carbonic anhydrase from *Brucella suis* with a series of water soluble glycosyl sulfanilamides. Bioorg Med Chem Lett 20:2178–2182
104. Wahle K, Caruso D, Ochoa J, Quiles J (2004) Olive oil and modulation of cell signaling in disease prevention. Lipids 39:1223–1231
105. Fu G, Pang H, Wong Y (2008) Naturally occurring phenylethanoid glycosides: potential leads for new therapeutics. Curr Med Chem 15:2592–2613
106. Davis RA, Watters D, Healy PC (2005) The isolation and synthesis of 3-chloro-4-hydroxyphenylacetamide produced by a plant-associated microfungus of the genus *Xylaria*. Tetrahedron Lett 46:919–921
107. Casero RA Jr, Marton LJ (2007) Targeting polyamine metabolism and function in cancer and other hyperproliferative diseases. Nat Rev Drug Discovery 6:373–390
108. Fleidervish IA, Libman L, Katz E, Gutnick MJ (2008) Endogenous polyamines regulate cortical neuronal excitability by blocking voltage-gated Na^+ channels. Proc Natl Acad Sci U S A 105:18994–18999
109. Wallace HM, Niiranen K (2007) Polyamine analogues – an update. Amino Acids 33:261–265

110. Soda K, Dobashi Y, Kano Y, Tsujinaka S, Konishi F (2009) Polyamine-rich food decreases age-associated pathology and mortality in aged mice. Exp Gerontol 44:727–732
111. Xu M, Davis RA, Feng Y, Sykes ML, Shelper T, Avery VM, Camp D, Quinn RJ (2012) Ianthelliformisamines A-C, antibacterial bromotyrosine-derived metabolites from the marine sponge *Suberea ianthelliformis*. J Nat Prod 75:1001–1005
112. Yin S, Davis RA, Shelper T, Sykes ML, Avery VM, Elofsson M, Sundin C, Quinn RJ (2011) Pseudoceramines A-D, new antibacterial bromotyrosine alkaloids from the marine sponge *Pseudoceratina* sp. Org Biomol Chem 9:6755–6760
113. Buchanan MS, Carroll AR, Fechner GA, Boyle A, Simpson MM, Addepalli R, Avery VM, Hooper JNA, Su N, Chen HW, Quinn RJ (2007) Spermatinamine, the first natural product inhibitor of isoprenylcysteine carboxyl methyltransferase, a new cancer target. Bioorg Med Chem Lett 17:6860–6863
114. Carta F, Temperini C, Innocenti A, Scozzafava A, Kaila K, Supuran CT (2010) Polyamines inhibit carbonic anhydrases by anchoring to the zinc-coordinated water Molecule. J Med Chem 53:5511–5522
115. Lopez M, Paul B, Hofmann A, Morizzi J, Wu QK, Charman SA, Innocenti A, Vullo D, Supuran CT, Poulsen S-A (2009) S-glycosyl primary sulfonamides – a new structural class for selective inhibition of cancer-associated carbonic anhydrases. J Med Chem 52:6421–6432
116. Rodriguez OM, Maresca A, Tempera CA, Bravo RD, Colinas PA, Supuran CT (2011) N-β-glycosyl sulfamides are selective inhibitors of the cancer associated carbonic anhydrase isoforms IX and XII. Bioorg Med Chem Lett 21:4447–4450
117. Lopez M, Trajkovic J, Bornaghi LF, Innocenti A, Vullo D, Supuran CT, Poulsen S-A (2011) Design, synthesis, and biological evaluation of novel carbohydrate-based sulfamates as carbonic anhydrase inhibitors. J Med Chem 54:1481–1489
118. Woo LWL, Purohit A, Malini B, Reed MJ, Potter BVL (2000) Potent active site-directed inhibition of steroid sulphatase by tricyclic coumarin-based sulphamates. Chem Biol 7:773–791
119. Winum J-Y, Vullo D, Casini A, Montero J-L, Scozzafava A, Supuran CT (2003) Carbonic anhydrase inhibitors: Inhibition of cytosolic isozymes I and II and the membrane-bound, tumor associated isozyme IX with sulfamates also acting as steroid sulfatase inhibitors. J Med Chem 46:2197–2204
120. Temperini C, Innocente A, Scozzafava A, Supuran CT (2008) Carbonic anhydrase inhibitors. Interaction of the antitumor sulfamate EMD 486019 with twelve mammalian carbonic anhydrase isoforms: kinetic and X-ray crystallographic studies. Bioorg Med Chem Lett 18:4282–4286
121. Ho YT, Purohit A, Vicker V, Newman SP, Robinson JJ, Leese MP, Ganeshapillai D, Woo LWL, Potter BVL, Reed MJ (2003) Inhibition of carbonic anhydrase II by steroidal and non-steroidal sulphamates. Biochem Biophys Res Commun 305:909–914
122. Lloyd MD, Pederick RL, Natesh R, Woo LW, Purohits A, Reed MJ, Acharya KR, Potter BVL (2005) Crystal structure of human carbonic anhydrase II at 1.95 Angstrom resolution in complex with 667-coumate, a novel anti-cancer agent. Biochem J 385:715–720
123. Abbate F, Winum J-Y, Potter BVL, Casini A, Montero J-L, Scozzafava A, Supuran CT (2004) Carbonic anhydrase inhibitors: X-ray crystallographic structure of the adduct of human isozyme II with EMATE, a dual inhibitor of carbonic anhydrases and steroid sulfatase. Bioorg Med Chem Lett 14:231–234

Chapter 17
Glaucoma and the Applications of Carbonic Anhydrase Inhibitors

Andrea Scozzafava and Claudiu T. Supuran

Abstract Inhibition of carbonic anhydrase (CA, EC 4.2.1.1) has pharmacologic applications in the treatment of glaucoma, a disease affecting a large number of people and characterized by an elevated intraocular pressure (IOP). At least three isoforms, CA II, IV and XII are targeted by the sulfonamide inhibitors, some of which are clinically used drugs. Acetazolamide, methazolamide and dichlorophenamide are first generation CA inhibitors (CAIs) still used as systemic drugs for the management of this disease. Dorzolamide and brinzolamide represent the second generation inhibitors, being used topically, as eye drops, with less side effects compared to the first generation drugs. Third generation inhibitors have been developed by using the tail approach, but they did not reach the clinics yet. The most promising such derivatives are the sulfonamides incorporating either tails with nitric oxide releasing moieties or hybrid drugs possessing prostaglandin (PG) F agonist moieties in their molecules. Recently, the dithiocarbamates have also been described as CAIs possessing IOP lowering effects in animal models of glaucoma. CAIs are used alone or in combination with other drugs such as adrenergic agonist/antagonists, or PG analogs, being an important component of the antiglaucoma drugs armamentarium.

Susan C. Frost and Robert McKenna (eds.). Carbonic Anhydrase: Mechanism, Regulation, Links to Disease, and Industrial Applications

A. Scozzafava
Laboratorio di Chimcia Bioinorganica, Universita degli Studi di Firenze, Florence, Italy
e-mail: andrea.scozzafava@unifi.it

C.T. Supuran (✉)
NEUROFARBA Department, Universita degli Studi di Firenze, Polo Scientifico, Florence, Italy
e-mail: claudiu.supuran@unifi.it

S.C. Frost and R. McKenna (eds.), *Carbonic Anhydrase: Mechanism, Regulation, Links to Disease, and Industrial Applications*, Subcellular Biochemistry 75,
DOI 10.1007/978-94-007-7359-2_17, © Springer Science+Business Media Dordrecht 2014

Keywords Carbonic anhydrase • Sulfonamide • Dithiocarbamate • Intraocular pressure • Vasodilation • Glaucoma • Carbonic anhydrase inhibitors

1 Introduction

Glaucoma consists of a group of eye diseases showing a broad spectrum of clinical presentation and aetiologies, which lead to a permanent loss of visual function due to the damage of the optical nerve [1–5]. Several different types of glaucoma are known, and they are characterized by an elevated intraocular pressure (IOP) [1–4]. The source of this abnormal increase is related to a misfunctioning of tissues of the trabecular meshwork, located in the anterior chamber angle of the eye, between the cornea and the iris. Their role is to maintain a balanced pressure in the anterior chamber of the eye by allowing the outflow of aqueous humor [1–5]. Aqueous humor is a transparent liquid, rich in bicarbonate, which fills the region between the cornea and the lens [5–11]. It is continuously secreted by the ciliary body around the lens, and constantly flowing from the ciliary body to the anterior chamber [5–7]. From the anterior chamber the aqueous humor passes through the trabecular meshwork into the Schlemm canal and from it into a multitude of aqueous veins which eventually merge with the blood carrying veins [5–11]. Glaucoma is very often characterized by an excessive retention/secretion of aqueous humor in the anterior chamber, whose effect leads to a steep increase of the IOP to values as high as 25–30 mmHg [1–5].

Glaucoma has been classified in two different categories, the closed angle and the open angle glaucoma [1, 2]. In the first case the closure of the anterior chamber angle is observed as a result of a contact between the iris and the surface of the trabecular meshwork [1, 2]. Closure of this angle prevents the drainage of the aqueous humor. The open angle form is represented by any kind of glaucoma in which the above angle remains open, but the drainage of the aqueous humor is decreased [1, 2]. The exact reasons of such an altered flow remains largely unknown [1].

The general method for treating glaucoma consists in IOP reduction therapy, through pharmacology, laser therapy or surgical operation [1–4]. It should be noted that some patients with glaucomatous vision loss can show relatively low IOP, but they represent a minority of the glaucoma patients [1, 2]. Also in such cases, therapies that control IOP are recommended [1, 2]. Drug therapies effective for IOP reduction include both agents that decrease aqueous humor production and agents that increase the outflow facility. Such therapies can be applied topically (directly on the eye) or administered systemically [1–4].

The relationship between glaucoma and carbonic anhydrase (CA, EC 4.2.1.1), the enzyme catalyzing the hydration of CO_2 to bicarbonate and protons [12–16] is rather straightforward and known for decades. The pioneering studies of Friedenwald [6], Kinsey [7] and Kinsey and Barany [8] on the chemistry and dynamics of aqueous humor showed that the main constituent of this secretion is sodium bicarbonate. The next step was the identification of CA, in the anterior uvea of the eye by Wistrand [9], who demonstrated that it is this enzyme (present mainly in the ciliary processes) responsible for the bicarbonate secretion, as a consequence of the hydration reaction of carbon dioxide, with formation of bicarbonate (and obviously, protons). Becker [10] then showed that the sulfonamide CA inhibitor (CAI) acetazolamide, **1**, produced a drop of IOP in experimental animals and humans, whereas Kinsey and Reddy [11] proved that this phenomenon is due to a reduced bicarbonate secretion as a consequence of CA inhibition by the sulfonamide drug. This was the beginning of a novel treatment for glaucoma, already in the 1950s [17]. As this condition is affecting an increasing number of the population, also being the leading cause of blindness in the Western countries, it is essential to explore novel medications for its treatment [1–4]. In fact, CAIs constitute one of the main drug classes used for the treatment of glaucoma [12–17].

2 First Generation Inhibitors: Systemically Acting Drugs (Acetazolamide, Methazolamide, Dichlorophenamide)

Discovered in the 1950s by Roblin's group [18], the heterocyclic sulfonamides such as acetazolamide **1**, methazolamide **2** and ethoxzolamide **3** [19] as well as the aromatic compound dichlorophenamide **4**, represent the first generation of clinically used CAIs [17]. They are very strong (typically low nanomolar) inhibitors of most CA isoforms of the 15 ones presently known in humans (see Ref. [13] for details). Except ethoxzolamide which has fewer clinical applications, acetazolamide **1**, methazolamide **2** and dichlorophenamide **4** are systemically used CAIs, mainly as antiglaucoma drugs, even if they were discovered decades ago, and even if they have a range of side effects (see discussion later in the text) [5, 17].

Systemic inhibitors are useful in reducing elevated IOP characteristic of many glaucoma forms, as they represent the most efficient physiological treatment of glaucoma. Indeed by inhibiting the ciliary process enzymes (the sulfonamide susceptible isozymes CA II, CA IV and CA XII [5, 12, 17, 20]), a reduced rate of bicarbonate and aqueous humor secretion is achieved, which leads to a 25–30 % decrease of IOP [5, 17]. However, as mentioned above, these compounds are promiscuous, strong inhibitors of all CA isoforms, and the inhibition of various CA isozymes present in other tissues than the eye leads to an entire range of side effects [5, 12, 17]. The most prominent ones are: numbness and tingling of extremities, metallic taste, depression, fatigue, malaise, weight loss, decreased libido, gastrointestinal irritation, metabolic acidosis, renal calculi and transient myopia [5, 12, 17]. Indeed, all these compounds (**1–4**) indiscriminately inhibit most of the CA isozymes (such as CA I, CA II, CA IV, CA VA, VB, CA VII, XIII and XIV) abundant in other organs than the eye, such as blood, kidneys, lungs, gastrointestinal tract, CNS, etc. [5, 12, 17]. As a consequence there are limitations of their use due to patience compliance. However, acetazolamide and dichlorophenamide are even nowadays components of regimens for the treatment of refractory glaucoma, which does not respond to adrenergic antagonists, or prostaglandin (PG) analogs [1, 5].

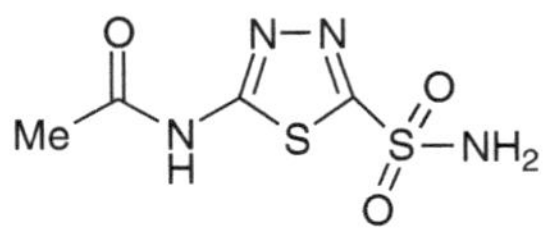

1

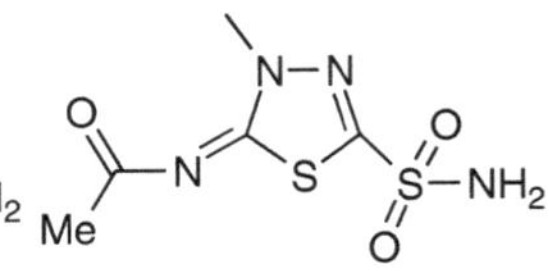

2

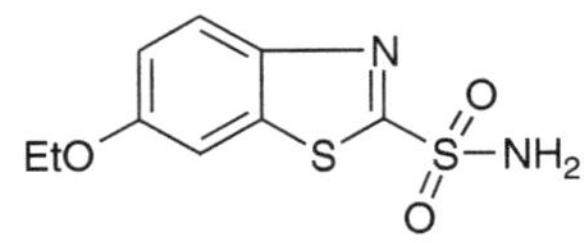

3

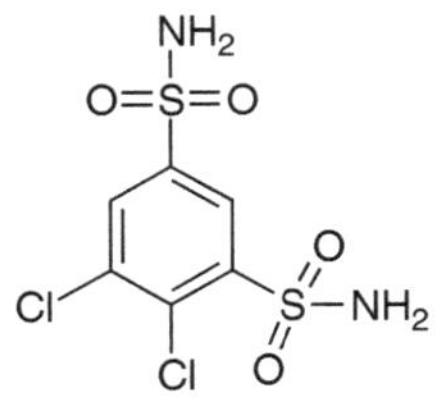

4

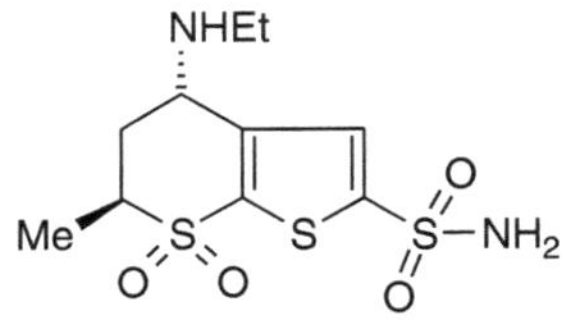

5

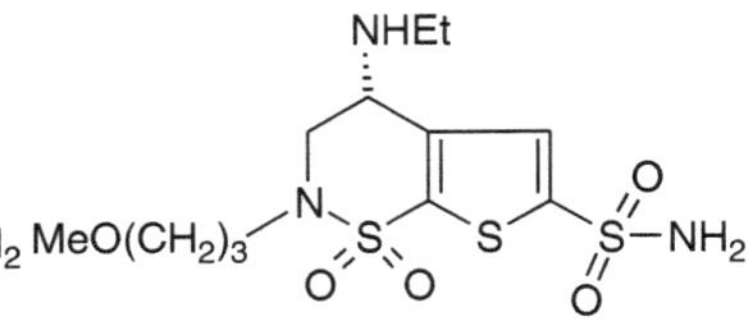

6

3 Second Generation Inhibitors: Topically Acting Sulfonamides (Dorzolamide and Brinzolamide)

The idea to administer topically, directly into the eye the sulfonamide CAI was already addressed by Becker in the 1950s [10]. This and other studies involving the clinically used compounds **1–4** which were administered as suspensions into the eye of experimental animals, only gave negative results, being concluded that sulfonamide CAIs were effective as antiglaucoma drugs only via the systemic route [10, 17]. The lack of efficiency of the first generation sulfonamide CAIs via the topical route was due to the fact that the drug was unable to arrive at the ciliary processes where CAs are present [5, 21]. The inadequate drug penetrability through the cornea was ascrivable to the inappropriate physico-chemical properties for such a route of administration of the sulfonamides **1–4**.

In 1983, in a seminal paper Maren's group [21] postulated that a water-soluble sulfonamide, possessing a relatively balanced lipid solubility (in order to be able to penetrate through the cornea) as well as strong enough CA inhibitory properties, would be an effective IOP lowering drug via the topical route, but at that moment no inhibitors possessing such properties existed, as the bio-organic chemistry of this class of compounds was rather unexplored at that time [19]. Water-soluble sulfonamide CAIs started to be developed in several laboratories soon thereafter, and by 1995 the first such pharmacological agent, dorzolamide **5** has been launched for clinical use by Merck, as 2 % eye drops [22, 23]. A second, structurally related compound, brinzolamide **6** (discovered at Alcon Laboratories), has then been approved for the topical treatment of glaucoma in 1999 [24]. These two compounds are still the only topically acting CAIs in clinical use at this moment.

Dorzolamide **5** and brinzolamide **6** are nanomolar CA II/CA XII inhibitors [13, 22], possess a good water solubility, are sufficiently liposoluble to penetrate through the cornea, and may be administered topically, directly into the eye, as a

2 % water solution (of the dorzolamide hydrochloride salt) or as 1 % suspension (as the brinzolamide hydrochloride salt) 2–3 times a day [5, 22, 24]. The two drugs are effective in reducing IOP and show fewer side effects as compared to the systemically applied drugs. The observed common side effects include stinging, burning or reddening of the eye, blurred vision, pruritus, bitter taste [5, 22, 24]. All but the last are probably due to the fact that dorzolamide (the best studied topical CAI) is the salt of a weak base with a very strong acid, so that the pH of the drug solution is rather acidic (generally around 5.5). The last side effect mentioned above is probably due to drug-laden lachrymal fluid draining into the oropharynx and inhibition of CA present in the saliva (CA VI) and the taste buds (CA II and CA VI), with the consequent accumulation of bicarbonate, and was observed both with systemic as well as topical CAIs [5, 12]. Brinzolamide produces less stinging but more blurred vision as compared to dorzolamide [5, 24]. Unfortunately, dorzolamide showed some more serious side effects, such as contact allergy, nephrolithiasis, anorexia, depression and dementia and irreversible corneal decompensation in patients who already presented corneal problems [25–28]. Thus, even if dorzolamide and brinzolamide represent indeed a major progress in the fight against glaucoma with therapies based on CAIs, novel types of topically effective inhibitors belonging to this class of pharmacological agents are still needed.

4 Third Generation Sulfonamide Inhibitors

An approach was reported [29] for obtaining novel types of sulfonamide CAIs with good hydrosolubility and IOP lowering effects, which has been nominated the "tail approach". It consists in attaching water-solubilizing functionalities to the molecules of aromatic/heterocyclic sulfonamides incorporating derivatizable moieties of the amino, imino or hydroxyl type. Such moieties included, among others, pyridine-carboximido; carboxypyridine-carboxamido, quinolinesulfonamido; picolinoyl, isonicotinoyl, perfluoroalkyl/arylsulfonyl-, as well as amino acyl groups, whereas ring systems which have been derivatized by using the above mentioned moieties included: 2-; 3- or 4-aminobenzenesulfonamides; 4-(ω-aminoalkyl)-benzenesulfonamides; 3-halogeno-substituted-sulfanilamides; 1,3-benzene-disulfonamides; 1,3,4-thiadiazole-2-sulfonamides; benzothiazole-2-sulfonamides as well as thienothiopyran-2-sulfonamides, and were chosen in such a way as to demonstrate that the proposed approach was a general one [29–32]. Compounds prepared by the tail approach showed 2–3 times more effective topical IOP lowering effects in rabbits as compared to dorzolamide **5** [29–32]. They possessed good water solubility (as hydrochlorides, triflates or trifloroacetates), inhibition in the low nanomolar range against hCA II and IV, good penetrability through the cornea, and very good IOP lowering properties in both normotensive and glaucomatous rabbits (widespread animal models of glaucoma) [29–32]. What is more important, this effect lasted for a prolonged period of time as compared to the similar effect of dorzolamide [29–32]. These promising compounds were not developed for clinical use mainly because the company which acquired rights on them was incorporated in another one, which was not interested in this type of applications.

The tail approach proved however to be a general and versatile modality to obtain a wide range of CAIs belonging to several classes of compounds, the sulfonamides being just one particular case (reviewed in Refs. [13, 33–35]).

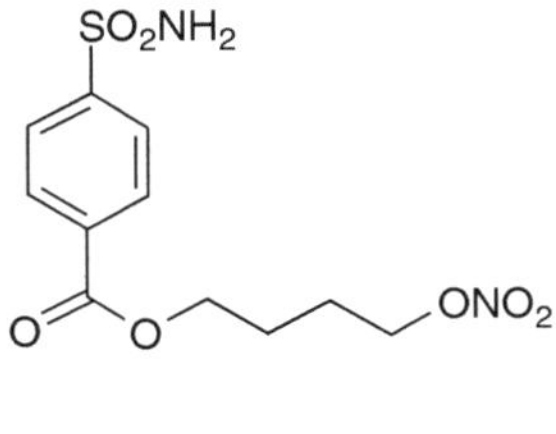

7

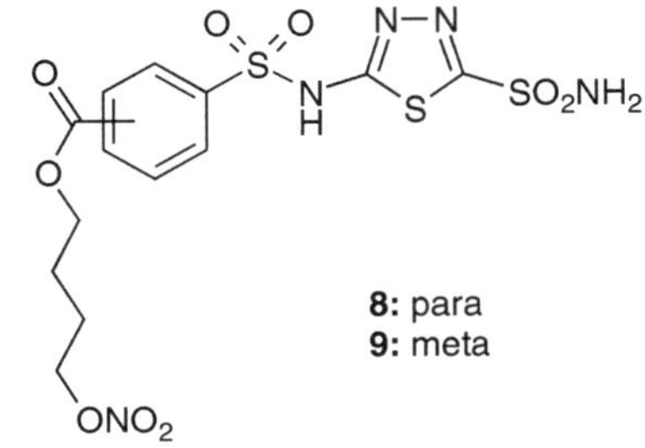

8: para
9: meta

An interesting development of the tail approach consisted in designing CAIs incorporating nitrate ester moieties, which led to hybrid drugs possessing a sulfonamide and an NO-donating moiety pharmacophore in their molecule [36–38]. Nitric oxide (NO), a radical gas produced by the enzyme nitric oxide synthase (NOS), is involved in vasodilation, aqueous humour outflow within the eye, local modulation of ocular blood flow and retinal ganglion cells death apoptosis [36–38]. It appeared thus of interest to combine these two pharmacophores, a sulfonamide CAI, and a moiety able to donate NO, in the molecule of the same compound [36–38]. In this way, a large number of sulfonamides with NO-donating properties were reported in the last several years, among which those of types **7–9** were among the most interesting ones [36–39]. Several aromatic/heterocyclic sulfonamide scaffolds have been used to synthesize compounds incorporating NO-donating moieties of the nitrate ester type [36–38]. Some of these compounds showed effective in vitro inhibition of the target isoforms involved in glaucoma (in the low nanomolar range) i.e., hCA II, IV and XII, and the X-ray crystal structure of some of them bound to the dominant isoform hCA II also revealed factors associated with this marked inhibitory activity (see Chap. 15 in this book) [36–38]. More importantly, in an animal model of ocular hypertension, one of these new compounds (derivative **7**), was twice more effective than dorzolamide **5** in reducing elevated IOP characteristic of this disease, anticipating its potential for the treatment of glaucoma [37]. A detailed pharmacologic study of **7** was recently reported [38]. Chronic administration of **7** as 2 % eyedrops to glaucomatous albino rabbits resulted in an important reduction in IOP (of 45–50 %) already after the first week of treatment, with a regular decreasing trend during the treatment period [38]. This reduction was much higher than that observed when dorzolamide at 2 % was administered in the same animal model and with an identical administration schedule [38]. Furthermore, in the ophthalmic artery of the treated rabbits, both systolic and diastolic velocities, were significantly reduced in eyes treated with the hybrid drug **7** in comparison to dorzolamide **5**, suggesting thus a beneficial effect of this class of CAIs on the blood supply to the optic nerve (in addition to the IOP reduction), which was not observed with dorzolamide or brinzolamide [38]. Thus, the sulfonamides incorporating NO-donating moieties represent good candidates for the next generation topically acting antiglaucoma agents: they may exert beneficial activity both by reducing elevated IOP and by supplying more blood (and thus oxygen) to the optic nerve.

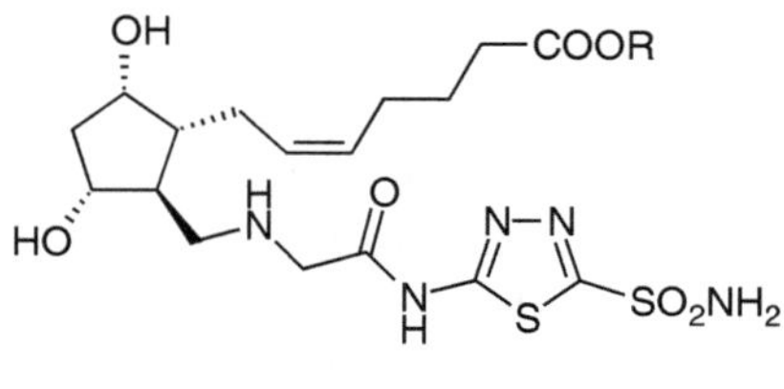

10a: R = i-Pr
10b: R = H

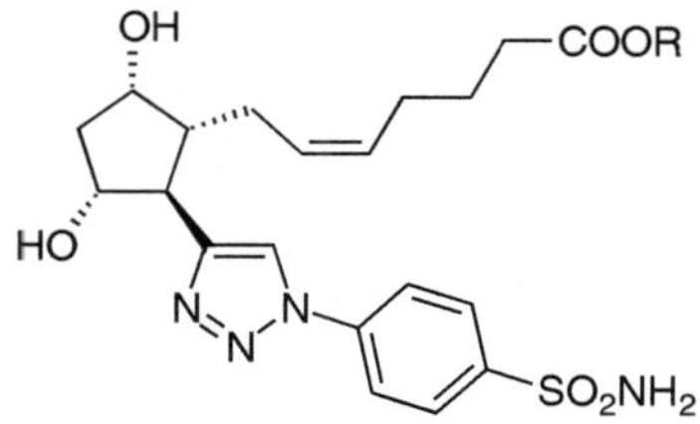

11a: R = Me
11b: R = H

Very recently [39], this hybrid drugs approach has been also used for obtaining compounds incorporating sulfonamide and PG moieties in the same molecule. The rationale of having two pharmacophores with well-known antiglaucoma effects, i.e., the aromatic/heterocyclic sulfonamide one associated with CA inhibition (and whence reduced bicarbonate secretion in the aqueous humor), and the PG F one associated with increased outflow of the aqueous humor, has led to molecules of types **10** and **11** [39]. Both heterocyclic (1,3,4-thiadiazole-2-sulfonamide, such as **10a,b**) and aromatic (benzenesulfonamides, such as **11a,b**) derivatives have been reported. They incorporate, in addition to the aromatic/heterocyclic sulfonamide moiety, the $PGF_{2\alpha}$ moiety, occurring in eicosanoid such as latanoprost, which are effective antiglaucoma drugs [5]. The most interesting compound seems to be **11b** which showed an inhibition constants of around 10 nM against hCA II and was also a good PG F receptor (FP) agonist (EC_{50} of 5.7 nM) [39]. Although the ocular permeability of this (and related) hybrid drugs was good, no in vivo IOP lowering data with these derivatives were reported so far. However, the hybrid drug approach is a valid one for obtaining CAIs with additional biological activity, which can add on to the efficacy of such agents as antiglaucoma drugs.

5 Dithiocarbamates as Topically Acting Antiglaucoma CAIs

Although the sulfonamides dominated the drug design landscape of CAIs for many years, recently, new chemotypes emerged that interact with these enzymes by a similar or different inhibition mechanism as the sulfonamides [13, 33–35]. Among them, the dithiocarbamates (DTCs) are undoubtedly the most interesting ones [40–43]. These compounds have been rationally discovered as CAIs after our report of trithiocarbonate (CS_3^{2-}) as an interesting (milli – micromolar) inhibitor [44]. In the X-ray crystal structure of this inorganic anion bound to CA II, it has been observed a monodentate coordination of the inhibitor via one sulfur atom to the zinc ion from the enzyme active site, and a hydrogen bond in which another sulfur and the OH of Thr199 were involved [44]. Thus, the CS_2^- was discovered as a new zinc-binding group (ZBG) for generating CAIs. As DTCs are simple compounds that incorporate this new ZBG, a rather large series of such compounds was prepared and evaluated for their inhibitory activity against several mammalian, fungal and bacterial CAs [40–43]. Several low nanomolar and subnanomolar CAIs were thus detected against all these enzymes [40–43]. X-ray crystal structures were also reported for several DTCs complexed to hCA II. Many DTCs inhibited hCA II in the low nanomolar range. Their binding mode to the enzyme is identical to that of trithiocarbonate, i.e. with one sulfur of the CS_2^- moiety coordinated to the metal ion, but the organic scaffold present in the DTC was observed to make extensive contacts with many amino acid residues from the active site, which explained the wide range of inhibitory power of these derivatives (from the subnanomolar to the micromolar, for the entire series of around 30 DTC reported so far [40–43]). Interestingly, the highly water soluble morpholine DTC **12** was also very effective in vivo as an antiglaucoma agent when administered topically, directly into the eye of hypertensive rabbits [41], a widely used animal model of glaucoma [38]. Although few such compounds were investigated up until now in detail, the DTCs, being readily available, easy to synthesize and with an excellent water solubility, may have a firm place in the antiglaucoma drugs armamentarium.

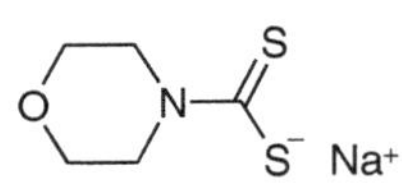

12

6 Conclusions

CAIs are widely used in the treatment of glaucoma, both systemically and as topically-acting agents. Three such drugs, acetazolamide, methazolamide and dichlorophenamide, are the first generation, systemic sulfonamides, in clinical use since the 1950s, and even if they possess a range of side effects they are still useful for the treatment of refractory forms of the disease. The newer, topically acting agents dorzolamide and brinzolamide, are used as eye drops and show fewer systemic side effects compared to acetazolamide or dichlorophenamide, but they show many ocular side effects. Both drugs are used alone or in combination with β-blockers or PG F2 receptor agonists. However, no new drugs from this pharmacological class appeared for more than 15 years. Several interesting approaches have been however reported, which led to many interesting compounds which were evaluated in animal models of the disease. Among them, the tail approach led to a high number of very effective sulfonamide CAIs, many of which with excellent water solubility, good activity against the target enzymes, and excellent preliminary activity in several animal models of glaucoma. Among these approaches, the hybrid drug one seems to have led to highly interesting compounds incorporating sulfonamides and NO-donating moieties on one hand, or sulfonamides and Prostaglandin F receptor agonist on the other hand. Such compounds, possessing a dual mode of action, were active in animal models of the disease. Finally, the dithiocarbamates have been reported as a new class of CAIs, and some representatives were also showed to possess significant antiglaucoma activity in animal models. Considering the lack of innovation in the ophthalmology field for the last 15 years, it would be desirable that some of these interesting compounds may arrive to clinics soon.

Acknowledgments Research from our laboratory was financed by several grants of the 6th and 7th Framework Programs of the European Union (DeZnIT, Metoxia and Dynano projects).

References

1. Quigley HA (2011) Glaucoma. Lancet 377:1367–1377
2. Quigley HA, Broman AT (2006) The number of people with glaucoma worldwide in 2010 and 2020. Br J Ophthalmol 90:151–156
3. Mincione F, Menabuoni L, Supuran CT (2004) Clinical applications of the carbonic anhydrase inhibitors in ophthalmology. In: Supuran CT, Scozzafava A, Conway J (eds) Carbonic anhydrase – its inhibitors and activators. CRC Press, Boca Raton, pp 243–254
4. Mincione F, Scozzafava A, Supuran CT (2009) Antiglaucoma carbonic anhydrase inhibitors as ophthalomologic drugs. In: Supuran CT, Winum JY (eds) Drug design of zinc-enzyme inhibitors: functional, structural, and disease applications. Wiley, Hoboken, pp 139–154
5. Carta F, Supuran CT, Scozzafava A (2012) Novel therapies for glaucoma: a patent review 2007–2011. Expert Opin Ther Pat 22:79–88
6. Friedenwald JS (1949) The formation of the intraocular fluid. Am J Ophthalmol 32:9–27
7. Kinsey VE (1953) Comparative chemistry of of aqueous humor in posterior and anterior chambers of rabbit eye. Arch Ophthalmol 50:401–417
8. Kinsey VE, Barany E (1949) The rate flow of aqueous humor. II. Derivation of rate of flow and its physiologic significance. Am J Ophthalmol 32:189–202
9. Wistrand PJ (1951) Carbonic anhydrase in the anterior uvea of the rabbit. Acta Physiol Scand 24:144–148
10. Becker B (1955) The mechanism of the fall in intraocular pressure by the carbonic anhydrase inhibitor Diamox. Am J Ophthalmol 39:177–183
11. Kinsey VE, Reddy DVN (1959) Turnover of total carbon dioxide in aqueous humors and the effect thereon of acetazolamide. Arch Ophthalmol 62:78–83
12. Supuran CT, Scozzafava A (2000) Carbonic anhydrase inhibitors and their therapeutic potential. Expert Opin Ther Pat 10:575–600
13. Supuran CT (2008) Carbonic anhydrases: novel therapeutic applications for inhibitors and activators. Nature Rev Drug Discov 7:168–181
14. Supuran CT (2010) Carbonic anhydrase inhibitors. Bioorg Med Chem Lett 20:3467–3474
15. Supuran CT (2010) Carbonic anhydrase inhibition/activation: trip of a scientist around the world in the search of novel chemotypes and drug targets. Curr Pharm Des 16:3233–3245

16. Supuran CT, Scozzafava A, Casini A (2003) Carbonic anhydrase inhibitors. Med Res Rev 23:146–189
17. Maren TH (1967) Carbonic anhydrase: chemistry, physiology and inhibition. Physiol Rev 47:595–781
18. Miller WH, Dessert AM, Roblin RO Jr (1950) Heterocyclic sulfonamides as carbonic anhydrase inhibitors. J Am Chem Soc 72:4893–4896
19. Supuran CT, Scozzafava A, Casini A (2004) Development of sulfonamide carbonic anhydrase inhibitors (CAIs). In: Supuran CT, Scozzafava A, Conway J (eds) Carbonic anhydrase – its inhibitors and activators. CRC Press, Boca Raton, pp 67–147
20. Liao SY, Ivanov S, Ivanova A, Ghosh S, Cote MA, Keefe K, Coca-Prados M, Stanbridge EJ, Lerman MI (2003) Expression of cell surface transmembrane carbonic anhydrase genes CA9 and CA12 in the human eye: overexpression of CA12 (CAXII) in glaucoma. J Med Genet 40:257–261
21. Maren TH, Jankowska L, Sanyal G, Edelhauser HF (1983) The transcorneal permeability of sulfonamide carbonic anhydrase innhibitors and their effect on aqueous humor secretion. Exp Eye Res 36:457–480
22. Sugrue MF (2000) Pharmacological and ocular hypotensive properties of topical carbonic anhydrase inhibitors. Prog Ret Eye Res 19:87–112
23. Maus TL, Larsson LI, McLaren JW, Brubaker RF (1997) Comparison of dorzolamide and acetazolamide as suppresors of aqueous humor flow in humans. Arch Ophthalmol 115:45–49
24. Silver LH (2000) Dose–response evaluation of the ocular hypotensive effect of brinzolamide ophthalmic suspension (Azopt). Brinzolamide dose–response study group. Surv Ophthalmol 44(suppl 2):147–153
25. Aalto-Korte K (1998) Contact allergy to dorzolamide eyedrops. Contact Dermatitis 39:206
26. Carlsen J, Durcan J, Swartz M, Crandall A (1999) Nephrolithiasis with dorzolamide. Arch Ophthalmol 117:1087–1088
27. Thoe Schwartzenberg GW, Trop GE (1999) Anorexia, depression and dementia induced by dorzolamide eyedrops (Trusopt). Can J Ophthalmol 34:93–94
28. Konowal A, Morrison JC, Brown SV, Cooke DL, Maguire LJ, Verdier DV (1999) Irreversible corneal decompensation in patients treated with topical dorzolamide. Am J Ophthalmol 127:403–406
29. Scozzafava A, Menabuoni L, Mincione F, Briganti F, Mincione G, Supuran CT (1999) Carbonic anhydrase inhibitors: synthesis of water-soluble, topically effective intraocular pressure lowering aromatic/heterocyclic sulfonamides containing cationic or anionic moieties: is the tail more important than the ring? J Med Chem 42:2641–2650
30. Scozzafava A, Briganti F, Mincione G, Menabuoni L, Mincione F, Supuran CT (1999) Carbonic anhydrase inhibitors. Synthesis of water-soluble, amino acyl/dipeptidyl sulfonamides possessing long lasting-intraocular pressure lowering properties via the topical route. J Med Chem 42:3690–3700
31. Scozzafava A, Menabuoni L, Mincione F, Briganti F, Mincione G, Supuran CT (2000) Carbonic anhydrase inhibitors: perfluoroalkyl/aryl-substituted derivatives of aromatic/heterocyclic sulfonamides as topical intraocular pressure lowering agents with prolonged duration of action. J Med Chem 43:4542–4551
32. Scozzafava A, Menabuoni L, Mincione F, Supuran CT (2002) Carbonic anhydrase inhibitors. A general approach for the preparation of water soluble sulfonamides incorporating polyamino-polycarboxylate tails and of their metal complexes possessing long lasting, topical intraocular pressure lowering properties. J Med Chem 45:1466–1476
33. Alterio V, Di Fiore A, D'Ambrosio K, Supuran CT, De Simone G (2012) Multiple binding modes of inhibitors to carbonic anhydrases: how to design specific drugs targeting 15 different isoforms? Chem Rev 112:4421–4468
34. Supuran CT (2011) Carbonic anhydrase inhibitors and activators for novel therapeutic applications. Future Med Chem 3:1165–1180
35. Supuran CT (2012) Structure-based drug discovery of carbonic anhydrase inhibitors. J Enzyme Inhib Med Chem 27:759–772

36. Steele RM, Batugo MR, Benedini F, Biondi S, Borghi V, Carzaniga L, Impagnatiello F, Miglietta D, Chong WK, Rajapakse R, Cecchi A, Temperini C, Supuran CT (2009) Nitric oxide-donating carbonic anhydrase inhibitors for the treatment of open-angle glaucoma. Bioorg Med Chem Lett 19:6565–6570
37. Mincione F, Benedini F, Biondi S, Cecchi A, Temperini C, Formicola G, Pacileo I, Scozzafava A, Masini E, Supuran CT (2011) Synthesis and crystallographic analysis of new sulfonamides incorporating NO-donating moieties with potent antiglaucoma action. Bioorg Med Chem Lett 21:3216–3221
38. Fabrizi F, Mincione F, Somma T, Scozzafava G, Galassi F, Masini E, Impagnatiello F, Supuran CT (2012) A new approach to antiglaucoma drugs: carbonic anhydrase inhibitors with or without NO donating moieties. Mechanism of action and preliminary pharmacology. J Enzyme Inhib Med Chem 27:138–147
39. Long DD, Frieman B, Hegde SS, Hill CM, Jiang L, Kintz S, Marquess DG, Purkey H, Shaw JP, Steinfeld T, Wilson MS, Wrench K (2013) A Multivalent approach towards linked dual-pharmacology prostaglandin F receptor agonist/carbonic anhydrase-II inhibitors for the treatment of glaucoma. Bioorg Med Chem Lett 23:939–943
40. Carta F, Aggarwal M, Maresca A, Scozzafava A, McKenna R, Supuran CT (2012) Dithiocarbamates: a new class of carbonic anhydrase inhibitors. *Crystallographic and kinetic investigations*. Chem Commun (Cambridge) 48:1868–1870
41. Carta F, Aggarwal M, Maresca A, Scozzafava A, McKenna R, Masini E, Supuran CT (2012) Dithiocarbamates strongly inhibit carbonic anhydrases and show antiglaucoma action in vivo. J Med Chem 55:1721–1730
42. Monti SM, Maresca A, Viparelli F, Carta F, De Simone G, Mühlschlegel FA, Scozzafava A, Supuran CT (2012) Dithiocarbamates strongly inhibit the beta-class fungal carbonic anhydrases from *Cryptococcus neoformans, Candida albicans* and *Candida glabrata*. Bioorg Med Chem Lett 22:859–862
43. Maresca A, Carta F, Vullo D, Supuran CT (2013) Dithiocarbamates strongly inhibit the beta-class carbonic anhydrases from *Mycobacterium tuberculosis*. J Enzyme Inhib Med Chem 28:407–411
44. Temperini C, Scozzafava A, Supuran CT (2010) Carbonic anhydrase inhibitors. X-Ray crystal studies of the carbonic anhydrase II – trithiocarbonate adduct – an inhibitor mimicking the sulfonamide and urea binding to the enzyme. Bioorg Med Chem Lett 20:474–478

Chapter 18
Carbonic Anhydrase Inhibitors and High Altitude Illnesses

Erik R. Swenson

Abstract Carbonic anhydrase (CA) inhibitors, particularly acetazolamide, have been used at high altitude for decades to prevent or reduce acute mountain sickness (AMS), a syndrome of symptomatic intolerance to altitude characterized by headache, nausea, fatigue, anorexia and poor sleep. Principally CA inhibitors act to further augment ventilation over and above that stimulated by the hypoxia of high altitude by virtue of renal and endothelial cell CA inhibition which oppose the hypocapnic alkalosis resulting from the hypoxic ventilatory response (HVR), which acts to limit the full expression of the HVR. The result is even greater arterial oxygenation than that driven by hypoxia alone and greater altitude tolerance. The severity of several additional diseases of high attitude may also be reduced by acetazolamide, including high altitude cerebral edema (HACE), high altitude pulmonary edema (HAPE) and chronic mountain sickness (CMS), both by its CA-inhibiting action as described above, but also by more recently discovered non-CA inhibiting actions, that seem almost unique to this prototypical CA inhibitor and are of most relevance to HAPE. This chapter will relate the history of CA inhibitor use at high altitude, discuss what tissues and organs containing carbonic anhydrase play a role in adaptation and maladaptation to high altitude, explore the role of the enzyme and its inhibition at those sites for the prevention and/or treatment of the four major forms of illness at high altitude.

Susan C. Frost and Robert McKenna (eds.). Carbonic Anhydrase: Mechanism, Regulation, Links to Disease, and Industrial Applications

E.R. Swenson (✉)
VA Puget Sound Health Care System and Department of Medicine,
University of Washington, Seattle, WA, USA
e-mail: eswenson@u.washington.edu

S.C. Frost and R. McKenna (eds.), *Carbonic Anhydrase: Mechanism, Regulation, Links to Disease, and Industrial Applications*, Subcellular Biochemistry 75,
DOI 10.1007/978-94-007-7359-2_18, © Springer Science+Business Media Dordrecht 2014

Keywords Hypoxia • Acid-base • Pulmonary edema • Cerebral edema • Erythrocytosis • Carbonic anhydrase inhibitors

1 Introduction

The initial use of a carbonic anhydrase (CA) inhibitor, acetazolamide, for high altitude acclimatization dates back to the 1960s; however, the underlying rationale was already evident in the 1930s with the initial experience arising from the use of sulfanilamide, the first effective oral antibiotic and serendipitously the first carbonic anhydrase inhibitor. This chapter will relate that history, discuss what tissues and organs containing carbonic anhydrase play a role in adaptation and maladaptation to high altitude, explore the role of the enzyme and its inhibition at those sites for the prevention and/or treatment of the four major forms of illness at high altitude; acute mountain sickness (AMS), high altitude cerebral edema (HACE), high altitude pulmonary edema (HAPE), and chronic mountain sickness (CMS). Most of what we understand about carbonic anhydrase inhibitors derives from cell, animal and human studies with acetazolamide and the results are interpreted as the consequence of CA inhibition. However, like any medication, CA inhibiting sulfonamides including acetazolamide may have effects separate from CA inhibition and evidence is emerging that these actions, either with or without concomitant CA inhibition may be useful at high altitude and in diseases associated with hypoxia and ischemia.

2 Acute Mountain Sickness and High Altitude Cerebral Edema

AMS is a constellation of symptoms experienced by many people in the first several days at high altitude. Although hypobaria accompanies any ascent to high altitude, poor acclimatization to acute high altitude is primarily intolerance to hypoxia, which manifests itself as headache, nausea, anorexia, gastrointestinal distress, poor sleep, generalized malaise and lassitude. At above 10,000 ft or 3,000 m, roughly 25–50 % percent of newcomers can be afflicted. AMS is not life-threatening and generally resolves spontaneously after several days. Very severe AMS, however, may evolve into HACE which can be lethal if not recognized and treated immediately. HACE is marked by brain swelling and increased intracranial pressure (ICP), which cause confusion, ataxia, convulsions and ultimately death.

Despite much investigation over the past 4 decades since it was first shown that acetazolamide is effective in AMS, our understanding of both the pathogenesis of AMS/HACE [1] and the actions of acetazolamide in these condition remains incomplete and more complicated than generally taught [2, 3]. This should come as no surprise considering the presence of carbonic anhydrase in many tissues relevant to high-altitude adaptation or maladaptation, and to the presence of enzyme there and elsewhere, the inhibition of which may contribute to the non-trivial incidence of side effects, some of which mimic the symptoms of AMS.

AMS is thought to be a consequence of non-lethal cerebral hypoxia and compensations invoked to return cerebral oxygen delivery to a more satisfactory level, such as by cerebral vasodilation and increase in cerebral blood flow (CBF). Some critical level of hypoxia differing among individuals (perhaps based upon genetic predisposition) may lead to changes in cerebrovascular permeability and in the integrity of the blood brain barrier (BBB). Increased CBF in association with changes in BBB permeability and impaired CBF autoregulation lead to mild vasogenic interstitial edema; with a possible increase in intracranial pressure that may be perceived as headache and the other symptoms of AMS. Although slight brain swelling (about 1 % increase in brain water) does occur in AMS it does not correlate with symptoms and persons with no AMS symptoms have equal swelling as those without AMS, making it unlikely that this small amount of swelling is pathogenic. Another theory for AMS is that cerebral hypoxia causes an increase in radical oxygen species (ROS) generation [4] which can cause irritation of pain fibers in the trigeminal nerve and increase BBB permeability for small molecules with irritant effects on neuronal function. Prophylactic strategies and treatments for AMS have aimed to increase cerebral oxygenation, reduce fluid accumulation, and decrease ROS generation. HACE is very likely the most severe expression of AMS, in which there is unequivocal brain swelling, breakdown of the blood brain barrier, and bleeding into some areas of the brain [5]. CA inhibitors act or may act positively on all of these factors in AMS, and possibly in preventing HACE by limiting the progression of AMS to HACE.

A brief history of CA and its inhibitors serves to give background perspective on acetazolamide use in AMS. Shortly after the discovery of CA in red blood cells in 1932 sulfanilamide, the first non-toxic oral antibiotic was introduced [6]. Almost immediately it was evident that animals and patients taking the drug develop a mild diuresis, a mild metabolic acidosis, and hyperventilation. These side effects were quickly recognized to be a consequence of renal CA inhibition. Following World War II, synthesis of stronger CA inhibitors yielded the 1,000-fold more potent sulfonamide, acetazolamide, which remains even to this day, the most commonly prescribed oral CA inhibitor. Although first used as a diuretic in heart failure patients and as a gastric acid suppressant, by the mid-1950s it found far greater efficacy in hydrocephalus and glaucoma, reducing both cerebrospinal fluid (CSF) and aqueous humor formation by 50 %. As an outgrowth of efforts to develop even more potent CA-inhibiting sulfonamides than acetazolamide, several other diuretics (high ceiling loop diuretics and the thiazides) acting on other ion transporters in the kidney were discovered. These have largely supplanted acetazolamide in diuretic therapy for heart and renal disease, unless a problematic metabolic alkalosis and alkalemia is present, for which the suppression of renal tubular bicarbonate reabsorption by acetazolamide helps to normalize acid–base status. Furosemide (and its analogs) and the thiazides are, in fact, weaker CA inhibitors, but equal or stronger in their diuretic action with fewer side effects, including no metabolic acidosis. Over the subsequent decades other uses of CA inhibitors have been identified for diseases in which some degree of fluid transport or acid–base alteration might be advantageous, including epilepsy, macular edema, sleep disordered breathing and other primary hypoventilation syndromes, obesity, and now most excitingly cancer.

Within the decade of its introduction, respiratory physiologists and clinicians began to explore whether acetazolamide-induced metabolic acidosis might be a useful respiratory stimulant for hypoxemic patients with chronic obstructive pulmonary disease (COPD) with the goal of improving arterial oxygenation [3]. Although it was effective in this regard for some patients with mild or moderate lung function impairment, many with moderate or severe disease could not tolerate the worsened dyspnea when forced to breathe more and in some cases the drug led to hypercapnic respiratory failure [3, 7]. While the increased work of breathing with hyperventilation is trivial in healthy persons, in those with limited lung function and weaker chronically fatigued respiratory muscles the added effort may not be possible or sustainable. Following a brief period of enthusiasm for use in COPD, this approach with few exceptions has been largely abandoned. In fact, the United States Food and Drug Administration lists severe COPD as a possible contraindication to its use.

Kronenberg and Cain, nonetheless, realized that such ventilatory stimulation might have a significant impact at high altitude in healthy persons, who can easily increase their breathing to significantly raise arterial PO_2 and blood oxygen content [8]. Indeed they showed better oxygenation and ventilation in subjects taking acetazolamide (5–10 mg/kg) undergoing simulation of high altitude in a hypobaric chamber. Shortly thereafter, Forwand et al. [9] demonstrated in a field study that this ventilatory stimulation by acetazolamide could reduce AMS symptoms, and its use for this purpose was quickly established and embraced. Since the 1970s over 200 studies with acetazolamide have shown it to be safe and 60–80 % effective in AMS, but still associated with side effects that some people cannot tolerate. It is interesting that some of the side effects experienced by those taking the drug at sea level (and thus not hypoxia-mediated) are some of the same symptoms of AMS- nausea, malaise, and loss of appetite [10]. Due to the overlap of side effects of acetazolamide with AMS symptoms, it may be that the effectiveness of acetazolamide in AMS is in fact underestimated.

Although an initial study of exercise in acute hypoxia found that acetazolamide taken for 1 day increased maximal oxygen consumption and work capacity by about 8 % [11], all subsequent studies, whether performed with acute hypoxia at sea level or after several days at high altitude showed either no improvement [12] or a slight decrease of about 5–10 % [13, 14]. Time to exhaustion at a high equivalent work rate also appeared to be slightly shorter after acetazolamide in two studies [12, 15, 16]. In all cases, the drug increases ventilation by about 10–15 % over control values and improves arterial oxygen saturation or arterial PO_2. Interestingly lactic acidosis (measured by elevation in blood lactate) is less with acetazolamide [17, 18] due not to decreased production but by decreased elimination out of muscle. In all the aforementioned studies the dosing of acetazolamide was of short duration (hours to 1 day). Why maximal exercise capacity appears to be slightly compromised with acetazolamide at high altitude despite better arterial oxygenation could represent the higher metabolic cost of increased ventilation that limits cardiac output distribution to the exercising muscle and/or the greater intramuscular accumulation of lactic acid. In contrast, chronic administration of acetazolamide at high altitude causes less decrease in near-maximal O_2 consumption than in placebo control subjects [19] and less total weight and muscle mass loss, suggesting that overall better oxygenation both at rest and during exertion over many days helps to preserve muscle mass and power output. One more possible benefit to acetazolamide at altitude is the better preservation of subendocardial oxygenation [20], which would be of advantage particularly to patients with coronary artery disease who travel to high altitude.

3 Mechanisms of Action by Acetazolamide and Other CA Inhibitors

3.1 Renal Carbonic Anhydrase Inhibition

Acetazolamide works by several actions most of which relate to CA inhibition, but there may be others independent of the enzyme inhibition. The single and most important action is inhibition of renal proximal and distal tubular CA which leads to a loss of bicarbonate in the urine and generation of a mild metabolic acidosis [2, 6]. In response to the limited oxygen availability of high altitude, ventilation is stimulated by the hypoxia sensitive peripheral chemoreceptors, located in the carotid bodies, which project information to the respiratory control centers in the brain stem. However, as a consequence of stimulated ventilation, arterial PCO_2 falls and blood pH rises and the resultant respiratory alkalosis limits the full ventilatory response to hypoxia by reducing input from the peripheral and central chemoreceptors, both of which are suppressed by hypocapnia. During the next several days at altitude, the kidneys respond by reducing bicarbonate reabsorption to partially counteract the respiratory alkalosis, in effect reversing the braking action of hypocapnia on the hypoxic drive to breathe. Thus in essence, acetazolamide at doses of 1–10 mg/kg simply accelerates the normal response of the kidneys that otherwise requires several days to occur once at high altitude. It is important to note that owing to the organic acid concentrating capacity of the kidney, these low doses achieve virtually complete CA inhibition in the kidney.

Another possible benefit to renal CA inhibition is the mild diuretic and natriuretic effect of acetazolamide. Although it is not exactly known if fluid retention precedes and contributes to AMS or is simply a consequence of AMS, it is conceivable that a preemptive loss of extracellular volume might be beneficial. This remains speculative and studies of other mild diuretics (without CA inhibiting effect) such as spironolactone have not been conclusive [21, 22].

3.2 Vascular Endothelial Cell CA Inhibition

The other tissue CA activity which may be fully inhibited at this low dose range is that of the vascular endothelium throughout the body, but most importantly in the brain. Here membrane bound isoforms of CA, such as CA IV and XII, with their extracellular orientation can be easily inhibited at low drug concentrations. Luminal brain vascular endothelial CA will be fully and immediately inhibited by even very low drug concentrations in blood, causing a small hindrance to normal "tissue-to-blood" transfer such that tissue PCO_2 will be elevated by 1–2 mmHg [23]. A slight CO_2 retention in the vicinity of both the central and peripheral chemoreceptors by this inhibition in combination with the renal metabolic acidosis will be sufficient to stimulate ventilation. Both cause elevations in H^+ concentration and counteract the effect of increased ventilation to washout CO_2 and raise pH. Acetazolamide (5–7 mg/kg iv) in normoxic humans induces a small increase in ventilation within minutes consistent with endothelial CA inhibition, well before any significant urinary bicarbonate loss occurs or red blood cell drug uptake reaches a critical inhibitory level [24]. At the level of the central chemoreceptors in the brain stem, this slight retention of CO_2 leads to greater efferent signaling to the respiratory control centers in the brain stem. While it might be logical to assume that red cell CA inhibition might also contribute to the efficacy of acetazolamide, it does not appear that sufficient red cell CA inhibition occurs at the low doses effective in AMS. Although inhibition of red cell carbonic anhydrase does cause ventilatory stimulation, the combination of high erythrocyte CA concentrations and in humans, two isozymes (CA I and CA II) requires much greater dosing ($>$15 mg/kg) to cause sufficient CO_2 retention and respiratory acidosis to stimulate ventilation [25].

3.3 CA Inhibition in the Central Nervous System

In the brain, three sites of CA might be involved in the protective action of acetazolamide; choroid plexus, cerebral vasculature and chemoreceptors. The choroid plexus secretes CSF and any reduction in CSF production will help to diminish intracranial pressure (ICP), which is set by the amount of water in the blood and interstitial fluid volumes, intracellular volume and CSF volume. In healthy conditions these volumes, although contained within the rigid confines of the skull are such that ICP is less than 20 mmHg. Yet, any increase in one or more of the component cranial volumes can quickly elevate ICP to values that begin to limit blood flow and or cause pain and neurological alterations. Complete choroid plexus CA inhibition reduces CSF production by 50 % and in principle could reduce overall intracranial volume and pressure in AMS. However, effective doses of acetazolamide which penetrate the BBB to reach the choroid plexus, and depress CSF flow are on the order of 20 mg/kg [26] and much greater than doses effective in AMS.

At the level of the cerebral vasculature, acetazolamide in high concentrations (>20 mg/kg) causes vasodilation, increases CBF and oxygenation [27], but at the lower doses used successfully in AMS, there is no increase in CBF [28, 29]. In fact, one study using 250 mg every 8 h for three doses found a slight fall of 10 % at simulated altitude, but enhanced autoregulation [29]. However, these changes were not correlated with AMS symptoms. Despite the lack of CBF increase or even slight fall, cerebral oxygenation was 3–5 % higher at rest and exercise in trekkers between 3,700 and 5,700 m while taking 375 mg bid [30]. These increases in cerebral oxygenation were greater than the corresponding improvements in arterial oxygenation generally observed with acetazolamide. Whether acetazolamide reduces cerebral oxygen consumption and so by this mechanim also increases cerebral oxygenation remains unexplored. The only caveat in wholly dismissing a possible benefit of increased CBF is that if regional brain blood flow changes are heterogeneous, then measurements of total CBF may not reflect increases in blood flow in some critical areas of the brain whose oxygenation will be improved by greater perfusion. There have been no studies of whether low dose acetazolamide alters regional CBF when total CBF is not increased.

The final site of CA in the central nervous system is the chemoreceptors, located in the brain stem (only CO_2 and pH sensitive) and the peripheral chemoreceptors of the carotid body (sensitive to O_2, as well as CO_2 and pH). Here it gets rather complicated in vivo, because the behavior and signaling of the two sites of chemoreception are clearly altered by systemic changes in acid–base status and oxygenation with high altitude, in addition to any possible consequence directly of their own enzyme inhibition. When the central chemoreceptors are studied in isolation it appears that the speed at which they respond to changes in PCO_2 or pH are slowed, but not the full response when CA is inhibited [31]. If local metabolically produced CO_2 is not as efficiently eliminated by loss of neurocellular and endothelial cell CA activity the slight tissue CO_2 retention will further decrease intracellular pH and stimulate ventilation in addition to the metabolic acidosis caused by renal CA inhibition.

When the peripheral chemoreceptors are studied in isolation inhibition of CA by methazolamide leads to a slowing of the response to a change in PCO_2 or pH, and also a reduction in the maximum response [32, 33]. In contrast, methazolamide delays, but does not decrease the response to hypoxia [32]. With respect to changes in ventilation or carotid body output with hypoxia, there is no increase in the acute isocapnic hypoxic ventilatory response (HVR) in vivo despite a renal metabolic acidosis. This indicates that local chemoreceptor CA inhibition abolishes the additive H^+-O_2 interaction in the peripheral chemoreceptors [24, 34, 35]. A lower peripheral chemoreceptor response to hypoxia has also been demonstrated during exercise [36]. This raises an interesting question as to what could be the advantage of taking a drug at altitude that has inhibitory effects on oxygen-sensing cells. The answer lies in the fact that the ventilatory response to high-altitude hypoxia is under partial suppression by the hypocapnic alkalosis. Consequently, any agent abolishing the H^+-O_2 interaction under these circumstances will blunt the action of a low PCO_2 on the hypoxic response and generate more ventilation than would otherwise occur [2, 3]. In humans at sea level, acetazolamide has no influence on the peripheral chemoreceptor contribution to the ventilation response on stepwise changes in end-tidal PCO_2 [37]. It cannot be excluded that, in part, H^+-O_2 and O_2 responses follow separate signal-transduction pathways in the carotid bodies. Future studies with acetazolamide are warranted to determine if, despite the absence of an H^+- O_2 interaction, the carotid bodies may retain their H^+ sensitivity. Predictably, the inhibitory effect of acetazolamide inhibition on HVR should be due to peripheral chemoreceptor CA inhibition. However, in the cat, even the more lipophilic inhibitor methazolamide does not reduce HVR [38], suggesting that acetazolamide and other CA inhibitors may act by a mechanism (s) other than CA inhibition as taken up below in the next section on HAPE.

The actions of acetazolamide on chemoreceptors and the vasculature suggest also how it reduces periodic breathing (PB) during sleep and in sleep apnea at high altitude [39, 40]. Repetitive apnea and hyperpnea leads to cyclical drops in arterial oxygenation that may disrupt sleep quality and add to the total hypoxic stress of high altitude. It has been proposed [41] with loss of higher cortical control on breathing during sleep that hypoxia-mediated high peripheral chemoreceptor output leads to a level of ventilation during sleep sometimes high enough to raise PaO_2 and lower $PaCO_2$ sufficiently to transiently suppress breathing. During such an apneic period, PaO_2 then falls and $PaCO_2$ rises enough to generate a combined strong stimulus to initiate overbreathing again which then repeats itself often 30–60 times an hour. During the hyper-ventilatory phase arterial PCO_2 may decrease enough to pass below the apneic threshold and exceed the "CO_2 reserve" (i.e., the difference in $PaCO_2$ during eupnea (normal breathing) and the apneic threshold). The CO_2 reserve when combined with "plant gain" (or the ventilatory increase required for a given decrease in $PaCO_2$) and "controller gain" (ventilatory responsiveness to CO_2 above eupnea), are the key determinants of breathing stability in sleep. In general, breathing during sleep is made more stable with increases in the CO_2 reserve, and reductions in plant and controller gain. In addition the rate at which the chemoreceptors respond with changes in their output to the respiratory control centers to varying and rapidly arterial O_2 and CO_2 levels may also be important. Acetazolamide alters all of these favorably. First, with elevated arterial PO_2 and decreased $PaCO_2$, plant gain will be reduced because larger changes in ventilation are needed to cause equivalent changes in blood gases. Second, a parallel left shift of the CO_2 response curve (ventilation vs. $PaCO_2$) will raise the difference between the prevailing $PaCO_2$ and the apneic threshold (CO_2 reserve) and decrease the propensity for apnea. A small rise in brain stem PCO_2 with inhibition of vascular CA will have a similar tonic stabilizing influence. Third, due to the higher prevailing PO_2, subjects are in a flatter region of the (hyperbolic) hypoxic response curve, resulting in a reduced O_2 controller gain. A further reason why acetazolamide reduces ventilatory controller gain in sleep is the abolishment of the CO_2-O_2 interaction so that responses to hypoventilation-induced combined hypoxia/hypercapnia may be reduced considerably. Lastly acetazolamide by slowing the rate at which the chemoreceptor afferent information arrives at the respiratory control centers could help dampen the magnitude of the ventilatory responses and in particular keep the arterial PCO_2 from going below the apneic threshold. Finally, it cannot be ruled out that acetazolamide increases the cerebrovascular response to combined hypercapnia/hypoxia [18], leading to a dampening influence on subsequent changes in brain stem PCO_2. Note that a lowered cerebrovascular CO_2 sensitivity may play a role in the genesis of periodic breathing at high altitude [42].

3.4 Non-CA Inhibiting Actions

As alluded to above CA inhibiting sulfonamides including acetazolamide, may have effects independent of CA inhibition and evidence is emerging that these actions, either with or without concomitant CA inhibition could be useful at high altitude and in other diseases associated with hypoxia and ischemia.

3.4.1 Aquaporin Modulation

Virtually all cells have specific membrane water channels that contribute to intracellular osmoregulation and extracellular water regulation, and in many organs they contribute to transepithelial fluid transport. Of the many members of the aquaporin (AQP) family, AQP-1 and AQP-4 are of potential interest with regard to acetazolamide and high altitude disease. In addition aquaporins may also serve as channels for small uncharged gas molecules, such as nitric oxide (NO) and CO_2 [43, 44]. Both AQP-1 and AQP-4 are expressed in the brain [45], particularly astrocytes and choroid plexus, where they are involved in CSF production [46], CBF [47], and brain extracellular fluid and water homeostasis [48]. Inhibition of AQP-4 and genetic deletion of AQP-4 is protective against some forms of cerebral edema [49, 50]. AQP-1 is the major isoform in red cells and the kidneys and intimately involved in whole body osmoregulation and possibly in CO_2/HCO_3 fluxes [51]. With regard to AQP-4 and to some extent AQP-1, some groups have found that acetazolamide in the uM range directly blocks water flux across the plasma membrane of cells in vitro [46, 52, 53]. In the study by Tanuimura et al. [53] interestingly, methazolamide was inactive, despite its very close structural similarity to acetazolamide; differing only by a methyl group substitution on the thiadiazole ring. Despite these several reports of AQP-1 and AQP-4 blockade by acetazolamide others have not been able to confirm these findings [54–56].

Although resolution of the question of whether acetazolamide directly blocks aquaporins is difficult, this does not necessarily rule out other means by which acetazolamide could alter brain water homeostasis involving aquaporin function. Acetazolamide has been reported to reduce vasogenic and cytotoxic forms of cerebral edema in animal models [57–61]. Although acetazolamide may reduce edema by other actions such as diuresis and vascular dilation, it also inhibits AQP-1 and AQP-4 gene and protein expression in models of brain and cardiac injury

[61, 62] and so by this action favorably alter brain water homeostasis. These actions of acetazolamide on cerebral edema and brain likely have less relevance in AMS than in HACE where frank edema occurs.

3.4.2 Radical Oxygen Species Modulation

The evidence that hypoxia causes increased radical oxygen species (ROS) generation, tissue injury or dysfunction, and vascular leakage is growing. Hypoxic exposure equivalent to typical high altitudes increases ROS formation [63, 64] and some studies show that AMS can be prevented by antioxidant administration [65]. If ultimately it is proven that increased ROS formation is responsible for AMS, then acetazolamide, a heterocyclic thiadiazole might perhaps work as an antioxidant given that other numerous compounds containing a 1-3-4 thiadiazole ring are ROS scavengers [66]. Intriguingly the depression of peripheral chemosensitivity to hypoxia by acetazolamide in humans can be overridden by antioxidants [67]. This somewhat paradoxical finding would suggest that acetazolamide may not be a ROS scavenger, but this remains to be tested directly. Natural defense against ROS includes a number of antioxidant proteins, such as superoxide dismutase, catalase, hemooxygenase and thioredoxins that are upregulated by the gene transcription factor, nuclear related factor-2 (Nrf-2). Recently it has been shown that methazolamide, but surprisingly not acetazolamide at clinically relevant dosing, activates Nrf-2 in the brain and decreases hypoxic-mediated cerebrovascular leakage in a rat model [68]. Whether this difference between the drugs just represents the greater lipophilicity of methazolamide over acetazolamide and great penetrance across the blood brain barrier, or some unique attribute of methazolamide will require more extensive pharmacological investigation.

3.4.3 Heat Shock Protein (HSP), Interleukin-1 Receptor Agonist (IL-1RA) and Hypoxia-Inducible Factor (HIF)

Heat shock protein-70 protects against cellular stress induced by hypoxia and IL-1RA is an anti-inflammatory cytokine. In a human study, subjects who did not develop AMS had higher blood levels of HSP-70 that those with AMS and acetazolamide increased HSP-70 in the AMS susceptible subjects who did not get AMS or had fewer symptoms [69]. In this same study, acetazolamide also raised IL-1RA levels. It is not clear whether these interesting findings are due to CA inhibition or to other actions of the drug.

The master hypoxic transcription factor, HIF-1, appears important in surviving hypoxic stress. When normoxic rats were given exceedingly large doses of acetazolamide (50–100 mg/kg) HIF-1 alpha was upregulated in brain tissue along with several of the genes known to have HIF-1 response elements in their promoter region [70]. However, another study of similar degree of acidosis (pH $\sim$ 7.0) in cultured cells showed moderate up-regulation of HIF-1 protein, but interestingly not all of the usual genes upregulated by HIF-1 with hypoxia were increased [71]. The relevance of these finding to high altitude is uncertain, because acetazolamide doses this high cause severe respiratory acidosis, but respiratory alkalosis occurs at high altitude. Whether the very slightly lower pH with acetazolamide at altitude compared to those not treated (usually about 0.05 units) causes any differences in HIF-1 metabolism remains unknown. Furthermore, two human studies have found no relationship between AMS susceptibility and numerous polymorphisms of HIF-1 [72, 73].

4 Clinical Aspects of Acetazolamide in AMS/HACE Prevention

Over 1800 subjects have been treated with acetazolamide in placebo-controlled trials and across all dosing regimens and altitudes examined the prophylactic efficacy of acetazolamide in preventing or reducing AMS severity is roughly 50–60 % [11, 74]. These meta-analyses found a slight dose response effectiveness increasing from 45 % with 125 mg bid to 50 % with 250 mg bid to 55 % with 375 mg bid. The calculated number of subjects needed to treat (NNT) to prevent one case of AMS is 7 with 125 mg bid, 6 with 250 mg bid and 3 with 375 mg bid. The only other CA inhibitors studied have been methazolamide and benzolamide. Methazolamide, a more lipophilic agent at equivalent CA inhibiting doses to acetazolamide appears to be equally effective [75, 76]. Benzolamide, with its ten-fold higher hydrophilicity and lower membrane permeance than acetazolamide was effective against AMS in a 72 h simulated altitude hypobaric chamber study [77] and in reducing periodic breathing during sleep at high altitude [78]. More recently, benzolamide at 100 mg bid was very effective at reducing AMS and maintaining better oxygenation in subjects in a Himalayan trekking expedition [79]. As treatment for AMS, one placebo controlled study of acetazolamide 250 mg bid showed that it hastened symptom relief and improved oxygenation within 24 h [80].

These studies of different CA inhibitors with varying intracellular penetrance (methazolamide > acetazolamide > > benzolamide) but similar efficacy in AMS prevention suggest strongly that the main impact of any CA inhibitor is by renal and endothelial cell CA inhibition, since benzolamide is largely excluded from the other sites of intracellular CA discussed above that might be involved in AMS pathogenesis. Because benzolamide [79] was found to reduce by roughly 50 % the side effects perceived by subjects and objective neuropsychological deficits (concentration, balance, memory recall) with acetazolamide 125 mg bid [81], it or similar impermeant inhibitors should be superior. Unfortunately, benzolamide has no patent protection and thus it is unlikely that any pharmaceutical company will market the drug. However, with an emerging realization that other plasma membrane-associated CA isozymes (IX and XII) are richly and almost uniquely expressed in cancer cells and critical to their growth and metastatic potential clinical development of other membrane-impermeant CA inhibitors which block CA IX are underway and have shown promising features. These drugs would be ideal for use at high altitude. Until that time, the use of acetazolamide at doses at or below 125 mg bid may come close to yielding the same results as benzolamide, and is worthy of further study up to moderate altitudes and with slower ascent rates. Lastly, it must be appreciated that among individuals heterogeneity in body size, drug clearance rates, which may be slowed at high altitude [82] and as yet uncovered genetic susceptibility to AMS may make it necessary to use higher doses with greater ascent rates and maximal altitude goals.

Finally, whether acetazolamide can prevent or treat HACE remains untested and unknown. HACE is very rare and thus controlled trials will never be practical. Dexamethasone, a glucocorticoid effective in many forms of cerebral edema, works effectively at early stages of HACE and remains the treatment of choice. Although some physicians use acetazolamide with dexamethasone in HACE the evidence is all anecdotal. Given the threat of impending lethal intracranial pressure (ICP) elevation, higher doses (>20 mg/kg) capable of inhibiting CSF production and perhaps reduce AQP-4 to lower ICP should be considered rather than the conventional 2–4 mg/kg. This could be tested formally in a rat model of HACE recently reported that demonstrated effectiveness of a derivative of gingko biloba [83].

Many questions remain to be investigated before a complete understanding of acetazolamide protection in AMS and HACE is possible. Although not an exhaustive list of experiments and methodological approaches, the following are several key directions for future research.

1. Regional brain blood flow, pH, metabolic rate, and oxygenation (such as with magnetic resonance imaging and spectroscopy) in areas of brain involved in respiratory control need to be performed in humans in conjunction with ventilation, ventilatory responses, and arterial blood gas measurements. Dose–response experiments with acetazolamide in normoxia and hypoxia, combined with plasma measurements of total and free drug levels, are needed to better calculate the degree of red blood cell and tissue CA inhibition. Although not presently available for humans, analogs of acetazolamide, lacking CA inhibiting activity, will be useful in animals to test whether and how much of the effect of acetazolamide is due to CA inhibition, or to some other action on O_2 and CO_2 sensing and responsiveness, redox state, ion channels or oxygen- and oxygen radical-dependent transcription factors such as HIF-1 or Nrf-2.
2. Studies of acetazolamide in conscious and sleeping chemoreceptor- intact animals are needed, both in normoxia and hypoxia, in which central and peripheral chemoreceptor contributions can be gauged by isolating the drug and/or systemic acid–base changes to either the central or peripheral chemoreceptors. This can be done by isolating the circulation to the carotid bodies and perfusing them with appropriately conditioned blood [84].
3. The contributions of the renal metabolic acidosis and diuresis of acetazolamide to the protection with acetazolamide in AMS need to be explored by using other mild diuretics to achieve the same magnitude of diuresis as acetazolamide and/or using acetazolamide but preventing bicarbonate and sodium losses by carefully titrated replacement.

4. Studies of membrane-impermeant CA inhibitors would be extremely interesting and likely would show more efficacy and fewer side effects than acetazolamide, as the limited data on benzolamide demonstrate. These ought to tested when they become available and approved for human use in cancer treatment.

5 High Altitude Pulmonary Edema

Despite the overwhelming evidence of efficacy of acetazolamide in AMS over many decades of use, the logical extension to HAPE prevention and treatment has received no attention in all of the many studies of acetazolamide at high altitude.

HAPE is the sudden development of pulmonary edema in otherwise healthy persons who have ascended within 1–5 days to high altitude. It may also occur in high altitude residents that have descended for more than several days and then reascend to their usual altitude of residence (reentry HAPE). It is caused by pressures high enough within the microvasculature and capillaries of the lung to overcome the structural integrity of the ultrathin alveolar capillary barrier that maintains a dry fluid-free airspace. This hydrostatic breach of the normal barrier leads to alveolar hemorrhage and fluid accumulation that prevent normal oxygen and carbon dioxide exchange. The edema causes even lower arterial oxygen levels than otherwise at that altitude, cough, breathlessness, fatigue, inability to do minimal exertion, and ultimately death if not treated [85]. Pulmonary artery pressure rises at high altitude due to hypoxic pulmonary vasoconstriction (HPV) and is a normal response of the lung to low oxygen levels in the alveolar gas that cause the surrounding blood vessels to constrict [86]. However, in some persons HPV is excessive and pressures can rise enough to lead to capillary stress failure [85, 87, 88]. Drugs which lower pulmonary artery pressure, oxygen and descent are used for HAPE treatment and certain of the same drugs used to treat HAPE can be used prophylactically to prevent its occurrence. Acetazolamide, in principle, should be beneficial in preventing HAPE since the ventilatory stimulation it induces will raise alveolar PO_2 and thus diminish the principle stimulus for HPV. Furthermore, its diuretic effect might lower the total amount of fluid available to leak into the lung.

On the basis of earlier studies into the role of CA in gas exchange, ventilation–perfusion heterogeneity and control of ventilation, I began to pursue the possibility that acetazolamide and other CA inhibitors might directly inhibit HPV and offer protection against HAPE. The first report of HPV inhibition by acetazolamide, however, was buried in a report on the effects of hypercapnia on the isolated perfused lung [89]. This novel and unprecedented finding of a CA inhibitor effect on a process not thought to involve acid–base exchange or a pH transduction signal went wholly unrecognized for more than a decade until CA mediation in the peripheral chemoreceptor response to hypoxia was demonstrated [32].

My first inkling that CA might be involved in reducing HPV arose from studies with the multiple inert gas elimination technique (MIGET). We found that acetazolamide (20–25 mg/kg) increased ventilation–perfusion heterogeneity in normoxic anesthetized dogs [90]. Using a global index of ventilation–perfusion (V/Q) mismatch (a summation of the alveolar-arterial differences for the six inert gases of MIGET), gas exchange efficiency decreased by roughly 25 %. We hypothesized that spontaneous minute to minute fluctuations of regional ventilation and/or perfusion lower or raise alveolar PCO_2 accordingly and evoke known pH-sensitive parenchymal, vascular, and airway bronchomotor responses to readjust the accompanying parallel gas or blood flow in a manner to return the local ventilation–perfusion ratio to its original state [90]. Since all of these CO_2 mediated changes are pH-dependent responses, slowing the rate of CO_2/HCO_3-$/H^+$ interconversion will reduce overall ventilation–perfusion matching. When we imposed known cyclic fluctuations in regional lobar perfusion of 60 s intervals to generate varying alveolar PCO_2 concentrations, overall ventilation–perfusion heterogeneity in dogs increased with acetazolamide [91], consistent with this hypothesis. Although our focus was on CO_2-pH sensitive mechanisms of V/Q matching in these studies, we reasoned that any fluctuation in regional ventilation or perfusion would, of course, also alter local alveolar PO_2 and thus the strength of hypoxic pulmonary vasoconstriction. In principle, our results were also consistent with a component of impairment and slowing of HPV by acetazolamide. It was this possibility that directed my focus to the pulmonary vasculature and the role, if any, of CA in pulmonary vascular smooth muscle physiology and hypoxic responses. In order to explore the question fully, my colleagues and I conducted work in isolated pulmonary artery smooth muscle cells, isolated perfused lungs and live animals. A comprehensive approach was necessary since CA is ubiquitously expressed throughout the body and consequences of CA inhibition might arise from direct pulmonary effects and/or secondary systemic effects.

6 Isolated Perfused Lung and Whole Animal Atudies

Our first studies used the isolated blood perfused rabbit lung, in which we measured both the rate of HPV development and its final magnitude [92]. The isolated perfused lung allowed us to examine HPV without the possible confounding effects of systemic hypercapnia and metabolic acidosis following CA inhibition in the whole animal, which in general are known to augment HPV [86]. We found that acetazolamide (30 uM in the perfusate) reduced HPV by roughly 50 % and reduced the rate of rise by 40 % as earlier shown in the same preparation [89]. Without providing any explanation for HPV inhibition, we found no rise in exhaled nitric oxide (NO) as a marker for increased NO production to account for HPV moderation. Whether acetazolamide might act via enhanced vasodilating NO generation remains a subject of some controversy. Although we found no increase in exhaled NO as a surrogate for increased NO in the vasculature, others have reported carbonic anhydrase has nitrite reductase activity and that this CA-mediated NO generation from nitrite can be enhanced by acetazolamide and dorzolamide [93]. It is not at all clear how CA might have a reductase activity at its active site or how ligating the zinc in the active site of the enzyme increases this activity. Because other hypoxia-sensitive nitrite reductases, such as xanthine oxidoreducatase and deoxyhemoglobin [94] generate more NO in the lung under hypoxic conditions, it is unlikely that acetazolamide works via increasing nitrite availability since renal CA inhibition leads to reduced nitrite reabsorption by the proximal tubule and greater urinary excretion [95].

Studies in the live animal are important to determine whether findings in the isolated perfused lung are reproducible in vivo, in which non-pulmonary effects of CA inhibition; systemic acidosis, effects on peripheral chemoreceptors and neural transmission might potentiate or oppose HPV at the lung and vascular level. In unanethetized beagles [96] with invasive monitoring of pulmonary hemodynamics, ventilation, lung gas exchange, and renal function, we found that acetazolamide (20 mg/kg) to achieve an equivalent concentration in blood to that used in the isolated perfused lung completely inhibited HPV when the dogs breathed 10 % oxygen gas (F_IO_2 of 0.10). Since conscious animals hyperventilate with CA inhibition (due to a combined respiratory and metabolic acidosis from red cell and renal CA inhibition) it was possible that the abolition of HPV may have resulted simply from the higher alveolar PO_2 with hyperventilation. In order to control for this potential reduced stimulus to HPV we lowered the F_IO_2 to 0.08 and achieved equivalent arterial and alveolar PO_2 as in the control animals breathing a F_IO_2 of 0.10. In this case, there was no difference in the effect on HPV. Lowering the dose of acetazolamide to 5 mg/kg, intravenous or oral (a dosing more relevant to human use) in a series of preliminary experiments also inhibited HPV, but not to the complete extent as 20 mg/kg [97]. We found in these experiments that inhibition of HPV could not be correlated with changes in plasma potassium, endothelin 1, or angiotensin II; factors that themselves alter HPV. Furthermore HPV suppression in the whole animal occurs despite the systemic acidosis with CA inhibition.

7 Pulmonary Vascular Smooth Muscle Studies

Altogether the data in the isolated perfused lung and whole animal clearly established that CA could be involved in the full expression of HPV, what process(es) it subserves and what cell type(s) in the lung (alveolar epithelial, vascular endothelial or arterial smooth muscle) are relevant could not be resolved by these experiments. Although HPV is a complex process [86] and is modulated by systemic acid–base status, peripheral chemoreceptor activity, the autonomic nervous system, and the pulmonary vascular endothelium, it is an inherent property of the pulmonary arterial and venous smooth muscle cells.

In the 60 years since its discovery CA has been found in virtually all cells including many in the lung; the alveolar epithelium, bronchial epithelium, vascular endothelium and in certain airway nerves [3]. However, up to this point it had never been demonstrated in the pulmonary vascular smooth muscle, although histochemical and immunocytochemical studies in other non-vascular smooth muscle sites and systemic vascular smooth muscle clearly suggested its presence [98]. Using rat pulmonary artery smooth muscle cells obtained from small to mid-sized resistance vessels, we found that acetazolamide had no effect on intracellular calcium (Ca^{2+}) in normoxia, but markedly slowed and reduced the magnitude of Ca^{2+} uptake upon exposure of these cells to 4 % O_2 [99] with an I_{50} of roughly 50 uM. While one cannot directly translate these pharmacological data in isolated rat arterial smooth muscle cells to the isolated perfused rabbit lung and live dog, the I_{50} in our cell work accords well with the dose–response effects in these other studies. Thus, in the rabbit lung the concentration was 30 uM [92] and we only achieved partial inhibition, while in the dog the dose of 20 mg/kg yielding a concentration of approximately 70 uM [100] was fully inhibitory [96] and 5 mg/kg yielding a concentration of about 15 uM only partially blocked HPV [97].

To explore the mechanism by which acetazolamide inhibits HPV, we undertook a number of experiments with agents or manipulations that alter intracellular Ca^{2+} content in combination with acetazolamide and hypoxia . Whether the effect of acetazolamide on hypoxia-induced Ca^{2+} responses is due to a non-specific inhibition of Ca^{2+} signaling, we studied the effect of acetazolamide on the response to 60 mM extracellular KCl. Potassium is a non-specific vasoconstrictor that induces an increase in intracellular Ca^{2+} in pulmonary artery smooth muscle cells by initiating membrane depolarization and Ca^{2+} influx through voltage-gated Ca^{2+} channels. KCl caused a significant increase in intracellular Ca^{2+} which was similar in magnitude to that induced by hypoxia, but was unaffected by pretreatment with 100 uM acetazolamide. These results indicate that acetazolamide has no effect on the activation of voltage-gated Ca^{2+}channels and suggest that the effect of acetazolamide on pulmonary artery smooth muscle cell Ca^{2+} signaling during hypoxia is not due to toxicity or a non-specific or generalized inability to increase intracellular Ca^{2+} by initiating extracellular Ca^{2+} uptake [99].

Acetazolamide in neural and systemic vascular smooth muscle has recently been proposed to alter the properties of Ca^{2+} activated K^+ channels, causing hyperpolarization [101, 102]. Since the hypoxia-induced increase in intracellular Ca^{2+} is believed to occur secondary to depolarization, hyperpolarization should prevent activation of voltage-gated Ca^{2+} channels and an increase in intracellular Ca^{2+} in response to hypoxia. We measured membrane potential (Em) using an Em-sensitive fluorescent dye and found that acetazolamide had no effect on basal normoxic Em and did not alter the KCl-induced depolarization. Acetazolamide in normoxic cells caused a very small statistically significant decrease in resting intracellular pH (pHi) from 7.26 to 7.22. While intracellular pH changes are known to affect intracellular Ca^{2+} concentrations and so might conceivably alter HPV, in order to show intracellular Ca^{2+} changes equivalent to the 100–150 nM changes we observed with hypoxia [99], the intracellular pH needs to be altered by greater than 0.5 pH units [103]. Exposure to 4 % O_2 had no significant effect on pHi and did not further alter the pHi of cells pretreated with acetazolamide. Lastly, we have preliminary evidence that acetazolamide does not inhibit Rho-kinase and so does not act by altering Ca^{2+} sensitivity of the smooth muscle contractile apparatus or by reducing ROS formation caused by endothelin (unpublished data).

These results taken as a whole suggest that acetazolamide has a specific action on hypoxia-mediated Ca^{2+} increases that are not due to changes in pHi or membrane potential and membrane L-type Ca^{2+} channels. We are presently pursuing the possibility that acetazolamide may alter the hypoxic-mediated sarcoplasmic reticulum (SR) release of Ca^{2+}, which has been shown to initiate the much larger influx of Ca^{2+} from the extracellular space via plasma membrane surface store operated Ca^{2+} channels [104].

It is a general rule in pharmacology that one should never base conclusions on a single drug or concentration and CA pharmacology is no exception [105]. Thus, we tested two other more potent CA inhibitors, benzolamide (a hydrophilic membrane impermeant inhibitor) and ethoxzolamide (a lipophilic cell membrane-permeant inhibitor) in both the isolated smooth muscle cells and the unanethetized dog. To our complete surprise, neither of these more powerful CA inhibitors [99] had any effect on the intracellular rise in Ca^{2+} of hypoxic pulmonary artery smooth muscle cells although they did slightly lower pHi as did acetazolamide, indicative of CA inhibition which lowers pHi by impairing CO_2 diffusion out of cells. Likewise, we have shown these two inhibitors are equally ineffective in the conscious dog since neither inhibits HPV [97]. Finally we methylated the sulfonamide nitrogen critical to the binding with zinc at the active site of the enzyme to yield an inactive drug, N-methyl acetazolamide (NMA). In both the pulmonary artery smooth muscle cells and in the dog, this drug with otherwise the exact same structure beyond the sulfonamide moiety, pK, water and lipid solubility, and intramolecular electronic charge distribution to acetazolamide, but no CA inhibiting activity was equipotent at inhibiting Ca^{2+} elevation with hypoxia and reducing HPV [99, 106]. In the cell studies NMA did not lower intracellular pH, and in the dogs it did not cause a diuresis, change urinary bicarbonate excretion or stimulate ventilation, all the expected findings of CA inhibition.

The relevance of the cellular and animal work has been established in humans. In a simulated 8 h exposure to 12 % oxygen, acetazolamide 250 mg bid given over 3 days reduced HPV by 80 % [35] and more recently 375 mg bid given 24 h before flying to 3,800 m also reduced pulmonary artery pressure over the next 2 days at high altitude [107]. These findings of acute effectiveness against HPV appear to be sustained for only several days at altitude or with hypoxic exposure because equivalent dosing in subjects at high altitude for more than 1–2 week does not lower pulmonary artery pressure [14, 108]. This suggests that over time other mechanisms resistant to acetazolamide such as smooth muscle hypertrophy causing pulmonary hypertension at altitude become more important in setting the elevated pulmonary vascular resistance than HPV.

At this juncture it is clearly apparent that acetazolamide inhibits HPV at the level of the pulmonary artery smooth muscle. It does so by a mechanism not dependent upon carbonic anhydrase and is not substantially altered by the effects of inhibition of carbonic anhydrase elsewhere in the body. In this case, the response of the pulmonary vasculature interestingly differs from that of the systemic vasculature in which the evidence is more compelling for a CA mediated role in vasomotor regulation. It has been shown that isolated porcine mesenteric arteries preconstricted with norepinephrine relaxed with acetazolamide but also to methazolamide and ethoxzolamide in a dose response manner consistent with their respective CA inhibitory potencies [109]. Intra-arterial infusions of acetazolamide into the human forearm reduce local vascular resistance and the response can be blocked by an inhibitor of Ca^{2+} activated K^{+} channels [110]. The molecular receptor for acetazolamide involved in HPV remains unknown.

As to whether acetazolamide itself prevents HAPE, we have shown acetazolamide is effective [111] in an rat model in which 20 mg/kg prevents the typical alveolar protein, red cell and lung water accumulation of human HAPE when rats are exposed to one-half atmosphere (∼18,000 ft) for 24 h. In humans the only supportive data are anecdotal unpublished reports by physicians in the mountains of Colorado who use acetazolamide to prevent reentry HAPE in children returning home from holidays at sea level. Placebo controlled studies are clearly needed to determine if acetazolamide or congeners of acetazolamide are efficacious and perhaps have been preventing many cases of HAPE concurrently with its use in AMS. Use of a non-CA inhibiting form of acetazolamide if proven effective would be superior to acetazolamide in that the many side effects of CA inhibition could be avoided.

8 Chronic Mountain Sickness

Presently more than 100 million people live permanently above 2,500 m in the Andean, Himalayan and North American Rocky Mountains. Most individuals are well adapted, particularly those descendants of many generations at high altitude. Nonetheless, some newcomers after weeks to months and resident natives develop excessive erythrocytosis or polycythemia beyond the normal slight increase in hematocrit and hemoglobin concentration which is a typical compensatory response to chronic hypoxia and high altitude residence. Hematocrits above 65 % cause increased blood viscosity and poor perfusion of all organs and tissues. Additionally the work of the heart is increased and leads to early cardiac dysfunction. Typical symptoms and findings of CMS include cyanosis, plethora, headache, fatigue, edema, systemic and pulmonary hypertension, relative hypoventilation and cor pulmonale (right heart failure). The pathogenesis of CMS and what drives excessive erythrocytosis are not known, but it is thought that either a hereditary lower drive to breathe (lower hypoxic and hypercapnic ventilatory responses) causes greater hypoxemia and more release from the kidneys of erythropoietin (EPO), the hormone that stimulates bone marrow production of red cells. Greater erythrocytosis leads to more hypoventilation and begets a vicious cycle [112]. Alternatively, the lower drive to breathe demonstrated in patients with CMS may be secondary to an initial extreme sensitivity of the kidney to hypoxia and greater EPO production that then causes secondary hypoventilation [113]. In either event, phlebotomy or drugs that stimulate breathing and secondarily suppress EPO release, or directly inhibit renal EPO production lower hematocrit, improve cardiac function and relieve symptoms of CMS [114].

Acetazolamide, by virtue of its ability to stimulate ventilation and prevent AMS has been explored as a low cost treatment option in CMS because phlebotomy is expensive and time consuming, while moving to low altitudes can be socially and economically disruptive. Two large randomized placebo-controlled studies performed in the Peruvian Andes found that chronic acetazolamide (250 mg daily) over 3 weeks or 6 months in patients with CMS and hematocrits above 65 % was quite effective and well tolerated. Roughly 40 % of patients had reductions in hematocrit below the definitional criterion for CMS, and there were improvements in arterial oxygenation, systemic and pulmonary blood pressures, CMS symptom scores, and reductions in EPO concentrations [115, 116]. Confirming the stimulation on ventilation, these authors found that both hypoxic and hypercapnic ventilatory responses were augmented with treatment [117]. In a rat model of CMS, acetazolamide either started before 3 weeks of hypobaric hypoxia to an equivalent of 5,500 m or started 10 days into the exposure reduced hematocrit, blood viscosity, pulmonary and systemic vascular resistance, and increased cardiac output [118, 119]. Relief of hypoxic pulmonary vasoconstriction by both an increase in alveolar PO_2 and a reduction in the stimulus for HPV as well as a direct inhibition of HPV [92, 96, 99] likely contribute to the higher cardiac outputs and amelioration of cor pulmonale.

The data and pathophysiological basis of CMS strongly support the concept that acetazolamide reduces EPO production by its ventilatory stimulation and it is unlikely that CA inhibitors work by any other mechanisms. However, it has been proposed that acetazolamide directly suppresses renal EPO production independent of improvements in oxygenation by decreasing the O_2 demands of sodium reabsorption in the proximal tubule where the majority of sodium bicarbonate reabsorption occurs and is inhibited almost 80 % by CA inhibitors [120]. In these normoxic rat experiments, the dosage of acetazolamide necessary to suppress EPO production was roughly 10–20 times higher than the dose yielding a natriuretic and bicarbonaturic effect. A better explanation for the suppressant effect of acetazolamide on EPO release at such high doses is the likely severe respiratory acidosis accompanying complete red cell CA inhibition [25] which is known to block normal EPO production [121].

9 Conclusions

Acetazolamide and other CA inhibitors have multiple actions that underlie their efficacy in the prevention of acute altitude illnesses. This is most clearly established for AMS, in which it is predominantly an effect of renal and vascular CA inhibition that generate a mild bicarbonaturia and slight tissue CO_2 retention in the central nervous system. These combined metabolic and respiratory acidoses counter the suppressive effects of the hypoxia-mediated hyperventilation-induced hypocapnic respiratory alkalemia, which acts to limit the full ventilatory response to hypoxia. These same actions also are protective against CMS. Actions of these inhibitors on other CA-dependent processes such as CBF and CSF regulation may play a role as well, but experiments sufficient to isolate these effects from the ventilatory stimulation will require sophisticated 'clamping' of ventilation changes. Likewise, investigation of non CA-inhibiting actions of acetazolamide will also entail further work, such as has been performed in studies of hypoxic pulmonary vasoconstriction as they may relate to HAPE. In fact, the several non-CA inhibiting actions of acetazolamide warrant a reconsideration of much work that has used acetazolamide as the definitive pharmacological probe for defining a role for CA in a physiological process or mechanism as so cogently outlined by Thomas Maren in his classic review of the subject [105]. Lastly, definitive clinical trials of acetazolamide and non-CA inhibiting analogues of acetazolamide (such as n-methyl acetazolamide) in HAPE prevention are needed. With the advent of topical CA inhibitors for the treatment of glaucoma, such as dorzolamide, and the promise of isozyme specific CA inhibitors in cancer therapy, acetazolamide now has its greatest application at high altitude and remains the drug of choice for AMS and CMS prophylaxis and treatment.

References

1. Bailey DM, Bärtsch P, Knauth M, Baumgartner RW (2009) Emerging concepts in acute mountain sickness and high-altitude cerebral edema: from the molecular to the morphological. Cell Mol Life Sci 66:3583–3594
2. Swenson ER (1998) Carbonic anhydrase inhibitors and ventilation: a complex interplay of stimulation and suppression. Eur Respir J 12:1242–1247
3. Swenson ER, Teppema L (2007) Prevention of acute mountain sickness: as yet an unfinished story. J Appl Physiol 102:1305–1307
4. Bailey DM, Taudorf S, Berg RM, Lundby C, McEneny J, Young IS, Evans KA, James PE, Shore A, Hulllin DA, McCord JM, Pedersen BK, Möller K (2009) Increased cerebral output of free radicals during hypoxia: implications for acute mountain sickness? Am J Physiol 297:R1283–R1292
5. Kallenberg K, Dehnert C, Dörfler A, Schellinger PD, Bailey DM, Knauth M, Bärtsch PD (2008) Microhemorrhages in nonfatal high-altitude cerebral edema. J Cereb Blood Flow Metab 28:1635–1642
6. Maren TH (1979) An historical account of CO2 chemistry and the development of carbonic anhydrase inhibitors. Pharmacologist 20:303–321
7. Coudon WL, Block AJ (1976) Acute respiratory failure precipitated by a carbonic anhydrase inhibitor. Chest 69:112–113
8. Kronenberg RS, Cain SM (1967) Effects of acetazolamide on physiologic and subjective responses of men to 14,000 feet. SAM-TR-67–81. Sch Aviat Med Tech Rep 67:1–10
9. Forwand SA, Landowne M, Follansbee JN, Hansen JE (1968) Effect of acetazolamide on acute mountain sickness. N Engl J Med 279:839–845
10. Ellsworth AJ, Larson EB, Strickland D (1987) A randomized trial of dexamethasone and acetazolamide for acute mountain sickness prophylaxis. Am J Med 83:1024–1030
11. Kayser B, Dumont L, Lysakowski C, Combescure C, Haller G, Tramèr MR (2012) Reappraisal of acetazolamide for the prevention of acute mountain sickness. Med Sci Sports Exerc 22:178–1841

12. Stager JM, Tucker A, Cordain L, Engebretsen BJ, Brechue WF, Matulich CC (1990) Normoxic and acute hypoxic exercise tolerance in man following acetazolamide. Med Sci Sports Exerc 22:178–184
13. Garske LA, Brown MG, Morrison SC (2003) Acetazolamide reduces exercise capacity and increases leg fatigue under hypoxic conditions. J Appl Physiol 94:991–996
14. Faoro V, Huez S, Giltaire S (2007) Effects of acetazolamide on aerobic exercise capacity and pulmonary hemodynamics at high altitudes. J Appl Physiol 103:1161–1165
15. McLellan T, Jacobs I, Lewis W (1998) Acute altitude exposure and altered **acid–base** states II. Effects on exercise performance and muscle and blood lactate. Eur J Appl Physiol Occup Physiol 57:445–451
16. Hackett PH, Schoene RB, Winslow RM, Paters RM Jr, West JB (1985) Acetazolamide and exercise in sojourners to 6,300 meters–a preliminary study. Med Sci Sports Exerc 17:593–597
17. Jonk AM, van den Berg IP, Olfert IM, Wray DW, Arai T, Hopkins SR, Wagner PD (2007) Effect of acetazolamide on pulmonary and muscle gas exchange during normoxic and hypoxic exercise. J Physiol 579:909–921
18. Scheuermann BW, Kowalchuk JM, Paterson DH, Cunningham DA (2000) Carbonic anhydrase inhibition delays plasma lactate appearance with no effect on ventilatory threshold. J Appl Physiol 88:713–721
19. Bradwell, A. R., Coote, J. H., Milles, J. J., Dykes, P. W., Forster, P. J. E., Chesner, I., and Richardson, N. V.; Birmingham Medical Research Expeditionary Society (1986) Effect of acetazolamide on exercise performance and muscle mass at high altitude. Lancet 327:1001–1005
20. Slavi P, Revera M, Faini A, Giullian A, Gregorini F, Agostoni P, Becerra CG, Blio G, Lombardi C, O'Rourke MF, Mancia G, Parati G (2013) Changes in subendocardial viability ratio with acute high-altitude exposure and protective role of acetazolamide. Hypertension 61:79–799
21. Larsen RF, Rock PB, Fulco CS (1986) Effect of spironolactone on acute mountain sickness. Aviat Space Environ Med 57:543–547
22. Basnyat, B., Holck, P. S., Pun, M., Halverson, S., Szawarski, P., Gertsch, J., Steif, M., Powell, S., Khanai, S., Shankar, R., Karambay, J., Alexander, H.D., Stone, A., Morrissey, C., Thompson, B. H., and Farrar, J.; SPACE Trial Group (2011) Spironolactone does not prevent acute mountain sickness: a prospective, double-blind, randomized, placebo-controlled trial by SPACE Trial Group (spironolactone and acetazolamide trial in the prevention of acute mountain sickness group). Wilderness Environ Med 22:15–22
23. Klocke RA (1996) Potential role of endothelial carbonic anhydrase in dehydration of plasma bicarbonate. Trans Am Clin Climatol Assoc 108:44–57
24. Swenson ER, Hughes JMB (1993) Effects of acute and chronic acetazolamide on resting ventilation and ventilatory responses in men. J Appl Physiol 74:230–237
25. Swenson ER, Maren TH (1978) A quantitative analysis of CO2 transport at rest and during maximal exercise. Respir Physiol 35:129–159
26. Vogh BP (1980) The relation of choroid plexus carbonic anhydrase activity to cerebro-spinal fluid formation: a study of three inhibitors in cat with extrapolation to man. J Pharmacol Exp Ther 213:321–331
27. Tachtsidis I, Tisdall M, Delpy DT, Smith M, Elwell CE (2008) Measurement of cerebral tissue oxygenation in young healthy volunteers during acetazolamide provocation: a transcranial Doppler and near-infrared spectroscopy investigation. Adv Exp Med Biol 614:389–396
28. Huang SY, Mccullough RE, Mccullough RG, Micco AJ, Manco-Johnson M, Well JV, Reeves JT (1998) Usual clinical dose of acetazolamide does not alter cerebral blood-flow velocity. Respir Physiol 72:315–326
29. Subudhi AW, Dimmen AC, Julian CG, Wilson MJ, Paneral RB, Roach RC (2011) Effects of acetazolamide and dexamethasone on cerebral hemodynamics in hypoxia. J Appl Physiol 110:1219–1225

30. Vuyk J, Van Den Bos J, Terhell K, De Bos R, Vietter A, Valk P, Van Beuzekom M, Kleef JV, Dahan A (2006) Acetazolamide improves cerebral oxygenation during exercise at high altitude. High Alt Med Biol 7:290–301
31. Coates LE, Li A, Nattie EE (1991) Acetazolamide on the ventral medulla of the cat increases phrenic output and delays the ventilatory response to CO2. J Physiol 441:433–451
32. Iturriaga R, Lahiri S, Mokashi A (1991) Carbonic anhydrase and chemoreception in the cat carotid body. Am J Physiol 261:C565–C573
33. Iturriaga R, Mokashi A, Lahiri S (1993) Dynamics of carotid body responses in vitro in the presence of CO2–HCO3-: role of carbonic anhydrase. J Appl Physiol 75:1587–1594
34. Bashir Y, Kann M, Stradling JR (1990) The effect of acetazolamide on hypercapnic and eucapnic/poikilocapnic hypoxic ventilatory responses in normal subjects. Pulm Pharmacol 3:151–154
35. Teppema LJ, Balanos GM, Steinback CD, Brown AD, Foster GE, Duff HJ, Leigh R, Poulin MJ (2007) Effects of acetazolamide on ventilatory, cerebrovascular, and pulmonary vascular responses to hypoxia. Am J Respir Crit Care Med 175:277–281
36. Scheuermann BW, Kowalchuk JM, Paterson DH, Cunningham DA (1999) Peripheral chemoreceptor function after carbonic anhydrase inhibition during moderat-intensity exercise. J Appl Physiol 86:1544–1551
37. Teppema LJ, Dahan A (1999) Acetazolamide and breathing. Does a clinical dose alter peripheral and central CO2 sensitivity? Am J Respir Crit Care Med 160:1592–1597
38. Teppema LJ, Bijl H, Mousavi-Gourabi B, Dahan A (2006) The carbonic anhydrase inhibitors methazolamide and acetazolamide have different effects on the hypoxic ventilatory response in the anaesthetized cat. J Physiol 574:565–572
39. Hackett PH, Roach RC, Harrison GL, Schoene RB, Mills WJ Jr (1987) Respiratory stimulants and sleep periodic breathing at high altitude Almitrine versus acetazolamide. Am Rev Respir Dis 135:896–898
40. Latshang TD, Nussbauemer-Ochsner Y, Heen RM, Ulrich CM, Lo Cascio CM, Ledergerber B, Kohler M, Bloch KE (2012) Effect of acetazolamide and autoCPAP therapy on breathing disturbances among patients with obstructive sleep apnea syndrome who travel to altitude: a ramdomized controlled trial. J Am Med Assoc 308:2390–2398
41. Dempsey JA (2005) Crossing the apnoeic threshold: causes and consequences. Exp Physiol 90:13–24
42. Ainslie PN, Burgess K, Subedi P, Burgess KR (2007) Alterations in cerebral dynamics at high altitude following partial acclimatization in humans: wakefulness and sleep. J Appl Physiol 102:658–664
43. Herrera M, Garvin JL (2011) Aquaporins as gas channels. Pflugers Arch 462:623–630
44. Itel F, Al-Samir S, Öberg F, Chami M, Kumar M, Supuran CT, Deen PM, Meier W, Hedfalk K, Gros G, Endeward V (2012) CO2 permeability of cell membranes is regulated by membrane cholesterol and protein gas channels. FASEB J 26:5182–5191
45. Benga O, Huber VJ (2012) Brain water channel proteins in health and disease. Mol Aspects Med 33:562–578
46. Ameli PA, Madan M, Chigurupati S, Yu A, Chan SL, Pattisapu JV (2012) Effect of acetazolamide on aquaporin-1 and fluid flow in cultured choroid plexus. Acta Neurochir Suppl 113:59–64
47. Igarashi H, Tsujita M, Suzuki Y, Kwee IL, Nakada T (2013) Inhibition of aquaporin-4 significantly increases regional brain blood flow. Neuroreport 24:324–328
48. Zanotto C, Abib RT, Batassini C, Tortorelli LS, Biasbetti R, Rodrigues L, Nardin P, Hansen F, Gottfried C, Leite MC, Goncalves CA (2013) Non-specific inhibitors of aquaporin-4 stimulate S100-B secretion in acute hippocampal slices of rats. Brain Res 1491:14–22
49. Igarashi H, Huber VJ, Tsujita M, Nakuda T (2011) Pretreatment with a novel aquaporin-4 inhibitor significantly reduces ischemic cerebral edema. Neurol Sci 32:113–116
50. Verkman AS, Binder DK, Bloch O, Auguste K, Papadopoulos MC (2006) Three distinct roles of aquaporin-4 in brain function revealed by knockout mice. Biochim Biophys Acta 1758:1085–1093

51. Zhang J, An Y, Gao J, Han J, Xueyang P, Pan Y, Tie L, Li X (2012) Aquaporin-1 translocation and degradation mediates the water transportation mechanism of acetazolamide. PLoS One 7:e45976
52. Huber VJ, Tsujita M, Kwee IL, Nakada T (2009) Inhibition of aquaporin-4 by antiepileptic drugs. Bioorg Med Chem 17:418–424
53. Tanimura Y, Hiroaki Y, Fujiyoshi Y (2009) Acetazolamide reversibly inhibits water conduction by aquaporin-4. J Struct Biol 166:16–21
54. Yamaguchi T, Iwata Y, Miura S, Kawada K (2012) Reinvestigation of drugs and chemicals as aquaporin-1 inhibitors using pressure-induced hemolysis in human erythrocytes. Biol Pharm Bull 35:2088–2091
55. Yang B, Zhang H, Verkmann AS (2008) Lack of aquaporin-4 water transport inhibition by antiepileptics and arylsulfonamides. Bioorg Med Chem 16:7489–7493
56. Sogaard R, Zeuthen T (2008) Test of blockers of AQP-1 water permeability by a high resolution method: no effects of tetraethylammonium ions or acetazolamide. Pflugers Arch 456:285–292
57. Harbaugh RD, James HE, Marshall LF, Shapiro HM, Laurin R (1979) Acute therapeutic modalities for experimental vasogenic edema. Neurosurgery 5:656–665
58. Czernicki Z, Kuroiwa T, Ohno K, Endo S, Ito U (1994) Effect of acetazolamide on early ischemic cerebral edema in gerbils. Acta Neurochir Suppl 60:329–331
59. Millson C, James HE, Shapiro HM, Laurin R (1981) Intracranial hypertension and brain edema in albino rabbits. Effects of acute therapy with diuretics. Acta Neurochir 56:167–181
60. Guo F, Hua Y, Wang J, Keep RF, Xi G (2012) Inhibition of carbonic anhydrase reduces brain injury after intracerebral hemorrhage. Transl Stroke Res 3:130–137
61. Katada R, Nishitani Y, Honmou O, Mizuo K, Okazaki S, Tateda K, Watanabe S, Matsumoto H (2012) Expression of aquaporin-4 augments cytotoxic barin edema after traumatic brain injury during acute ethanol exposure. Am J Pathol 180:17–23
62. Ran X, Wang H, Chen Y, Zeng Z, Zhou Q, Zheng R, Sun J, Wang B, Lv X, Liang Y, Zhang K, Liu W (2010) Aquaporin-1 expression and angiogenesis in rabbit chronic myocardial ischemia is reduced by acetazolamide. Heart Vessels 25:237–247
63. Bailey DM, Dehnert C, Luks AM, Menold E, Castell C, Schendlev G, Faoro V, Gutoski M, Evans KA, Taudorf S, James PE, McEneny J, Young IS, Swenson ER, Mairbäurl H, Bärtsch P, Bergef MM (2010) High-altitude pulmonary hypertension is associated with a free radical-mediated reduction in pulmonary nitric oxide bioavailability. J Physiol 588:4837–4847
64. Bailey DM, Roukens R, Knauth M, Kallenberg K, Christ S, Mohr A, Genius J, Storch-Hagenlocher B, Meisel F, McEneny J, Young IS, Steiner T, Hess K, Bärtsch P (2006) Free radical-mediated damage to barrier function is not associated with altered brain morphology in high-altitude headache. J Cereb Blood Flow Metab 26:99–111
65. Bailey DM, Davies B (2001) Acute mountain sickness; prophylactic benefits of antioxidant vitamin supplementation at high altitude. High Alt Med Biol 2:21–29
66. Prouillac C, Vicendo P, Garrigues JC, Poteau R, Rima G (2009) Evaluation of new thiadiazoles and benzothiazoles as potential radioprotectors: free radical scavenging activity in vitro and theoretical studies (QSAR, DFT). Free Radic Biol Med 46:1139–1148
67. Teppema LJ, Bijl H, Romberg R, Dahan A (2006) Antioxidants reverse depression of the hypoxic ventilatory response by acetazolamide in man. J Physiol 572:849–856
68. Lisk C, McCord J, Bose S et al (2013) Irwin1Nrf-2 activation: a potential strategy for the prevention of acute mountain sickness. Free Rad Biol Med 63:264–273
69. Julian GJ, Subudhi AW, Wilson MJ, Dimmen AC, Pecha T, Roach RC (2011) Acute mountain sickness, inflammation, and permeability: new insights from a blood biomarker study. J Appl Physiol 111:392–399
70. Xu J, Peng Z, Li R, Dou T, Xu W, Gu G, Liu Y, Kang Z, Tao H, Zhang JH, Ostrowski RP, Lu J, Sun X (2009) Normoxic induction of cerebral HIF-1alpha by acetazolamide in rats. Neurosci Lett 451:274–278
71. Willam C, Warnecke C, Schefold JC, Kügler J, Koehne P, Frei U, Wiesener M, Eckardt KU (2006) Inconsistent effects of acidosis on HIF-alpha protein and its target genes. Pfluegers Arch 451:534–543

72. Droma Y, Ota M, Hanaoka M, Katsuyama Y, Basnyat B, Neupane P, Arjyal A, Pandit A, Sharma D, Ito M, Kubo K (2008) Two hypoxia sensor genes and their association with symptoms of acute mountain sickness in Sherpas. Avait Space Environ Med 79:1056–1060
73. Hennis PJ, Bussell C, Darlison MG (2010) The lack of associations between alleles at the hypoxia-inducible factor 1A C1772T loci and response to acute hypoxia. Wilderness Environ Med 21:219–228
74. Low EV, Avery AJ, Gupta V, Schedlbauer A, Grocott MP (2012) Identifying the lowest effective dose of acetazolamide for the prophylaxis of acute mountain sickness: systematic review and meta-analysis. Br Med J 345:e6779
75. Wright AD, Bradwell AR, Fletcher RF (1983) Methazolamide and acetazolamide in acute mountain sickness. Aviat Space Environ Med 54:619–621
76. Forster P (1982) Methazolamide in acute mountain sickness. Lancet 8283:1254
77. Kronenberg RS, Cain SM (1968) Hastening respiratory acclimatisation to altitude with benzolamide (CL 11366). Aerospace Med 39:296–300
78. Swenson ER, Leatham KL, Roach RC, Schoene RB, Mills WJ Jr, Hacket HP (1991) Renal carbonic anhydrase inhibition reduces high altitude sleep periodic breathing. Respir Physiol 86:333–343
79. Collier D, Swenson ER, Hedges AM et al (2013) Benzolamide improves oxygenation and reduces acute mountain sickness during a high altitude trek and has less side effects than acetazolamide. Brit J Clin Pharmacol (in press)
80. Grissom CK, Roach RC, Sarnquist FH, Hackett PH (1992) Acetazolamide in the treatment of acute mountain sickness: clinical efficacy and effect on gas exchange. Ann Intern Med 116:461–465
81. Wang J, Ke T, Zhang X, Chen Y, Liu M, Chen J, Luo W (2013) Effects of acetazolamide on cognitive performance during high-altitude exposure. Neurotoxicol Teratol 35:28–33
82. Vij A, Kishore K, Dey J (2012) Effect of intermittent hypoxia on efficacy and clearance of drugs. Indian J Med Res 135:211–216
83. Botao Y, Ma J, Xiao W, Xiang Q, Fan K, Hou J, Wu J, Jing W (2013) Protective effects of gingkolide B on high altitude cerebral edema of rats. High Alt Med Biol 14:61–64
84. Smith CA, Rodman JR, Chenuel BJ, Henderson KS, Dempsey JA (2006) Response time and sensitivity of the ventilatory response to CO2 in unanesthetized dogs: central vs. peripheral chemoreceptors. J Appl Physiol 100:13–19
85. Bartsch P, Mairbaurl H, Maggiorini M, Swenson ER (2005) Physiological aspects of high-altitude pulmonary edema. J Appl Physiol 93:1101–1110
86. Swenson ER (2013) Hypoxic pulmonary vasoconstriction. High Alt Med Biol 14:101–110
87. Maggiorini M, Melot C, Pierre S, Pfeiffer F, Greve I, Sartori C, Lepori M, Hauser M, Scherrer U, Naeije R (2001) High-altitude pulmonary edema is initially caused by an increase in capillary pressure. Circulation 103:2078–2083
88. Swenson ER, Maggiorini M, Mongovin S, Gibbs JS, Greve I, Mairbäurl H, Bärtsch P (2002) Pathogenesis of high altitude pulmonary edema. Inflammation is not a pathogenic factor. J Am Med Assoc 287:2228–2235
89. Emery CJ, Sloan PJ, Mohammed FH, Barer GR (1977) Action of hypercapnia during hypoxia on pulmonary vessels. Bull Eur Physiopath Respir 13:763–776
90. Swenson ER, Robertson HT, Hlastala MP (1993) Effects of carbonic anhydrase inhibition on ventilation–perfusion matching in the dog lung. J Clin Invest 92:702–709
91. Swenson ER, Graham MM, Hlastala MP (1995) Acetazolamide slows ventilation-perfusion matching after changes in regional blood flow. J Appl Physiol 78:1312–1318
92. Deem S, Hedges RG, Kerr ME, Swenson ER (2000) Acetazolamide reduces hypoxic pulmonary vasoconstriction in isolated perfused rabbit lung. Respir Physiol 123:109–119
93. Aamand R, Dalsgaard T, Jensen FB, Simonsen U, Roepstorff A, Fago A (2009) Generation of nitric oxide from nitrite by carbonic anhydrase: a possible link between metabolic activity and vasodilation. Am J Physiol 297:H2068–H2074
94. Pickerodt PA, Emery MJ, Zarndt R, Martin W, Francis RC, Boemke W, Swenson ER (2012) Sodium nitrite mitigates ventilator-induced lung injury in rats. Anesthesiol 117:592–601

95. Chobanyan-Jürgens K, Schwarz A, Böhmer A, Beckmann B, Gutzki FM, Michaelsen JT, Stichtenoth DO, Tsikas D (2012) Renal carbonic anhydrases are involved in the reabsorption of endogenous nitrite. Nitric Oxide 26:126–129
96. Hohne C, Krebs MO, Seiferheld M, Boemke W, Kaczmarczyk G, Swenson ER (2004) Acetazolamide prevents hypoxic pulmonary vasoconstriction in conscious dogs. J Appl Physiol 97:515–521
97. Höhne C, Pickerodt PA, Boemke W, Swenson ER (2007) Pulmonary vasodilation by acetazolamide during hypoxia is not related to carbonic anhydrase inhibition. Am J Physiol 292:L178–L184
98. Berg JT, Ramanathan S, Gabrielli MG, Swenson ER (2004) Carbonic anhydrase in mammalian vascular smooth muscle. J Histochem Cytochem 52:1101–1106
99. Shimoda LA, Luke T, Sylvester JT, Shih HW, Jain A, Swenson ER (2007) Inhibition of hypoxia-induced calcium responses in pulmonary arterial smooth muscle by acetazolamide is independent of carbonic anhydrase inhibition. Am J Physiol 292:L1002–L1012
100. Maren TH (1967) Carbonic anhydrase: chemistry, physiology and inhibition. Physiol Rev 47:595–781
101. McNaughton NC, Davies CH, Randall A (2004) Inhibition of alpha (1E) Ca (2+) channels by carbonic anhydrase inhibitors. J Pharmacol Sci 95:240–247
102. Tricarico D, Barbieri M, Mele A, Carbonara G, Camerino DC (2004) Carbonic anhydrase inhibitors: specific openers of skeletal muscle BK channel of K + −deficient rats. FASEB J 18:760–761
103. Farrukh IS, Hoidal JR, Barry WH (1996) Effect of intracellular pH on ferret pulmonary arterial smooth muscle cell calcium homeostasis and pressure. J Appl Physiol 80:496–505
104. Wang J, Shimoda LA, Weigand L, Wang W, Sun D, Sylvester JT (2005) Acute hypoxia increases intracellular [Ca2+] in pulmonary arterial smooth muscle by enhancing capacitative Ca2+ entry. Am J Physiol 288:L1059–L1069
105. Maren TH (1977) Use of inhibitors in physiological studies of carbonic anhydrase. Am J Physiol 232:F291–F297
106. Pickerodt PA, Francis RC, Neubert F et al (2013) Pulmonary vasodilation by acetazolamide during hypoxia: impact of methyl-group substitution and route of administration in the conscious, spontaneously breathing dog. J Pharmacol Exp Therap (in press)
107. Ke T, Wang, J, Swenson ER (2013) Effect of acetazolamide and gingko biloba on the human pulmonary vascular response to an acute altitude ascent. High Alt Med Biol 14:162–167
108. Basnyat B, Hargrove J, Holck PS, Srivastav S, Alekh K, Ghimire LV, Pandey K, Griffiths A, Shankar R, Kaul K, Paudyal A, Stasiuk D, Basnyat R, Davis C, Southard A, Robinson C, Shandley T, Johnson DW, Zafren K, Williams S, Weiss EA, Farrar JJ, Swenson ER (2008) Acetazolamide fails to decrease pulmonary artery pressure at high altitude in partially acclimatized humans. High Alt Med Biol 9:209–216
109. Pickkers P, Garcha RS, Schachter M, Smits P, Hughes AD (1999) Inhibition of carbonic anhydrase accounts for the direct vascular effects of hydrochlorothiazide. Hypertension 33:1043–1048
110. Pickkers P, Hughes AD, Russel FG, Thien T, Smits P (2001) In vivo evidence for K(Ca) channel opening properties of acetazolamide in the human vasculature. Br J Pharmacol 132:443–450
111. Berg JT, Ramanathan S, Swenson ER (2004) Inhibition of hypoxic pulmonary vasoconstriction prevents high altitude pulmonary edema (HAPE) in rats. Wilderness Environ Med 15:32–37
112. Leon-Velarde F, Richalet JP (2006) Respiratory control in residents at high altitude; physiology and pathophysiology. High Alt Med Biol 7:125–137
113. Swenson ER (2012) Normal exercise capacity in chronic mountain sickness: how high can the hematocrit go without consequence? Chest 142:823–825
114. Rivera-Ch M, Leon-Velarde F, Huicho L (2007) Treatment of chronic mountain sickness; critical reappraisal of an old problem. Respir Physiol Neurobiol 158:251–265

115. Richalet JP, Rivera M, Bouchet P, Chirinos E, Onnen I, Petitjean O, Bienvenu A, Lasne F, Moutereau S, León-Velarde F (2005) Acetazolamide: a treatment of chronic mountain sickness. Am J Respir Crit Care Med 172:1427–1433
116. Richalet JP, Rivera-Ch M, Maignan M, Privat C, Pham I, Macarlupu JL, Petitjean O, León-Velarde F (2008) Acetazolamide for Monge's disease: efficacy and tolerance of 6-month treatment. Am J Respir Crit Care Med 177:1370–1376
117. Rivera-Ch M, Huicho L, Bouchet P, Richalet JP, León-Velarde F (2008) Effect of acetazolamide on ventilatory response in subjects with chronic mountain sickness. Respir Physiol Neurobiol 162:184–189
118. Pichon A, Connes P, Quidu P, Marchant D, Brunet J, Levy BI, Vilar J, Safeukui I, Cymbalista F, Maignan M, Richalet JP, Favret F (2012) Acetazolamide and chronic hypoxia: effects on haemorheology and pulmonary haemodynamics. Eur Respir J 40:1401–1409
119. Leon-Velarde F, Villafuerte FC, Richalet JP (2010) Chronic mountain sickness and the heart. Prog Cardiovasc Dis 52:540–549
120. Eckardt KU, Kurtz A, Bauer C (1989) Regulation of erythropoietin production is related to proximal tubular function. Am J Physiol 256:F942–F947
121. Miller ME, Rorth M, Parving H, Howard D, Reddington I, Valeri CR, Stohlman F Jr (1973) pH effect on erythropoietin response to hypoxia. N Engl J Med 288:706–710

Chapter 19
Thermal-Stable Carbonic Anhydrases: A Structural Overview

Vincenzo Alterio, Simona Maria Monti, and Giuseppina De Simone

Abstract The potential of carbonic anhydrase (CA) family as target for the drug design of inhibitors with various medicinal chemistry applications has been recognized from long time, whereas the industrial interest in using these enzymes as biocatalysts for carbon dioxide sequestration and biofuel production is only recently emerging. However, an efficient utilization in these processes often requires stable enzymes, able to work in the harsh conditions typical of the CO_2 capture process. In this context CAs active at very high temperatures are of extreme interest. In this chapter we have summarized in a comparative manner all existing data on thermostable CAs both isolated by extremophiles and obtained by protein engineering studies. Among the five CA-classes, the biochemical and structural features of thermostable α-, β- and γ-CAs have been discussed. Data show that so far α-CAs isolated from thermophilic organisms are the best candidates to be used in biotechnological processes, even if plenty of work can be still done in this field also with help of protein engineering.

Keywords Thermostability • Carbonic anhydrase • Protein engineering • Thermophiles • CO2 capture process • β-carbonic anhydrases

---

Susan C. Frost and Robert McKenna (eds.). Carbonic Anhydrase: Mechanism, Regulation, Links to Disease, and Industrial Applications

V. Alterio • S.M. Monti • G. De Simone (✉)
Istituto di Biostrutture e Bioimmagini-CNR, Naples, Italy
e-mail: vincenzo.alterio@cnr.it; marmonti@unina.it; gdesimon@unina.it

S.C. Frost and R. McKenna (eds.), *Carbonic Anhydrase: Mechanism, Regulation, Links to Disease, and Industrial Applications*, Subcellular Biochemistry 75,
DOI 10.1007/978-94-007-7359-2_19,

1 Introduction

It is becoming always more evident that CO_2 release from the combustion of fossil fuels is the principal cause of the recent global warming [1]. Thus, the development of efficient technologies for CO_2 capture and sequestration is becoming one of the most important challenge of the twenty-first century [2]. Among CO_2 capture techniques the reversible chemical absorption by alkaline aqueous solvents and subsequent desorption with a heated stripper column, to give pure CO_2 for compression and storage, represents one of the most mature applied technology [1, 3]. However, the application of this kind of technologies raises several technical challenges, with the main problem related to the high energy cost necessary for CO_2 desorption [1]. Recent studies have demonstrated that these high energy costs can be lowered by using aqueous solvents with lower heat of desorption, coupled with Carbonic Anhydrases (CAs), enzymes able to accelerate absorption of CO_2 [1].

CAs are ubiquitous enzymes involved in the reversible hydration of carbon dioxide to the bicarbonate ion and proton ($CO_2 + H_2O \leftrightarrows HCO_3^- + H^+$). They represent an excellent example of convergent evolution, as five distinct genetic families have been discovered so far: the α-CAs, which are mainly found in vertebrates, bacteria, algae and cytoplasm of green plants, the β-CAs, present in bacteria, algae and chloroplasts, the γ-CAs, present in archaea and some bacteria and the δ-CAs and ζ-CAs, which have been discovered only in some marine diatoms [4]. There are no significant sequence homologies between representatives of the different CA families, but all members contain a metal ion in their active site which is essential for catalytic reaction (Zn(II) for α- and β-CAs [5–9], Fe(II), Co(II) or Zn(II) for γ-CAs [10–13], Co(II) or Zn(II) for δ-CAs [14–22] and Cd(II) or Zn(II) for ζ-CAs [15, 23]).

As outlined above, the recent years have seen an increasing industrial interest in using CAs as biocatalysts for carbon dioxide capture from the flue gas of coal-fired power plants [1, 24]. At the same time these enzymes have also shown a great potential in the development of technologies aimed to improve CO_2 storage rates as an inorganic salt since carbonate minerals represent the largest reserve of CO_2 on Earth [25].

Unfortunately, the use of CAs in carbon dioxide capture procedures is generally strongly limited by the poor stability and activity that such enzymes present in the harsh conditions required in these processes, especially the very high temperatures. Thus two different approaches can be efficiently used to overcome these limitations: (a) employment of enzymes isolated from microorganisms living at high temperature, i.e., thermophiles, (b) use of protein engineering techniques to produce themostable enzymes. Aim of this review is to provide an exhaustive description of the thermostable CAs so far reported. In particular, the first part of the review will be dedicated to the CAs isolated from thermophilic organisms, focusing the attention on the structural features of such enzymes, in order to derive for each CA class possible structural determinants responsible for enzyme thermal stability. The final part of the paper will be instead focused on thermostable CAs obtained by protein engineering techniques highlighting the most recent advances in such field.

2 Carbonic Anhydrases Isolated from Thermophilic Organisms

In the last years an increasing interest has been shown by the scientific community in the use of proteins isolated from thermophilic organisms for understanding enzyme evolution and molecular mechanisms for protein thermostability [26]. Thermophilic organisms grow optimally at temperatures which range from 50 °C to more than 90 °C and their "extremozymes" are usually active at very high temperatures [27–30], offering unique biotechnological advantages over mesophilic (optimally active at 25–50 °C) or psychrophilic enzymes (optimally active at 5–25 °C). Once expressed in mesophilic hosts, these enzymes conserve most of their features being still active at high temperatures and showing high resistance to chemical denaturants [31]. Moreover, performing enzymatic reactions at high temperatures, they can work with high substrate concentrations, low viscosity, and present fewer risks of microbial contaminations, and often high reaction rates [27–32].

In the past, reports on CAs isolated from thermophilic organisms were limited to proteins belonging to the β- and γ-class [33–36], but such enzymes presented limited application in the aforementioned industrial processes due to their low catalytic activity when compared to the enzymes of the α-class. Fortunately, the last 2 years have seen an increase in the studies on thermostable CAs belonging to the α-class [32, 37–41], while so far no δ- and ζ-CA from thermophilic organisms has been described. In the next paragraphs the thermostable CAs so far reported will be described and grouped on the basis of their belonging class. Particular emphasis will be dedicated to the structural features of such enzymes and to the possible determinants of thermal stability in order to obtain information that could be used in subsequent protein engineering studies.

2.1 *Thermostable α-Carbonic Anhydrases*

Only few studies have been so far reported on themostable CAs belonging to the α-class and, until last year, many of them were described in patents rather than in publications in the open scientific literature. In particular, heat-stable α-CAs from *Bacillus clausii* [42], *Bacillus halodurans* [42], *Caminibacter* [43] and *Persephonella marina* [44] and their use for the extraction of CO_2 were described in three recent patents.

During 2012 there has been an increasing amount of literature data on such enzymes. In particular two extremely thermostable α-CAs from *Sulfurihydrogenibium yellowstonense* YO3AOP1 [37, 38] and *Sulphurihydrogenibium azorense* [39, 45, 46], respectively, have been extensively studied from a biochemical, kinetic and a structural point of view.

Sulfurihydrogenibium yellowstonense YO3AOP1 is a novel thermophilic bacterium, isolated in the Yellowstone National Park, USA [47], which belongs to the chemosynthetic bacterial communities living in hot springs, at temperatures of up to 110 °C. A translated genome inspection led to the identification in this bacterium of a novel α-CA, named SspCA [32, 37, 38, 40, 41]. After cloning and expression in *E. coli* this enzyme was characterized for its catalytic activities. In particular, it showed esterase activity with an optimal working temperature of 95 °C and a very high catalytic activity for the CO_2 hydration reaction ($k_{cat} = 9.35 * 10^5\ s^{-1}$; $k_{cat}/K_M = 1.1 * 10^8\ M^{-1}\ s^{-1}$ at 20 °C and pH 7.5), comparable to that of human (h) isoform hCA II [38]. The CO_2 hydration activity was also measured after incubation at different temperatures and compared with that of two mammalian α-CAs, namely hCA I and hCA II, revealing even more interesting biotechnological properties. Indeed, these studies showed that while mammalian enzymes were inactivated at temperature higher than 70 °C for all incubation times, SspCA was still active after incubation at 100 °C for 3 h, highlighting an exceptional resistance to high temperature with respect to its mammalian homologues [32]. Moreover, after immobilization within a polyurethane foam SspCA was unalterably active and stable up to 50 h at 100 °C [37].

A series of inhibition and activation studies were also performed on this enzyme [38, 40, 41], showing that SspCA was not substantially inhibited by bicarbonate and carbonate, making again this enzyme a very interesting candidate for biotechnological applications in the CO_2 capture processes [38].

The same group of authors reported also the characterization of another α-CA, named SazCA, identified by translated genome inspection in *Sulfurihydrogenibium azorense*, a thermophilic bacterium living in the terrestrial hot springs of the Azores [39, 45, 46]. The kinetic parameters of SazCA, were determined by a stopped-flow assay method using CO_2 as substrate, showing that this enzyme was the most active among all CAs known to date ($kcat = 4.40 * 10^6\ s^{-1}$; $k_{cat}/K_M = 3.5 * 10^8\ M^{-1}\ s^{-1}$ at 20 °C and pH 7.5), being 2.3 times more efficient than hCA II [46]. Thermoactivity and thermostability studies showed that SazCA was active in the temperature range from 0°C to 100 °C, with an optimal working temperature for the esterase activity at about 80 °C. The thermal stability of SazCA was also compared with that of SspCA. In particular the CO_2 hydration activity of both enzymes was measured after incubation at different temperatures/times. Interestingly, both enzymes were active after incubation at the temperature of 100 °C, even if SazCA was slightly less stable than SspCA [46]. Finally, also in this case anion inhibition studies revealed that bicarbonate, carbonate and sulfate were ineffective SazCA inhibitors [39].

Many studies have been performed in recent years to derive information on structural determinants of enzyme thermostability, in order to develop strategies for enzyme stabilization. At this aim the comparison of high resolution X-ray structures of thermostable proteins with their mesophilic counterparts can provide useful information. Structural studies on thermostable α-CAs are so far available only for SspCA [32]; indeed, the enzyme was recently crystallized in complex with an inhibitor molecule. Analysis of its structure revealed the typical fold of other human α-CAs, characterized by a central ten-stranded mostly parallel β-sheet (Fig. 19.1). As observed in the structure of other α-CAs [4], the active site is located in a deep conical cavity, which extends from the protein surface to the

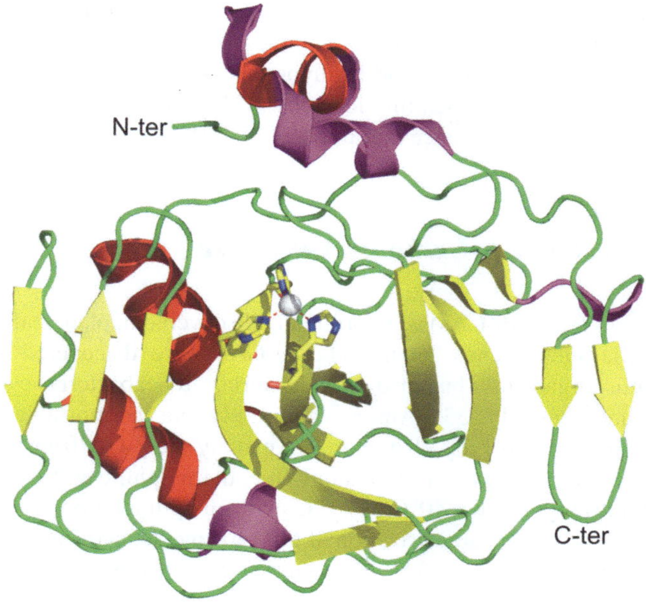

Fig. 19.1 SspCA monomer overall fold (PDB code 4G7A). β-strands are reported in *yellow*, α-helices in *red* and 3_{10}-helices in *magenta*. Secondary structure assignments were obtained from PROMOTIF [48]

centre of the molecule. The catalytic zinc ion is located at the bottom of this cavity, tetrahedrally coordinated by three histidine residues (His89, His91 and His108) and by the sulfonamide group of the inhibitor. Interestingly, the enzyme crystallized forming a dimer characterized by a large interface area and stabilized by a large number of hydrogen bonds and hydrophobic interactions. Since the active sites of both monomers forming the dimer were completely accessible to the substrate, and the dimeric arrangement was also evidenced by light scattering experiments, the authors hypothesized that SspCA works as a physiological dimer [32]. Furthermore, the high structural similarity of such dimer with that formed by another bacterial α-CA, namely *Neisseria gonorrhoeae* CA (NgCA) [49], strongly suggested that this dimeric arrangement could be typical of all bacterial α-CAs [32].

The availability of structural data on the thermostable SspCA and on several mesophilic homologues provided the authors the opportunity to identify possible structural determinants responsible for thermostability in the α-class CAs. In particular, a number of parameters which in literature were reported to be important for themostability, such as the percentage of secondary structural elements, the number of hydrogen bonds, ionic interactions, ionic networks, and others [50, 51] were compared between SspCA, and two its mesophilic homologues, namely hCA I and hCA II, for which thermal denaturation data were available [32]. The analysis of such parameters clearly showed that a higher content of secondary structural elements, associated to an increased number of charged residues on the protein surface and of large ionic networks, were the main distinguishing features of SspCA with respect to its mesophilic homologues. These findings are of extreme importance since they provide a working hypothesis on the mechanisms responsible for increased protein thermostability in α-class CAs, thus offering a tool to exploit in engineering α-CAs to enhance thermostability [32].

2.2 *Thermostable β-Carbonic Anhydrases*

β-carbonic anhydrases (β-CAs) were originally discovered in plant leaf chloroplasts in 1939 [52]. Subsequently, many other β-CAs were discovered not only in photosynthetic organisms, but also in eubacteria, archeabacteria, cyanobacteria, microalgae and yeast [11, 53]. Among these, only one thermostable enzyme has been so far isolated and fully characterized, namely the β-CA from *Methanobacterium thermoautotrophicum,* a lithoautotrophic, thermophilic archaeon that grows at temperatures ranging from 40°C to 70 °C, with an optimal growth temperature at 65 °C [54]. The kinetic characterization of this enzyme, named Cab, by means of a stopped flow assay method at 25 °C, showed that it is a less efficient catalyst for the CO_2 hydration reaction ($k_{cat} = 1.7*10^4\ s^{-1}$; $k_{cat}/K_M = 5.9 * 10^6\ M^{-1}\ s^{-1}$) [34] with respect to the α-class thermostable enzymes above reported, thus indicating a lower applicability of this enzyme in CO_2 capture processes with respect to SspCA [32, 37, 38, 40, 41] and SazCA [39, 45, 46]. However, the biochemical characterization revealed also in this case a very high thermostability, showing that Cab maintains nearly all catalytic activity after 15 min incubation at temperatures up to 75 °C, even if only very little activity was recovered when incubated at temperatures of 90 °C or higher [34].

The crystal structure of Cab was also solved at 2.1 Å resolution by a multiwavelength anomalous diffraction experiment [55]. Analysis of the structure, as observed for other β-CAs structurally characterized, showed that the core of the molecule consisted of a five-stranded β-sheet surrounded by several α-helices (Fig. 19.2) [55]. Interestingly, the enzyme crystallizes as a dimer [55], although previous analytical ultracentrifugation experiments and gel filtration chromatography were indicative of a tetrameric structure [34, 56]. These data were in agreement with the observation that β-CAs can exists in different oligomeric states, ranging from dimers to tetramers to octamers [11, 34, 57–59]. The dimer, as already observed for several others Cab mesophilic homologs [7, 60], consists of two identical monomers, related by a twofold rotation axis (Fig. 19.3), which interact through a large interface area (2110 Å^2/subunit) [55]. Residues in strand β2 and helices α2, α3, α4, and α5 contribute to the dimer stabilization. In particular, hydrogen bonds between residues from strand β2 in both subunits result in the formation of a 10-stranded β-sheet, whereas helices α4 and α5 protrude from one monomer and make extensive contacts with the other monomer (Fig. 19.3) [55]. The two active sites are located at the dimeric interface, each containing one Zn(II) ion coordinated by two Cys and one

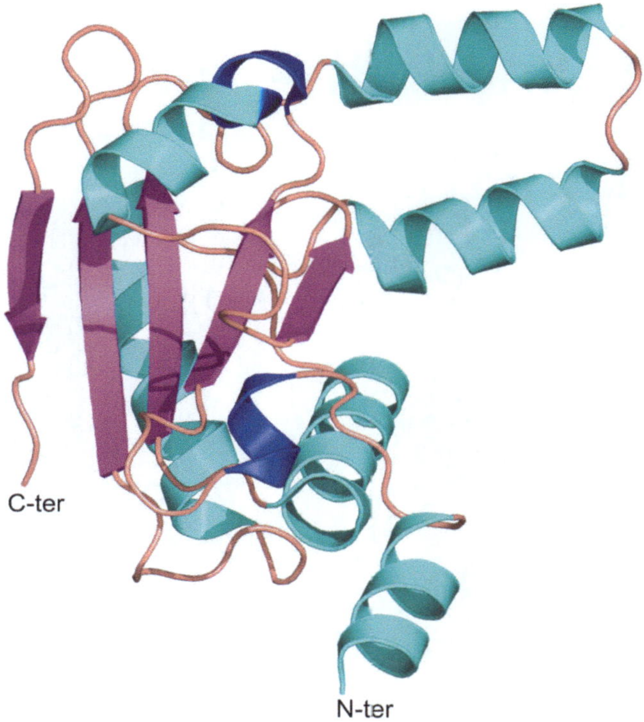

Fig. 19.2 Cab monomer overall fold (PDB code 1G5C). β-strands are reported in *magenta*, α-helices in *cyan* and 3_{10}-helices in *blue*. Secondary structure assignments were obtained from PROMOTIF [48]

His residues belonging to the same monomer. A water molecule completes in both cases the tetrahedral Zn(II) coordination [55]. The comparison of Cab active site with that of other β-CAs structurally characterized revealed that these enzymes can be classified in two major classes, type I and type II, on the basis of the fourth ligand of the Zn(II) ion [11]. Cab belongs to the first class, which is characterized by the presence of an exchangeable fourth ligand (i.e. water, acetate, acetic acid), which is hydrogen bonded to the aspartic residue of a conserved Asp-Arg dyad. A residue which could act as hydrogen bond donor to a zinc-bound bicarbonate ion in the exchangeable ligand position and a narrow hydrophobic active site cleft are the other main features of this β-CA class. On the contrary in the type II β-CAs the fourth coordination position is occupied by the aspartic residue which in the first class is part of the Asp-Arg diad. Moreover, the residue which could act as hydrogen bond donor to a bicarbonate ion bound in the position occupied by the aspartic residue is missing [11].

Even though Cab enzyme was discovered and characterized more than 10 years ago [34, 54], a systematic study aimed at identifying possible structural

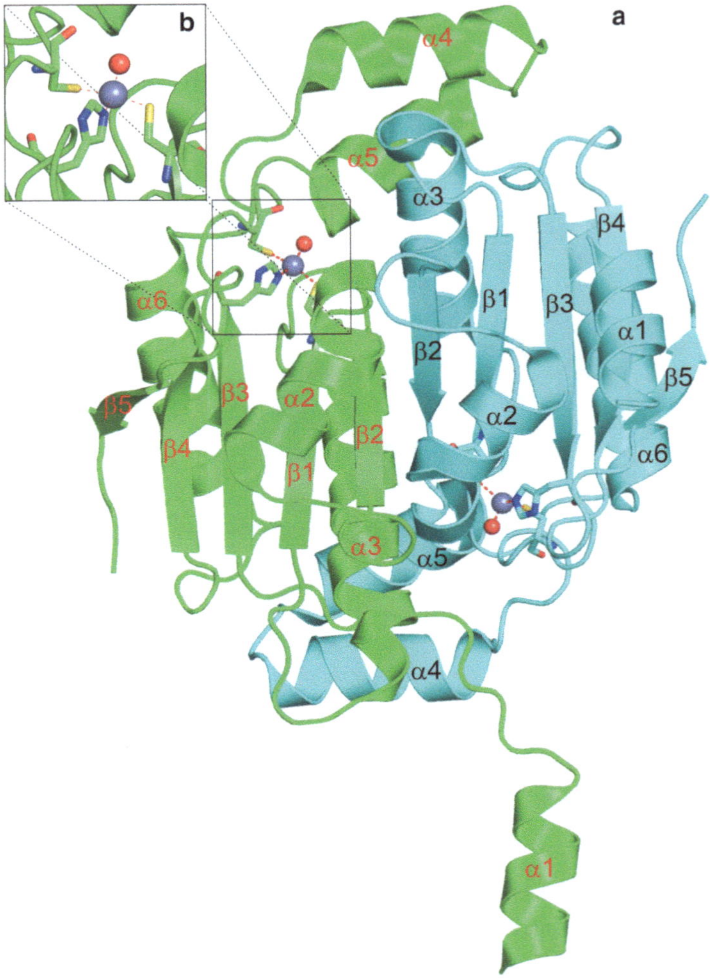

Fig. 19.3 (**a**) Dimeric structure of Cab (PDB code 1G5C) with one monomer colored in *green*, the other in *cyan*. Zn^{2+} ions are shown as *spheres* and their coordinating residues are shown in *ball-and-stick* representation. Secondary structure assignments were obtained from PROMOTIF [48]. (**b**) Enlarged view of the active site of Cab

determinants of thermostability, as in the study above reported for SspCA, is so far still missing. The absence of such a study is likely attributable to the rather high diversity of the structurally characterized members of the β-CA family, which makes rather difficult to evaluate the effect of a single parameter on enzyme thermal stability. Indeed, apart the similarity of the central β-sheet core, such enzymes largely differ for sizes, oligomeric arrangement, domain composition of the monomeric unit etc., showing that like no other CA class, β-CAs exhibit the broadest structural and functional diversity [11]. It would be highly desirable that future characterization of other members of this enzyme class could be performed, in order to lead to a more detailed comparative study, which could give the basis for the rational design of mutated proteins with improved biotechnological features.

2.3 *Thermostable γ-Carbonic Anhydrases*

Even if the γ-class of carbonic anhydrases (γ-CA) was first identified in 1994 [61], so far the overall knowledge of such enzymes is still limited. Indeed, only few members of the class have been fully characterized from a biochemical and structural point of view [10], with the enzyme Cam, from the moderately thermophilic methanoarchaeon *Methanosarcina thermophila* (MtCam) [33, 61] being the best studied. Biochemical studies, performed on this enzyme, heterologously produced in *E. coli,* showed that it contained a zinc ion tightly bound in the active site and displayed catalytic activity for the CO_2 hydration reaction but it did not present esterase activity [33]. Stopped-flow spectroscopy at 25 °C was utilized for the determination of kinetic parameters for the CO_2 hydration reaction ($k_{cat} = 6.8*10^4\ s^{-1}$; $k_{cat}/K_M = 3.1 * 10^6\ M^{-1}\ s^{-1}$) [62, 63], indicating, as already observed for Cab from *Methanobacterium thermoautotrophicum* [34], a catalytic efficiency lower than that showed by the α-CAs SspCA [32, 37, 38, 40, 41] and SazCA [39, 45, 46]. Finally, thermostability studies showed that MtCam is a moderately thermostable enzyme, since even if it retains nearly all catalytic activity when incubated for 15 min at temperature up to 55 °C, the optimal temperature for growth of the source organism, it maintains only little activity when incubated above 75 °C [33]. All these data suggest a possible use of this enzyme in CO_2 capture processes, even if with lower success with respect to the corresponding thermostable α-CA homologues.

Interestingly, subsequent biochemical studies showed that, although MtCam contains Zn^{2+} in the active site when it is purified aerobically [33], it contains Fe^{2+} when purified anaerobically [62] or overproduced in the strictly related species *M. acetivorans* [64]. The enzyme containing Fe^{2+} in the active site shows a significant increase in CA activity that is rapidly lost after exposure to air, as a result of oxidation to Fe^{3+} and loss of the metal from the active site. These studies suggested that Fe^{2+}-Cam is the predominant in vivo enzyme form [62].

The crystal structure of zinc containing MtCam was solved [65, 66] providing very interesting results on the γ-CA family, which shows features completely different from those of the enzymes of the α- and β-class. The first important difference is that MtCam crystallizes as a trimer, in agreement with ultracentrifugation experiments that also indicated the trimer as the predominant form in solution [65]. The structure of each monomer in the trimer is characterized by a seven-turn left-handed β-helix, which has a short α-helix on its top and a long C-terminal α-helix running antiparallel to the axis of the β-helix (Fig. 19.4). An extensive interaction

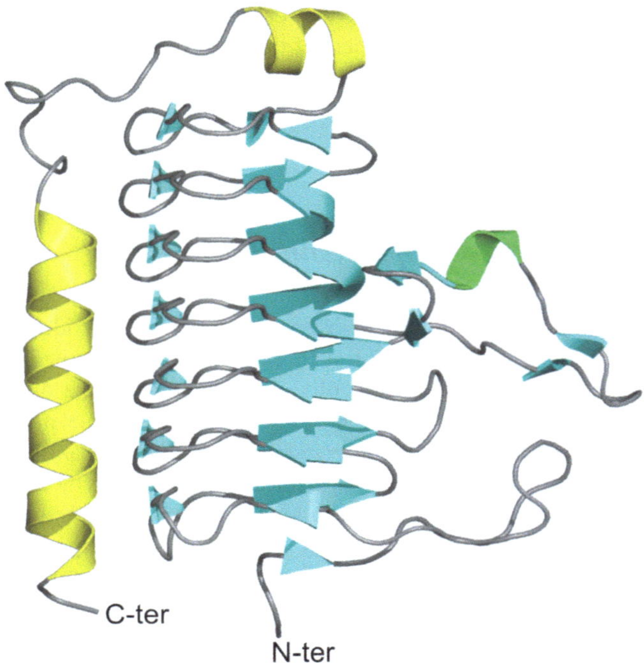

Fig. 19.4 MtCam monomer overall fold (PDB code 1QRG). β-strands are reported in *cyan*, α-helices in *yellow* and the 3_{10}-helix in *green*. Secondary structure assignments were obtained from PROMOTIF [48]

of the three β-helix structures originate the trimer, characterized by a large interface between two adjacent monomers as well as a great number of H-bonds, salt bridges and hydrophobic interactions [65]. The most interesting feature of the trimerization is that the three active site zinc ions are located at the interface between two adjacent monomers. As in the α-class CAs, these ions are coordinated by three histidine residues but not all derive from the same monomer. In particular, one monomer provides one histidine ligand and the adjacent monomer provides the remaining two [65]. Two additional water molecules complete the coordination sphere originating a trigonal bipiramydal geometry (Fig. 19.5) [66]. The substitution of the Zn(II) ion with Co(II) ion, by unfolding and refolding in the presence of Co(II), leads to a distorted octahedral geometry of the metal ion [66]. The analysis of these structures, together with mutagenesis experiments [67], also pointed out an important role in the proton transfer process, the rate limiting step of the catalytic reaction, for residues Glu62 and Glu84 [66].

Further structural studies on γ-CAs were subsequently published. In particular two full length enzymes, isolated respectively from the hyperthermophilic archaeon *Pyrococcus horikoshii* (PhCamH) [35], and the mesophilic bacterium *Escherichia Coli* (EcYrdA) [68], and a MtCam homolog consisting of the N-terminal domain

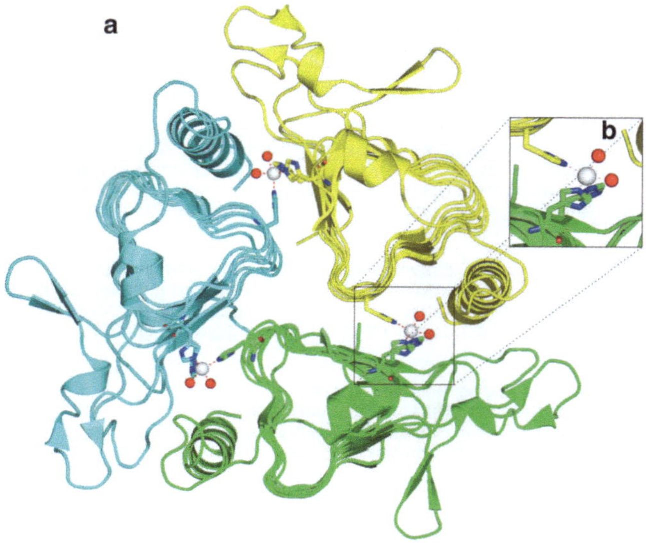

Fig. 19.5 (**a**) MtCam trimer (PDB code 1QRG). The three monomers are reported in *green*, *yellow* and *cyan*. Zinc ions are shown as *spheres* and their coordinating residues are shown in *ball-and-stick* representation. Secondary structure assignments were obtained from PROMOTIF [48]. (**b**) Enlarged view of the active site of MtCam

of the carboxysomal protein CcmM from the thermophilic β-cyanobacterium *Thermosynechococcus elongates* (TeCcmM209) [36] were crystallized and structurally characterized. The structural superposition of these three enzymes with MtCam, the prototype of the γ-class, revealed that they present a very similar overall architecture, with the most important differences located at the N- and C-terminus and in an acidic loop, located after β10 (MtCam structure numbering), which in MtCam contains the proton shuttle residue Glu84 (Fig. 19.6). Indeed, as also visible in the structure based sequence alignment reported in Fig. 19.6, this loop is very short in PhCamH [35] and EcYrdA [68], and is longer in MtCam [65] and TeCcmM209 [36]. In all three proteins PhCamH [35], EcYrdA [68], and TeCcmM209 [36] Glu84 is not conserved. These structural data, together with sequence analyses of MtCam homologues derived from the databases, showed that γ-class CAs can be divided in two main subclasses: a first subclass, whose members are strictly homologues to MtCam, and a second subclass, named CamH, which comprises the majority of γ-CAs, characterized by a general conservation of all residues essential for MtCam catalytic activity with the exception of Glu62 and the long acidic loop containing the fundamental proton shuttle residue Glu84 (Fig. 19.6). The absence of these residues is quite surprising in view of their role in catalytic reaction (see above), and could be indicative of a different function of the members of this subclass; however, two enzymes from this sub-class have been characterized and shown to have CA activity [36, 69].

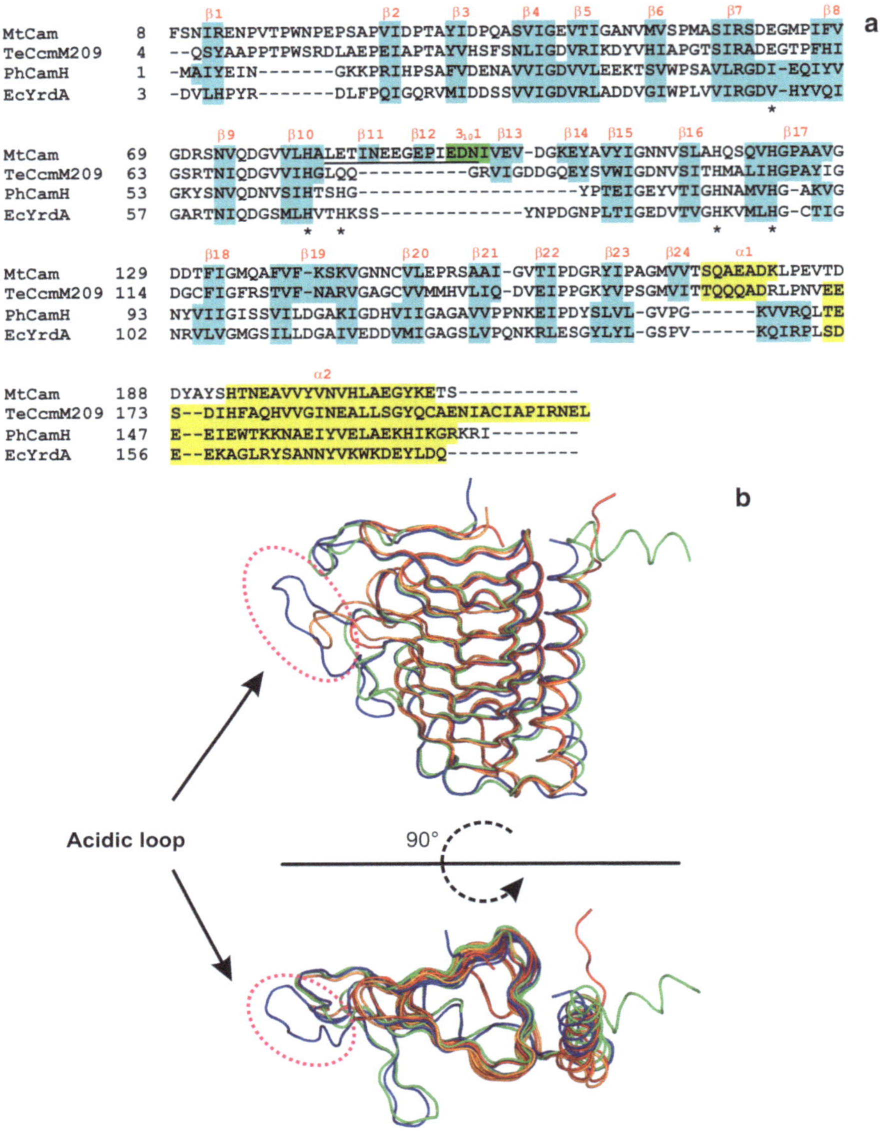

Fig. 19.6 (**a**) Structure-based sequence alignment of γ-CAs with known 3D structure. MtCam secondary structure elements are indicated and named according to PROMOTIF analysis [48]. α-helices, 3_{10}-helices and β-strands for MtCam (PDB code 1QRG), TeCcmM209 (PDB code 3KWC), PhCamH (PDB code 1V3W) and EcYrdA (PDB code 3TIO) are highlighted in *yellow*, *cyan* and *green*, respectively. Catalytic histidines, Glu84 and Glu62 of MtCam are indicated with *asterisks*, while the "acidic loop" (Leu83-Asn96) is underlined. (**b**) Ribbon diagram showing the superposition of MtCam (*blue*), TeCcmM (*green*), PhCamH (*red*) and EcYrdA (*orange*) monomers

While structural information is available for several γ-CA proteins, there are only limited data in the literature on their catalytic activity or thermostability. Indeed, a complete biochemical characterization has been reported only for MtCam [33, 69, 70], while no data at all were reported on PhCamH activity. EcYrDA has been described as a not active enzyme [68], while TeCccM209 presents CA catalytic activity [36]. On the basis of these very few and controversial data it has not been possible to perform a detailed comparative study on the structural determinants of thermostability for this enzyme class, as that above reported for the α-class. Moreover, another great lack in this context is represented by the absence of data on the catalytic activity of PhCamH. Indeed, since this enzyme is isolated from an archaeon that grows optimally at the temperature of 98 °C, it is expected to present a very high thermostability. Thus, the capability to efficiently catalyze the CO_2 hydration reaction could make this enzyme a good candidate to be used in CO_2 capture process.

3 Thermostable Carbonic Anhydrases Obtained by Protein Engineering

A large number of experimental and theoretical studies have been reported aimed at understanding factors responsible for protein thermal stability, leading to enormous progresses in such field [50, 51]. However, even if the packing of the hydrophobic core, disulfide and salt bridges, hydrogen bonding, and overall rigidity have been shown to have a role in stabilization, it is becoming always more evident that a generalizable mechanism of thermostabilization does not exist and that each protein family adopts a different strategy to cope with high temperature. As a consequence, the production of thermostabilized enzyme still constitutes an important challenge which can be realized using a variety of different approaches [24, 71–73]. Among these, rational design approaches based on the optimization of existing polar, ionic, hydrophobic and hydrogen bond interactions, revealed by an X-ray crystallographic analysis, are among the most utilized.

Such kind of approach has also been recently utilized by McKenna's group to produce thermostabilized mutants of hCA II [24]. hCA II is the best known and characterized member of the CA family and seems to be an ideal candidate to be used in biotechnological processes. Indeed, it is a very soluble protein which can be easily over-expressed and purified in *Escherichia coli* and it is one of the faster and more efficient members of the family with a k_{cat} of $1.4 * 10^6\ s^{-1}$ and a K_{cat}/K_M of $1.5 \times 10^8\ M^{-1}\ s^{-1}$ [24, 74]. Unfortunately, its use in CO_2 capture processes is strongly limited by the low thermal stability; indeed, hCA II becomes completely inactive after incubation at temperature higher than 70 °C [32]. For this reason, hCA II variants with enhanced thermostability that still preserve high solubility and catalytic activity are of great interest to be used in the harsh conditions of carbon dioxide sequestration processes [24].

The McKenna group recently reported data on a group five hCA II mutants. These mutants were designed starting from the results reported in a previous patent [75], and coupling mutations of some surface leucine residues (Leu100, Leu224 and Leu240) to enhance thermostability, with strategic mutations of some active site residues (Tyr7, Asn62 and Asn67), to create variants that could have at the same time improved thermal stability and kinetics. In particular, the first mutant, named TS1, contained simultaneously the three mutation Leu100His, Leu224Ser and Leu240Pro. Then this triple mutant was used to generate four other mutants where to the three Leu substitutions the following mutations were added: Tyr7Phe (TS2), Tyr7Phe + Asn62Leu (TS3), Tyr7Phe + Asn67/Gln (TS4) and Tyr7Phe + Asn62/Leu + Asn67/Gln (TS5). The best variant obtained by such study, namely TS4, showed an enhancement in thermal stability by about 6 °C and in rate constants of proton transfer by about 6-fold with respect to the wild type enzyme [24]. This study is of a particular interest since it showed the potentiality to combine suitable replacement of surface residues with mutations of active site residues to obtain significantly improved kinetic and stability features [24].

4 Conclusions

In this paper existing data on thermostable CAs both isolated by thermophilic organisms and obtained by protein engineering studies have been summarized and described in a comparative manner. Studies here reported show that thermophilic organisms are a good source of thermostable CAs belonging to the three classes α, β and γ. Moreover, such studies have also highlighted that thermostable CAs belonging to the α-class, such as SspCA and SazCA, seem to be at the moment, both for thermostability features and kinetic properties, the most interesting candidates to be used in the harsh condition of the CO_2 capture processes. Interesting results have also been reported for the rational protein engineering of the well know enzyme hCA II, showing that with suitable mutations of surface and active site residues improved variants both for kinetics and thermostability can be achieved. However, it should be also noted that such variants present features still far from those required in biotechnological processes and thus for the moment enzymes isolated form thermophiles still constitute the primary choice in such processes.

References

1. Savile CK, Lalonde JJ (2011) Biotechnology for the acceleration of carbon dioxide capture and sequestration. Curr Opin Biotechnol 22:818–823
2. Sahoo PC, Jang YN, Lee SW (2012) Immobilization of carbonic anhydrase and an artificial Zn(II) complex on a magnetic support for biomimetic carbon dioxide sequestration. J Mol Catal B Enzymatic 82:37–45

3. Puxty G, Rowland R, Allport A, Yang Q, Bown M, Burns R, Maeder M, Attalla M (2009) Carbon dioxide postcombustion capture: a novel screening study of the carbon dioxide absorption performance of 76 amines. Environ Sci Technol 43:6427–6433
4. Alterio V, Di Fiore A, D'Ambrosio K, Supuran CT, De Simone G (2012) Multiple binding modes of inhibitors to carbonic anhydrases: how to design specific drugs targeting 15 different isoforms? Chem Rev 112:4421–4468
5. Satoh D, Hiraoka Y, Colman B, Matsuda Y (2001) Physiological and molecular biological characterization of intracellular carbonic anhydrase from the marine diatom *Phaeodactylum tricornutum*. Plant Physiol 126:1459–1470
6. Nishimori I, Onishi S, Takeuchi H, Supuran CT (2008) The α and β classes carbonic anhydrases from *Helicobacter pylori* as novel drug targets. Curr Pharm Des 14:622–630
7. Schlicker C, Hall RA, Vullo D, Middelhaufe S, Gertz M, Supuran CT, Mühlschlegel FA, Steegborn C (2009) Structure and inhibition of the CO_2-sensing carbonic anhydrase Can2 from the pathogenic fungus *Cryptococcus neoformans*. J Mol Biol 385:1207–1220
8. Isik S, Kockar F, Aydin M, Arslan O, Guler OO, Innocenti A, Scozzafava A, Supuran CT (2009) Carbonic anhydrase inhibitors. Inhibition of the β-class enzyme from the yeast *Saccharomyces cerevisiae* with sulfonamides and sulfamates. Bioorg Med Chem 17:1158–1163
9. Carta F, Maresca A, Covarrubias AS, Mowbray SL, Jones TA, Supuran CT (2009) Carbonic anhydrase inhibitors. Characterization and inhibition studies of the most active β-carbonic anhydrase from *Mycobacterium tuberculosis*, Rv3588c. Bioorg Med Chem Lett 19:6649–6654
10. Ferry JF (2010) The gamma-carbonic anhydrases. Biochim Biophys Acta 1804:374–381
11. Rowlett RS (2010) Structure and catalytic mechanism of the β-carbonic anhydrases. Biochim Biophys Acta 1804:362–373
12. Smith KS, Jakubzick C, Whittam TS, Ferry JG (1999) Carbonic anhydrase is an ancient enzyme widespread in prokaryotes. Proc Natl Acad Sci U S A 96:15184–15189
13. Elleuche S, Pöggeler S (2010) Carbonic anhydrases in fungi. Microbiology 156:23–29
14. Lane TW, Morel FMM (2000) A biological function for cadmium in marine diatoms. Proc Natl Acad Sci U S A 97:4627–4631
15. Xu Y, Feng L, Jeffrey PD, Shi Y, Morel FM (2008) Structure and metal exchange in the cadmium carbonic anhydrase of marine diatoms. Nature 452:56–61
16. Cox EH, McLendon GL, Morel FM, Lane TW, Prince RC, Pickering IJ, George GN (2000) The active site structure of *Thalassiosira weissflogii* carbonic anhydrase 1. Biochemistry 39:12128–12130
17. Park H, Song B, Morel FM (2007) Diversity of the cadmium-containing carbonic anhydrase in marine diatoms and natural waters. Environ Microbiol 9:403–413
18. Lane TW, Saito MA, George GN, Pickering IJ, Prince RC, Morel FM (2005) Biochemistry: a cadmium enzyme from a marine diatom. Nature 435:42
19. McGinn PJ, Morel FMM (2008) Expression and regulation of carbonic anhydraseses in the marine diatom *Thalassiosira pseudonana* and in natural phytoplankton assemblages from Grat Bay, New Jersey. Physiol Plantarum 133:78–91
20. Lane TW, Morel FMM (2000) Regulation of carbonic anhydrase expression by Zinc, Cobalt, and carbon dioxide in the marine diatom *Thalassiosira weissflogii*. Plant Physiol 123:345–352
21. Roberts SB, Lane TW, Morel FMM (1997) Carbonic anhydrase in the marine diatom *Thalassiosira weissflogii* (Bacillariophyceae). J Phycol 33:845–850
22. Yee D, Morel FMM (1996) In vivo substitution of zinc by cobalt in carbonic anhydrase of a marine diatom. Limnol Oceanogr 41:573–577
23. Alterio V, Langella E, Viparelli F, Vullo D, Ascione G, Dathan NA, Morel FM, Supuran CT, De Simone G, Monti SM (2012) Structural and inhibition insights into carbonic anhydrase CDCA1 from the marine diatom *Thalassiosira weissflogii*. Biochimie 94:1232–1241
24. Fisher Z, Boone CD, Biswas SM, Venkatakrishnan B, Aggarwal M, Tu C, Agbandje-McKenna M, Silverman D, McKenna R (2012) Kinetic and structural characterization of thermostabilized mutants of human carbonic anhydrase II. Protein Eng Des Sel 25:347–355

25. Mignardi S, De Vito C, Ferrini V, Martin RF (2011) The efficiency of CO_2 sequestration via carbonate mineralization with simulated wastewaters of high salinity. J Hazard Mater 191:49–55
26. Unsworth LD, van der Oost J, Koutsopoulos S (2007) Hyperthermophilic enzymes–stability, activity and implementation strategies for high temperature applications. FEBS J 274:4044–4056
27. Adams MWW (1993) Enzymes and proteins from organisms that grow near and above 100 degrees C. Ann Rev Microbiol 47:627–658
28. Adams MWW, Kelly RM (1995) Enzymes from microorganisms in extreme environments. Chem Eng News 73:32–42
29. Adams MWW, Perler FB, Kelly RM (1995) Extremozymes: expanding the limits of biocatalysis. Biotechnol (NY) 13:662–668
30. Gomes J, Steiner W (2004) The biocatalytic potential of extremophiles and extremozymes. Food Technol Biotechnol 42:223–235
31. Vieille C, Zeikus GJ (2001) Hyperthermophilic enzymes: sources, uses, and molecular mechanisms for thermostability. Microbiol Mol Biol Rev 65:1–43
32. Di Fiore A, Capasso C, De Luca V, Monti SM, Carginale V, Supuran CT, Scozzafava A, Pedone C, Rossi M, De Simone G (2013) X-Ray structure of the first "extremo-α-carbonic anhydrase", a dimeric enzyme from the thermophilic bacterium, *sulfurihydrogenibium yellostonense YO3AOP1*. Acta Crystallogr D (In press)
33. Alber BE, Ferry JG (1996) Characterization of heterologously produced carbonic anhydrase from *Methanosarcina thermophila*. J Bacteriol 178:3270–3274
34. Smith KS, Ferry JG (1999) A Plant-Type (β-Class) Carbonic anhydrase in the thermophilic methanoarchaeon *Methanobacterium thermoautotrophicum*. J Bacteriol 181:6247–6253
35. Jeyakanthan J, Rangarajan S, Mridula P, Kanaujia SP, Shiro Y, Kuramitsu S, Yokoyama S, Sekar K (2008) Observation of a calcium-binding site in the gamma-class carbonic anhydrase from *Pyrococcus horikoshii*. Acta Crystallogr D 64:1012–1019
36. Peña KL, Castel SE, de Araujo C, Espie GS, Kimber MS (2010) Structural basis of the oxidative activation of the carboxysomal gamma-carbonic anhydrase, CcmM. Proc Natl Acad Sci U S A 107:2455–2460
37. Capasso C, De Luca V, Carginale V, Cannio R, Rossi M (2012) Biochemical properties of a novel and highly thermostable bacterial α-carbonic anhydrase from *Sulfurihydrogenibium yellowstonense YO3AOP1*. J Enzyme Inhib Med Chem 27:892–897
38. De Luca V, Vullo D, Scozzafava A, Carginale V, Rossi M, Supuran CT, Capasso C (2012) Anion inhibition studies of an α-carbonic anhydrase from the thermophilic bacterium *Sulfurihydrogenibium yellowstonense YO3AOP1*. Bioorg Med Chem Lett 22:5630–5634
39. Vullo D, De Luca V, Scozzafava A, Carginale V, Rossi M, Supuran CT, Capasso C (2012) Anion inhibition studies of the fastest carbonic anhydrase (CA) known, the extremo-CA from the bacterium *Sulfurihydrogenibium azorense*. Bioorg Med Chem Lett 22:7142–7145
40. Vullo D, De Luca V, Scozzafava A, Carginale V, Rossi M, Supuran CT, Capasso C (2012) The first activation study of a bacterial carbonic anhydrase (CA). The thermostable α-CA from *Sulfurihydrogenibium yellowstonense YO3AOP1* is highly activated by amino acids and amines. Bioorg Med Chem Lett 22:6324–6327
41. Vullo D, De Luca V, Scozzafava A, Carginale V, Rossi M, Supuran CT, Capasso C (2013) The alpha-carbonic anhydrase from the thermophilic bacterium *Sulfurihydrogenibium yellowstonense YO3AOP1* is highly susceptible to inhibition by sulfonamides. Bioorg Med Chem 21:15341538
42. Borchert M, Saunders P (2008) Heat stable carbonic anhydrases and their use. WO2008095057
43. Borchert M, Saunders P (2010) Heat stable carbonic anhydrases and their use. WO2010151787
44. Borchert MS (2012) Heat stable persephonella carbonic anhydrases and their use. WO2012025577
45. Akdemir A, Vullo D, De Luca V, Scozzafava A, Carginale V, Rossi M, Supuran CT, Capasso C (2013) The extremo-α-carbonic anhydrase (CA) from *Sulfurihydrogenibium azorense*, the fastest CA known, is highly activated by amino acids and amines. Bioorg Med Chem Lett 23:1087–1090

46. De Luca V, Vullo D, Scozzafava A, Carginale V, Rossi M, Supuran CT, Capasso C (2013) An α-carbonic anhydrase from the thermophilic bacterium *Sulphurihydrogenibium azorense* is the fastest enzyme known for the CO_2 hydration reaction. Bioorg Med Chem 21:14651469
47. Nakagawa S, Shtaih Z, Banta A, Beveridge TJ, Sako Y, Reysenbach AL (2005) *Sulfurihydrogenibium yellowstonense* sp. nov., an extremely thermophilic, facultatively heterotrophic, sulfur-oxidizing bacterium from Yellowstone National Park, and emended descriptions of the genus *Sulfurihydrogenibium*, *Sulfurihydrogenibium subterraneum* and *Sulfurihydrogenibium azorense*. Int J Syst Evol Microbiol 55:2263–2268
48. Hutchinson EG, Thornton JM (1996) PROMOTIF–a program to identify and analyze structural motifs in proteins. Protein Sci 5:212–220
49. Huang S, Xue Y, Sauer-Eriksson E, Chirica L, Lindskog S, Jonsson BH (1998) Crystal structure of carbonic anhydrase from *Neisseria gonorrhoeae* and its complex with the inhibitor acetazolamide. J Mol Biol 283:301–310
50. Matsui I, Harata K (2007) Implication for buried polar contacts and ion pairs in hyperthermostable enzymes. FEBS J 274:4012–4022
51. Haney PJ, Stees M, Konisky J (1999) Analysis of Thermal Stabilizing Interactions in Mesophilic and Thermophilic Adenylate Kinases from the *Genus Methanococcus*. J Biol Chem 274:28453–28458
52. Neish AC (1939) Studies on chloroplasts: Their chemical composition and the distribution of certain metabolites between the chloroplasts and the remainder of the leaf. Biochem J 33:300–308
53. Moroney JV, Bartlett SG, Samuelsson G (2001) Carbonic anhydrases in plants and algae. Plant Cell Environ 24:141–153
54. Smith DR, Doucette-Stamm LA, Deloughery C, Lee H, Dubois J, Aldredge T, Bashirzadeh R, Blakely D, Cook R, Gilbert K, Harrison D, Hoang L, Keagle P, Lumm W, Pothier B, Qiu D, Spadafora R, Vicaire R, Wang Y, Wierzbowski J, Gibson R, Jiwani N, Caruso A, Bush D, Safer H, Patwell D, Prabhakar S, McDougall S, Shimer G, Goyal A, Pietrokovski S, Church GM, Daniels CJ, Mao J, Rice P, Nölling J, Reeve JN (1997) Complete genome sequence of *Methanobacterium thermoautotrophicum* ΔH: functional analysis and comparative genomics. J Bacteriol 179:7135–7155
55. Strop P, Smith KS, Iverson TM, Ferry JG, Rees DC (2001) Crystal structure of the "cab"-type β class carbonic anhydrase from the archaeon *Methanobacterium thermoautotrophicum*. J Biol Chem 276:10299–10305
56. Smith KS, Cosper NJ, Stalhandske C, Scott RA, Ferry JG (2000) Structural and kinetic characterization of an archaeal β-class carbonic anhydrase. J Bacteriol 182:6605–6613
57. Kimber MS, Pai EF (2000) The active site architecture of *Pisum sativum* beta-carbonic anhydrase is a mirror image of that of alpha-carbonic anhydrases. EMBO J 19:1407–1418
58. Mitsuhashi S, Mizushima T, Yamashita E, Yamamoto M, Kumasaka T, Moriyama H, Ueki T, Miyachi S, Tsukihara T (2000) X-ray structure of beta-carbonic anhydrase from the red alga, *Porphyridium purpureum*, reveals a novel catalytic site for CO_2 hydration. J Biol Chem 275:5521–5526
59. Guilloton MB, Korte JJ, Lamblin AF, Fuchs JA, Anderson PM (1992) Carbonic anhydrase in *Escherichia coli*. A product of the cyn operon. J Biol Chem 267:3731–3734
60. Teng YB, Jiang YL, He YX, He WW, Lian FM, Chen Y, Zhou CZ (2009) Structural insights into the substrate tunnel of *Saccharomyces cerevisiae* carbonic anhydrase Nce103. BMC Struct Biol 9:67
61. Alber BE, Ferry JG (1994) A carbonic anhydrase from the archaeon *Methanosarcina thermophila*. Proc Natl Acad Sci U S A 91:6909–6913
62. Tripp BC, Bell CB 3rd, Cruz F, Krebs C, Ferry JG (2004) A role for iron in an ancient carbonic anhydrase. J Biol Chem 279:6683–6687
63. Alber BE, Colangelo CM, Dong J, Stålhandske CM, Baird TT, Tu C, Fierke CA, Silverman DN, Scott RA, Ferry JG (1999) Kinetic and spectroscopic characterization of the gamma-carbonic anhydrase from the methanoarchaeon *Methanosarcina thermophila*. Biochemistry 38:13119–13128

64. MacAuley SR, Zimmerman SA, Apolinario EE, Evilia C, Hou YM, Ferry JG, Sowers KR (2009) The archetype γ-class carbonic anhydrase (Cam) contains iron when synthesized in vivo. Biochemistry 48:817–819
65. Kisker C, Schindelin H, Alber BE, Ferry JG, Rees DC (1996) A left-hand beta-helix revealed by the crystal structure of a carbonic anhydrase from the archaeon *Methanosarcina thermophila*. EMBO J 15:2323–2330
66. Iverson TM, Alber BE, Kisker C, Ferry JG, Rees DC (2000) A closer look at the active site of gamma-class carbonic anhydrases: high-resolution crystallographic studies of the carbonic anhydrase from *Methanosarcina thermophila*. Biochemistry 39:9222–9231
67. Zimmerman SA, Ferry JG (2006) Proposal for a hydrogen bond network in the active site of the prototypic gamma-class carbonic anhydrase. Biochemistry 45:5149–5157
68. Park HM, Park JH, Choi JW, Lee J, Kim BY, Jung CH, Kim JS (2012) Structures of the γ-class carbonic anhydrase homologue YrdA suggest a possible allosteric switch. Acta Crystallogr D 68:920–926
69. Zimmerman SA, Tomb JF, Ferry JG (2010) Characterization of CamH from *Methanosarcina thermophile*: founding member of a subclass of the γ class of carbonic anhydrases. J Bacteriol 192:1353–1360
70. Zimmerman S, Domsic JF, Tu C, Robbins AH, McKenna R, Silverman DN, Ferry JG (2013) Role of Trp19 and Tyr200 in catalysis by the γ-class carbonic anhydrase from *Methanosarcina thermophila*. Arch Biochem Biophys 529:11–17
71. Jochens H, Aerts D, Bornscheuer UT (2010) Thermostabilization of an esterase by alignment-guided focused directed evolution. Protein Eng Des Sel 23:903–909
72. Reetz MT, Carballeira JD, Vogel A (2006) Iterative saturation mutagenesis on the basis of B factors as a strategy for increasing thermostability. Angew Chem 118:7909–7915
73. Mrabet NT, Van den Broeck A, Van den brande I, Stanssens P, Laroche Y, Lambeir AM, Matthijssens G, Jenkins J, Chiadmi M, van Tilbeurgh H, Rey F, Janin J, Quax WJ, Lasters I, Maeyer M, Wodak SJ (1992) Arginine residues as stabilizing elements in proteins. Biochemistry 31:2239–2253
74. Supuran CT (2008) Carbonic anhydrases: novel therapeutic applications for inhibitors and activators. Nat Rev Drug Discov 7:168–181
75. Daigle R, Desrochers M (2009) Carbonic anhydrase having increased stability under high temperature conditions. US 7,521,217, B2

Chapter 20
Carbonic Anhydrases in Industrial Applications

Javier M. González and S. Zoë Fisher

Abstract Carbonic anhydrases (CAs) catalyze a fundamental reaction: the reversible hydration and dehydration of carbon dioxide (CO_2) and bicarbonate (HCO_3^-), respectively. Current methods for CO_2 capture and sequestration are harsh, expensive, and require prohibitively large energy inputs, effectively negating the purpose of removing CO_2 from the atmosphere. Due to CA's activity on CO_2 there is increasing interest in using CAs for industrial applications such as carbon sequestration and biofuel production. A lot of work in the last decade has focused on immobilizing CA onto various supports for incorporation into CO_2 scrubbing applications or devices. Although the proof of principle has been validated, current CAs being tested do not withstand the harsh industrial conditions. The advent of large-scale genome sequencing projects has resulted in several emerging efforts seeking out novel CAs from a variety of microorganisms, including bacteria, micro-, and macro-algae. CAs are also being investigated for their use in medical applications, such drug delivery systems and artificial lungs. This review also looks at possible downstream uses of captured and sequestered CO_2, from using it to enhance oil recovery to incorporating it into useful and financially viable products.

Keywords Carbon sequestration • Green house gas • Biomimetic • Biofuel • Thermophiles • Medical applications

Susan C. Frost and Robert McKenna (eds.). Carbonic Anhydrase: Mechanism, Regulation, Links to Disease, and Industrial Applications

J.M. González • S.Z. Fisher (✉)
Bioscience Division, Los Alamos National Laboratory, Los Alamos, NM, USA
e-mail: javierg@lanl.gov; zfisher@lanl.gov

S.C. Frost and R. McKenna (eds.), *Carbonic Anhydrase: Mechanism, Regulation, Links to Disease, and Industrial Applications*, Subcellular Biochemistry 75,
DOI 10.1007/978-94-007-7359-2_20, © Springer Science+Business Media Dordrecht 2014

1 Introduction

CAs rank among the best known enzyme catalysts, able to convert CO_2 to bicarbonate with a maximal turnover rate (k_{cat}) approaching 10^6 s^{-1}, as in the case of human CA II [1] (Table 20.1). Due to the highly efficient CO_2 hydration as catalyzed by CAs, there is considerable interest in using CAs for biofuel and carbon sequestration applications, in both research and industry.

CA is an attractive biomimetic CO_2 sequestering agent as it can be produced fairly inexpensively, has very good kinetics (i.e. less enzyme is needed), is re-usable, and as a biological catalyst it can work at ambient temperatures and under mild experimental conditions [4]. Unfortunately, most of the industrial processes that seek to employ CA in CO_2 capture and sequestration are considered to be quite harsh by biological standards; i.e. they occur under elevated temperatures and/or extreme pH [5–7]. The pressing need for the development of industrial CAs is manifested through the number of patents filed and biotech companies that have emerged in this field. Several of these, including Codexis, Novozymes, CO2 Solution Technologies, Carbozyme, etc., are looking at ways to either improve CA for various applications or by using the enzyme in engineering platforms to extract CO_2 from the air.

1.1 CO_2 Production, Capture, and Sequestration

Carbon dioxide (CO_2) is one of the most abundant greenhouse gases (GHGs), along with water vapor (H_2O), nitrous oxide (N_2O), methane (CH_4), and ozone (O_3). In addition to these naturally occurring GHGs, there are also a number of man-made gases, such as hydrofluorocarbons (HFCs), and perfluorocarbons (PFCs) [8]. Anthropogenic atmospheric CO_2 levels are increasing rapidly and are considered one of the largest contributors to man-made climate change [9]. The net rise in atmospheric CO_2 can be partly attributed to increased emissions due to population rise and increased economic growth rates. Aggravating this situation is the observation that the airborne fraction of CO_2 is also increasing, indicating a

Table 20.1 Summary of known kinetic constants for CO_2 hydration activity for selected CAs. *BCA* bovine CA, *HCA I & II* human CA I & II, *Hp-alpha CA Helicobacter pylori* CA, *SspCA* CA from *Sulfurihydrogenibium Yellowstonense*

CA type	k_{cat}	k_{cat}/K_M	Reference
BCA	8.3×10^5	9.2×10^7	[2]
HCA I	2.0×10^5	5.0×10^7	[3]
HCA II	1.4×10^6	1.5×10^8	[1]
Hp-CA	2.5×10^5	1.5×10^7	[3]
SspCA	9.4×10^5	1.1×10^8	[3]

decrease in ability of natural sinks (land and oceans) to absorb this excess CO_2. The decline in natural CO_2 sinks is thought to be due to alterations in land use for agricultural activities as well as deforestation [9]. The current concentration of CO_2 in the earth's atmosphere is about 390 parts per million (ppm), a significant increase from $\sim$315 ppm since the 1960s [10, 11]. Investigation of the National Oceanic and Atmospheric Administration (NOAA) records on atmospheric CO_2 content shows that levels are increasing at about 2 ppm per year for the last decade. The Intergovernmental Panel of Climate Change (IPCC) calculates that we will see a rise of between 1.2 °C and 6.4 °C by the end of the twenty-first century. To avoid this we should reduce our output of greenhouse gases, including CO_2, by 50–80 % by 2050 to avoid catastrophic climate change [12]. The IPCC also calculates that a seemingly modest increase of only 2 °C is sufficient to cause serious problems.

Due to this larger-than-expected increase in CO_2, there are very large collaborative national and international efforts underway to develop methods to remove atmospheric CO_2 in a safe and cost-effective way (http://www.fossil.energy.gov). The consequences of climate change are under intense study and its dramatic effects on wildlife, global water supply, weather patterns, and ecosystems, have just recently been realized and observed [13, 14]. In order to make these efforts financially attractive to industry, additional research is underway to improve downstream conversion of captured and sequestered CO_2 to products with added value.

Extracting CO_2 from the air can be very expensive and requires harsh chemical processes and elevated temperatures [15]. Typical industrial fumes contain anywhere from 10 % to 20 % CO_2. In post-combustion CO_2 capture, the gas is separated from the waste stream in a variety of ways. These include amine scrubbing, mineral carbonation, pressure storage, absorption of CO_2 onto solids, or taking it up into liquids. It is technically very challenging, and costly, to preferentially capture CO_2 out of mixed waste gas that also may contain nitrogen, sulfur, and other compounds [16]. A very attractive approach to enhanced CO_2 capture and sequestration is to use biomimetic or biological/enzymatic methods that are by their nature environmentally benign, renewable, selective, and relatively inexpensive [4]. The National Energy Technology Lab of the Department of Energy (NETL DOE) has determined that if downstream storage or usage of immobilized CO_2 (Sect. 1.5) is to be feasible, then catalysts have to be developed with the ability to dramatically increase sequestration rates while performing adequately under reduced thermal and pressure conditions (http://www.netl.doe.gov). Earlier studies have shown that CA works better to facilitate CO_2 transport across a membrane as bicarbonate than amine or alkali-based chemical methods that are commonly proposed to make bicarbonate [17].

For most applications discussed here, CO_2 has to undergo multiple steps (1–5 below) and transformations to get it from the atmosphere to an inert product, in this example $CaCO_3$ [4, 18].

CO_2 has limited solubility in water, and the uncatalyzed hydration to bicarbonate is very slow (0.037 s^{-1}) [19, 20]. Calcium is shown in this example, but it should be noted that $CO_3{}^{2-}$ can readily react with Li^+, Mg^{2+}, and Na^+ to form ion pairs

[20]. This is an area where carbonic anhydrases (CAs) can make a large impact due to their ability to work very well under ambient conditions.

$$CO_2(gas) \Leftrightarrow CO_2(aqueous) \tag{20.1}$$

$$CO_2(aqueous) + H_2O \Leftrightarrow H_2CO_3 \tag{20.2}$$

$$H_2CO_3 \Leftrightarrow HCO_3{}^- + H^+ \tag{20.3}$$

$$HCO_3{}^- \Leftrightarrow CO_3{}^{2-} + H^+ \tag{20.4}$$

$$Ca^{2+} + CO_3{}^{2-} \Leftrightarrow CaCO_3 \tag{20.5}$$

The focus of this paper is to report on these efforts that include engineering improvements designed to separate the enzyme from harsh conditions, bio-engineered CA for improved stability and/or kinetics, immobilization of the enzyme for stability and reusability, as well as looking for new CAs from many different organisms (Figs. 20.1 and 20.2), and also a brief look at some of the potential medical applications of CA in artificial lung technology. A major bottleneck is what to do with the captured CO_2, and a summary is provided with the proposed downstream processes and products that will convert CO_2 into useful commodity products.

2 Mammalian CA for Biomimetic Carbon Sequestration

While early work in this field showed the proof of concept of using both human and bovine CAs in industrial settings, there are now major efforts underway to seek improved CA isoforms for this purpose [4, 24]. Figure 20.1 shows the structure and surface representation of selected α- and γ-CA isoforms that have been used in these industrial settings. Table 20.1 gives a comparison of the kinetic constants for some of these CA isoforms. The reason why these mammalian enzymes are so attractive is that they can be easily overexpressed in bacteria or commercially purchased. Also, as Table 20.1 shows, they represent the fastest kinetics for these applications. Under the right conditions, temperature, and ionic strength, HCA II can catalyze the hydration of CO_2 at about 10^6 per second [1]. In one case, a team obtained large volumes of bovine blood from a slaughterhouse and were able to prepare fairly pure bovine CA (BCA) with just simple ammonium sulfate precipitation [25]. However, the isoforms that have really desirable kinetic and solubility properties are not typically stable enough to withstand high temperatures and other industrial conditions. HCA II is irreversibly denatured at around 58 °C and is also susceptible to inhibition by small anions, such as sulfate, cyanate, and azide [7, 24].

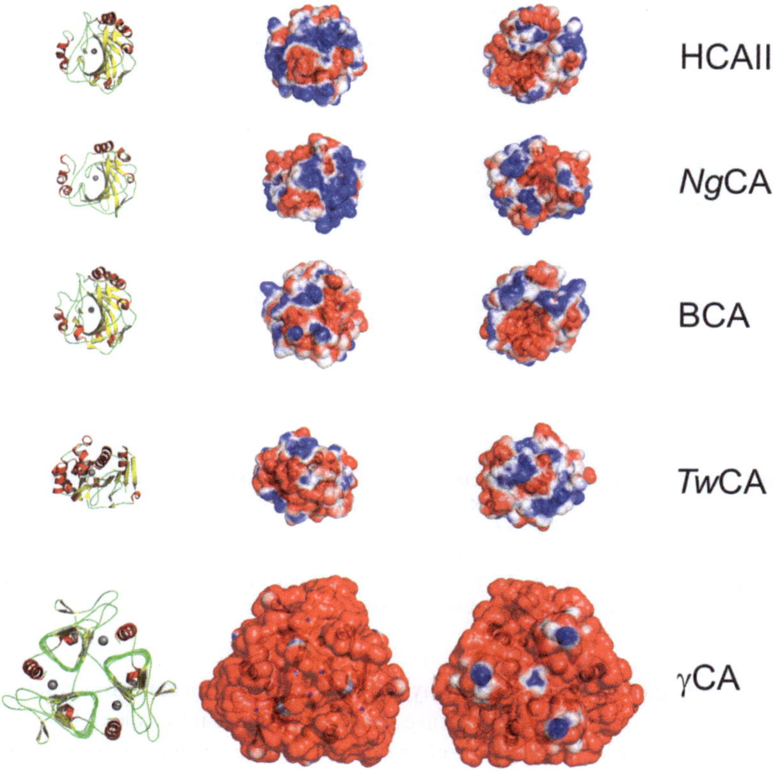

Fig. 20.1 Three-dimensional fold and solvent-accessible surface charge distribution of selected CAs that have been used for industrial purposes. Human CA II (HCA II, PDB 3KS3), *Neisseria gonorrhoeae* (*Ng*CA, PDB 1KOQ), bovine CA (BCA, PDB 1V9E), *Thalassiosira weissflogii* (*Tw*CA, PDB 3BOE), and *Methanosarcina thermophila* (γ-CA, PDB 1QRE). Images were produced with PYMOL 4.2 [21], and charge distributions were calculated with APBS 1.2 [22]

3 CA and Algal Biofuel Production

There are many reports on the effects of CA on carbon flux and fixation reactions in cyanobacteria and algae [26]. A 1982 US patent [27] describes how to boost algal growth (*Dunaliela sp.*) by simply lysing a portion of a algal culture, liberating the endogenous cytoplasmic CA in the process, and then adding it back to the rest of culture to give a significant boost in cell growth [27]. Two research teams at Los Alamos National Laboratory are investigating the effect of adding extracellular CA to cyanobacteria and algal culture media. In certain cyanobacterial species there is a significant increase in growth rates. Another team has expressed HCA II in the chloroplast of algae and has measured improved carbon flux in these cells, showing the proof of concept that growth enhancements are possible (Richard Sayre, personal communication).

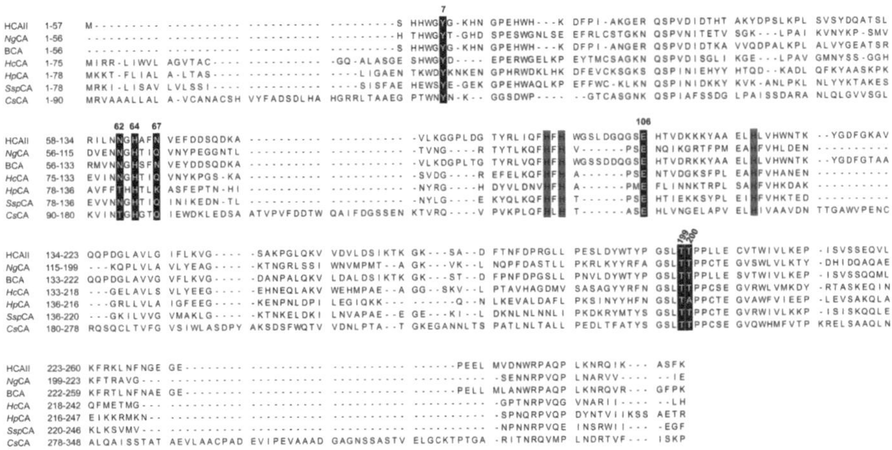

Fig. 20.2 Sequence alignment of selected α-class CAs: human (HCA II, UniProt P00918), *Neisseria gonorrhoeae* (*Ng*CA, UniProt Q50940), bovine (BCA, UniProt P00921), *Hahella chejuensis* strain KCTC 2396 (*Hc*CA, UniProt Q2SNS4), *Helicobacter pylori* strain J99 (*Hp*CA, UniProt Q9ZK30), *Sulfurihydrogenibium* sp. strain YO3AOP1 (*Ssp*CA, UniProt B2V8E3), and *Chlorella sorokiniana* (*Cs*CA, UniProt O64462). *Gray*-shaded residues indicate Zn ligands. *Black*-shaded residues indicate important active site residues. Alignment was calculated with PROBCONS [23]

CAs role in carbon fixation pathways in photosynthetic organisms (plants, algae, and cyanobacteria) is also under intense study as a potential avenue to boost biofuel production. Of course, plants and other photosynthetic organisms have evolved to efficiently extract the most CO_2 out of a fairly dilute source: regular air. The uptake of CO_2 into cells as bicarbonate is the rate-limiting step in biomass production in many organisms. These organisms have evolved carbon-concentrating mechanisms that include CA, to overcome these issues and deliver inorganic carbon directly to Rubisco, a notoriously inefficient enzyme [28, 29]. In cyanobacteria, the carbon concentrating mechanism has evolved and exhibits numerous parts, including active bicarbonate transporters, CA, and structural shell proteins that make up the protein microcompartment called the carboxysome. The carboxysome consists mostly of Rubisco and CA and is finely tuned to boost carbon fixation reactions by Rubisco. CO_2 is produced locally by CA from bicarbonate that was actively imported into the cytosol. Rubisco incorporates CO_2 and produces 3-phosphoglycerate and is highly sensitive to O_2. The structure of the carboxysome and the packaging of the enzyme inside also serves as barrier to CO_2 leaking back out to the cytosol [28]. Rubisco is one of the slowest enzymes and can only incorporate between 3 and 10 molecules of CO_2 per second. Indeed, numerous research efforts are being conducted to improve this bottleneck in natural carbon fixation pathways for improved food crop cultivation and biomass production [30].

Microalgae, or cyanobacteria, are widely regarded as a very promising source for biofuel production. Different strains under the right conditions, which may include nutrient deprivation, can produce lipids and/or hydrocarbons (e.g. Biodiesel). This avenue is environmentally sustainable as the cells will use CO_2, need only sunlight and limited micronutrients to grow, and the byproducts can be used in food production, fertilizers, and other commodity products [31, 32]. Using endogenous algal CA indirectly, in either calcite deposition or for fixing CO_2 as biofuels, is also actively being pursued by many different international and national groups [33]. Recently, studies have shown that CO_2 capture and sequestration measured by calcite deposition, was boosted by the presence of algal species *Chlorella* and *Spirulina*. Calcite precipitation is an active function in microalgae with many proposed purposes, including its buffer capabilities against a pH rise, or to protect the action of the bicarbonate transporting pumps, which are very sensitive to carbonate ions. Researchers have looked at candidate species to simultaneously enhance lipid production while boosting their carbon intake through CO_2 or bicarbonate [34]. Fulke et al. looked at several different species in photobioreactors and concluded that *Chlorella* was the most successful at producing large amounts of lipids, up to 29 % of cell mass, on 3 % bubbled CO_2 gas. They were also able to identify production of calcite during cell growth [35]. Another group observed something similar in simulated raceway ponds and saw that enhanced capture was negatively impacted by the addition of acetazolamide, a commonly used CA inhibitor [7, 34]. This data strongly implicates the role of CA in bicarbonate production in these mini-raceway systems [34]. Using bubbled CO_2 directly in algal or cyanobacterial cultures is not the best option as most of the gas is lost due to low solubility. CA can control this by producing bicarbonate, which is not only soluble but it also the carbon source of choice for a variety of algal strains [36, 37]. Rost et al. studied the extra- and intra-cellular CA and its activity at different pH values in the red tide dinoflagellates. Certain strains preferred bicarbonate uptake at elevated pH at or above pH 9. One species increased its extracellular CA activity upon increasing pH to 9, but it seems that in these species there is a very delicate balance between pH and substrate availability, and their ability to adapt to these conditions [37]. Clearly there is an interesting natural phenomenon at work in these different organisms that can be maximized and exploited for simultaneous biofuel production, while using CO_2/bicarbonate and possibly making significant amounts of calcite. Of course, each organism has its drawbacks, such as slow growth rates, working in a narrow pH range, or needing debilitating nutrient deprivation to make lipids. A possible solution will be to find a synergistic combination of the right organism, paired with the right CA, which will be able to produce fuels while reducing carbon, in a cost effective way. With the increasing interest in algal biofuels, we can expect a lot more work in this area.

4 Various Techniques and Advantages of CA Immobilization

Several techniques and materials have been applied for the immobilization of CA and not all of these will be covered in this review. The benefit to immobilization of the enzyme is that it increases the stability of CA, allows repeated use, and eases enzyme recovery, driving costs down during the process. Some CAs, like

HCA II, are amenable to immobilization as they have numerous lysine residues on the surface, in addition to free hydroxyl groups, thus facilitating immobilization chemistry [38]. Besides the usual matrices used for the immobilization of enzymes, a group has developed a chimeric enzyme between CA and a cellulose binding domain protein (CBD). In this, case a bacterial α-CA from *Neisseria gonorrhoeae* was fused to the CBD from *Clostridium thermocellum* and showed satisfactory binding affinities to inexpensive cellulose without CA activity loss [39]. In addition to numerous reports on CA immobilization on solid supports, there are also reports of polymeric materials being used, such as poly(acrylic acid-co-acrylamide) hydrogels [40] and Sepharose 4B [41]. All these methods have certain limitations: limited diffusion of substrate/product, leaching and loss of enzyme, no gain in stability or performance, and cost of material. A summary of some of the techniques and materials used to immobilize CA is presented below, with varied success in terms of activity, reusability, and stability.

4.1 Chitosan and Alginate

Positive results were obtained from several groups using chitosan-based nanoparticles or hydrogels [42, 43]. Chitosan is composed of a linear polysaccharide composed of D-glucosamine and N-acetyl-D-glucosamine. It is prepared from the treatment of crustacean shells with sodium hydroxide. Alginate is prepared from seaweed and is also a polysaccharide, with D-mannuronate and D-guluronate. Both chitosan and alginate are biocompatible and are used in many biological applications [44, 45]. As such, they are a safe and environmentally benign choice, since chitosan-coated alginate beads are non-toxic and biodegradable. The feasibility of trapping CA in/on chitosan was first established by Simsek-Ege et al. in the early 2000s, and subsequent studies focused on issues with enzyme retention and length of cross-linking conditions [46]. Single-enzyme nanoparticles composed of CA and chitosan did not show better esterase activity than the free enzyme, however, they performed significantly better in $CaCO_3$ precipitation reactions (140 vs. 30 mg precipitate/mg of CA) [43]. Others have also looked at immobilizing CA from a variety of bacteria, such a *Pseudomonas fragi*, *Micrococcus lylae*, *Micrococcus luteus 2*, and *Bacillus pumilus*. These enzymes performed quite satisfactorily, retaining up to 50 % of their original activity after 30 days of storage [47]. Further studies with the *B. pumilus* CA attached to chitosan beads showed it to have better esterase activity kinetics than the free enzyme [48]. Another group has forgone the enzyme purification step and has immobilized live *P. fragi* cells to chitosan and was able to observe $CaCO_3$ precipitation (a measure of catalyzed conversion of CO_2 to bicarbonate) [49]. Similar studies with *B. pumilus* also showed improved esterase activity for immobilized whole cells [50]. The surface expression of *Helicobacter pylori* CA on *E. coli* was also successful and the merging of these advances may lead to the ability to immobilize intact bacteria with surface-expressed CA to surfaces or beads for biomineralization of CO_2 [51].

4.2 Metal-Based Nanoparticles

BCA was also covalently attached to octa(aminophenyl)silsesquioxane-functionalized Fe-O-SiO nanoparticles. Investigators were able reuse the nanoparticles after a simple separation using magnets. They also looked at CO_2 capture efficiency, by measuring calcite precipitation, and these performed better than free enzyme. The enzyme nanoparticles withstood 30 cycles showed, with complete reusability. Also, the immobilized enzyme retained over 80 % activity after 30 days of storage [52].

Nanobiocatalysts were synthesized by electrostatic immobilization of HCA II onto Ag or Au nanoparticles that were assembled over an SBA-15 core. The Ag particles performed very well, with ~ 87 % enzyme activity retained after 30 days of storage. The Au nanoparticles stored adequately at 25 °C and were re-used 20 times with full activity retention as determined by the esterase assay. Unfortunately, these have similar kinetics as the free enzyme [52, 53].

4.3 Mesoporous Silica

Several groups have looked at BCA immobilized onto some form of mesoporous silica SBA-15 spheres/beads. A micelle is formed during self-assembly and has hydrophobic block at its core, with an external hydrophilic layer. After assembly under acidic conditions there is a hydrolysis and subsequent condensation of the silica component that results in a silica network [54]. A group of researchers looked at different methods to attach the enzyme, including covalent attachment, enzyme adsorption, and cross-linked enzyme aggregation. Overall, the enzyme activity was the same as that of the free enzyme, but showed additional desirable characteristics such as stability, reusability, and storage endurance [55]. Wanjari et al. also looked at mesoporous aluminosilicate as a support for CA. This material is well suited to enzyme immobilization due to its large surface area and pore size. Interestingly, the K_M for the immobilized enzyme was higher compared to the free form, indicating decreased affinity of the enzyme for the substrate due to suboptimal substrate/product exchange. The immobilized enzyme was still over 50 % active after 25 days at 25 °C, while the free enzyme was only ~ 20 % active after the same treatment. Unfortunately, the enzyme appeared to either leach out or denature, and was only operating at 20 % carbonation capacity after 7 cycles [54].

4.4 Polyurethane Foam

In addition to enzyme adsorption and covalent attachment, it is also possible to trap enzymes in porous materials. The original irreversible enzyme trapping protocol in polyurethane (PU) foam was developed in the 1980s by Wood et al. [56]. For CO_2 sequestration applications, this work was further developed to include CA. This protocol involves mixing the HYPOL prepolymer with the enzyme solution. HYPOL is polyethylene glycol with isocyanate end groups, polymerization is initiated through nucleophilic attack by hydroxide of carbonyl groups, following a protonation/deprotonation event with the release of CO_2. This results in isocyanate being converted to an amine that reacts rapidly with a nearby isocyanate group, thereby cross-linking two adjacent prepolymer chains. The generated CO_2 escapes and in the process forms a porous, spongy PU foam. Through the enzyme's surface amine and hydroxyl groups it cross-links to the isocyanate group on the prepolymer. This process is quite fast and a high percentage of active enzyme is covalently obtained in the final PU [38]. In contrast to other materials, after seven cycles there was no detectable enzyme leaching or a reduction in CA activity. After 45 days of storage of the CA-PU foam at room temperatures, it retained 100 % of its activity, while the free enzyme was completely inactive after the same period at 4 °C [38].

4.5 *Thin Liquid Membranes*

Numerous groups have attempted the immobilization of CA between thin liquid membranes for CO2 extraction from flue gas. There are several patents in this arena (eg. US patents 20080003662, 6143556, and 7132090), with a concomitant low number of primary research articles. The setup comprises three zones, namely, a thin liquid layer is packed between two membranes typically consisting of some polypropylene derivative, which is strengthened to prevent curving of the pliable membrane [17]. The very thin aqueous layer sandwiched between the membranes contains CA. This arrangement is then designed to have the CO_2-rich air flow on one side, allowing diffusion of CO_2 into the aqueous layer, and dissolving to become the substrate of CA to form bicarbonate, a soluble ion. On the desorption side there is usually a sweep gas (inert), plain air, or vacuum with lower CO_2 concentrations than the feed side. This pushes the equilibrium to favor formation of CO_2 on the desorption side of the liquid membrane, with subsequent gas escape. In the correct configuration this setup can support adsorption and desorption of CO_2 and the only driving force is the partial CO_2 pressure. This directionality has been exploited and also shows very good selectivity, with negligible contamination by oxygen, sulfur, etc. [17, 57]. There are many technological and practical problems with the design, and drying of the membrane, pH of the CA phase, and costs, are among the main issues. Tests with an enzyme-based CO_2 scrubber performed reasonably well and there are reports of such a system tested continuously at an Alcoa plant for a month. This setup was able to capture ~ 80 % of the CO_2 being emitted [16].

5 Emerging Novel CAs from Thermophiles and Other Organisms

In the last decade, a number of novel CAs have been isolated from a variety of organisms, including halo- and thermo-tolerant microorganisms. This has been largely possible through advanced genome sequencing techniques. A few of these newly discovered CAs are from the β class, but the most that have been investigated for their CO_2-capture industrial applications, are from the α class (Fig. 20.2). Unlike the mammalian α-class CAs, some of these seem to have much better intrinsic temperature stability, and in some cases display increased salt tolerance. These attributes are highly suitable for industrial applications, where elevated temperatures and small molecule contaminants are commonplace.

5.1 Bacterial CAs

A novel α-class CA from the marine bacterium *Hahella chejuensis* was identified [58]. Genome sequencing revealed that, besides CA, this organism also expresses lytic enzymes that may be active against marine algal bloom organisms. Subsequently, a group cloned and codon-optimized the CA gene and overexpressed it in *E. coli*. The purified CA has a melting temperature of ∼ 60 °C and showed maximal activity at 50 °C. More importantly, the enzyme showed very high pH stability and activity at pH 10.0. In addition, this CA is highly salt tolerant and for the esterase activity assays it showed twice the activity at 2.0 M NaCl, compared to 0.67 M NaCl. This CA also showed calcite precipitation from Ca^{2+} ions [59].

A few recent reports described the purification and characterization of a new plant-like β-class CA from *Bacillus subtilis* VSG-4. Ramanan and coworkers reported the purification and characterization of this novel CA [60]. For the purified, free enzyme they showed that unlike most other β-class CAs, this one had esterase activity and they were able to use an inhibitor affinity column to purify it. The enzyme also seemed to work well between pH 7.0 and 11.0. Subsequently, this enzyme was entrapped in chitosan beads and compared to the free enzyme for pH and temperature performance [42]. The enzyme had average performance compared to mammalian isoforms, working optimally at pH 8.2 and 37 °C. The authors note that the immobilized enzyme withstands storage better than the free form. Interestingly, this form was very good at calcite precipitation, with 480 mg $CaCO_3$ per mg of protein [42].

As briefly described above, another group was able to express an α-class CA from *Helicobacter pylori* on the surface of *E. coli* using a surface-anchoring system derived from ice nucleation protein from *Pseudomonas syringae* [51] (Figs. 20.1 and 20.2). They showed that the shortest linker led to the best expression and activity levels of the surface CA. Also, the anchoring of the CA seemed to stabilize the enzyme compared to the free form. This combination of technologies, like the CA fused to a cellulose binding domain, opens the way for more innovative design of unique way to express, display, and immobilize CA for industrial applications [39, 51].

Recently, the first report using inexpensive recombinant CA from *Neisseria gonorrhoeae* to precipitate calcite was published [61]. The investigators were able to obtain high expression levels and demonstrated that the enzyme worked well. Also, these authors tested a crude cell lysate (without purifying the CA from *E. coli* expression lysate) and showed significant ability to form calcite.

Sharma et al. [49] cloned, expressed, and purified CAs from multiple bacterial species to test their commercial potential against the more commonly used BCA. They looked at CAs purified from *Pseudomonas fragi*, *Micrococcus lylae*, *Micrococcus luteus 2*, and BCA. At 35–45 °C and pH 8–9 the CA from *M. luteus 2* outperformed all the others and also precipitated the most calcium as calcite compared to BCA [62]. This shows that bacterial and recombinant CAs can be very useful for incorporation into CO_2 scrubbers.

An exciting CA isolated from a the chemolithotropic thermophilic bacterium *Sulfurihydrogeninium yellowstonense*, SspCA, has been recently reported [63]. SspCA is another bacterial α-class CA, but it is smaller than HCA II (26 and 29 kDa, respectively) (Fig. 20.2). *S. yellowstonense* was isolated from Yellowstone National Park in 2003, where this organism thrives in hot springs at high concentrations of hydrogen sulfide [3]. This CA is strikingly thermostable, retaining 100 % activity after 2 h incubation at 100 °C. SspCA has also been immobilized in polyurethane foam and, like the free enzyme, retained 100 % of its activity after 48 h incubation at 100 °C [63]. SspCA shares all of the α-CA characteristics, for instance, it shows esterase and hydratase activities, it binds to a CA inhibitor-affinity column for purification, and displays HCA II-like kinetics (Table 20.1) [3]. Inspection of a sequence alignment (Fig. 20.2), shows the SspCA retains all the important active-site residues for efficient catalysis, such as Tyr7, Asn62 and 67, Thr199 (HCA II numbering). A structural comparison of SspCA and HCA II will be very interesting to better understand what determines the large difference in thermal stability between these two highly related CAs.

5.2 CAs from Plants and Algae

A recent report described the cloning, expression, and characterization of a CA from saltwater algae *Dunaliella*. This organism is under intense study for algal biofuel production, since up to 37 % of its weight can be lipids, displaying high CO_2 uptake levels, making it a good candidate for CO_2 reduction while producing fuels. This CA was overexpressed in *E. coli* and showed a melting temperature of 47 °C, making it less stable than HCA II, but it had a very broad pH stability range between pH 7.6–10.0 [64]. Another study showed that it was possible to extract a very crude

CA fraction from freshwater algae *Chlorella vulgaris* and that it used calcium to make calcite from dissolved CO_2. However, this enzyme seemed to work best at 30–40 °C, and pH 7.5–8.5 [65].

6 Useful Products Made of Sequestered CO_2

Once CO_2 gas is captured it can be chemically converted to stable compounds or pressurized to make liquid CO_2. Liquid CO_2 can then easily be transported via pipelines or container trucks and pumped underground or into the ocean where it will form long-lived stationary pools. The storage capacity of the ocean is huge, an estimated thousand billion tons. However, there are many concerns about this method, including cost, stability, and unknown long-term biological impact. The other storage mechanism that is actively and successfully being implemented in different countries (Canada, Algeria) involves the geological long-term storage of captured CO_2. For underground storage, the CO_2 is injected into depleted oil and gas reservoirs, and inaccessible coal seams, and is even being used to enhance oil recovery [15]. Another option that geologists are actively pursuing is the production of magnesite or calcite ($MgCO_3$ or $CaCO_3$) with subsequent burial of the solid carbonates, a process called geosequestration. There are concerns on the effect of acid rain on these deposits, mainly because of the possible sudden release of CO_2. Certain geographic regions will be intrinsically better suited to geosequestration as the naturally occurring serpentinite (for magnesite) or wollastonite (for calcite) can be directly carbonated and stored underground [66, 67]. This may also be a more suitable option where rainfall is low, for example in the southwest of the United States [67].

There is strong government and academic interest in what to do with the captured or sequestered CO_2, besides burying it. There are many approaches that essentially convert captured CO_2 (whether in liquid or gas form) to useful commodity products. These products include materials like polycarbonates, acrylates, methane, stable carbonate storage polymers, stronger cement for building materials, or other solid materials with added value [68–70].

Carbonates, such as calcite ($CaCO_3$), can be derived from the simple reaction of $CaCl_2$ with bicarbonate. It has very low solubility and is stable and durable, as evinced by its ubiquitous use as a main constituent of shells in marine organisms [71, 72]. Calcite also occurs as the key component of limestone and marble and is very abundant in the earth's crust. Interestingly, there has been a report that nacrein – an integral protein component of the mollusk shell matrix – contains a proteinaceous component identified as CA and is thought to be directly involved with shell formation [73]. There are many industrial uses for calcite as a pigment for paint formulation, an acid-neutralizer (either as a stomach antacid or to neutralize acidic mining run-off), and to make cement and other building materials.

Sasol, based in South Africa, has implemented the largest scale gas to liquid fuel (Syngas) production plants that use Fischer-Tropsch chemistry. This process involves the production of synthetic liquid hydrocarbons from coal, biomass, and/or natural gas, and the reaction between H_2 and carbon monoxide (CO) mixtures to produce Syngas [74]. To enable the use of CO_2 for this application there has to be an efficient catalyst to reduce CO_2 to CO and this has not yet been possible at scale or at a reasonable cost [75]. Methane is also present in Syngas up to 15 %, and there are reports of the electrochemical reduction of CO_2 to methane in bicarbonate mixtures applying Cu electrodes, with an impressive 65 % efficiency [76].

Glycerol is emerging as an abundant and low cost by-product of biodiesel production from the transesterification of fatty acids. Glycerol is amenable to many chemical modifications such as selective oxidation, pyrolysis, gasification, and carbonation to produce glycerol carbonate [77, 78]. Glycerol carbonate has many potential uses, including as a substitute for ethylene and polypropylene carbonate, and in its polymeric form it could be used for the production of polycarbonates and polyurethanes. These chemicals are widely used as flexible hosing, foam insulation, and as plastics for food and beverage containers [79].

The US Department of Energy (DOE), through the National Energy Technology Labs (NETL) and academic partners, is funding a spectrum of research projects investigating closed-cycle catalysts for solar, chemical, and electrochemical methods to convert CO_2 to commodity products that are ideally energy and carbon neutral. A major interest is to carboxylate materials for use in road and building construction. One of these companies, CCS Materials, is developing an inorganic CO_2-binder for use in concrete production instead of Portland cement. The added feature of this material is that is does not require being kiln-fired, thus dramatically reducing the amount of energy input. Detailed descriptions of these projects can be found at www.netl.doe.gov under the Carbon Sequestration topic. An important driver for this research is that geologic storage of large amounts of CO_2 may not be feasible in many areas and this avenue of research represents a way to supplement the method of simply burying CO_2.

7 Medical Applications of CA

7.1 Artificial Lungs

Respiratory failure affects hundreds of thousands of patients in the US every year, and the average wait time for lung transplants are 1–1.5 years. Patient mortality is rather high during the waiting period, typically around 11 %. During this time many patients will also be kept alive with mechanical ventilators, but these have negative effects by causing baro- or volu-traumas by over pressurizing or distending lung tissue, causing massive damage in the long term [80, 81]. There is a medical need to overcome these issues and help patients that are waiting for lungs. An attractive option is an extracorporeal lung assistant that will facilitate the blood gas exchange. Current models are good for gas exchange but can only be used for patients in intensive care units and are not implantable because of their large size. Therefore, there is a need for smaller and more efficient artificial lungs, to be effective at comfortably assisting such patients [81]. The natural concentration gradient driving O_2 versus CO_2 exchange shows that CO_2 removal is much harder to achieve. Artificial lungs constructed of hollow fiber membranes (HFM) have been around for a while and a lot of work has centered on making these better for patient use. They offer a decent platform for CO_2 removal from blood, but early models suffered from suboptimal performance due to the requirement for a very large blood contacting interface (1 m^2), as well as issues with hemo- and bio-compatibility [82]. The HFM constructs also suffer from a lot of platelet aggregation, necessitating the development of thromboresistant surfaces. Several of the issues can be overcome by the use of CA in artificial lung constructs [80].

Early bioactive HFM models involved the immobilization of active BCA on conventional HFMs, thus enabling facilitated diffusion of bicarbonate. This setup naturally mimics the function of CA natively found in lung capillaries [80]. To prepare the HFM for CA attachment, the surface hydroxyls were activated by plasma deposition. The HFMs were then activated further by the addition of cyanogen bromide and subsequently reacted with CA to promote linkages between surface lysines and the membrane. CA-modified fibers contained close to 90 % of the theoretical maximum with no further improvement observed for increased enzyme loading. It was observed that the presence of CA did not impair gas permeability, showing that the intrinsic properties were not negatively affected. Tests with buffer were conducted, as using blood causes nonspecific binding of plasma protein, platelet aggregation, and digestion of CA by proteases [82]. Doping the buffer with free CA had an additive effect on CO_2 exchange. The results were remarkable, with a boost of 75 % for CO_2 exchange and zero enzyme leaching over time. This has dramatic implications for size reduction of artificial lung devices [80].

To overcome biocompatibility issues, such as platelet aggregation, several alternatives or modifications have been proposed or tested. Kaar et al. suggested using the commonly used heparin as a way to block thrombus formation. Such coatings would also require amine chemistry and is compatible with CA immobilization, allowing simultaneous adhesions of both heparin and CA from a mixed solution [80]. Another report focused on siloxane modification of commercial polypropylene Celgard HFM surface. Interestingly, attachment of the siloxane alone did nothing to ablate platelet binding and activation. In contrast, the CA-HFM system showed a significant reduction in platelet aggregation and activation. Later studies focused more on the phenomenon mechanics and quantitated it as a 95 % reduction in thrombus formation [83]. This is thought to occur through a general physical blocking of platelet binding rather than a mechanism involving CA catalytic activity [82]. This technology will certainly benefit from the upcoming improved CAs, like highly active HCA mutants and new thermostable CAs.

7.2 CO_2 Sequestration for Confined Spaces

As expected, elevated CO_2 levels have negative effects on humans, such as impaired judgment and, in extreme cases, death. CO_2 control is important in confined spaces where there is little buffering ability to absorb this gas. The NASA was initially involved in developing CA-based CO_2 capturing technology for use in confined spaces inhabited by humans, such as spacecraft or submarines. In this application a small amount of CA is dissolved in thin aqueous buffered films and compressed between porous polypropylene membranes [16, 84]. Compared to coal-fired power plant flue gas or industrial fumes, the concentration of CO_2 is very low, only 0.1 %. Under these conditions, CA is very good at selectively capturing the CO_2. Analysis of the produced (scrubbed) gas shows that the CA-containing setup selectively lets N_2 and O_2 through, with ratios of 1400:1 and 866:1, respectively [84, 85]. The relatively low concentration of CO_2 readily dissolves in the thin layer of enzyme containing buffer, once it diffuses across the membrane, it is removed on the back end by either a vacuum or a carrier gas. CA outperformed purely chemical methods using diethylamine solutions, with much higher selectivity, 442:1 and 270:1 for N_2 and O_2, respectively. In addition, the presence of CA decreased the resistance of CO_2 transport by $\sim$70 %. An added attractive feature to this setup is that it is very efficient at ambient pressure and temperature and worked with dry air, with no further energy inputs required [85]. While this technology is exciting, there are concerns with long-term operation and possible drying of the films. The need to keep the membranes wet, or at least humid, will add cost and operational difficulties to their practical use. This initial work led to an alternative setup involving hollow microporous fibers with better applications in the flue gas-scrubbing environment [17].

7.3 Miscellaneous

One interesting study worth mentioning here is the use of CA in CO_2-responsive cationic hydrogels for antidote delivery to treat analgesic overdose. This elegant delivery system will limit the harmful effects of analgesic overdose without losing therapeutic levels of drug and involves feedback-regulated antidote delivery [86]. While opioids like morphine have very potent analgesic effects, overdoses can cause respiratory hypoventilation, leading to increased CO_2 levels, decreased O_2 levels, and can lead to acidosis-induced death. Naltrexone and naloxone are morphine antidotes, and one study looked at devices where the antidote is packaged in a polymer, that is in turn put inside a cellulose dialysis tube containing regulated enzymes. When morphine reaches high levels, it diffuses into the device and activates the enzymes, causing a release of the antidote [87]. An alternative strategy, instead of relying on the morphine levels to active antidote release, is to design a responsive system that reacts to a toxicity biomarker, such as high CO_2 levels or changing pH [86]. Cationic hydrogels based on N,N-dimethylaminoethyl methacrylate (DMAEMA) have been used to make glucose sensitive insulin releasing systems [88]. Modified DMAEMA cationic hydrogels display an appropriate pKa of 7.5 that makes them good candidates to "sense" changes in blood pH. Including CA as a CO_2 sensor improved the function of these antidote-delivery systems and provided proof of concept that such elegant and finely targeted delivery systems can work, without negating the positive analgesic effects of the opioid [86]. A potentially complementary study showed the design of a switchable co-block polymer-based hydrogel that underwent a gel to sol transition upon exposure to CO_2, with no additives such as acid or base. This process was reproducible and they were able to show CO_2-induced release of an encapsulated protein [89]. There are many potential applications of CA in stimuli-triggered delivery of drugs, especially if sensing of CO_2, bicarbonate, or changes in pH are applicable.

Another potential medical use for CA has been reported as a component of artificial blood. There is always more demand than supply when it comes to donated blood for use in trauma injuries or major surgery. As such, there has been a number of blood substitutes developed and some of these are currently in phase III clinical trials. These mostly consist of 4–5 cross-linked stroma-free hemoglobin (polySFHb) molecules that have adequate oxygen transport capability. However, there is a push to develop agents that have more blood-like qualities and that can act like complete red blood cells (RBCs) [90, 91]. Besides the transportation of oxygen, RBCs are also responsible for removing CO_2 from the tissues and delivering it to the lungs. Increased CO_2 in the body leads to acidosis, and if left untreated it will end up in coma and death. Most of the CO_2 produced in the body diffuses with the concentration gradient into the RBCs, here a small fraction is carried as dissolved gas, but most of it is converted by CA and carried as the soluble bicarbonate ion. The proton and bicarbonate produced by CA also helps to buffer small acids produced through metabolism. In some artificial blood products, catalase (CAT) and superoxide dismutase (SOD) have been added to polyHb to confer antioxidant activity to these products. These may be especially protective during ischemia reperfusion injuries and can help recovery. Combining all of these elements led to the development of a novel blood substitute called PolySFHb-SOD-CAT-CA. The process to make this product proved to be very successful and reports show that all the cross-linked components work very well even though the assays comparing free CA to the cross-linked form showed lower activity for the bound CA. This could be overcome by adding more CA to the cross-linking reaction. Similarly, the CAT was shown to work in the complex, but there is no mention of SOD activity when cross-linked. These kinds of products are very promising and better than using transfused whole blood as it can be sterilized, stored for long periods, and contains no blood antigens [90].

8 Conclusions and Future Directions

The future is promising for CA in industrial applications, from boosting biofuel production to incorporation into artificial lungs. With the advent of fast and cost-effective genome sequencing, we can expect many new and different CAs to be discovered and characterized. The proof of principle for using free or immobilized CA in algae or for biomineralization of calcite has been definitively established. Certain challenges remain, such as a way to make large quantities of low cost CA or to get more out of known CAs by improving their chemical and thermal stability. A Brazilian group has looked at using bovine blood, which is produced in large quantities due to slaughtering activities in the meat industry, as a large source of bovine CA [25]. With additional research looking at minimal processing of such raw products for carbon sequestration efforts, such recycling and usage of by-products may be very advantageous. It is likely that a combination of Molecular Biology techniques to boost overexpression of protein, coupled with Directed Evolution techniques will result in abundant, highly active and stable CA for future applications.

References

1. Silverman DN, Lindskog S (1988) The catalytic mechanism of carbonic-anhydrase – implications of a rate-limiting protolysis of water. Acc Chem Res 21:30–36
2. Pocker Y, Bjorkquist DW (1977) Comparative studies of bovine carbonic anhydrase in H_2O and D_2O. Stopped-flow studies of the kinetics of interconversion of CO_2 and HCO3. Biochemistry 16:5698–5707
3. Vullo D, Luca VD, Scozzafava A, Carginale V, Rossi M, Supuran CT, Capasso C (2012) The alpha-carbonic anhydrase from the thermophilic bacterium Sulfurihydrogenibium yellowstonense YO3AOP1 is highly susceptible to inhibition by sulfonamides. Bioorg Med Chem 15:1534–1538
4. Bond GM, Stringer J, Brandvold DK, Simsek FA, Medina MG, Egeland G (2001) Development of integrated system for biomimetic CO(2) sequestration using the enzyme carbonic anhydrase. Energy Fuel 15:309–316
5. Lee SW, Park SB, Jeong SK, Lim KS, Lee SH, Trachtenberg MC (2010) On carbon dioxide storage based on biomineralization strategies. Micron 41:273–282
6. Savile CK, Lalonde JJ (2011) Biotechnology for the acceleration of carbon dioxide capture and sequestration. Curr Opin Biotechnol 22:818–823
7. Fisher SZ, Aggarwal M, Kovalevsky AY, Silverman DN, McKenna R (2012) Neutron diffraction of acetazolamide-bound human carbonic anhydrase II reveals atomic details of drug binding. J Am Chem Soc 134:14726–14729
8. Intergovernmental Panel on Climate Change (IPCC) (2007) http://www.ipcc.ch/publications_and_data/ar4/syr/en/main.html
9. Canadell JG, Le Quere C, Raupach MR, Field CB, Buitenhuis ET, Ciais P, Conway TJ, Gillett NP, Houghton R, Marland G (2007) Contributions to accelerating atmospheric CO(2) growth from economic activity, carbon intensity, and efficiency of natural sinks. Proc Natl Acad Sci U S A 104:18866–18870
10. Thoning KW, Tans PP, Komhyr WD (1989) Atmospheric carbon-dioxide at Mauna Loa observatory.2. Analysis of the Noaa Gmcc Data, 1974–1985. J Geophys Res Atmos 94:8549–8565

11. Tans PP (2013) http://www.esrl.noaa.gov/gmd/ccgg/trends/
12. Intergovernmental Panel on Climate Change (IPCC) (2001) http://www.grida.no/climate/ipcc_tar/
13. Post E, Forchhammer MC, Bret-Harte M, Callaghan TV, Christensen TR, Elberling B, Fox AD, Gilg O, Hik DS, Hoye TT, Ims RA, Jeppesen E, Klein DR, Madsen J, McGuire A, Rysgaard S, Schindler DE, Stirling I, Tamstorf MP, Tyler NJ, van der Wal R, Welker J, Wookey PA, Schmidt NM, Aastrup P (2009) Ecological dynamics across the arctic associated with recent climate change. Science 325:1355–1358
14. Clarke G, Leverington D, Teller J, Dyke A (2003) Superlakes, megafloods, and abrupt climate change. Science 301:922–923
15. Benson SM, Surles T (2006) Carbon dioxide capture and storage: an overview with emphasis on capture and storage in deep geological formations. Proc IEEE 94:1795–1805
16. Pierre AC (2012) Enzymatic carbon dioxide capture. ISRN Chemical Engineering 2012:753687, p. 22
17. Bao L, Trachtenberg MC (2006) Facilitated transport of CO2 across a liquid membrane: comparing enzyme, amine, and alkaline. J Membr Sci 280:330–334
18. Dreybrodt W, Lauckner J, Liu ZH, Svensson U, Buhmann D (1996) The kinetics of the reaction $CO_2 + H_2O \rightarrow H^+ + HCO_3^-$ as one of the rate limiting steps for the dissolution of calcite in the system H_2O–CO_2–$CaCO_3$. Geochim Cosmochim Acta 60:3375–3381
19. Dodds WS, Stutzman LF, Sollami BJ (1956) Carbon dioxide solubility in water. Ind Eng Chem 1:92–95
20. Boron WF (2010) Evaluating the role of carbonic anhydrases in the transport of HCO3–related species. Biochim Biophys Acta 1804:410–421
21. The PyMOL molecular graphics system, version 1.5.0.4 Schrödinger, LLC. 2013
22. Baker NA, Sept D, Joseph S, Holst MJ, McCammon JA (2001) Electrostatics of nanosystems: application to microtubules and the ribosome. Proc Natl Acad Sci U S A 98:10037–10041
23. Do CB, Mahabhashyam MS, Brudno M, Batzoglou S (2005) ProbCons: probabilistic consistency-based multiple sequence alignment. Genome Res 15:330–340
24. Fisher Z, Kovalevsky AY, Mustyakimov M, Silverman DN, McKenna R, Langan P (2011) Neutron structure of human carbonic anhydrase II: a hydrogen-bonded water network "switch" is observed between pH 7.8 and 10.0. Biochemistry 50:9421–9423
25. da Costa OJ, Sala L, Cerveira GP, Kalil SJ (2012) Purification of carbonic anhydrase from bovine erythrocytes and its application in the enzymic capture of carbon dioxide. Chemosphere 88:255–259
26. Service RF (2011) Algae's second try. Science 333:1238–1239
27. Bloch MR, Sasson J, Ginzburg ME, Goldman Z, Ginzburg BZ, Garti N, Porath A (1982) Oil products from algae. US patent 4,341,038, 1982
28. Badger MR, Price GD (2003) CO_2 concentrating mechanisms in cyanobacteria: molecular components, their diversity and evolution. J Exp Bot 54:609–622
29. Cannon GC, Heinhorst S, Kerfeld CA (2010) Carboxysomal carbonic anhydrases: structure and role in microbial CO_2 fixation. Biochim et Biophys Acta-Proteins Proteomics 1804:382–392
30. Ellis R (2010) Biochemistry tackling unintelligent design. Nature 463:164–165
31. Pires J, Alvim-Ferraz M, Martins F, Simoes M (2012) Carbon dioxide capture from flue gases using microalgae: engineering aspects and biorefinery concept. Renew Sustain Energy Rev 16:3043–3053
32. Gonzalez-Fernandez C, Ballesteros M (2012) Linking microalgae and cyanobacteria culture conditions and key-enzymes for carbohydrate accumulation. Biotechnol Adv 30:1655–1661
33. Shekh AY, Krishnamurthi K, Mudliar SN, Yadav RR, Fulke AB, Devi SS, Chakrabarti T (2012) Recent advancements in carbonic anhydrase-driven processes for CO_2 sequestration: minireview. Crit Rev Environ Sci Technol 42:1419–1440
34. Ramanan R, Kannan K, Deshkar A, Yadav R, Chakrabarti T (2010) Enhanced algal CO_2 sequestration through calcite deposition by *Chlorella* sp and *Spirulina platensis* in a mini-raceway pond. Bioresour Technol 101:2616–2622

35. Fulke AB, Mudliar S, Yadav R, Shekh A, Srinivasan N, Ramanan R, Krishnamurthi K, Devi SS, Chakrabarti T (2010) Bio-mitigation of CO_2, calcite formation and simultaneous biodiesel precursors production using *Chlorella* sp. Bioresour Technol 101:8473–8476
36. Chi Z, O'Fallon JV, Chen S (2011) Bicarbonate produced from carbon capture for algae culture. Trends Biotechnol 29:537–541
37. Rost B, Richter KU, Riebesell U, Hansen PJ (2006) Inorganic carbon acquisition in red tide dinoflagellates. Plant Cell Environ 29:810–822
38. Ozdemir E (2009) Biomimetic CO_2 sequestration: 1. Immobilization of carbonic anhydrase within polyurethane foam. Energy Fuels 23:5725–5730
39. Liu Z, Bartlow P, Dilmore RM, Soong Y, Pan Z, Koepsel R, Ataai M (2009) Production, purification, and characterization of a fusion protein of carbonic anhydrase from *Neisseria gonorrhoeae* and cellulose binding domain from *Clostridium thermocellum*. Biotechnol Prog 25:68–74
40. Cheng LH, Zhang L, Chen HL, Gao CJ (2008) Hollow fiber contained hydrogel-CA membrane contactor for carbon dioxide removal from the enclosed spaces. J Membr Sci 324:33–43
41. Hosseinkhani S, Szittner R, Nemat-Gorgani M, Meighen EA (2003) Adsorptive immobilization of bacterial luciferases on alkyl-substituted Sepharose 4B. Enzyme Microb Technol 32:186–193
42. Oviya M, Giri SS, Sukumaran V, Natarajan P (2012) Immobilization of carbonic anhydrase enzyme purified from *Bacillus subtilis* Vsg-4 and its application As CO_2 sequesterer. Prep Biochem Biotechnol 42:462–475
43. Yadav R, Satyanarayanan T, Kotwal S, Rayalu S (2011) Enhanced carbonation reaction using chitosan-based carbonic anhydrase nanoparticles. Curr Sci 100:520–524
44. Machida-Sano I, Ogawa S, Ueda H, Kimura Y, Satoh N, Namiki H (2012) Effects of composition of iron-cross-linked alginate hydrogels for cultivation of human dermal fibroblasts. Int J Biomater 2012:820513
45. Zhai P, Chen XB, Schreyer DJ (2013) Preparation and characterization of alginate microspheres for sustained protein delivery within tissue scaffolds. Biofabrication 5:015009
46. Simsek-Ege FA, Bond GM, Stringer J (2002) Matrix molecular weight cut-off for encapsulation of carbonic anhydrase in polyelectrolyte beads. J Biomater Sci Polym Ed 13:1175–1187
47. Prabhu C, Wanjari S, Puri A, Bhattacharya A, Pujari R, Yadav R, Das S, Labhsetwar N, Sharma A, Satyanarayanan T, Rayalu S (2011) Region-specific bacterial carbonic anhydrase for biomimetic sequestration of carbon dioxide. Energy Fuel 25:1327–1332
48. Wanjari S, Prabhu C, Yadav R, Satyanarayana T, Labhsetwar N, Rayalu S (2011) Immobilization of carbonic anhydrase on chitosan beads for enhanced carbonation reaction. Process Biochem 46:1010–1018
49. Sharma A, Bhattacharya A, Shrivastava A (2011) Biomimetic CO_2 sequestration using purified carbonic anhydrase from indigenous bacterial strains immobilized on biopolymeric materials. Enzyme Microb Technol 48:416–426
50. Prabhu C, Wanjari S, Gawande S, Das S, Labhsetwar N, Kotwal S, Puri AK, Satyanarayana T, Rayalu S (2009) Immobilization of carbonic anhydrase enriched microorganism on biopolymer based materials. J Mol Catal B: Enzym 60:13–21
51. Fan LH, Liu N, Yu MR, Yang ST, Chen HL (2011) Cell surface display of carbonic anhydrase on *Escherichia coli* using ice nucleation protein for CO_2 sequestration. Biotechnol Bioeng 108:2853–2864
52. Vinoba M, Bhagiyalakshmi M, Jeong SK, Nam SC, Yoon Y (2012) Carbonic anhydrase immobilized on encapsulated magnetic nanoparticles for CO_2 sequestration. Chem A Eur J 18:12028–12034
53. Vinoba M, Lim KS, Lee SH, Jeong SK, Alagar M (2011) Immobilization of human carbonic anhydrase on gold nanoparticles assembled onto amine/thiol-functionalized mesoporous SBA-15 for biomimetic sequestration of CO_2. Langmuir 27:6227–6234
54. Wanjari S, Prabhu C, Satyanarayana T, Vinu A, Rayalu S (2012) Immobilization of carbonic anhydrase on mesoporous aluminosilicate for carbonation reaction. Micropor Mesopor Mater 160:151–158

55. Vinoba M, Bhagiyalakshmi M, Jeong SK, Yoon YI, Nam SC (2012) Immobilization of carbonic anhydrase on spherical SBA-15 for hydration and sequestration of CO_2. Colloids Surf B Biointerfaces 90:91–96
56. Wood LL, Hartdegen FJ, Hahn PA (1982) Enzymes bound to polyurethane, US Patent 4,342,834, 1982
57. Trachtenberg MC, Tu CK, Landers RA, Willson RC, McGregor ML, Laipis PJ, Kennedy JF, Paterson M, Silverman DN, Thomas D, Smith RL, Rudolph FB (1999) Carbon dioxide transport by proteic and facilitated transport membranes. Life Support Biosph Sci 6:293–302
58. Jeong H, Yim JH, Lee C, Choi SH, Park YK, Yoon SH, Hur CG, Kang HY, Kim D, Lee HH, Park KH, Park SH, Park HS, Lee HK, Oh TK, Kim JF (2005) Genomic blueprint of *Hahella chejuensis*, a marine microbe producing an algicidal agent. Nucleic Acids Res 33:7066–7073
59. Ki MR, Min K, Kanth BK, Lee J, Pack SP (2012) Expression, reconstruction and characterization of codon-optimized carbonic anhydrase from *Hahella chejuensis* for CO(2) sequestration application. Bioprocess Biosyst Eng 36:375–381
60. Ramanan R, Kannan K, Vinayagamoorthy N, Ramkumar KM, Sivanesan SD, Chakrabarti T (2009) Purification and characterization of a novel plant-type carbonic anhydrase from *Bacillus subtilis*. Biotechnol Bioprocess Eng 14:32–37
61. Kim IG, Jo BH, Kang DG, Kim CS, Choi YS, Cha HJ (2012) Biomineralization-based conversion of carbon dioxide to calcium carbonate using recombinant carbonic anhydrase. Chemosphere 87:1091–1096
62. Sharma A, Bhattacharya A (2010) Enhanced biomimetic sequestration of CO_2 into $CaCO_3$ using purified carbonic anhydrase from indigenous bacterial strains. J Mol Catal B: Enzym 67:122–128
63. Capasso C, De LV, Carginale V, Cannio R, Rossi M (2012) Biochemical properties of a novel and highly thermostable bacterial alpha-carbonic anhydrase from Sulfurihydrogenibium yellowstonense YO3AOP1. J Enzyme InhibMed Chem 27:892–897
64. Kanth BK, Min K, Kumari S, Jeon H, Jin ES, Lee J, Pack SP (2012) Expression and characterization of codon-optimized carbonic anhydrase from *Dunaliella* species for CO(2) sequestration application. Appl Biochem Biotechnol 167:2341–2356
65. Li L, Fu ML, Zhao YH, Zhu YT (2012) Characterization of carbonic anhydrase II from *Chlorella vulgaris* in bio-CO_2 capture. Environ Sci Pollut Res Int 19:4227–4232
66. Zevenhoven R, Eloneva S, Teir S (2006) Chemical fixation of CO_2 in carbonates: routes to valuable products and long-term storage. Catal Today 115:73–79
67. Allen DJ, Brent GF (2010) Sequestering CO(2) by mineral carbonation: stability against acid rain exposure. Environ Sci Technol 44:2735–2739
68. Sakakura T, Choi JC, Yasuda H (2007) Transformation of carbon dioxide. Chem Rev 107:2365–2387
69. Arakawa H, Aresta M, Armor JN, Barteau MA, Beckman EJ, Bell AT, Bercaw JE, Creutz C, Dinjus E, Dixon DA, Domen K, DuBois DL, Eckert J, Fujita E, Gibson DH, Goddard WA, Goodman DW, Keller J, Kubas GJ, Kung HH, Lyons JE, Manzer LE, Marks TJ, Morokuma K, Nicholas KM, Periana R, Que L, Rostrup-Nielson J, Sachtler WM, Schmidt LD, Sen A, Somorjai GA, Stair PC, Stults BR, Tumas W (2001) Catalysis research of relevance to carbon management: progress, challenges, and opportunities. Chem Rev 101:953–996
70. Beckman EJ (1999) Polymer synthesis – making polymers from carbon dioxide. Science 283:946–947
71. Suzuki M, Saruwatari K, Kogure T, Yamamoto Y, Nishimura T, Kato T, Nagasawa H (2009) An acidic matrix protein, Pif, is a key macromolecule for nacre formation. Science 325:1388–1390
72. Astachov L, Nevo Z, Brosh T, Vago R (2011) The structural, compositional and mechanical features of the calcite shell of the barnacle *Tetraclita rufotincta*. J Struct Biol 175:311–318
73. Miyamoto H, Miyoshi F, Kohno J (2005) The carbonic anhydrase domain protein nacrein is expressed in the epithelial cells of the mantle and acts as a negative regulator in calcification in the mollusc *Pinctada fucata*. Zoolog Sci 22:311–315
74. Fischer F, Tropsch H (1930) Process for the production of paraffin hydrocarbons with more than one carbon atom. US patent 1,746,464, 1930

75. Centi G, Perathoner S (2009) Opportunities and prospects in the chemical recycling of carbon dioxide to fuels. Catal Today 148:191–205
76. Hori Y, Kikuchi K, Murata A, Suzuki S (1986) Production of methane and ethylene in electrochemical reduction of carbon-dioxide at copper electrode in aqueous hydrogencarbonate solution. Chem Lett 15:897–898
77. Guerrero-Peacuterez M, Rosas J, Bedia J, Rodriacuteguez-Mirasol J, Cordero T (2009) Recent inventions in glycerol transformations and processing. Recent Patents Chem Eng 2:11–21
78. Hu J, Li J, Gu Y, Guan Z, Mo W, Ni Y, Li T, Li G (2010) Oxidative carbonylation of glycerol to glycerol carbonate catalyzed by PdCl2(phen)/KI. Appl Catal Gen 386:188–193
79. Nguyen N, Demirel Y (2011) A novel biodiesel and glycerol carbonate production plant. Int J Chem Reactor Eng 9:A108
80. Kaar JL, Oh HI, Russell AJ, Federspiel WJ (2007) Towards improved artificial lungs through biocatalysis. Biomaterials 28:3131–3139
81. Sreenivasan R, Bassett EK, Hoganson DM, Vacanti JP, Gleason KK (2011) Ultra-thin, gas permeable free-standing and composite membranes for microfluidic lung assist devices. Biomaterials 32:3883–3889
82. Oh HI, Ye SH, Johnson CA Jr, Woolley JR, Federspiel WJ, Wagner WR (2010) Hemocompatibility assessment of carbonic anhydrase modified hollow fiber membranes for artificial lungs. Artif Organs 34:439–442
83. Arazawa DT, Oh HI, Ye SH, Johnson CA Jr, Woolley JR, Wagner WR, Federspiel WJ (2012) Immobilized carbonic anhydrase on hollow fiber membranes accelerates CO(2) removal from blood. J Memb Sci 404–404:25–31
84. Ge J, Cowan RM, Tu C, McGregor ML, Trachtenberg MC (2002) Enzyme-based CO2 capture for advanced life support. Life Support Biosph Sci 8:181–189
85. Cowan RM, Ge JJ, Qin YJ, McGregor ML, Trachtenberg MC (2003) CO2 capture by means of an enzyme-based reactor. Ann N Y Acad Sci 984:453–469
86. Satav SS, Bhat S, Thayumanavan S (2010) Feedback regulated drug delivery vehicles: carbon dioxide responsive cationic hydrogels for antidote release. Biomacromolecules 11:1735–1740
87. Roskos KV, Fritzinger BK, Tefft JA, Nakayama GR, Heller J (1995) Biocompatibility and in vivo morphine diffusion into a placebo morphine-triggered naltrexone delivery device in rabbits. Biomaterials 16:1235–1239
88. Hassan CM, Doyle FJ, Peppas NA (1997) Dynamic behavior of glucose-responsive poly(methacrylic acid-g-ethylene glycol) hydrogels. Macromolecules 30:6166–6173
89. Han D, Boissiere O, Kumar S, Tong X, Tremblay L, Zhao Y (2012) Two-way CO_2-switchable triblock copolymer hydrogels. Macromolecules 45:7440–7445
90. Bian Y, Rong Z, Chang TM (2012) Polyhemoglobin-superoxide dismutase-catalase-carbonic anhydrase: a novel biotechnology-based blood substitute that transports both oxygen and carbon dioxide and also acts as an antioxidant. Artif Cells Blood Substit Immobil Biotechnol 40:28–37
91. Gould SA, Moore EE, Hoyt DB, Ness PM, Norris EJ, Carson JL, Hides GA, Freeman IH, DeWoskin R, Moss GS (2002) The life-sustaining capacity of human polymerized hemoglobin when red cells might be unavailable. J Am Coll Surg 195:445–452

Index

S.C. Frost and R. McKenna (eds.), *Carbonic Anhydrase: Mechanism, Regulation, Links to Disease, and Industrial Applications*, Subcellular Biochemistry 75,
DOI 10.1007/978-94-007-7359-2, © Springer Science+Business Media Dordrecht 2014

D

E

MIX
Papier aus verantwortungsvollen Quellen
Paper from responsible sources
FSC® C105338

If you have any concerns about our products,
you can contact us on
ProductSafety@springernature.com

In case Publisher is established outside the EU,
the EU authorized representative is:
Springer Nature Customer Service Center GmbH
Europaplatz 3, 69115 Heidelberg, Germany

Printed by Libri Plureos GmbH
in Hamburg, Germany